SIMULATION of ECOLOGICAL and ENVIRONMENTAL MODELS

SIMULATION of ECOLOGICAL and ENVIRONMENTAL MODELS

MIGUEL F. ACEVEDO

CRC Press is an imprint of the
Taylor & Francis Group, an **informa** business

CRC Press
Taylor & Francis Group
6000 Broken Sound Parkway NW, Suite 300
Boca Raton, FL 33487-2742

First issued in paperback 2019

© 2013 by Taylor & Francis Group, LLC
CRC Press is an imprint of Taylor & Francis Group, an Informa business

No claim to original U.S. Government works

ISBN-13: 978-1-4398-8506-2 (hbk)
ISBN-13: 978-0-367-86680-8 (pbk)

This book contains information obtained from authentic and highly regarded sources. Reasonable efforts have been made to publish reliable data and information, but the author and publisher cannot assume responsibility for the validity of all materials or the consequences of their use. The authors and publishers have attempted to trace the copyright holders of all material reproduced in this publication and apologize to copyright holders if permission to publish in this form has not been obtained. If any copyright material has not been acknowledged please write and let us know so we may rectify in any future reprint.

Except as permitted under U.S. Copyright Law, no part of this book may be reprinted, reproduced, transmitted, or utilized in any form by any electronic, mechanical, or other means, now known or hereafter invented, including photocopying, microfilming, and recording, or in any information storage or retrieval system, without written permission from the publishers.

For permission to photocopy or use material electronically from this work, please access www.copyright.com (http://www.copyright.com/) or contact the Copyright Clearance Center, Inc. (CCC), 222 Rosewood Drive, Danvers, MA 01923, 978-750-8400. CCC is a not-for-profit organization that provides licenses and registration for a variety of users. For organizations that have been granted a photocopy license by the CCC, a separate system of payment has been arranged.

Trademark Notice: Product or corporate names may be trademarks or registered trademarks, and are used only for identification and explanation without intent to infringe.

Visit the Taylor & Francis Web site at
http://www.taylorandfrancis.com

and the CRC Press Web site at
http://www.crcpress.com

Contents

Part I Introduction, Mathematical Review, and Software Fundamentals

Contents

Part I Introduction, Mathematical Review, and Software Fundamentals

Preface

This book evolved from lecture notes and laboratory manuals that I have written over many years to teach modeling and simulation to graduate and undergraduate students in the United States, Spain, and Latin America. I developed the vast majority of this material over 15 years to teach a first-year graduate course in environmental modeling for the environmental science and applied geography programs at the University of North Texas. In that course, we focus on problem solving in geographical and environmental sciences with an emphasis on simulation hands-on experience. More recently, I have expanded and used this material for blended first-year graduate and senior-level undergraduate engineering students.

Modeling and simulation methods are the same in a broad range of disciplines in science and engineering, and so are the computational tools that we can use. Methods vary by discipline either because of academic tradition or because of the priority given to certain problems. However, methods not traditionally employed in a discipline sometimes become part of its arsenal as priorities shift and methods are "imported" from other fields where they have shown to be effective.

This book aspires to educate twenty-first century scientists and engineers in modeling and simulation that require a unified presentation of methods, including modeling ecological systems subject to disturbances or stresses due to human–nature interactions. It also looks to provide practical training on performing analysis using computers. Furthermore, given the importance of interdisciplinary work in sustainability, this book attempts to bring together methods that are applicable to a variety of science and engineering disciplines dealing with earth systems, the environment, ecology, and human–nature interactions. Therefore, this book contributes to undergraduate and graduate education in biology, environmental science, geography and earth science, and engineering.

OVERVIEW

I have organized this textbook into three major parts:

- Part I: Chapters 1 and 2—introduction, mathematical review, and software fundamentals
- Part II: Chapters 3 through 8—one-dimensional models and fundamentals of modeling methodology
- Part III: Chapters 9 through 16—multidimensional models, focusing on structured populations, communities, and ecosystems, including cycles and flow of energy

A major difference between Parts II and III is the use of matrix algebra in Part III to address multidimensional problems. To make this possible, Chapter 9 covers the fundamentals of matrices and linear algebra.

Although this organization may seem unconventional, it allows flexibility in using the book in various countries, types of curricula, and levels of student progress in any curriculum. In the United States, for example, most undergraduate students in the sciences do not take a linear algebra course, and in some engineering programs, linear algebra is not required until the third (junior) year upon completion of a second calculus class. Therefore, I have left the multidimensional material for last after a substantial review of matrix algebra, allowing the students to become familiar with modeling methodology and major issues of nonlinearity, stability, and disturbances at an earlier stage.

USE OF THE BOOK

There are several ways to use this book. For example, a junior-level (third-year) course for undergraduate students can cover Parts I and II at a rhythm of about two chapters per week during a typical 15-week semester by emphasizing examples. A more challenging or honors section could include the review of matrices (Chapter 9) and a couple of chapters from Part III. A senior-level and first-year graduate course can be based on the entire book. My experience has been that last-year (senior) undergraduate and first-year graduate students in the sciences are unfamiliar with matrix algebra and would need Chapter 9. However, a senior-level undergraduate or graduate engineering course may not need coverage of this chapter, except for the computer session.

PEDAGOGY

Each chapter starts with conceptual and theoretical material covered with enough mathematical detail to serve as a firm foundation to understand how the methods work. Over the many years that I have used this material, I have confirmed my belief that students rise to the challenge of understanding the mathematical concepts and develop a good understanding of simulation that facilitates their future learning of the more advanced and specialized methods needed in their professions or research areas. To facilitate learning these concepts, I have included examples and exercises that illustrate the applications and how to go from concepts to problem solving.

In each chapter, a hands-on computer session follows the theoretical foundations, which helps the student **learn by doing**. In addition, this computer session allows the reader to grasp the practical implications of the theoretical background. This book is not a software manual, but the computer examples are developed with sufficient detail so that the students can do the work on their own either in an instructor-supervised computer classroom, or laboratory environment, or unassisted and on their own time. This design gives maximum flexibility to the instructor and the student. In a similar fashion to the theoretical section, the computer session includes exercises to practice comprehension of the material.

I have based the computer examples and exercises on functions I have built in a package for R (see next section). Using this package, named **seem**, the learner is able to quickly produce simulations and explore the effect of changing conditions in the model without many programming complications. This way, most of the learner's focus is on understanding the model dynamics and not on the mechanics of producing the simulation. However, I include a detailed explanation of the R functions as optional material at the end of each chapter in a section called *Seem Functions Explained*. The more advanced and motivated learners are encouraged to understand these functions and to feel free to modify them as necessary. The instructor may assign material from this section depending on the background of the students. In addition, starting with Chapter 6, I have included several *Build Your Own* sections where all readers are challenged to build a model using the concepts explained.

COMPUTER EXAMPLES AND EXERCISES

I have organized the computer examples using the **R** system, which is **open source** and very simple to download, install, and run. As some authors put it, R has evolved into the *lingua franca* of statistical computing. R competes with major systems of scientific computing, yet because it is open source, it is free of commercial license cost while having access to thousands of packages to perform a tremendous variety of analysis; at the time of this writing, there are 3398 packages available. Even students with no prior knowledge of programming are quickly acquainted with the basics of programming in R. For those users who still prefer a graphical user interface (GUI), there is a diversity of GUIs also available as open source.

R is a **GNU project** system. The GNU project includes Free Software and General Public License. GNU stands for "Gnu's Not Unix." R is available from the Comprehensive R Archive

Network (CRAN), the major repository mirrored all over the world. The simplest approach is to download the precompiled binary distribution for your operating system (Linux, Mac, or Windows). In this book, we will assume a Windows installation because it is a very common situation in university environments.

In addition to the R GUI, there are several other GUIs available to simplify or extend R. For example, a **Web** GUI can be used to enter commands over a Web browser, and therefore from smart phones and pads with Web access, and the **R Commander** GUI and its several plug-in packages can be used to simplify entering commands.

As mentioned before, students can execute the computer examples and exercises in the classroom and at their own pace using their computers. Over the years, I have tested this material in both modes. I conduct a weekly instructor-supervised session in the computer classroom, where I run demonstrations from the instructor's machine equipped with a projector or through systematic instructions followed by students in their assigned computers, or I simply let the students follow the instructions given in the book and ask for help as needed. Students can go to the computer lab to work on their assigned exercises or complete the assignments off-campus by installing R on their home computers.

HOW TO USE THE BIBLIOGRAPHY

I have included citations pointing to several textbooks that cover the topics at similar levels and are written for different audiences; these books can help students read tutorial explanations from other authors. Many times, reading the same thing in different words or looking at different figures helps understand a topic better. In addition, the references point to several introductory texts that can be used for review. I have also included reference items that provide entry points or hooks to specialized books and articles on some topics. This helps advanced students access the bibliography for their research work.

SUPPLEMENTARY MATERIAL

The **seem** package available from CRAN provides all the data files and scripts used here. The package is available via CRAN and also by links provided at the Texas Environmental Observatory (TEO) website, www.teo.unt.edu, and the author's website, which can be reached from his departmental affiliation website, www.ee.unt.edu. The publisher also offers supplementary materials, available with qualifying course adoption. These include a solutions manual and PowerPoint® slides with figures and equations to help in preparing lectures.

Acknowledgments

I am very grateful to the many students who took classes with me and used preliminary versions of this material. Their questions and comments helped shape the content and improve the presentation. My sincere thanks go to several students who have worked as teaching assistants for classes taught with this material and helped improve successive drafts over the years—in particular, H. Goetz and K. Anderle, who kindly made many suggestions. I also thank students whom I have guided on their theses and dissertation research. Working with them provided insight about the type of methods that would be useful to cover in this textbook.

Many colleagues have been inspirational—to name just a few: T. W. Waller and K. L. Dickson of the UNT environmental science program, now emeritus faculty; M. A. Harwell, Harwell Gentile & Associates, LC; H. H. Shugart, University of Virginia; T. G. Hallam, University of Tennessee, retired; D. L. Urban, Duke University; J. Raventós, Universidad de Alicante; M. Ataroff, P. Soriano, and M. Ablan, Universidad de Los Andes.

I am very grateful to Irma Shagla-Britton, environmental science and engineering editor at CRC Press, who was enthusiastic from day one, and Laurie Schlags, project coordinator, who helped immensely in the production process. Several reviewers provided excellent feedback that shaped the final version and approach of the manuscript, i.e., M. Uzcategui, Universidad de Los Andes, who provided suggestions for improvements.

Special thanks to my family and friends, who were so supportive and willing to postpone many important things until I completed this project. Last, but not least, I give special thanks to the open source community for making R such a wonderful tool for research and education.

Miguel Acevedo
Denton, Texas

Author

Miguel F. Acevedo has 38 years of academic experience, the last 20 of these as faculty member of the University of North Texas (UNT). His career has been interdisciplinary, especially at the interface of science and engineering. He has served UNT in the department of geography, the graduate program in environmental sciences of the biology department, and more recently in the electrical engineering department. He obtained his PhD in biophysics from the University of California Berkeley in 1980 and master degrees in electrical engineering and computer science from Berkeley (ME, 1978) and the University of Texas at Austin (MS, 1972). Prior to UNT, he was at the Universidad de Los Andes in Merida, Venezuela, where he taught for 18 years in the School of Systems Engineering, the graduate program in tropical ecology, and the Center for Simulation and Modeling (CESIMO). He has served on the Science Advisory Board of the U. S. Environmental Protection Agency and on many review panels of the U. S. National Science Foundation. He has received numerous research grants and written many journal articles, book chapters, and proceedings articles. UNT has recognized him with the Regent's Professor rank, the Citation for Distinguished Service to International Education, and the Regent's Faculty Lectureship.

Part I

Introduction, Mathematical Review, and Software Fundamentals

1 Introduction

1.1 MODELING AND SIMULATION

This book introduces the theory and practice of models and simulation applied in ecology, geography, environmental science, and engineering, presented in a unified framework. Integrating ecological and environmental models helps understand the science and management of ecosystems and environmental problem solving. Models and their simulation are useful tools to make explicit our assumptions of how ecosystems work and to incorporate our current scientific understanding of these systems. The complexity of ecosystem management and environmental problem solving requires more quantitative approaches, such as those provided by models and their simulation using computers.

Because management and problem solving often demand the prediction of environmental changes, the emphasis of the models covered in this book is on **dynamics**, that is to say changes over time. The book describes mathematical fundamentals to understand and analyze models and the methodology to simulate models. We start with simple models to explain modeling and simulation concepts. Once understood, it is easier to apply these concepts to more complex models.

Hands-on exercises in computer sessions are included in every chapter. Practice is an effective way to learn modeling. We use data sets in conjunction with models, for model calibration and evaluation, in order to emphasize the importance of data in the modeling process. Computer exercises cover a wide range of model applications and emphasize simulations with simple programs to explain basic concepts. This foundation will lead to a better understanding of more complicated models readily available today by virtue of the websites at universities and government agencies such as the U.S. Environmental Protection Agency and the U.S. Geological Survey.

1.2 PROCESS-BASED DYNAMIC MODELS

A model is a simplified representation of reality, based on concepts, hypotheses, and theories of how the real system works. In this book, we are interested in models that represent reality as a set of mathematical equations based on the **processes** at work; for example, a differential equation representing tree growth over time based on the increment of its diameter. For this purpose, we use the concept that diameter increases when the tree is small and that growth decreases when the tree is large. This method is in contrast to **empirical** models that build a quantitative relationship between variables based on data without an explicit consideration of the process yielding that relation. For example, using regression we can derive a predictor of tree height as a function of tree diameter based on measured data from 20 trees of different heights and diameters.

This book emphasizes **process-based** or **mechanistic** models as opposed to empirical models. I have written a related book emphasizing empirical models (Acevedo, 2012). However, we use empirical models throughout this book to estimate parameters of the process-based models based on data from field and laboratory experiments. For example, we can use a mechanistic model to calculate the flow of a stream using the water velocity and cross-sectional area, but we estimate velocity using an empirical relation of velocity to water depth. In addition, we will use empirical models to convert output variables of process-based models to other variables. For example, we will predict tree diameter increments from a process-based model of tree growth and then convert the diameter to height using an empirical relation of height versus diameter.

Temporal dynamics and spatial gradients make the concept of **rate of change** of paramount importance in this book. Most model equations in this book represent a balance of quantities according to rates of change. Derivatives, integrals, and differential equations are excellent tools for representing dynamic models. This book assumes some background knowledge of these mathematical techniques, but we present these concepts from an intuitive point of view for those students who may not have this background.

Exercise 1.1

We want to build a model of bacteria population growth through time. Work with population density, that is, the number of cells per unit volume. Assume that you could establish a rate of growth as a function of population density. Conceptually, propose a process-based model. How would you develop an empirical model?

Exercise 1.2

We want to build a model of sunlight vertical transmission through a forest canopy. Assume that light attenuation is proportional to the density of foliage at various heights and that we can measure light at various heights. This is an example of spatial gradients. Conceptually, propose a process-based model. How would you develop an empirical model?

1.3 APPLICATIONS OF ENVIRONMENTAL MODELING

The models covered in this book have a broad range of applications. Instead of being very comprehensive, we will cover some selected applications that focus on those dealing with how populations, ecological communities, and ecosystems respond to various environmental drivers, both natural and human-caused. In the latter, we include not only chemical compounds, determining an exposure of biological components to stress, but also other disturbances, such as harvesting. We will then employ the models to help predict or understand the effect of the exposure or disturbance on the population or ecosystem.

Figure 1.1 summarizes how this book approaches these applications to ecological systems and environmental problems. This figure can serve as a road map for the topics covered in this book and

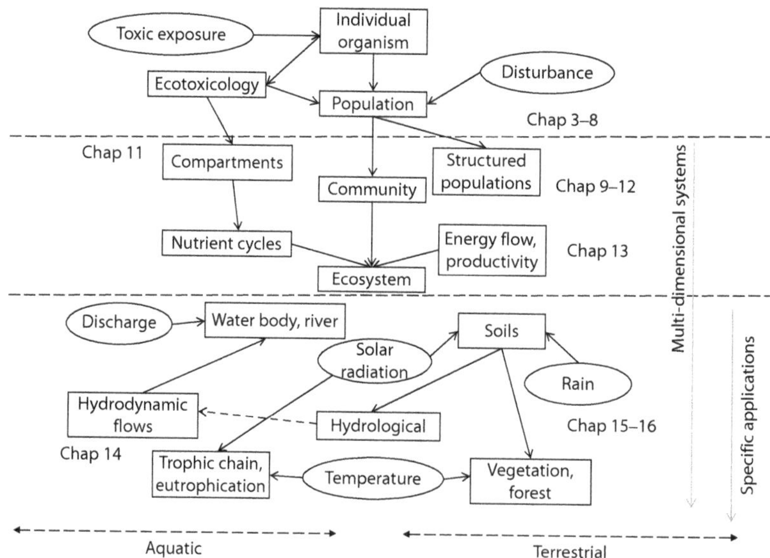

FIGURE 1.1 Modeling: themes and applications covered in this book.

their interrelationships. We can come back to this diagram as we go along for guidance on how a particular theme fits in this overall framework.

Exercise 1.3

Name and discuss one application of environmental models. Name and discuss one application of ecological models. Discuss one example of the integration of ecological and environmental models.

1.4 MODELING METHODOLOGY

Terms of model methodology will become clear as we go through the different chapters. The following list, together with Figure 1.2, summarizes this methodology:

- Conceptual: system description and problem definition
  - Define the questions to be answered by the model
  - Verbal descriptions, assumptions, questions
  - Graphical block diagrams and arrows
  - Identify biological (species, organisms) and physical and chemical (*abiotic*) components
  - Identify scales in time and space
- Mathematical: model and components
  - Equations
  - Parameters, variables, units
  - Preliminary analysis
- Simulation: numerical analysis
  - Use available data for calibration or parameter estimation
  - Perform sensitivity analysis
  - Use available data for model evaluation
  - Define scenarios to be simulated
- Analysis: generation of results
  - Answer the questions that motivated using the model
  - Limits of these answers
  - Next steps and model extension

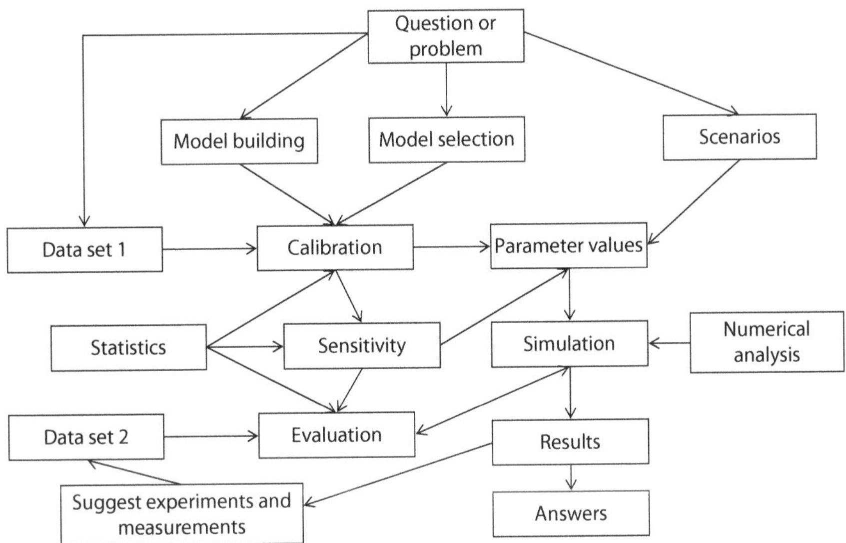

FIGURE 1.2 Modeling methodology.

Exercise 1.4

What are the first steps to take when considering the application of the modeling method?

1.5 BOOK ORGANIZATION

I have organized this textbook in three major parts comprising 16 chapters.

1.5.1 Part I, Chapters 1–2

The chapter you are now reading and the next are devoted to an introduction to the subject. This chapter provides an orientation to the book, introduction to R software fundamentals, and overview of the bibliography. Chapter 2 contains a tutorial-style mathematical review and a gentle introduction to the use of R functions in connection with these mathematical concepts. Very informally, we give a working concept of derivatives, integrals, ordinary differential equations (ODEs), and random variables.

1.5.2 Part II, Chapters 3–8

In these chapters, we go through the fundamentals of modeling **methodology**. To accomplish this, we will limit ourselves to the simplest **one-dimensional** models (one variable ODE) so that we focus on the methods and not on model complexity. In Chapter 3, we introduce the simplest model, the **exponential**, and use it to discuss the concept of model **calibration** using linear and nonlinear regression to estimate parameter values. In Chapter 4, we learn computer **simulation** by covering simple numerical methods to integrate an ODE. In addition, we show how to generate multiple instances of a random process.

In Chapter 5, we dwell on model **evaluation**, a method of great importance in simulation. We talk about how to calculate differences between model results and measured data and how to analyze the sensitivity of the model to changes in the parameter values. Chapter 6 is devoted to nonlinear models; for this, we exploit the advantage of examining a simple model so that we can easily relate the theoretical dynamics with the computer simulation. In that chapter, we see a glimpse of what would happen when we couple two models, one for a resource and the other for a consumer, that obey the equations covered at those preliminary stages.

Chapter 7 covers the basics of the concepts of stability and forced models. Here we establish how to make a difference in the model between what we consider **endogenous** aspects of the system, and the effects of what we deem to be **exogenous** or external drivers. This turns out to be an important aspect of applications, because it will allow us to consider various aspects of disturbances and stress. Chapter 8 completes the second part of the book by covering simulation scenarios and more advanced concepts of sensitivity.

1.5.3 Part III, Chapters 9–16

In this third part, we discuss multidimensional models. To be able to do that, we take the time in Chapter 9 to cover linear algebra, matrices, and the basics of linear systems. We end that chapter by learning about the importance of eigenvalues and eigenvectors. Once the foundation of Chapter 9 has been laid out, we can embark on studying modeling applications to structured populations (Chapter 10), ecotoxicology (Chapter 11), communities (Chapter 12), and ecosystems (Chapters 13–16). Prevalent throughout all these chapters is the issue of analyzing more than one variable.

The subject matter of Chapters 10–16 is very broad. We will be concerned with only some selected aspects to keep our treatment relatively simple. One of the things that I hope you grasp is that similar simple models are used to analyze various aspects of ecological interactions and responses to external drivers. In a manner of speaking, a set of tools is used repetitively to build more complicated models of different processes.

We divide the last four chapters (13–16) into two major application domains: **aquatic** and **terrestrial** ecosystems. We demonstrate that the concepts of energy flow (mainly productivity) and material cycling (mainly nutrients) prevail in these applications (Chapter 13). However, we are required to pay attention to the specificities of each domain. For example, in one case we deal with how nutrients and producers disperse and move in a water body (Chapter 14), and in the other case how water from rain is dispersed and stored in soil for use by producers (Chapter 15). A common feature of both domains is that sunlight and temperature condition various processes. In aquatic systems, the temperature of concern is that of water, whereas in terrestrial systems we deal with the temperature of air and soil. In addition, rainfall and evapotranspiration impose important conditions on the terrestrial domain (Chapter 15). Our final chapter illustrates how simple models are hooked together to generate more comprehensive models.

1.6 SIMULATION PLATFORMS

We can code model simulations in a variety of computer programming languages, for example, FORTRAN, C, C++, Java, Basic, Visual Basic, and Pascal. Furthermore, there are user-friendly interfaces that facilitate model building. For example, Stella supports code development by using flow diagrams. Many models developed over the last 40 years are now readily available on the Internet. Some of these have user-friendly interfaces and extensive documentation. Others require more expertise to use and adapt to the problem at hand.

Seem, an R package that I developed specifically for this book, allows the user to access a collection of preprogrammed models, change the input data, perform simulations, and plot output. It also offers the opportunity to modify the preprogrammed models so that you can build your own. This package uses R, a **GNU** (Gnu's Not Unix) program supported by the R project. The GNU project includes free software and a general public license. We introduce the R program later in this chapter. The computer sessions in this book assume a Windows installation of R. However, the R programs provided can run under other operating systems, such as Linux. We start our study of R in the next section of this chapter.

1.7 COMPUTER SESSION

1.7.1 Working Directory

The computer sessions of this book assume that you have access to write and read files in a **working directory**, which may be a folder in the local hard disk, **c:**, a network home drive, **h:**, or a removable drive (flash drive), like **e:**. For the purpose of following the exercises in this book, we denote the working directory by **seem**. Later, when we extract the downloaded **seem** archive, a folder or directory for each chapter will be created within the working directory.

Exercise 1.5

Make or create a folder (or directory) with the name **seem** in your working drive (e.g., **c:**, **h:**, or **e:**).

1.7.2 Installing R

Download R from the Comprehensive R Archive Network (CRAN) repository, http://cran.us.r-project.org/, by looking for the **precompiled binary distribution** for your operating system (Linux, Mac, or Windows). In this book, we will assume a Windows installation. Thus, for Windows, select the **base** and then the executable download for the current release; for example, at the time this chapter was last updated, the release was R-2.14.2. Save it in your disk and run this program following the installation steps. It takes just a few minutes. During installation, it is convenient to choose the option to install manuals in Portable Document Format (PDF).

Exercise 1.6

Install the current version of R.

1.7.3 Running R: Change dir and R Console

You could just double-click on the shortcut created during installation and you would get the R GUI that opens up with the > prompt on the **R Console** (Figure 1.3). However, to facilitate following the exercises in the book, I recommend that you first go to **File | Change dir ...**, browse to your working directory **seem**, and select it so that all subsequent references to files are made from this location. The R Console is where you type commands upon the > prompt and receive text output.

1.7.4 Setup

It is also possible to customize the launch or work directory by including it in the **Start in** option of the shortcut. To do this, right-click on the shortcut, go to **Properties**, then type, for example, **c:\seem** as your launch or work directory (Figure 1.4). Double-click to Run; R will start from this folder named **seem**. This remains valid unless it is edited.

When working on machines that are shared by many individuals (e.g., a university computer lab), a more permanent solution is to create a new personalized shortcut in your working directory. Find the shortcut on the desktop; right-click, select **Create Shortcut**, and browse to the desired location (folder **seem**) as shown in Figure 1.5. Then, right-click on this new shortcut and select **Properties**, and edit **Start in** as shown before. Thereafter, run R by double-clicking on this new personalized shortcut.

1.7.5 Help and Reference

R Manuals are available, via the **Help** menu, in PDFs that are viewable, for example, with Acrobat Reader (Figure 1.6), and hypertext markup language (HTML) that is viewed using a web browser (Figure 1.7).

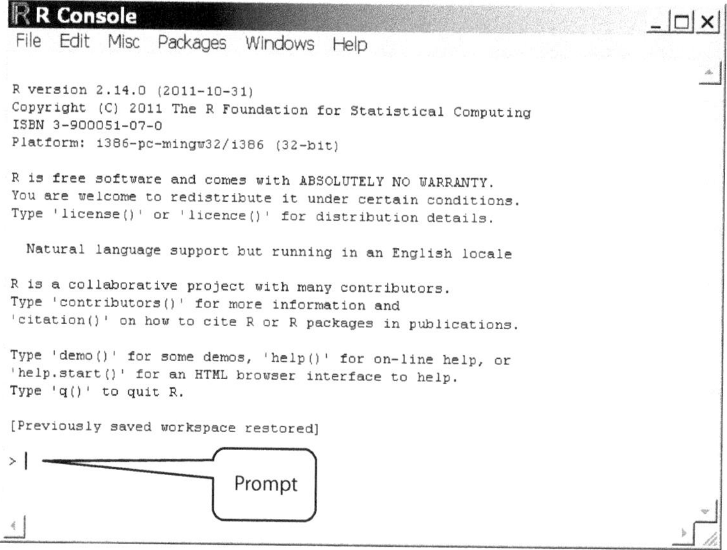

FIGURE 1.3 Start of the R GUI and R Console.

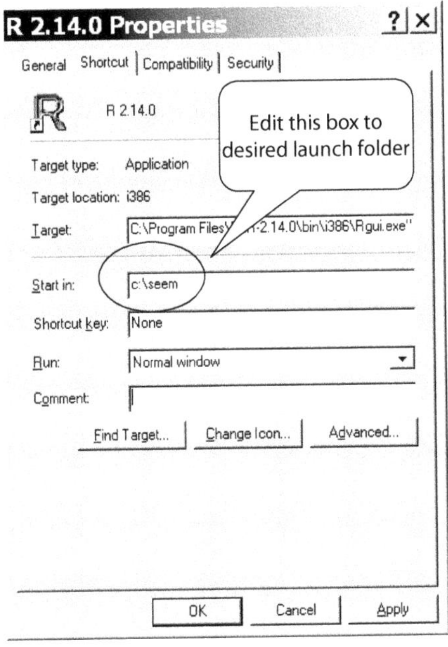

FIGURE 1.4 Modifying the **Start in** option in the R shortcut properties box.

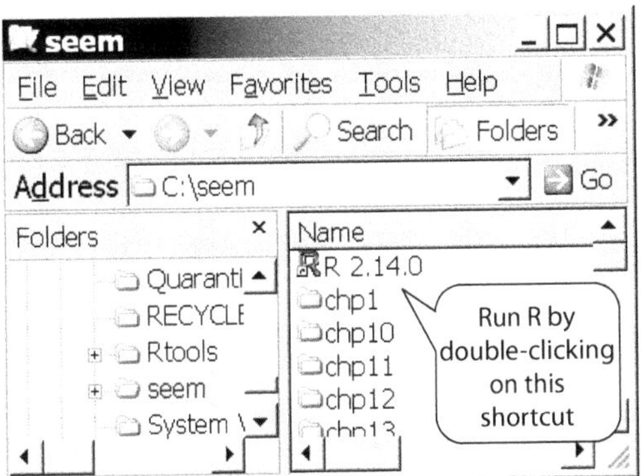

FIGURE 1.5 New R shortcut to reside in your working folder (e.g., **c:\seem**).

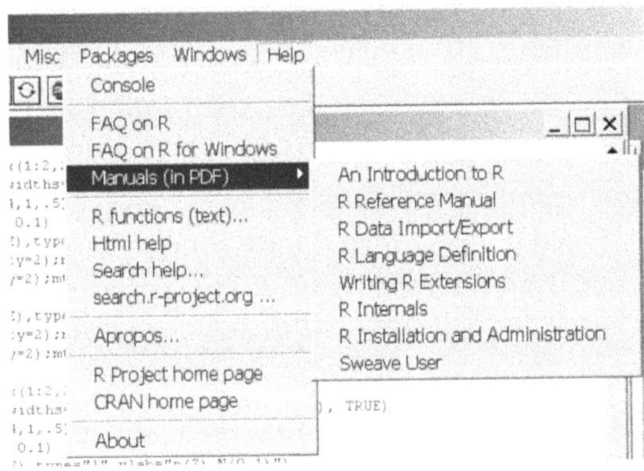

FIGURE 1.6 Finding R manuals in PDF from the Help menu.

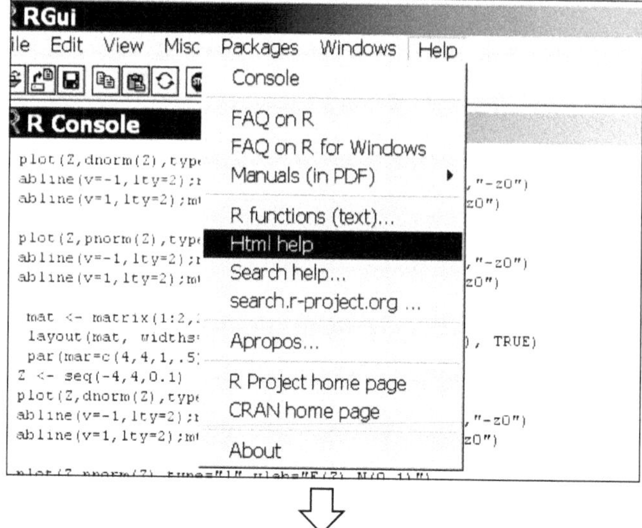

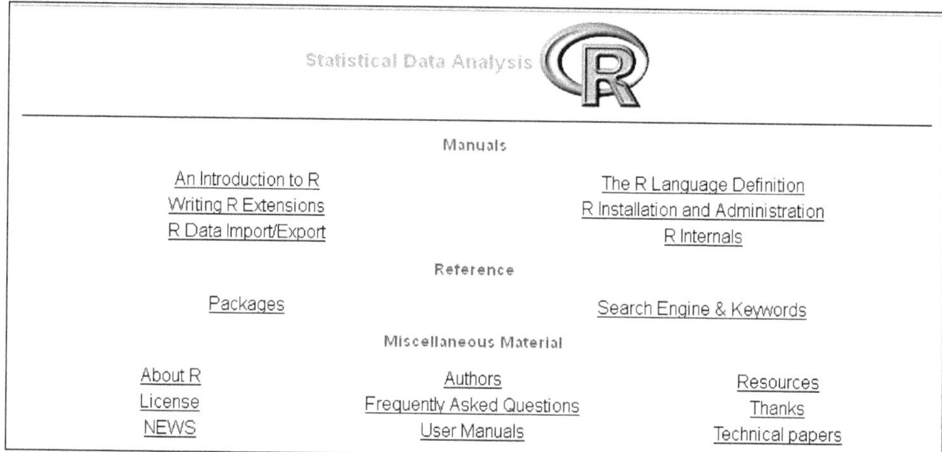

FIGURE 1.7 Help from a browser in HTML format.

You can also obtain help on specific functions from the Help menu; select **R functions (text) …** as shown in Figure 1.8 and then type the function name in the dialog box. In addition, you can obtain help on specific functions directly from the R Console by typing `help` at the prompt, for example, `help(plot)`. It is also possible to just type the question mark followed by the function name, for example, `?plot`. You can also launch the Help menu with the `help.start(·)` function.

1.7.6 Graphics Device

This is where you receive graphical output (Figure 1.9). From an active graphics window select **File | Print** from the menu to print a graph or **File | Save as** to save in a variety of graphics formats, including metafile, postscript, BMP, PDF, and JPEG. You could simply use copy and paste from the graphics window, or **File | Copy to clipboard** and then paste to a desired application.

1.7.7 Store Your Objects

The workspace contains "objects," which are data, functions, and results. They are stored in a file, **.Rdata**. Use the **File | Save workspace** menu to store an image of your objects. The **.Rdata** file

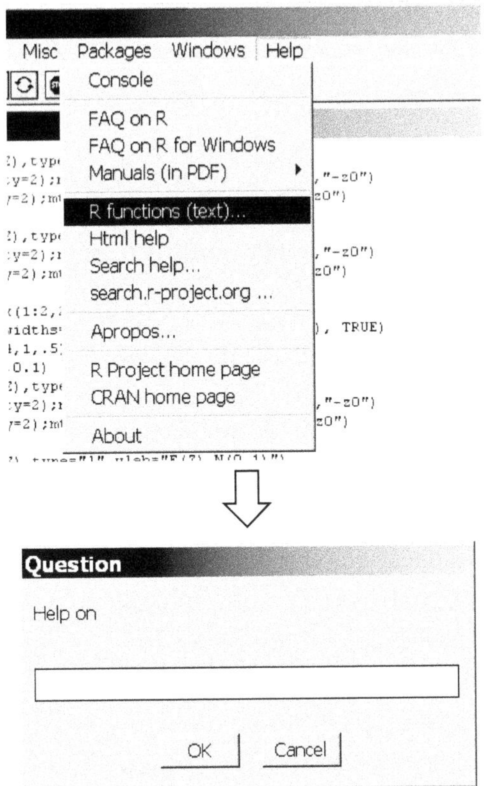

FIGURE 1.8 Help on specific functions.

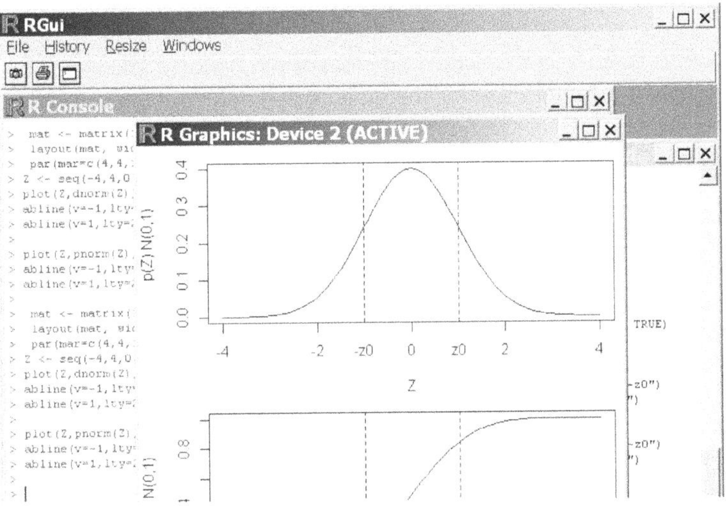

FIGURE 1.9 R graphics windows.

resides in the working directory, which was selected by the **File | Change dir ...** option or specified in the **Start In** field of the R shortcut. We will use this **.Rdata** file for all computer sessions to facilitate access to functions created by various exercises. After launching the program, you can load the workspace using **File | Load Workspace** and browse to find the desired **.Rdata** file. Alternatively, you can also simply double-click on the **.Rdata** file to launch the program and load the workspace.

To list objects in your workspace, type `ls()` or `objects()`; if this is your first run, then you would not have existing objects and you will get `character(0)`.

```
> ls()
character(0)
> objects()
character(0)
```

Just to see how this works, create an object by simply typing

```
>a <- 10
```

where the operator "<-" assigns value 10 to object **a**; objects are not stored until they are assigned a name. Equivalently you can write the same using the equals sign ("="). However, the equals sign is reserved for other purposes, like giving values to arguments of functions. Double-check that you indeed have the newly created object by using `ls()`:

```
> ls()
[1] "a"
```

The number in brackets on the left-hand side is the position of the object in the list. We can check the contents of **a** by typing its name:

```
> a
[1] 10
```

Exercise 1.7

To make sure you understand the workspace, save your workspace **.Rdata** file. Then, close R and start a new R session. Load the workspace and make sure you have created the object before.

You may want to have control of where the **.Rdata** file is stored and store objects in different **.Rdata** files. For now I recommend using the same **.Rdata** folder in the launch folder for all computer sessions to facilitate access to the functions and objects you will create using the various examples of this book.

1.7.8 Scripts

When editing a long sequence of commands, a convenient tool is to write a **script** to sort and visualize all the previous commands at once. From **File | New Script** you get an editor window to type your code, which can be saved using **File | Save as** with the extension **.R**, for example, **myscript.R**. It is easy to execute script lines: just right-click on the line or selection of lines and choose **Run line or selection**. To run the entire script from the active script window go to the **Edit** menu and select **Run all**. Later, to use the script, go to **File | Open script** and select the desired file. In addition, you can type **source("myscript.R")** or use **File | Source** from the menu to execute the entire script. Another way of editing scripts is to use a separate text editor and then copy and paste the desired lines to the R Console for execution.

Exercise 1.8

Create a simple text file, **test.txt**, using a text editor and save within the **seem** folder. Then, remove this file.

1.7.9 Close R Session

You can use **q()** or **File | Exit** to finalize R and close the session. When you do this, a message window asks whether you want to save the workspace (Figure 1.10). Reply "Yes," so that you will have the created objects available for the next time you use R.

1.7.10 Basic R Skills

The R material in this chapter is only a very brief tutorial guide to start using R interactively from the RGui in Windows. There is a lot more to R than summarized in these notes. In later chapters, I will provide more details on R; for example, Chapter 2 explains how to do simple calculations and graphics, and Chapter 3 describes how to read data from files.

1.7.11 Data Files and Functions

One of the advantages of the R system is the availability of packages to perform various methods of analysis. I have organized the data sets and functions for this book as a package named **seem** written to support this book. The package is available from the R repository. From the RGui, select **Packages | Install packages(s)**, select a mirror site near you, browse to find **seem**, and select. Once installed and successfully unpacked, use **Packages | Load package** and select **seem**. You do not need to repeat the package installation as long as you do not re-install the R software or want to update the package.

Also, the package is available as an archive, **seem_*.tar.gz**, from the websites given in the preface. Download and install from **Packages | Install packages(s) from local zip files**. Once installed and successfully unpacked, you are ready to load it by using **Packages | Load package** and selecting **seem**. We will start to learn how to use **seem**'s functions in Chapter 4. For now, we want to unpack the data file, which should have been stored as a **datafiles.zip** file in **c:\Program Files\R\R-2.14.2\library\seem\datafiles**. You can copy this file to your working directory, **seem**.

The reason for using archives is that download time is reduced by using a compressed format. You can unzip the **datafiles.zip** file using Windows or an archiving application and this will extract several folders, each containing several files. Each chapter corresponds to a folder; for example, **chp1**, **chp2**, and so on are extracted within **seem**. We can examine the contents of each folder with the explorer or file manager. It is convenient to configure the file manager or explorer to show all the file extensions; otherwise, you will not see the **.txt** part of the filename.

In addition, to load a package from the Rgui you can run the function `library` from the R console, i.e. `library(seem)`. One way to load the package automatically every time you start R is to add the following segment of code to the **Rprofile.site** file in R's **etc** directory, which typically is **C:\Program Files\R\R-2.14.2\etc**. The **Rprofile.site** is a text file and can be edited with a text editor.

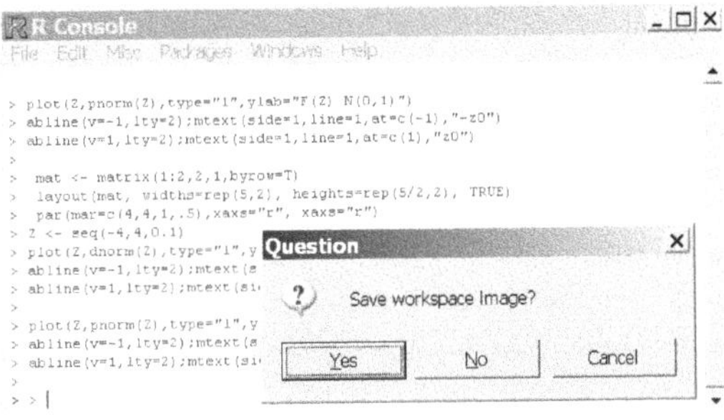

FIGURE 1.10 A pop-up appears when you are ready to close a session.

```
local({
  old <- getOption("defaultPackages")
  options(defaultPackages = c(old, "seem"))
})
```

In Chapters 10, 11, 14, and 16 we will use programs written in Fortran and C and executed from an interface R function. These are part of library **seem.dll** of seem. Therefore, this dll must be loaded before using those programs by executing function `dyn.load()` in the following manner:

```
>dyn.load(paste(.libPaths(),"/seem/libs/i386/seem.dll",sep=""))
```

You may want to postpone this last part until Chapter 10, but it is convenient then to add this line of code to the **Rprofile.site** file as well. Therefore, you may add the following to the file to load seem and the seem.dll automatically every time you start R:

```
local({
  old <- getOption("defaultPackages")
  options(defaultPackages = c(old, "seem"))
  dyn.load(paste(.libPaths(),"/seem/libs/i386/seem.dll",sep=""))
})
```

There are further instructions on the book's website on how to avoid loading the package every session and execute the models given in the book, along with an additional interface, **Rseem**, and other functions and data files.

1.8 SUPPLEMENTARY READING

Several books are useful for a first introduction to ecological and environmental models. They employ a variety of approaches: system dynamics that use Stella to implement numerical methods (Ford, 1999); tutorial explanations of basic concepts with numerical methods developed in Excel (Hardisty et al., 1993); and system analysis and programs in BASIC that emphasize earth science (Huggett, 1993) or biology (Keen and Spain, 1992). Although Hemond and Fechner's book (1994) is not a modeling book, it provides an introduction to the major concepts underpinning exposure models in water, air, and land. Jorgensen's classic book (1988) covers a wide range of topics. The book by Swartzman and Kaluzny (1987) covers a wide range of ecological models and techniques for programming simulations. Kot's book (2001) is a comprehensive introduction to mathematical ecology with abundant examples. Jopp et al. (2011) offer a broad perspective of modeling applications to ecological systems with practical implications.

Many engineering students are not familiar with ecology and may need to read from some introductory books, such as those by Ricklefs and Miller (2000), Colinvaux (1993), and Gotelli (2008). In addition, these books present some of the fundamentals by using mathematical models; therefore, they serve as a transition to the simulation material covered in this book. For those readers unfamiliar with earth science and environmental science, some basic readings are recommended (Kaufman and Franz, 2000).

Zeigler et al. (2000) have formalized the theory of simulation and searched for ways to integrate a variety of simulation approaches. A comprehensive treatment of scientific programming and simulation using R is now available (Jones et al., 2009). In the last few years, several ecological modeling books have used R (Bolker, 2008; Ellner and Guckenheimer, 2006; Soetaert and Herman, 2009; Stevens, 2009).

2 Review of Basic Mathematical Concepts and Introduction to R

This chapter is a brief and informal review of some basic concepts from calculus and probability and an introduction to basic calculations using R. If you are not familiar with mathematical notions, you may also have to read introductory textbooks on calculus and statistics.

2.1 VARIABLES

A **variable** represents a varying quantity, for example, time t. It can take one of many values; for example, time $t = 2$ days denotes that t takes the value 2 in units of days. We will work mostly with variables that can take **real** values, for example, $t = 1.533$. Also, we define an interval of possible values; for example, t could take values in the interval from zero to 10 days, including the interval ends 0 and 10. It is common to denote a close interval like this by [0, 10]. We refer to this type of variable as **continuous**.

Exercise 2.1

Use a variable X to denote human population on Earth. Explain why it varies and give an example of a value.

2.2 FUNCTIONS

A **function** is a map that assigns a value of a dependent variable to each value of an independent variable. For example, $X(t)$ is a function X of t, where X is the **dependent** variable and t is the **independent** variable. For a more specific example, take

$$X(t) = at \tag{2.1}$$

which is a function expressing that variable X is proportional to the variable t, where the constant of proportionality is the coefficient a. This is an example of a **linear** function (Figure 2.1).

As another example, consider the function

$$X(t) = at^b \tag{2.2}$$

where the independent variable has been raised to a power b. When $b = 2$, this is a quadratic equation and it is called a **parabola**. We obtain a **nonlinear** function when the exponent is different from one, $b \neq 1$; see Figure 2.2.

In this book, we will use a wide variety of nonlinear functions. An example is given as

$$X(t) = \frac{a}{t+b} \tag{2.3}$$

which corresponds to a **hyperbolic** form. Using Equation 2.3 with $a = 2$ and $b = 1$, we obtain the graph shown in Figure 2.3.

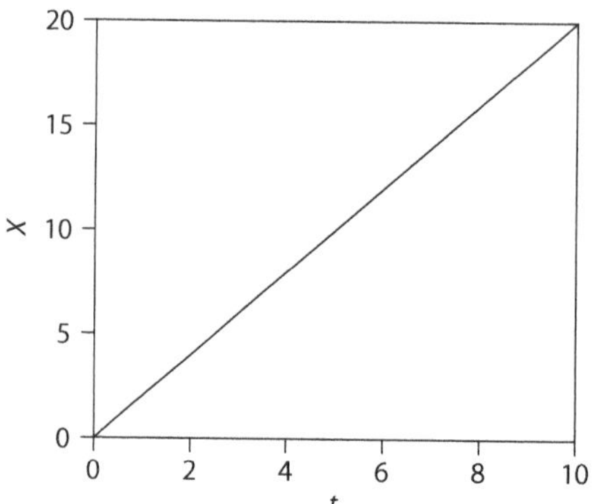

FIGURE 2.1 A linear function.

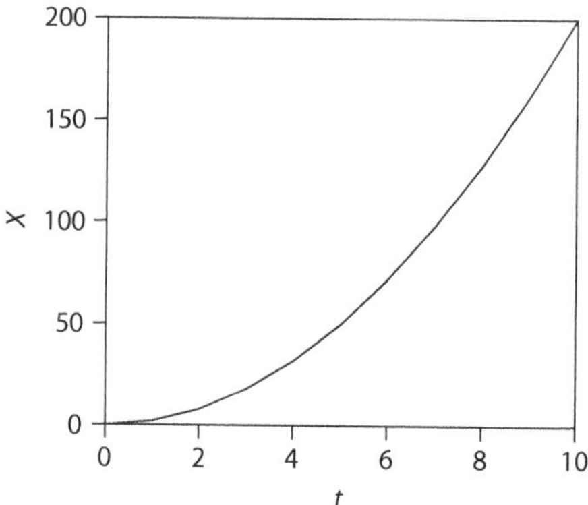

FIGURE 2.2 A parabolic function.

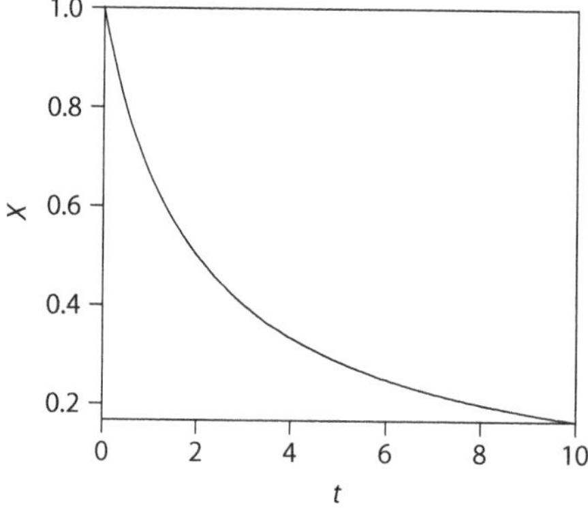

FIGURE 2.3 A hyperbolic function.

Another example is given as

$$X(t) = \exp(at) = e^{at} \tag{2.4}$$

which corresponds to an **exponential** function. In this last equation, the number e = 2.71828 is the base of natural **logarithms**. Recall that a logarithm is the power to which a base is raised to obtain a given number. For example, using base 10, log(100) = 2 or $10^2 = 100$. We use the exponential function often in this book to express that variable *X* has a rate of change that increases or decreases linearly with *X*. We will cover the exponential function in detail in the next chapter.

2.3 DERIVATIVES AND OPTIMIZATION

A concept often used in environmental models and employed throughout this book is the **derivative** of a function. For example, the derivative of $X(t)$ with respect to time t is denoted by

$$\frac{dX(t)}{dt} \tag{2.5}$$

This derivative represents a **rate of change** of X with time t and is equal to the **gradient** or **slope** of X with respect to t. This assumes that X varies continuously along t. You can think of a derivative as a ratio of very small change of two variables, for example, a very small change ΔX of X divided by a very small change Δt of t. The derivative is approximately equal to the slope obtained from the ratio $\Delta X/\Delta t$ (Figure 2.4). Therefore,

$$\frac{dX}{dt} \sim \frac{\Delta X}{\Delta t} \tag{2.6}$$

when deltas ΔX and Δt are very small or **infinitesimal** and can be referred to as **differentials** dX and dt of X and t.

There are rules to calculate the derivative of a function. The simplest ones are for polynomials or sums of power functions, like the one in Equation 2.2. The derivative is the product of the exponent and the variable raised to the power minus one. So the derivative of X with respect to t for the X given in Equation 2.2 is

$$\frac{dX}{dt} = abt^{b-1} \tag{2.7}$$

Exercise 2.2

Assume that $a = 1$ and $b = 2$. Evaluate the derivative of X with respect to t for the X given in Equation 2.2 and plot it. Is the derivative a constant with respect to t? Is the derivative a linear function with respect to t?

One application of derivatives is finding an **optimum** (minimum or maximum) of a function with respect to a variable. At an optimum, the derivative takes a value of zero. For example, the quadratic function or the parabola

$$X(t) = a + (t - b)^2 \tag{2.8}$$

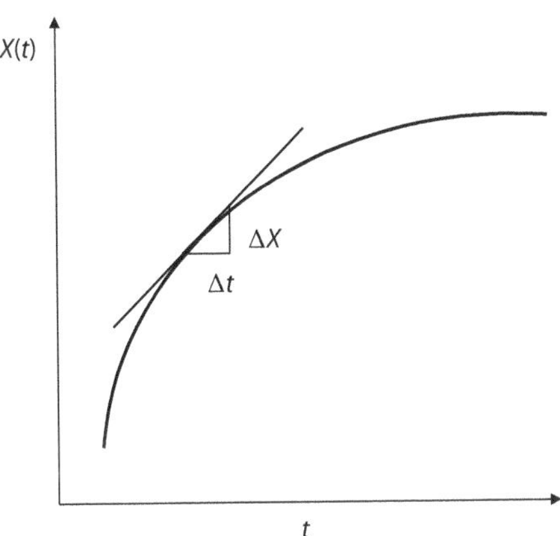

FIGURE 2.4 Derivative as the slope of the variable X as a function of time, t.

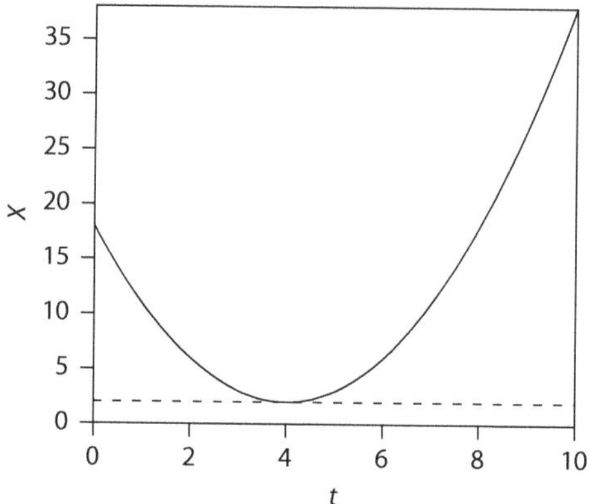

FIGURE 2.5 The derivative is zero at an optimum.

has a derivative

$$\mathrm{d}X/\mathrm{d}t = 2(t - b) \tag{2.9}$$

At an optimum, this derivative must be zero, and this will happen when the term in parentheses becomes zero, which occurs when $t = b$. Therefore, an optimum occurs at this value of t. Substituting in Equation 2.8, we see that the function takes the value

$$X(b) = a + (b - b)^2 = a$$

For $a = 2$ and $b = 4$, we can see that this optimum is a minimum with a value $a = 2$ and it occurs at $b = 4$ (Figure 2.5).

Exercise 2.3

Plot Equation 2.8 when $b = -4$ and $a = 3$. Find the values of the function and its derivative at $t = b$.

2.4 INTEGRALS AND AREA

You can think of the **integral** of a function as the inverse of the derivative and as the area under the curve representing the function within a defined interval. The integral is denoted by the integration symbol, ∫, applied to the function

$$\int_{t_1}^{t_2} X(t)\mathrm{d}t \tag{2.10}$$

where the term after the integration symbol is called the integrand and t_1 and t_2 are called the limits of the integration interval. The differential dt indicates that the integration is performed over the variable t. The area is calculated as a summation of many small rectangles of width dx and height $X(t)$ when dx is infinitesimal (Figure 2.6) (Carr, 1995).

There are rules to calculate integrals; some of the commonly used ones follow. The integral of the inverse of X is the logarithm:

$$\int \frac{1}{X}\mathrm{d}X = \ln(X)$$

The integral of a power function such as that in Equation 2.2 is

$$\int at^b \mathrm{d}t = \begin{cases} a\dfrac{t^{b+1}}{b+1} & \text{for } b \neq -1 \\ a\ln(t) & \text{for } b = -1 \end{cases}$$

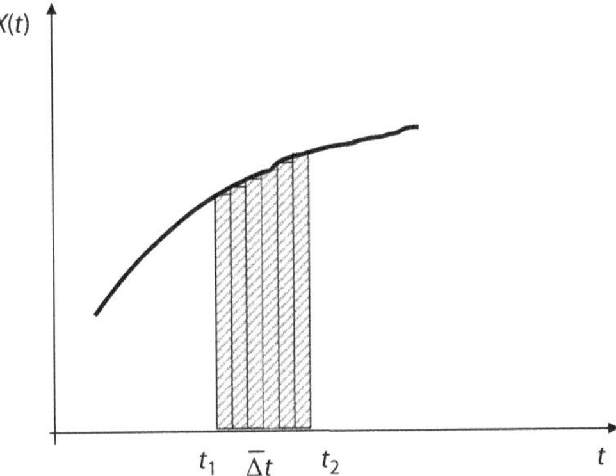

FIGURE 2.6 Integral as an area under the curve obtained by the summation of the area of many small rectangles of width Δt.

To evaluate for the given limits, one subtracts the integral evaluated at the lower limit from the integral evaluated at the upper limit. For example, if the limits for the integration of the term in Equation 2.2 were $t_1 = t_0 > 0$, $t_2 = t > t_0$, then

$$\int_{t_0}^{t} as^b \mathrm{d}s = \begin{cases} \dfrac{a}{b+1}\left(t^{b+1} - {t_0}^{b+1}\right) & \text{for } b \neq -1 \\ a\left(\ln(t) - \ln(t_0)\right) & \text{for } b = -1 \end{cases}$$

where we have used a dummy variable, s, for the integrand to distinguish it from the variable t used for the upper limit.

Exercise 2.4

Assume that $a = b = 1$ for the term in Equation 2.2. What is the area under the curve between $t = 2$ and $t = 6$? Illustrate the area using a graph like the one given in Figure 2.6.

2.5 ORDINARY DIFFERENTIAL EQUATIONS

Derivatives are used to build **differential equations** or **dynamic models** that state that the **rate of change** of X at any given time t, is a function $f(\cdot)$ of the value of X at that time t. These are **ordinary differential equations** (or **ODE** for brevity):

$$\frac{\mathrm{d}X(t)}{\mathrm{d}t} = f(X(t)) \tag{2.11}$$

When the function $f(\cdot)$ is linear, that is, the rate of change of a variable = coefficient × variable, we get a linear ODE. Using the derivative of X with respect to time t for the rate of change of X, we write

$$\frac{\mathrm{d}X(t)}{\mathrm{d}t} = aX(t) \tag{2.12}$$

where the coefficient a is a **per unit** rate of change. We will discuss this type of linear ODE with more depth in the next chapter.

The solution of the ODE given in Equation 2.11 is a function, X, that satisfies the ODE. A solution of Equation 2.11 is found by separating the terms in X and t on different sides of the equation,

$$\frac{1}{f(X)}\mathrm{d}X = \mathrm{d}t \tag{2.13}$$

and by integrating both sides,

$$\int_{X(t_0)}^{X(t_1)} \frac{1}{f(X)} \mathrm{d}X = \int_{t_0}^{t_1} \mathrm{d}t \tag{2.14}$$

which requires the limits of integration for t and X. Here we define the initial time t_0 and the time t_1 at which we want to evaluate the solution. To evaluate the solution $X(t_1)$, we need to know the initial value $X(t_0)$ of X or the value of X at the initial time t_0. Depending on the complexity of $f(X)$, the ODE can be solved analytically using Equation 2.14. Often, we resort to a numerical calculation as explained in Chapter 4.

An extremely simple example will be the one where $f(X)$ is a constant a. In this case, the ODE is given as $\frac{\mathrm{d}X}{\mathrm{d}t} = a$. To solve it, apply Equation 2.14,

$$\int_{X(t_0)}^{X(t_1)} \frac{1}{a} \mathrm{d}X = \int_{t_0}^{t_1} \mathrm{d}t$$

which yields

$$\frac{1}{a}\left(X(t_1) - X(t_0)\right) = t_1 - t_0$$

and after solving for $X(t_1)$,

$$X(t_1) = a(t_1 - t_0) + X(t_0)$$

we obtain a straight line with slope a starting at the initial condition $X(t_0)$.

Exercise 2.5

Assume that $t_1 = t$ and $t_0 = 0$. Write the solution for the ODE of the example immediately above, using the initial condition $X(0) = 1$.

It is important to start getting familiar with the terminology used in modeling. We can do this at this early stage by recognizing how some of the terms apply to this simple model as follows:

- $\mathrm{d}X/\mathrm{d}t = aX$ is a dynamic **model**. In this case, it is given by an **ODE**.
- $X(t)$ is the **state variable** at time t. In this case, the state could also be called the **output**. In other cases, the output is a function of the state.
- a is a **parameter**.
- $X(t_0)$ is the **initial condition** or **initial state** (technically it is also a parameter).

2.6 FUNCTIONS OF SEVERAL VARIABLES

In many instances, a variable is dependent on more than just one independent variable. For example, the variable X can be a function of time t and position z:

$$X(t, z) = at + bz \tag{2.15}$$

In this case, we can look at two-dimensional (2-D) graphs to visualize the function. For example, with $a = 2$ and $b = 1$, Equation 2.15 yields the results shown in Figures 2.7 and 2.8. Figure 2.7 represents the results as isolines or contour lines (or lines of equal value of X) and Figure 2.8 uses an image representation.

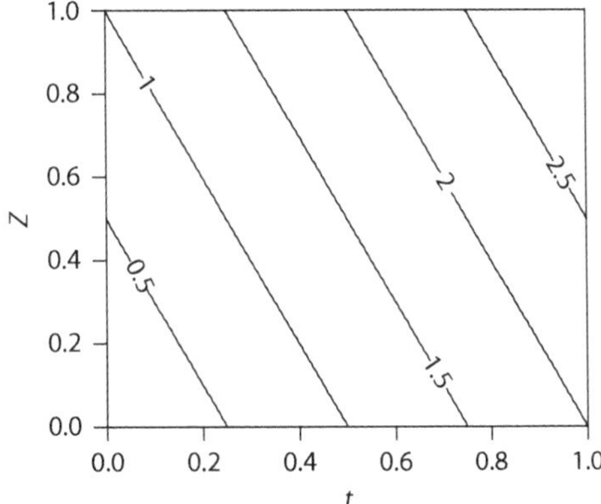

FIGURE 2.7 A linear function of two variables: an isoline or contour line view.

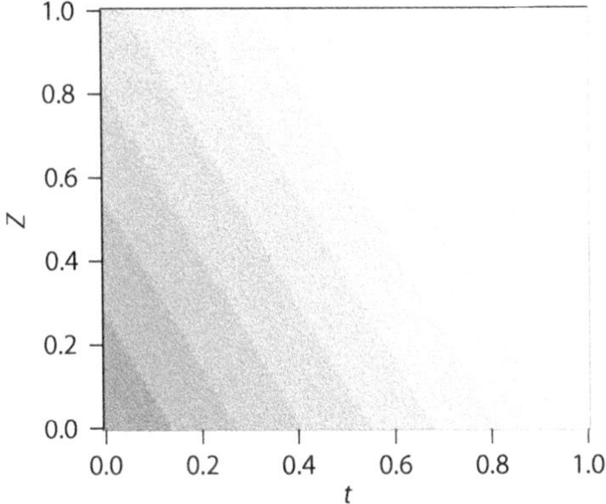

FIGURE 2.8 Function of two variables: image representation. The lighter the gray, the higher the value of *X*.

When we have more than one independent variable, we can take the derivative of the function with respect to each one of the variables. These are the **partial derivatives** of the function. In the example above, we can have two partial derivatives, one with respect to *t* and another with respect to *z*:

$$\frac{\partial X}{\partial t} = a \qquad \frac{\partial X}{\partial z} = b \tag{2.16}$$

Note that the symbol ∂/∂ is used for the partial derivative, which is different from the one used for ordinary derivatives, d/d.

One application of multivariable functions and partial derivatives is finding an **optimum** (minimum or maximum) of a function with respect to a set of parameters. At an optimum, all the partial derivatives take a value of zero. We will use this concept in the next chapter.

2.7 RANDOM VARIABLES AND DISTRIBUTIONS

A random variable (RV) is a rule or a map that associates a **probability** to each **event** in a sample space. The events can be defined from the intervals present in a range of real values; in this case, the values of an RV, *X*, are continuous in this range. We call this type of RV **continuous**. Consider the following example: measurements of concentration of a mineral (in parts per million [ppm]) at a given location can take values between 0 and 10,000 ppm. We may be interested in an event defined as that occurring when the measured concentration is in the interval 10–15 ppm.

A probability density function (pdf) $p(X)$ is based on the intervals; the probability of a value being in an infinitesimal interval of X between x and $x + \mathrm{d}x$ (Figure 2.9) is given by

$$p(x)\mathrm{d}x = P[x < X \le x + \mathrm{d}x] \tag{2.17}$$

where P denotes probability. Here, $p(x)$ is always positive or zero, that is, $p(x) \ge 0$.

The probability of a value being in an interval of X between a and b can be found using the integral

$$P[a < X \le b] = \int_a^b p(x)\mathrm{d}x \tag{2.18}$$

which is the area under the curve $p(x)$ for a given interval from a to b (Figure 2.10). When the interval is the whole range of values of X, then the value of the integral should be 1:

$$\int_{-\infty}^{+\infty} p(x)\mathrm{d}x = 1 \tag{2.19}$$

We have indicated the entire range of real values by selecting the limits from minus infinity ($-\infty$) to plus infinity ($+\infty$) or in practical terms from a very large negative value to a very large positive value.

Consider, for example, a uniform continuous RV. The density has the same value over the range $[a,b]$,

$$U_{a,b}(x) = \begin{cases} \dfrac{1}{b-a} & \text{for} \quad a \le x \le b \\ 0 & \text{otherwise} \end{cases} \tag{2.20}$$

as shown in Figure 2.11.

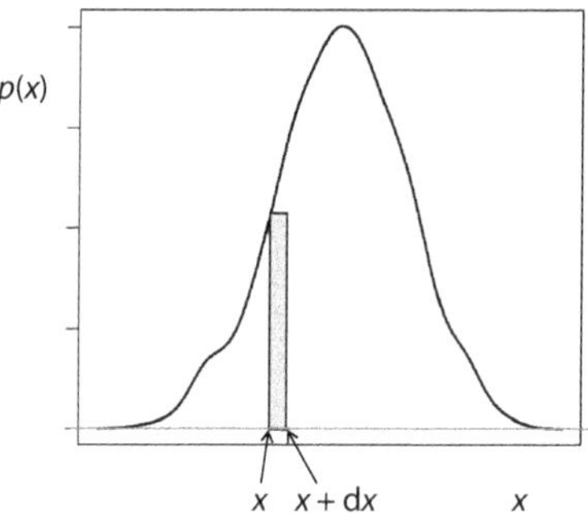

FIGURE 2.9 Probability density function of a continuous random variable. Probability is area under the curve between two values.

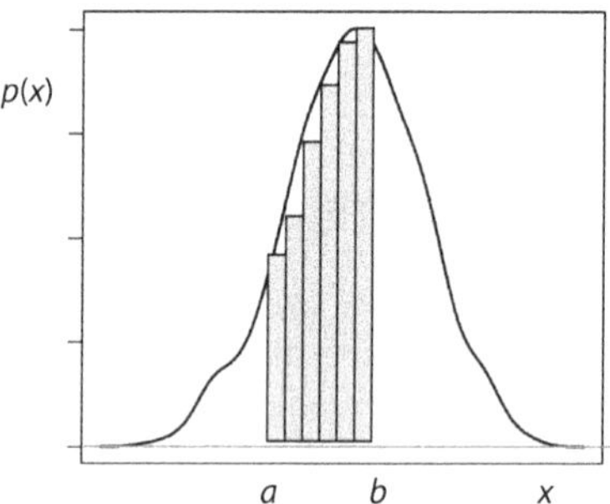

FIGURE 2.10 Probability of X having a value between a and b.

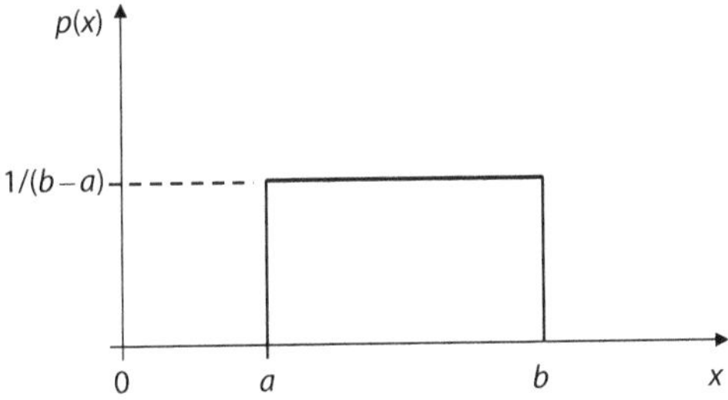

FIGURE 2.11 Probability density function of a uniform random variable.

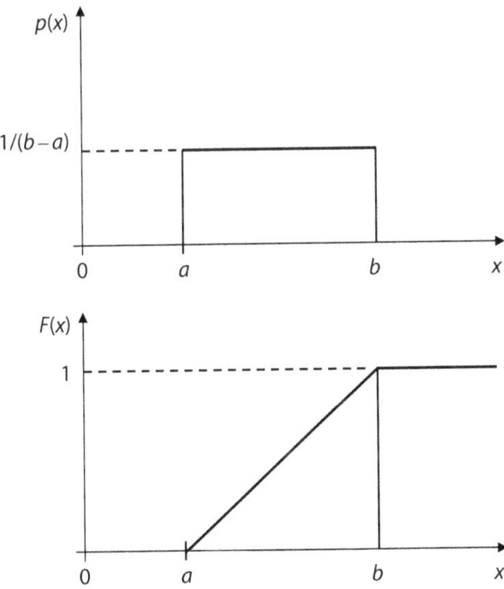

FIGURE 2.12 Probability density function and cumulative density function for a uniform random variable. Integration of a constant yields a linear increase (ramp function).

The "cumulative" distribution function (cdf) at a given value is defined by "accumulating" all probabilities up to that value:

$$F(x) = P[X \leq x] = \int_{-\infty}^{x} p(s)\mathrm{d}s \tag{2.21}$$

Note that the value at which we evaluate the cdf is the upper limit of the integral. Variable s is a dummy variable to avoid confusion with x. The cdf $F(x)$ at a value x is the area under the density curve up to that value. The value of the cdf for the largest value of X is its largest value and should be equal to 1. For example, for the uniform continuous RV, $U_{a,b}(x)$ is a ramp with slope $1/(b - a)$; see Figure 2.12.

2.8 RANDOM VARIABLES AND MOMENTS

The **first moment** of X is the **expected value** of X denoted by operator $E[\,]$ applied to X, that is $E[X]$. This is the same as the **mean** of X. When X is continuous,

$$\mu_X = E[X] = \int_{-\infty}^{+\infty} xp(x)\mathrm{d}x \tag{2.22}$$

Note that the integration is over all values of X.

As an example, consider an RV uniformly distributed in [0,1]. In this case, $b = 1$ and $a = 0$. We know that $p(x) = 1/(b - a) = 1$:

$$\mu_X = E[X] = \int_0^1 x\text{d}x = \frac{1}{2}x^2\Big|_0^1 = 1/2 \tag{2.23}$$

The expected value or mean is a theoretical concept. To calculate it, we need the pdf of the RV. The mean is not the same as the **statistic** known as the **sample mean**, which is the arithmetic average of n data values x_i comprising a sample:

$$\bar{X} = \frac{1}{n}\sum_{i=1}^{n} x_i \tag{2.24}$$

Note that the mean (first moment) is denoted with the Greek letter μ, whereas the sample mean or average is denoted with a bar on top of X, that is to say, $\bar{X}$.

The second **central** (i.e., with respect to the mean) moment is the **variance** or the expected value of the square of the difference with respect to the mean:

$$\sigma_X^2 = E[(X - \mu_X)^2] \tag{2.25}$$

If X is continuous, then

$$\sigma_X^2 = E[(X - \mu_X)^2] = \int_{-\infty}^{+\infty} (x - \mu_X)^2 p(x)\text{d}x \tag{2.26}$$

The expectation $E[.]$ is calculated by an integral when the RV is continuous. The variance is a theoretical concept. To calculate it, we need the pdf of the RV. The **standard deviation** is the square root of this variance:

$$\sigma_X = \sqrt{\sigma_X^2}$$

From the definition of variance given in Equation 2.25, we can derive a more practical expression by substituting $\mu_X = E[X]$ and expanding the square of a sum to obtain

$$\sigma_X^2 = E[(X - E(X))^2] = E[X^2 - 2XE(X) + E(X)^2]$$

Then, take the expected value of each term and use the fact that the expected value of a constant is the same constant to obtain

$$\sigma_X^2 = E[X^2] - 2E(X)E(X) + E(X)^2 = E[X^2] - 2E(X)^2 + E(X)^2$$

and finally

$$\sigma_X^2 = E[X^2] - E(X)^2 = E[X^2] - \mu_X^2 \tag{2.27}$$

The variance or second central moment is not the same as the **statistic** known as the **sample variance**, which is the variability measured relative to the arithmetic average of n data values, x_i, comprising a sample:

$$\text{var}(X) = s_X^2 = \frac{1}{n}\sum_{i=1}^{n}(x_i - \bar{X})^2 \tag{2.28}$$

This is the average of the square of the deviations from the sample mean. Alternatively,

$$\text{var}(X) = s_X^2 = \frac{1}{n-1}\sum_{i=1}^{n}(x_i - \bar{X})^2 \tag{2.29}$$

where $n - 1$ is used to account for the fact that the sample mean was already estimated from the n values. We write s_X^2 to denote the sample variance to differentiate from the variance σ_X^2.

This equation can be converted to a more practical one by using Equation 2.24 in Equation 2.29 and doing algebra to obtain

$$\text{var}(X) = s_X^2 = \frac{1}{n-1}\left[\sum_{i=1}^{n} x_i^2 - \frac{1}{n}\left(\sum_{i=1}^{n} x_i\right)^2\right] \tag{2.30}$$

This is easier to calculate because we can sum the squares of x_i and subtract the square of the sum of the x_i (Carr, 1995; Davis, 2002). Standard deviation is the square root of the variance. In this case, it refers to the standard deviation of a sample, which is different from the theoretical standard deviation.

The first and second central moments (that is, mean and variance) are also referred to as **parameters** of the RV (Carr, 1995; Davis, 2002). You have to be careful to avoid confusion with the term parameter as applied to an equation or a model. The mean and variance are different from the **statistics**, which are associated with the **sample** (Davis, 2002).

Another way of looking at this is to think of the pdf as a theoretical model expressing the underlying probability structure of the RV. These functions allow for the calculation of the moments. However, the statistics are calculated from the observed data and are used to **estimate** the moments.

2.9 SOME IMPORTANT RANDOM VARIABLES

The **uniform** RV has a pdf as given in Equation 2.20. The mean and variance are

$$\mu_X = \frac{b+a}{2} \qquad \sigma_X^2 = \frac{(b-a)^2}{12} \tag{2.31}$$

As an example, consider an RV uniformly distributed between a = 0, b = 1. The mean is 1/2 and the variance is 1/12.

The **normal**, or **Gaussian** RV, has a pdf given as

$$N_{\mu,\sigma}(x) = \frac{1}{\sqrt{2\pi}\sigma}\exp\left[-\frac{(x-\mu)^2}{2\sigma^2}\right] \quad \text{for} \quad -\infty < x < +\infty \tag{2.32}$$

with mean μ and variance σ^2. This is a symmetrical pdf, that is, the area under the curve left of the mean is the same as the area under the curve to the right of the mean. The area of the curve on both sides of the mean increases with distance: at one standard deviation on both sides ($\mu \pm \sigma$), the area is 0.68, at two standard deviations ($\mu \pm 2\sigma$), the area is 0.95, and at three standard deviations ($\mu \pm 3\sigma$), the area is 0.99 (Davis, 2002).

As another example, consider an RV normally distributed with mean = 1 and variance = 0.25. What is the probability of obtaining a value in between 0.5 and 1.5? The standard deviation is 0.5. This interval is one standard deviation away from the mean on each side. Therefore, the probability is 0.68.

A **standard normal** RV is a normal RV with a mean equal to zero ($\mu = 0$) and variance equal to one ($\sigma^2 = 1$). To obtain a standard normal RV from a normal RV, we subtract the mean and divide it by the standard deviation:

$$Z = \frac{X - \mu_X}{\sigma_X} \tag{2.33}$$

The new RV Z is standard normal. Its mean is 0 and variance is 1. All values to the left of the mean are negative ($z < 0$) and all values to the right of the mean are positive ($z > 0$). Note that z is scaled in units of standard deviation. Because the normal is symmetric, calculating the area under the standard pdf curve from $-\infty$ up to a value $-z_0$ (left of the mean) is the same as calculating the area under the curve from that value $+z_0$ to $+\infty$ (right of the mean); see Figure 2.13.

Exercise 2.6

Define an RV from the outcome of soil moisture measurements in the range of 20–100% in volume. Give an example of an event. Assuming that it can take values in [20,100] uniformly, plot the pdf and cdf. Calculate the mean and variance.

Exercise 2.7

At a site, the monthly air temperature is normally distributed. It averages to 20°C with a standard deviation of 4°C. What is the probability that the value of the air temperature in a given month exceeds 24°C? What is the probability that it is below 16°C or above 24°C?

The exponential pdf is given by

$$p(x) = a\exp(-ax) \tag{2.34}$$

and the cdf is given by

$$F(x) = 1 - \exp(-ax) \tag{2.35}$$

It has mean and variance equal to $\mu_X = 1/a$ and $\sigma_X^2 = 1/a^2$, respectively.

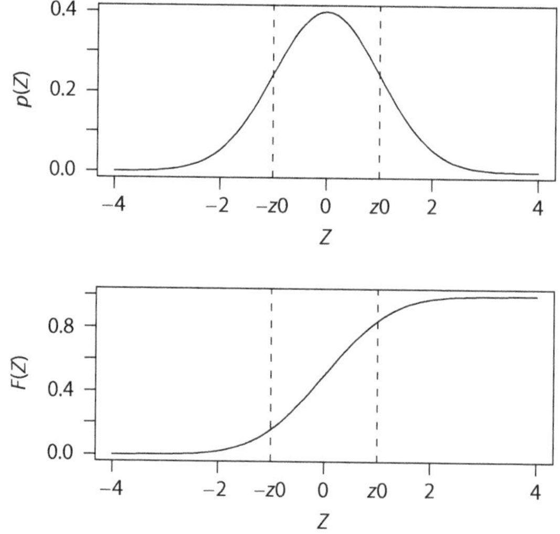

FIGURE 2.13 Standard normal pdf and cdf.

2.10 COVARIANCE AND CORRELATION

In many cases, we are interested in how variables relate to each other. The **bivariate** or simplest case is that of **two** RVs, X and Y. An important concept is the joint variation or the expected value of the product of the two variables, where each one is centered at the mean. This is called the **covariance** and can be written as

$$\text{cov}(X,Y) = E[(X-\mu_X)(Y-\mu_Y)] \tag{2.36}$$

Please note this is a theoretical concept, since the expectation operator implies using the distribution of the product. Therefore, we require the joint pdf of X and Y.

A derived concept is the **correlation coefficient** obtained by scaling the covariance to values less than or equal to 1. The scaling factor is the product of the two independent standard deviations:

$$\rho = \frac{\text{cov}(X,Y)}{\sigma_X \sigma_Y} \tag{2.37}$$

Because this product is always larger than the expected value of the product, the ratio is always less than 1. This fact can also be seen by calculating the correlation coefficient for maximum covariance, which occurs when X and Y are identical:

$$\rho = \frac{\text{cov}(X,Y)}{\sigma_X \sigma_Y} = \frac{\text{cov}(X,X)}{\sigma_X \sigma_X} = \frac{\sigma_X^2}{\sigma_X^2} = 1 \tag{2.38}$$

Expanding Equation 2.36, and since the expectation of a constant is the same constant, we obtain

$$\begin{aligned} \text{cov}(X,Y) &= E[(XY - Y\mu_X - X\mu_Y + \mu_X\mu_Y)] \\ &= E[XY] - E[Y]\mu_X - E[X]\mu_Y + \mu_X\mu_Y \\ &= E[XY] - \mu_Y\mu_X - \mu_X\mu_Y + \mu_X\mu_Y \\ &= E[XY] - \mu_X\mu_Y \end{aligned} \tag{2.39}$$

The same idea can be applied to a sample to obtain the **sample correlation coefficient**, r, where the covariance is a **sample covariance** and the denominator corresponds to the product of the standard deviations of the sample.

2.11 COMPUTER SESSION

2.11.1 Variables

When we use a continuous variable on a computer, we can only use a discrete approximation based on a set of values. Having more values in the interval improves the approximation. There are many ways to assign a set of values to a variable in R. For example, we can use a sequence from zero to 10 in steps of one:

```
> t <- seq(0,10,1)
```

Recall that the operator "<-" is used for assignment. Double-check that you have the newly created object by using `ls()`:

```
> ls()
[1] "t"
```

Object t is temporarily stored in **.Rdata** (workspace). We can double-check the contents of the object by typing its name:

```
> t
 [1] 0 1 2 3 4 5 6 7 8 9 10
```

We can see that this object is a one-dimensional (1-D) array. The number in brackets on the left-hand side is the position of the entry first listed in that row. For example, the entry in position 1 is zero. Object t should have 11 values, and this can be verified using the function `length`:

```
> length(t)
 [1] 11
```

We can obtain a better approximation to a continuous variable by changing the step size to 0.1:

```
> t <- seq(0,10,0.1)
> t
  [1] 0.0 0.1 0.2 0.3 0.4 0.5 0.6 0.7 0.8 0.9 1.0 1.1 1.2 1.3 1.4 1.5
 [17] 1.6 1.7 1.8 1.9 2.0 2.1 2.2 2.3 2.4 2.5 2.6 2.7 2.8 2.9 3.0 3.1
 [33] 3.2 3.3 3.4 3.5 3.6 3.7 3.8 3.9 4.0 4.1 4.2 4.3 4.4 4.5 4.6 4.7
 [49] 4.8 4.9 5.0 5.1 5.2 5.3 5.4 5.5 5.6 5.7 5.8 5.9 6.0 6.1 6.2 6.3
 [65] 6.4 6.5 6.6 6.7 6.8 6.9 7.0 7.1 7.2 7.3 7.4 7.5 7.6 7.7 7.8 7.9
 [81] 8.0 8.1 8.2 8.3 8.4 8.5 8.6 8.7 8.8 8.9 9.0 9.1 9.2 9.3 9.4 9.5
 [97] 9.6 9.7 9.8 9.910.0
>
```

Important tip: you can recall previously typed commands using the up arrow key. For example, you can use the up arrow key and recall the line already used to apply length:

```
> length(t)
[1] 101
```

We can refer to specific entries of an array using brackets or square braces. For example, the entry in position 24 of t is 2.3:

```
> t[24]
[1] 2.3
```

A colon symbol (":") declares a sequence of entries. For example, the first 10 positions of t, can be listed as follows:

```
> t[1:10]
[1] 0.0 0.1 0.2 0.3 0.4 0.5 0.6 0.7 0.8 0.9
```

To remove entries, use the minus sign; for example, use the following to remove the first entry of t:

```
> t[-1]
[1] 0.1 0.2 0.3 0.4 0.5 0.6 0.7 0.8 0.9 1.0 1.1 1.2 1.3 1.4 1.5 1.6
…. etc
```

As we can see, we have removed entry 0.0. In addition, we can confirm by using the length:

```
> length( t[-1])
[1] 100
```

A blank inside the brackets addresses all the entries. For example,

```
> t[]
[1] 0.0 0.1 0.2 0.3 0.4 0.5 0.6 0.7 0.8 0.9 1.0 1.1 1.2 1.3 1.4 1.5
... etc
```

This is the same as if we would have used just `t`.

2.11.2 Plots

As an example, let us use t in the interval [0,10] and $a = 2$ in Equation 2.1. To evaluate a function in R, first define the independent variable, next give a value to the parameter, and then apply the function. For example, define t as a sequence from 0 to 10 in steps of 1 and then apply a multiplication:

```
t <- seq(0,10,1)
a <- 2
X <-a*t
```

All values of array t are multiplied by a, and therefore, the new array X contains 11 values of variable X:

```
> X
[1] 0 2 4 6 8 10 12 14 16 18 20
```

A function can be represented as a graph in the x-y plane, where the x-axis or horizontal axis is used for the independent variable and the y-axis or vertical axis is used for the dependent variable. To visualize the result, we can run a simple plot function:

```
>plot(t,X)
```

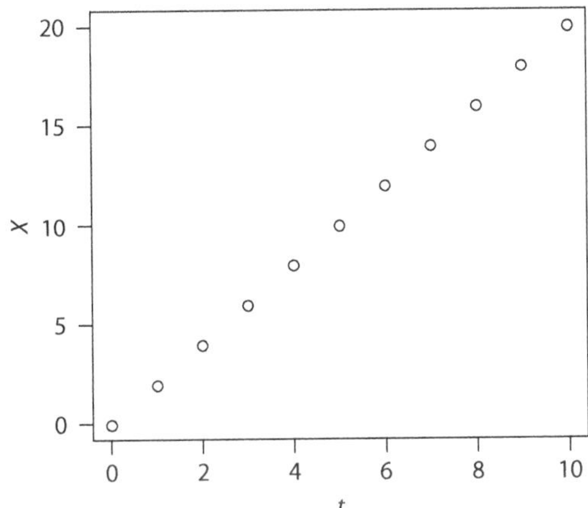

FIGURE 2.14 A plot of a linear function.

A graphics function will send the information to the active graphics window or create a new graphics window if needed. We can see that we obtain Figure 2.14, where the graph shows a circle symbol for each pair of values t, X. We can run the plot function using a line graph, obtained with the argument `type = "l"`. Here "l" is a letter character for "line." Do not confuse it with the number one:

```
>plot(t,X, type="l")
```

gives Figure 2.15.

Using the arguments `xaxs = "i"` and `yaxs = "i"`, the plot function skips the margins inside the plot area:

```
>plot(t,X, type="l", xaxs="i", yaxs="i")
```

as shown in Figure 2.1. Instead of including these arguments in the call to **plot**, it is often practical to first set graphics parameters using `par` before calling the `plot` function. This way all subsequent calls to **plot** will use these settings while the graphics device remains open. For example, the following will produce both plots with internal style "i" for both axes:

```
>par(xaxs="i",yaxs="i")
>plot(t,X)
>plot(t,X, type="l")
```

The contents of a graphics window can be saved in image format using **File|Save as**. For example, it can be saved as JPEG in one of three quality values, or TIFF, or PDF. You could also simply copy or save with a right-click on the graphics window or **File|Copy to clipboard** and then paste the contents into an application.

Exercise 2.8

Draw line graphs of the function given in Equation 2.1 for several values of coefficient a. Use $a = 0.1, 1, 10$. Produce three graphs.

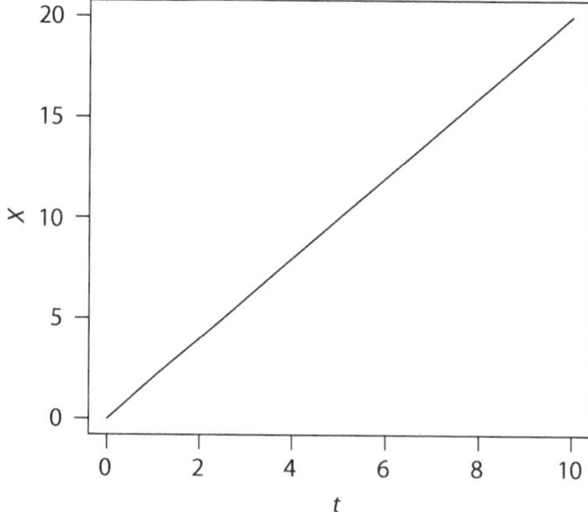

FIGURE 2.15 A line graph plot of a linear function.

Figure 2.2 can be produced with the following script where we used $a = 2$, $b = 2$. A semicolon is used to separate statements in the same line, and the symbol ^ is used for the power operator:

```
t <- seq(0,10,1)
a <- 2; b <-2
X <- a*t^b
plot(t,X, type="l")
```

Exercise 2.9

Plot Equation 2.2 using $a = 2$ and three different values of b: 0.1, 1, and 3.

With a script just like the previous one, we can produce Figure 2.3 except that we use `X <- a/(t+b)` and increment the number of values in the set for X by using time step of 0.1 so that the graph looks more continuous.

2.11.3 Defining Functions in R

We will use the hyperbolic function to introduce how to define functions in R. The following line assigns f as a function with arguments t, a, and b:

```
f <- function(t, a, b) X <- a/(t+b)
```

Now, to call the function, first give values to the arguments and use the name of the function:

```
t <- seq(0,10,0.1)
a <- 2; b <-2
X <- f(t,a,b)
plot(t,X, type="l")
```

Once we store a function in the workspace, we can write a call to it with proper arguments. It is available whenever we need it as long as we save the workspace before we exit the session. You can verify that the function, f, is stored in your workspace using `ls()`.

Exercise 2.10

Generate values for a function $y = ax + b$, $a = 0.1$, $b = 0.1$. Plot y for values of x in 0 to 1.

2.11.4 Exponential

For example, first define the independent variable as a sequence from 0 to 10 in steps of 0.1, then give a value to the rate coefficient, calculate variable `X` for all t using function `exp`, and plot:

```
t <- seq(0,10,0.1)
r<-0.1
X <- exp(r*t)
plot(t,X, type="l")
```

See the result in Figure 2.16. The limits of the x and y axes of the graph can be changed using `xlim=c(,)` and `ylim=c(,)`, where c(,) denotes a 1-D array with two entries. For example, `xlim=c(0,5)` has two elements, a minimum of 0 and a maximum of 5, and `ylim=c(0,2)` has two elements, a minimum of 0 and a maximum of 2.

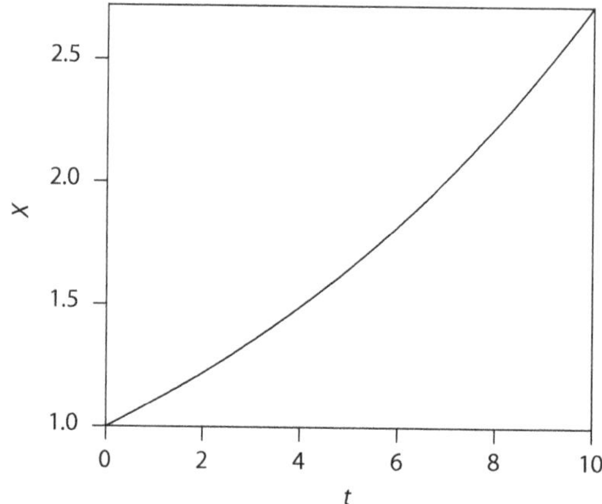

FIGURE 2.16 An exponential function with t taking values in [0, 10] and $r = 0.1$.

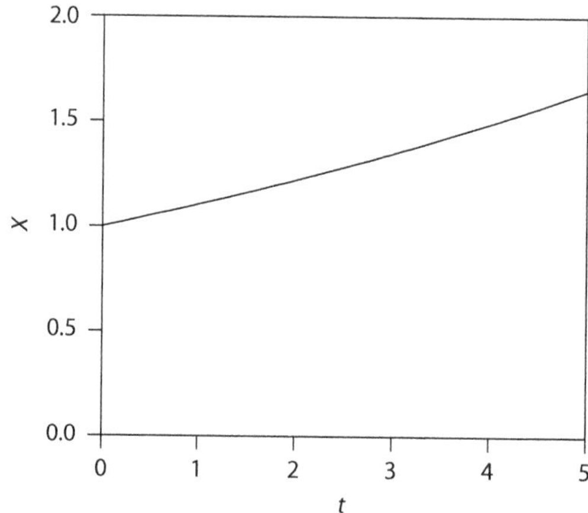

FIGURE 2.17 Changing the limits of the x and y axes of the graph.

```
>plot(t,X, type="l", xlim=c(0,5), ylim=c(0,2))
```

See Figure 2.17.

Exercise 2.11

Generate values for an exponential function with r = −0.1. Plot X for values of t in 0 to 10. Then limit the t-axis to interval [0,1].

2.11.5 Derivatives

A computer cannot take the exact derivative but it can calculate an approximation using Equation 2.6. Using R, we have the function `diff`, which can be used to calculate ΔX when X is given by an array. For example, assume we calculate `X` as a linear function as in the previous section:

```
> dt <- 0.1 # set time step
> t <- seq(0,10,dt)
> a <- 2; X <-a*t  # parameter and function
```

Now we take the difference and approximate the derivative by the ratio $\Delta X/\Delta t$:

```
> dX <- diff(X)
> dX.dt <- dX/dt
> dX.dt
 [1] 2 2 2 2 2 2 2 2 2 2 2 2 2  etc
```

We see that the derivative is a constant 2, which is the constant slope of the linear function. Note that the resulting array is shorter than the original because the first difference is the second entry minus the first entry of X.

```
> length(dX)
[1] 100
> length(X)
[1] 101
```

Exercise 2.12

Assume that $a = 1$ and $b = 2$. Calculate the derivative of X with respect to t for X as given in Equation 2.2 using **diff** and plot it. Compare this with Exercise 2.2. Hint: The first entry of t should be removed so that dX.dt and t have the same length and are compatible for the function plot.

We can evaluate the minimum or maximum of an array or a matrix using **min** and **max** functions. For example, define a function f for Equation 2.8 and evaluate it at a = 2, b = 4.

```
f <- function(t,a,b) {
  X <- a +(t-b)^2
  }
t <- seq(0,10,0.1)
X <- f(a=2,b=4,t)
```

We can use min to find the minimum and which to find the entry number at which the minimum occurs:

```
>min(X); t[which(X==min(X))]
[1] 2
[1] 4
```

2.11.6 Reading Data Files

As an example, let us work with a data set of 100 numbers drawn from a normal distribution of mean = 40 and standard deviation = 10 that are stored in a file **chp2\test100.txt**. Use a text editor (e.g., Notepad, Vim) to look at the file. Again, it is convenient to configure the list of files to show all the file extensions. Otherwise, you will not see the **.txt** part of the filename.

Because the working directory is **seem**, the path to the file is relative to this folder, for example, in this case **chp2\test100.txt**. Therefore, you could use this name to scan the file. Use forward slash "/" to separate folder and filename. Next, create an object, X, by reading or scanning this data file:

```
> X <- scan("chp2/test100.txt")
```

Values read or scanned from the file are assigned to object `X`. Double-check that you have the newly created object by using `ls()`:

```
> ls()
[1] "X"
```

Note: Object `X` will be stored in **seem\.Rdata** (i.e., in the workspace), but file **test100.txt** resides in **seem\chp2**. Double-check the object contents by typing its name.

```
> X
  [1] 48 38 44 41 56 45 39 43 38 57 42 31 40 56 42 56 42 46 35 40 30 49 36 28 55
 [26] 29 40 53 49 45 32 35 38 38 26 38 26 49 45 30 40 38 38 36 45 41 42 35 35 25
 [51] 44 39 42 23 44 42 52 55 46 44 36 26 42 31 44 49 32 39 42 41 45 50 39 55 48
 [76] 49 26 50 46 56 31 54 26 29 32 34 40 53 37 27 45 37 34 32 33 35 50 37 74 44
```

We can see that this object is an array. As we mentioned already, the number in brackets on the left-hand side is the position of the entry first listed in that row. For example, the entry in position 26 is 29 and the entry in position 51 is 44. Since this object is a 1-D array, we can check the size of this object by using command `length`:

```
> length(X)
[1] 100
```

Plot a histogram by using the function `hist` applied to a single variable (univariate or 1-D) object. Example:

```
>hist(X)
```

On the window corresponding to the graphics device, you will see a histogram like that shown in Figure 2.18. You could use a probability density scale (the bar height times the width would be a probability, i.e., from 0 to 1) on the vertical axis by using `hist` with the option `freq=F`:

```
>hist(X, freq=F)
```

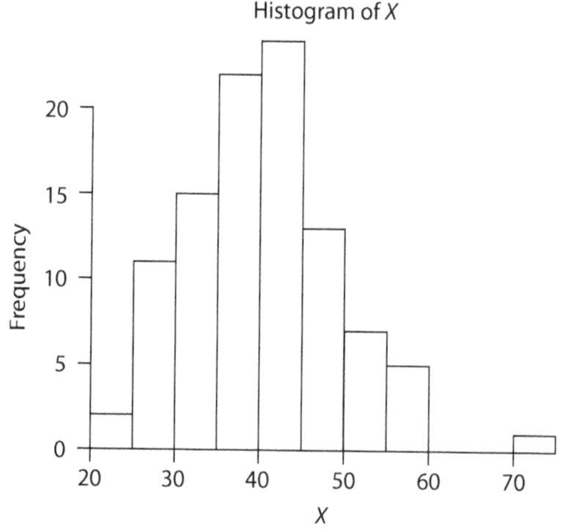

FIGURE 2.18 Histogram in frequency units.

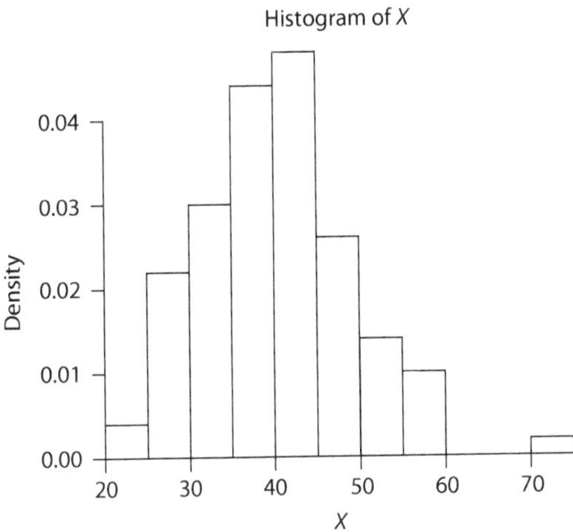

FIGURE 2.19 Histogram using probability units.

R handles many options in a similar manner: it gives a logical variable the value `T` (for **T**rue) and `F` (for **F**alse by default). Here, by default, `T` corresponds to the bar height that is equal to the count. The result in this case is given in Figure 2.19.

We can calculate sample mean, variance, and standard deviation using the functions `mean`, `var`, and `sd`.

```
> round(mean(X),2);round(var(X),2); round(sd(X),2)
[1] 40.86
[1] 81.62
[1]  9.034
```

The sample mean and standard deviation approach the population mean and standard deviations of 40 and 10, respectively.

Part II

One-Dimensional Models and Fundamentals of Modeling Methodology

3 Exponential Model

The exponential model is widely used in ecology and environmental science, for example, in population growth, mass loss, depuration of a toxicant from an organism, and decay of pollutants in water. In addition, the exponential model is a common building block of many models. In this chapter, we explain the fundamentals of the exponential model and expand on two simple applications in environmental modeling. The first is in population dynamics, for example, the growth or decline of populations (Hallam, 1986b). The second is in chemical fate, for example, the decay of chemicals (Hemond and Fechner, 1994).

3.1 FUNDAMENTALS

The simplest and most common ecological and environmental model is the **exponential model**. An exponential law of change of a variable $X(t)$ with time t is obtained by raising the number e to an exponent related to elapsed time:

$$X(t) = X(0)\mathrm{e}^{kt} \tag{3.1}$$

Equation 3.1 is often written using the function exp(·) as

$$X(t) = X(0)\exp(kt) \tag{3.2}$$

Equation 3.1, or equivalently Equation 3.2, describes the changes of a variable X with time t, beginning at the initial value $X(0)$ at time zero ($t = 0$).

This law results from solving an ordinary differential equation (ODE) or **model** stating that the **rate of change** of X at any given time t is proportional to the value of X at that time t. In other words, the rate of change of a variable = coefficient × variable. Using the derivative of X with respect to time t for the rate of change of X, we have

$$\frac{\mathrm{d}X(t)}{\mathrm{d}t} = kX(t) \tag{3.3}$$

where the coefficient k is a **per unit** rate of change or **rate coefficient**. When X multiplies this rate, we get the **net or total** rate of change, which will vary according to the value of X at any particular time t.

When the coefficient k does not depend on X, the model is **linear**. This occurs when k is either dependent only on time t or a constant. In this case, the exponential model is very simple and easy to work with. Figure 3.1 illustrates the linear rate for three values of the rate coefficient: one negative, another one positive, and the other equal to 0.

Why is Equation 3.2 the solution to Equation 3.3? To see this, we use the method described in the previous chapter to integrate Equation 3.3. Move X to the left-hand side (divide both sides by X) to obtain the per unit rate of change:

$$\frac{1}{X}\frac{\mathrm{d}X}{\mathrm{d}t} = k \tag{3.4}$$

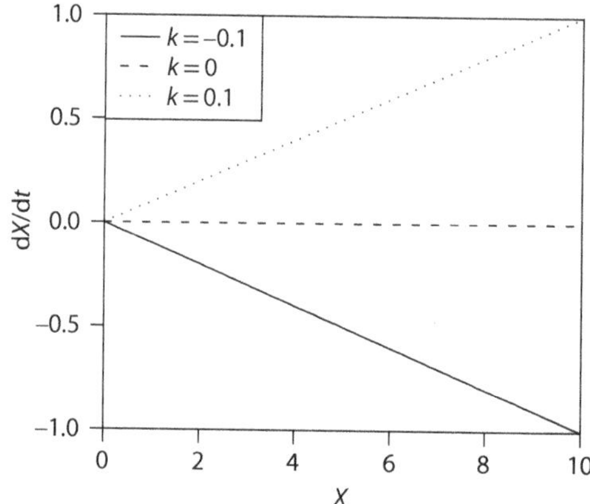

FIGURE 3.1 Exponential model for different values of the constant k. The slope of the line is the coefficient k.

Then, move dt to the right-hand side (multiply both sides by dt) and integrate both sides between the initial time, t_0, and the final time, t_f:

$$\int_{x(t_0)}^{x(t_f)} \frac{1}{X} \mathrm{d}X = \int_{t_0}^{t_f} k \mathrm{d}t \tag{3.5}$$

Recall that the integral on the left is the natural log to get

$$\ln[X(t_f)] - \ln[X(t_0)] = \ln\left[\frac{X(t_f)}{X(t_0)}\right] = k(t_f - t_0) \tag{3.6}$$

and now invert the log to obtain the exponential function for X at the final time, t_f,

$$X(t_f) = X(t_0)\exp\left[k(t_f - t_0)\right] \tag{3.7}$$

which can be calculated once we have the initial condition $X(t_0)$. It is common practice to let $t_0 = 0$ and t_f as variable t. In this manner, we get $X(t) = X(0)\exp(kt)$, which is precisely the solution given by Equation 3.1 or 3.2.

An easy way to check that Equation 3.1 is indeed the solution of Equation 3.3 is to take the derivative of Equation 3.1 with respect to t and plug it into the left-hand side of Equation 3.3 (Davis, 2002).

Exercise 3.1

Demonstrate that the derivative of Equation 3.1 satisfies the ODE given in Equation 3.3.

3.2 UNITS

Units are very important in modeling because variables represent physical, chemical, and biological properties and quantities, and therefore, to specify the amount of each variable we need to provide the units. The units of both sides of an equation must match. For example, the units of the left-hand side of Equation 3.3 are $\mathrm{d}X(t)/\mathrm{d}t$ = [units of X][units of t]$^{-1}$ and the units of the right-hand side are $kX(t)$ = [units of k][units of X]. For the units of both sides to match, it follows that the units of the coefficient k must be the inverse of the unit of time; that is to say, [units of k] = [units of t]$^{-1}$.

Recall that abbreviations to denote powers of ten are very useful for shorthand. Some common ones are **k** of kilo for 10^3 (e.g., kg, read as kilogram), **m** of milli for 10^{-3} (e.g., mg, read as milligram), μ of micro for 10^{-6} (e.g., μg read as microgram), and **p** of pico for 10^{-12} (e.g., pg read as picogram).

The **concentration** is given as a ratio of solute to solvent or of solute to solution. There are different ways of expressing this ratio. Two basic ones are per volume of solution or by weight of solution. Take, for example, percent by volume, milligrams per liter (mg/liter), and parts per million (ppm).

A useful convention to specify units is to give **dimensions** of length, mass, and time with the letters L, M, and T, regardless of the actual units. For example, mass in g or kg will both be specified with M; concentration in g/liter or mg/liter by $M \cdot L^{-3}$. This type of unit notation is very useful to perform a check on units without all the details of the actual units. For example, for Equation 3.3, when X is the concentration of a chemical $[M \cdot L^{-3}][T^{-1}] = [\text{units of } k][M \cdot L^{-3}]$, then we can see that $[\text{units of } k] = T^{-1}$.

Exercise 3.2

Suppose we are using an exponential model to describe the changes of light intensity as a function of the depth, z, in a water column. What are the units of coefficient k? Use the dimensional notation and then specify the units if z is given in meters.

Molar mass is the sum of all the atomic masses in a chemical formula. It is the mass corresponding to one mole. For example, the molar mass of silver chloride (AgCl) is 143.32 g/mole. Another important concentration unit is **molarity**, which is the ratio of moles of solute to the volume of solution in liters. That is to say,

$$M = \frac{\text{moles of solute}}{\text{liter of solution}}$$

Therefore, 1 M (read as one **molar**) is 1 mole per liter. Do not confuse M used here for molarity with the M used for mass in the dimensional notation L, M, and T explained earlier. Converting from mass of solute to moles of solute requires knowing the molar mass. For example, if you have 0.14332 g = 143.32 mg of AgCl in 1 liter, then the molarity is 1 milliM (one millimolar) or 1 mM because

$$\frac{143.32\,\text{mg/liter}}{143.32\,\text{g/mol}} = 1 \times 10^{-3}\,\text{mol/liter} = 1\,\text{mM}$$

Exercise 3.3

What would be the molarity if you have 1.4332 ng = 1433.2 pg of AgCl in 1 liter?

3.3 POPULATION DYNAMICS

A population could be defined as a collection of organisms of the same species in a given region (Ford, 1999; Hallam, 1986b; Keen and Spain, 1992). The exponential model applies to nonstructured, spatially homogeneous populations and assumes **density-independent** growth, which is to say the per capita rate is independent of population density.

By population density we mean the number of individuals per unit area or per unit volume, depending on what type of population we are considering. For example, for trees we may use individuals per unit area, and for zooplankton we may use individuals per unit volume (of water). Therefore, units of population density are expressed in $[\text{individuals}] \cdot L^{-2}$ or L^{-3}. For simplicity we will use [ind] or [indiv] to denote population density regardless of whether we are dealing with area or volume.

The main assumption of the exponential population model is (Hallam, 1986b) that the net rate of change is proportional to the density, X:

$$\text{rate of change } = r \times \text{density}$$

Mathematically, the ODE is given by

$$\frac{dX}{dt} = rX \tag{3.8}$$

The constant r is the **per capita rate of change** or the rate coefficient. The letter r commonly denotes the rate coefficient in population dynamics. The net rate of change, rX, will vary according to the value of X at any particular time, t.

The units are: for density [ind] and for rate [ind]$\cdot T^{-1}$. For example, if the unit of t is [days] then the unit of r is [days^{-1}] and the unit of dX/dt is [ind days^{-1}].

Exercise 3.4

A population has a density of $X(t) = 1$ [ind] at time $t = 2$ [years]. This population grows with a rate coefficient of $r = 0.5$ [year^{-1}]. What is the net rate of change of the population at time $t = 2$ [years]? What are the units?

The solution of the exponential model is

$$X(t) = X_0 \exp(rt) \tag{3.9}$$

where $X_0 = X(0)$, that is, the initial density. This solution is the same as Equation 3.2. It is a simple, "closed form" solution. We do not really need a computer simulation for this model.

The exponential model can generate three main outcomes (Figure 3.2) according to the value of the coefficient r as follows:

- When $r > 0$, the net rate of change is positive, X increases, and the population grows.
- When $r < 0$, the net rate of change is negative, X decreases, and the population declines.
- When $r = 0$, the net rate of change is zero and the population does not change.

This simple exponential model and its solution can help answer questions like this: How long does it take for a population to double if the intrinsic rate of growth is known? That is to say, how long does it take for X to double if r is known? To answer this question, denote t_d for the unknown doubling time and substitute in the solution (Equation 3.9) for the exponential model

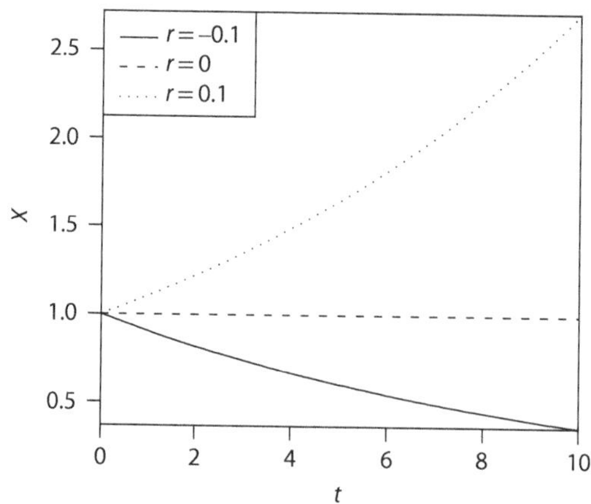

FIGURE 3.2 Exponential solution for the three ranges of r.

$$X(t_d) = X_0 \exp(rt_d) \tag{3.10}$$

Now, when the density doubles, we must have $X(t_d) = 2X_0$. Therefore, substitute $2X_0$ for $X(t_d)$ in the left-hand side to obtain

$$2X_0 = X_0 \exp(rt_d) \tag{3.11}$$

and after eliminating X_0 on both sides, we have

$$2 = \exp(rt_d) \tag{3.12}$$

we need to solve for t_d. Recall that the natural logarithm function ln(.), or log with base e, is the inverse of the exponential function. Take the natural log of both sides to obtain

$$\ln(2) = \ln(\exp(rt_d)) = rt_d \tag{3.13}$$

Now solve for t_d to obtain

$$t_d = \frac{\ln(2)}{r} = \frac{0.693}{r} \tag{3.14}$$

This is the final expression for the doubling time.

Example: A bacterial population has a density of $X(t) = 10^3$ ind or cells · liter^{-1} at time $t = 0$ days. In this case [ind] is a density in per volume basis. This population grows with a rate coefficient of $r = 1$ d^{-1}. The doubling time is given by

$$t_d = \frac{\ln(2)}{r} = \frac{0.693}{r} = 0.693 \text{ d}$$

What is the net rate of change of the population after 2 days? To answer this question, we calculate the rate coefficient × density after 2 days

$$\frac{dX(2)}{dt} = rX(2)$$

First, calculate density after 2 days using the exponential model solution at $t = 2$ d with $r = 1$ d^{-1}:

$$X(2) = X(0)\exp(rt) = 10^3 \exp(1 \times 2) = 7.39 \times 10^3 \text{ cells} \cdot \text{liter}^{-1}$$

Therefore, the rate after 2 days is given by

$$\frac{dX(2)}{dt} = rX(2) = 1 \times 7.39 \times 10^3 = 7.39 \times 10^3 \text{ cells} \cdot \text{liter}^{-1} \cdot \text{d}^{-1}$$

Exercise 3.5

A population has a density of $X(t) = 100$ ind at time $t = 0$ days. This population grows with a rate coefficient of $r = 0.1$ d^{-1}. What is the doubling time? What is the net rate of change of the population after 4 days?

Exercise 3.6

We know that the doubling time is 2 days. What is the rate coefficient? Include the units for the rate coefficient in your answer.

Let us make a linkage to a **difference equation** often used in population modeling. Go back to Equation 3.9 and define T as the duration of a cycle or season of reproduction. Using the population at a cycle, k, as the initial condition for the next cycle, $k + 1$, we can write

$$X((k+1)T) = \exp(rT)X(kT) \qquad k = 0, 1, 2, \ldots \tag{3.15}$$

Now define $\lambda = \exp(rT)$ as a net rate of growth and drop the T for brevity, then we have

$$X(k+1) = \lambda X(k) \qquad k = 0, 1, 2, \ldots \tag{3.16}$$

Equation 3.16 is a commonly used model of population growth in discrete time.

Exercise 3.7

Calculate λ for $r = 0.02$ year^{-1}, and $T = 1$ year is a reproductive season. For a population $X(k) = 100$, calculate the population at the end of the next season, that is to say $X(k + 1)$.

The solution of Equation 3.16 is found by iterating for $k = 1, 2, 3, \ldots$

$$\begin{aligned} X(1) &= \lambda X(0) \\ X(2) &= \lambda X(1) = \lambda^2 X(0) \\ X(3) &= \lambda X(2) = \lambda^3 X(0) \end{aligned} \tag{3.17}$$

which can be generalized as

$$X(k) = \lambda^k X(0) \tag{3.18}$$

The sequence of the population values $X(k)$ grows without bound when $\lambda > 1$, declines to 0 when $\lambda < 1$, and remains constant when $\lambda = 1$; see Figure 3.3.

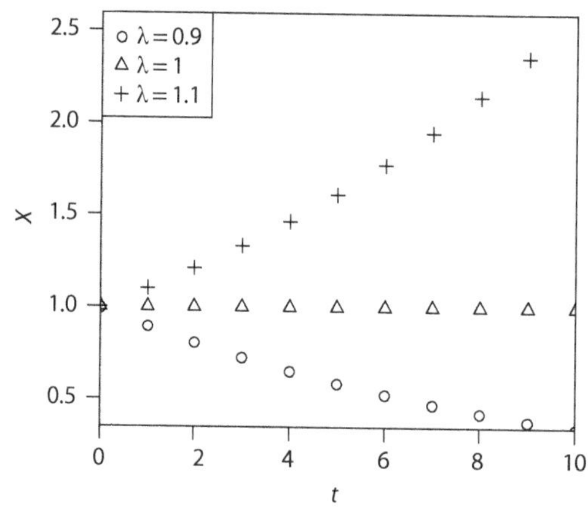

FIGURE 3.3 Discrete-time solution of the difference equation for the three ranges of λ.

Exercise 3.8

Does the population described in the previous exercise grow or decline?

3.4 CHEMICAL FATE

In this case, the dependent variable X denotes the concentration of a compound in some environmental medium, for example, in water, sediments of a stream, or soil (Ford, 1999, pp. 29–30; Keen and Spain, 1992, pp. 17–22).

The simplest assumption is linearity or first-order **kinetics** that leads to an exponential model stating that the rate of decay or degradation or breakdown is proportional to the concentration:

$$\mathrm{d}X/\mathrm{d}t = -kX \tag{3.19}$$

where k is the degradation or decay rate. Usually there are different chemical, physical, or biological mechanisms and agents to drive the reaction, for example, oxidation by chemicals, photolysis by light, and biodegradation by bacteria.

The model and its solution are the same as in the population example, except that we always assume a negative sign. In addition, we use a different letter, k, for the coefficient because k commonly denotes rate coefficient in kinetics.

The concentration unit for X is $\mathrm{M \cdot liter^{-3}}$, and therefore, the left-hand side has unit $\mathrm{M \cdot liter^{-3} \cdot T^{-1}}$ and the right-hand side has [units of k] $\mathrm{M \cdot liter^{-3}}$. Therefore [units of k] = $\mathrm{T^{-1}}$ as explained earlier. For example, X could be in $\mathrm{mg \cdot liter^{-1}}$ and k in $\mathrm{hour^{-1}}$ or $\mathrm{h^{-1}}$.

We can answer questions like how long does it take for the concentration, X, of a compound to reduce to one-half? The calculation process is the same as in Equations 3.10 through 3.14 but by using ½ instead of 2. Use t_h to denote the half-life (time that it takes for the concentration to degrade to one-half of the initial concentration) and therefore we have

$$\frac{1}{2}X_0 = X_0 \exp(-kt_\mathrm{h}) \tag{3.20}$$

that is,

$$1/2 = \exp(-kt_\mathrm{h}) \tag{3.21}$$

Take the natural log of both sides as before, and recall that the log of a ratio is the log of the numerator minus the log of the denominator to obtain

$$\ln(1) - \ln(2) = \ln(\exp(-kt_\mathrm{h})) \tag{3.22}$$

and that the log of 1 is zero,

$$-\ln(2) = -kt_\mathrm{h} \tag{3.23}$$

The negative sign cancels on both sides and solves for the half-life,

$$t_\mathrm{h} = \frac{\ln(2)}{k} = \frac{0.693}{k} \tag{3.24}$$

as a function of the degradation rate coefficient, k.

As an example, assume that a fungicide degrades at a rate of $k = 0.0693\ \text{d}^{-1}$. What is the half-life? The half-life is given by

$$t_h = \frac{\ln(2)}{k} = \frac{0.693}{0.0693} = 10\ \text{d}$$

Exercise 3.9

Assume that the decay rate coefficient is $0.5\ \text{h}^{-1}$; what is the half-life?

Assume that a chemical has $k = 0.0693\ \text{d}^{-1}$. How long does it take to degrade to 1% of the initial value? Use the solution of the exponential model:

$$\frac{1}{100}X(0) = X(0)\exp(kt_{1\%})$$

Cancel $X(0)$ on both sides, take logarithms of both sides, and simplify to obtain

$$t_{1\%} = \frac{\ln(100)}{k} = \frac{4.60}{0.0693} = 66.45\ \text{d}$$

Therefore, it takes 66.45 days to degrade to 1% of the initial value.

Exercise 3.10

A pesticide degrades with a rate coefficient of $k = 0.693\ \text{d}^{-1}$. What is the half-life? How long does it take to degrade to 5% of the initial value?

R can be used for quick calculations. For example, if the doubling time is 4 days, then what is r?

```
> 0.693/4
[1] 0.17325
```

If the decay rate is $2\ \text{d}^{-1}$, what is the half-life?

```
> 0.693/2
[1] 0.3465
```

3.5 MODEL TERMS

It is important to reiterate the terminology used in modeling. We can do this at this stage by recognizing how some of the terms apply to the exponential model as follows:

- $dX/dt = rX$ is a dynamic **model**. In this case, it is given by an **ODE**.
- $X(t)$ is the **state variable** at time t. In the simple exponential model that we are considering in this chapter, the state is also the **output**. In other cases, the output is a function of the state.
- r is a parameter.
- X_0 is the **initial condition** or **initial state** (technically it is also a parameter).

3.6 REVIEW OF SIMPLE LINEAR REGRESSION

In the next section, we will explain the concept of parameter estimation or model calibration, which requires a basic understanding of **regression**. Most likely, you know this from a statistics course. However, in this section, we review some basic notions of simple linear regression in case you are not familiar with this technique.

Let Y be a random variable defined as the "dependent" variable and X another random variable defined as the "independent" variable. This is a **bivariate** or two-variable situation. Assume that we have a joint sample x_i, y_i, $i = 1, \ldots, n$, or measurements of X and Y. These data pairs display in the X-Y plane as a **scatter** plot. See the example in Figure 3.4.

Denote by $\hat{Y}$ the **linear least squares estimator** of Y from X, then

$$\hat{Y} = b_0 + b_1 X \tag{3.25}$$

This is the equation of a straight line with **intercept** b_0 and **slope** b_1. For each data point, i, we have the **estimated** value of Y at the specific points, x_i,

$$\widehat{y_i} = b_0 + b_1 x_i \tag{3.26}$$

The **error** (**residual**) for data point i is given as

$$e_i = \widehat{y_i} - y_i \tag{3.27}$$

Take the square and sum over all observations to obtain the **total squared error**

$$q = \sum_{i=1}^{n} e_i^2 = \sum_{i=1}^{n} (\widehat{y_i} - y_i)^2 \tag{3.28}$$

We want to find the **coefficients** (intercept and slope) b_0 and b_1, which minimize the sum of squared errors (over all $i = 1, \ldots, n$), that is to say,

$$\min_b q = \min_b \sum_{i=1}^{n} e_i^2 = \min_b \sum_{i=1}^{n} (\widehat{y_i} - y_i)^2 \tag{3.29}$$

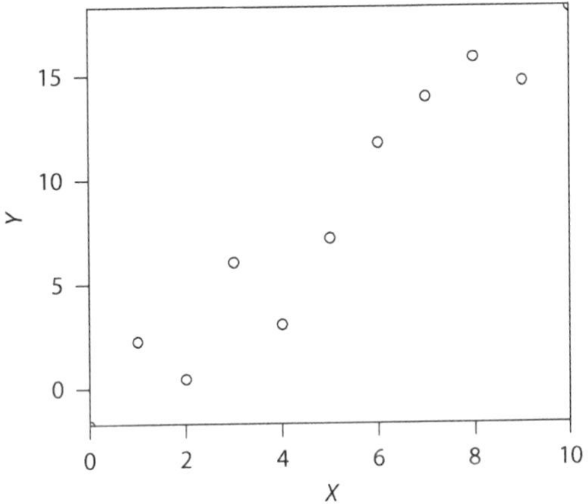

FIGURE 3.4 Scatter plot of X and Y.

How do we find which values of b_0 and b_1 minimize q? We express q as a function of b_0 and b_1 and then find the values of each one that make the slope or derivative of q zero. Those should be the values of b_0 and b_1 that minimize q.

To accomplish this, substitute Equation 3.26 in Equation 3.28 to obtain the sum of the squared errors as a function of the coefficients:

$$q = \sum_{i=1}^{n} e_i^2 = \sum_{i=1}^{n} (\hat{y}_i - y_i)^2 = \sum_{i=1}^{n} (b_0 + b_1 x_i - y_i)^2$$

Expand the square of the sum of the three terms in the summand:

$$q = \sum_{i=1}^{n} (b_0 + b_1 x_i - y_i)^2 = \sum_{i=1}^{n} (b_0^2 + b_1^2 x_i^2 + y_i^2 + 2b_0 b_1 x_i - 2b_1 x_i y_i - 2b_0 y_i)$$

Find partial derivatives of q with respect to b_0 and b_1. Start with the intercept b_0:

$$\begin{aligned}\frac{\partial q}{\partial b_0} &= \frac{\partial}{\partial b_0} \sum_{i=1}^{n} (b_0^2 + b_1^2 x_i^2 + y_i^2 + 2b_0 b_1 x_i - 2b_1 x_i y_i - 2b_0 y_i) \\ &= \sum_{i=1}^{n} (2b_0 + 2b_1 x_i - 2y_i) = 2(nb_0 + b_1 \sum_{i=1}^{n} x_i - \sum_{i=1}^{n} y_i)\end{aligned} \tag{3.30}$$

Set this derivative to zero:

$$\frac{\partial q}{\partial b_0} = 2\left(nb_0 + b_1 \sum_{i=1}^{n} x_i - \sum_{i=1}^{n} y_i \right) = 0 \tag{3.31}$$

Note that the term inside parenthesis in the right-hand side must be equal to zero and solving for b_0, we have

$$b_0 = -\frac{b_1}{n} \sum_{i=1}^{n} x_i + \frac{1}{n} \sum_{i=1}^{n} y_i \tag{3.32}$$

Now take the derivative with respect to b_1 and repeat as explained above

$$\begin{aligned}\frac{\partial q}{\partial b_1} &= \frac{\partial}{\partial b_1} \sum_{i=1}^{n} (b_0^2 + b_1^2 x_i^2 + y_i^2 + 2b_0 b_1 x_i - 2b_1 x_i y_i - 2b_0 y_i) \\ &= \sum_{i=1}^{n} (2b_1 x_i^2 + 2b_0 x_i - 2x_i y_i) = 2(b_1 \sum_{i=1}^{n} x_i^2 + b_0 \sum_{i=1}^{n} x_i - \sum_{i=1}^{n} x_i y_i)\end{aligned} \tag{3.33}$$

Make it zero:

$$\frac{\partial q}{\partial b_1} = 2\left(b_1 \sum_{i=1}^{n} x_i^2 + b_0 \sum_{i=1}^{n} x_i - \sum_{i=1}^{n} x_i y_i \right) = 0 \tag{3.34}$$

Note that the term in parenthesis must be zero and also we can substitute our earlier result in Equation 3.32 for b_0 to obtain

$$b_1 \sum_{i=1}^{n} x_i^2 + \left[-\frac{b_1}{n} \sum_{i=1}^{n} x_i + \frac{1}{n} \sum_{i=1}^{n} y_i \right] \sum_{i=1}^{n} x_i - \sum_{i=1}^{n} x_i y_i = 0 \tag{3.35}$$

Now solving for b_1, we have

$$b_1 = \frac{\sum_{i=1}^{n} x_i y_i - \frac{1}{n} \sum_{i=1}^{n} y_i \sum_{i=1}^{n} x_i}{\sum_{i=1}^{n} x_i^2 - \frac{1}{n} (\sum_{i=1}^{n} x_i)^2} \tag{3.36}$$

Once b_1 is calculated using Equation 3.36, we can calculate b_0 using Equation 3.32.

We can rearrange Equation 3.36 in terms of the sample mean of X and Y:

$$b_1 = \frac{\sum_{i=1}^{n} (x_i - \bar{X})(y_i - \bar{Y})}{\sum_{i=1}^{n} (x_i - \bar{X})^2} = \frac{s_{\text{cov}}(X,Y)}{s_x^{\ 2}} \tag{3.37}$$

Here the numerator is the sample covariance of X and Y, whereas the denominator is the sample variance of X. We repeat Equation 3.32 here for easy reference and take note that the components are the sample means of Y and X:

$$b_0 = \frac{1}{n} \sum_{i=1}^{n} y_i - \frac{b_1}{n} \sum_{i=1}^{n} x_i = \bar{Y} - b_1 \bar{X} \tag{3.38}$$

In summary, Equation 3.36 or 3.37 and Equation 3.38 are used to calculate the coefficients b_0 and b_1, which determine the regression line (Figure 3.5).

Rewriting Equation 3.38 as

$$\bar{Y} = b_0 + b_1 \bar{X}$$

we note that the regression line goes through the sample means of X and Y. Using the correlation coefficient in Equation 3.37, we can rewrite as

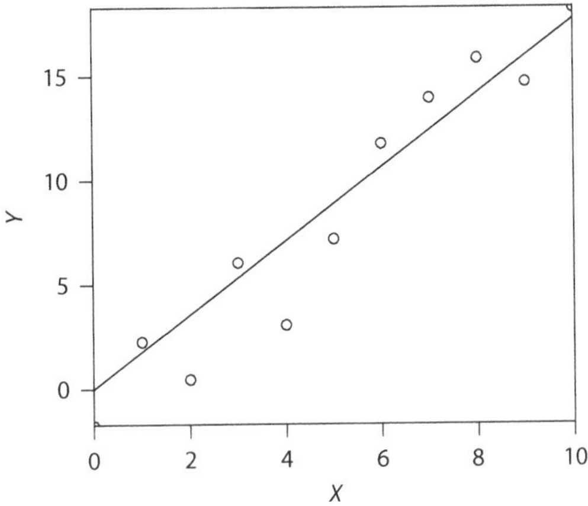

FIGURE 3.5 Regression line plotted on top of the scatter plot.

$$b_1 = \frac{s_{\text{cov}}(X,Y)}{s_X^{\,2}} = \frac{r s_X s_Y}{s_X^{\,2}} = r\frac{s_Y}{s_X} \tag{3.39}$$

In other words, the slope is the sample correlation coefficient multiplied by the ratio of sample standard deviations of Y over X.

There are three important error terms in regression. In the following, SS denotes sum of squares and MS denotes mean squares. The "total error" SS_T in Y is the sum of the squared differences of sample points minus the sample mean:

$$\text{SS}_\text{T} = \sum (y_i - \bar{Y})^2 \tag{3.40}$$

The "model" or "explained" error SS_M is the sum of the squared differences of the estimated points minus the sample mean:

$$\text{SS}_\text{M} = \sum (\hat{y}_i - \bar{Y})^2 \tag{3.41}$$

The "residual" or "unexplained" error SS_E is the sum of the squares of the difference between the estimated points and observations:

$$\text{SS}_\text{E} = \sum (\hat{y}_i - y_i)^2 \tag{3.42}$$

Now the total error is the sum of the model error and the residual (Acevedo, 2012), that is to say,

$$\text{SS}_\text{T} = \text{SS}_\text{M} + \text{SS}_\text{E} \tag{3.43}$$

Please note that when these quantities are divided by n, the number of observations, we get for SS_T the sample variance of Y $\text{MS}_\text{T} = s_Y^2$, for SS_M the average or mean squared model error MS_M, and for SS_E the average or mean squared residual error MS_E.

$$s_Y^{\,2} = \text{MS}_\text{E} + \text{MS}_\text{M} \tag{3.44}$$

A common measure of the goodness-of-fit is the ratio of the model error to the total error:

$$R^2 = \frac{\text{MS}_\text{M}}{\text{MS}_\text{T}} = \frac{\text{MS}_\text{M}}{\text{MS}_\text{E} + \text{MS}_\text{M}} = \frac{\text{MS}_\text{M}}{s_Y^{\,2}} \tag{3.45}$$

When MS_M (which is minimized by the least squares procedure) is very small, then R^2 approaches 1. Note that R^2 is the fraction (or percent) of the variance of Y explained by the regression model. Also,

$$1 - R^2 = \frac{\text{MS}_\text{E}}{\text{MS}_\text{T}} = \frac{\text{MS}_\text{E}}{\text{MS}_\text{E} + \text{MS}_\text{M}} = \frac{\text{MS}_\text{E}}{s_Y^{\,2}} \tag{3.46}$$

that is, the difference of R^2 with respect to 1 is the fraction of the variance of Y unexplained by the regression model.

Note that

$$R^2 = \frac{\text{MS}_\text{M}}{s_Y^{\,2}} = \frac{(1/n)\sum\left(b_0 + b_1 X - \bar{Y}\right)^2}{s_Y^{\,2}} = \frac{(1/n)\sum\left(\bar{Y} - b_1\bar{X} + b_1 X - \bar{Y}\right)^2}{s_Y^{\,2}} \tag{3.47}$$

where the sample mean of Y cancels, b_1 is a common factor, and recognizing the sample variance of X, we obtain

$$R^2 = \frac{(1/n)\sum b_1^2\left(X-\bar{X}\right)^2}{s_Y^2} = \frac{b_1^2(1/n)\sum\left(X-\bar{X}\right)^2}{s_Y^2} = \frac{b_1^2 s_X^2}{s_Y^2} \tag{3.48}$$

and by recalling the expression for b_1, we get

$$R^2 = \frac{\left(\dfrac{s_{\text{cov}}(X,Y)}{s_X^2}\right)^2 s_X^2}{s_Y^2} \tag{3.49}$$

therefore

$$R^2 = \left(\frac{s_{\text{cov}}(X,Y)}{s_X s_Y}\right)^2 = r^2 \tag{3.50}$$

The square root of R^2 is equal to r, which is the correlation coefficient.

We do not just look at the R^2 to evaluate a regression. It is necessary to check the assumptions of the method and to examine the significance of the regression, the confidence interval, and the trend in the unexplained or residual error. For example, we examine the following diagnostics.

The statistical significance of the trend, that is, whether the slope is nonzero. Soundness of the linearlty assumption; by examining the scatter plot of Y versus X to see how good the linear assumption is. If the y_i points seem to follow a definite nonstraight pattern or curve, then the linearity is suspicious even when getting a good R^2. Randomness of the residual error by looking at a plot of the residuals as a function of the estimated or predicted y. The residuals should be scattered up and down around zero (i.e., "noise"), telling us that the error is independent of the position in the regression line. The residuals should also be examined for normality. In addition, we can examine their standard error. The residual standard error is the standard deviation of the residuals:

$$s_e = \sqrt{\frac{\sum(y_i-\hat{y}_i)^2}{n-2}} \tag{3.51}$$

where we have used $n - 2$ because two parameters (slope and intercept) were estimated. For more details on regression diagnostics see Acevedo (2012).

3.7 MODEL CALIBRATION: PARAMETER ESTIMATION

The process of parameter estimation, referred to in some cases as model calibration, consists of finding the value of a given parameter. In this chapter, we explain the method using as an example the parameter k in the exponential model discussed in the previous chapter.

To estimate the parameter value, we need data about the change of X with time t. The simplest case is to have the values $X_1 = X(t_1)$ and $X_2 = X(t_2)$ for two points in time t_1 and t_2.

$$X_2 = X_1 \exp\left[k\left(t_2-t_1\right)\right] \tag{3.52}$$

Now take the logarithm of both sides and solve for k,

$$k = \frac{\ln(X_2/X_1)}{t_2-t_1} \tag{3.53}$$

Example: If the half-life is known, then $t_2 - t_1$ is the t_h and $X_2 = 0.5X_1$. Therefore,

$$k = \frac{\ln(2)}{t_h} \tag{3.54}$$

Parameter estimation from a data set usually involves empirical modeling (Swartzman and Kaluzny, 1987, pp. 1–11). If you have n data points, $X(t_i)$, $i = 1,\ldots,n$, then regression of $X(t_i)$ versus t_i yields k. We assume that you have a basic understanding of the regression method (see previous section).

A simple way of performing regression in this case is to apply the linear regression after logarithmic transformation of $X(t_i)$ in the following manner. For each t_i, we must have

$$X(t_i) = X_0 \exp(kt_i) \tag{3.55}$$

or dividing both sides by X_0 and taking the natural logarithm,

$$\ln\left(X(t_i)/X_0 \right) = \ln(\exp(kt_i)) = kt_i \tag{3.56}$$

We can see that a linear regression of $\ln(X(t_i)/X_0)$ versus t_i yields a slope that should correspond to k. In this case, we want the regression with **zero intercept** ($b_0 = 0$) since the first position of $\ln(X(t_i)/X_0)$ is zero because $\ln(1) = 0$.

3.8 NONLINEAR REGRESSION

In many cases, we need to estimate the parameters of a nonlinear equation relating Y to X of the general form

$$Y = f(X, p) \tag{3.57}$$

where $f(\ldots)$ is a function and p are the parameters. For example, the exponential model with parameters k and Y_0 is a nonlinear function. As we know from the previous section, sometimes we can transform Equation 3.57 into a linear regression problem. However, this is not always possible and so we can apply the process of nonlinear regression. This consists of postulating the function that may fit the data, for example, an exponential curve, and then using an optimization algorithm to minimize the square error (residuals) with respect to the coefficients. In other words, find the value of the coefficients that would yield a minimum square error.

The **error (residual)** for data point i is given by

$$e_i = y_i - \widehat{y}_i = y_i - f(x_i, p) \tag{3.58}$$

Take the square and sum over all observations to obtain the **total squared error**

$$q = \sum_{i=1}^{n} e_i^2 = \sum_{i=1}^{n} (y_i - f(xi, p))^2 \tag{3.59}$$

We want to find the values of the **coefficients** p that minimize the sum of the squared errors (over all $i = 1,\ldots,n$), that is to say, find p such that

$$\min_p q = \min_p \sum_{i=1}^{n} e_i^2 = \min_p \sum_{i=1}^{n} (y_i - f(x_i, p))^2 \tag{3.60}$$

An optimization algorithm works in the following manner. First, it reads an initial guess of the values of the coefficients. Then, it recursively changes the parameter values by a small amount and moves down the gradient (derivative) in the q surface until changes in parameter values no longer yield a decrease in q. Sometimes, we can obtain the initial guess of the parameter values by means of a linear regression performed on some approximation or transform of the nonlinear function.

It is often difficult to determine a functional relationship or a model. This is usually possible by knowing or postulating how the system works. For example, if we are trying to find a coefficient of light attenuation in the water column of a lake, we know that light attenuation follows an exponential law, because the rate of decay is linear with depth.

3.9 COMPUTER SESSION

Let us use an example from exponential population growth with two columns: t[years] and X[ind]. Use the file **chp3/expodata.txt**. Verify the contents using a text editor (e.g., notepad or Vim).

```
t[years]  X[ind]
0         1.088
1         1.156
2         1.240
3         1.383
4         1.500
5         1.709
6         1.851
7         2.075
8         2.250
9         2.528
10        2.770
```

We want to make an object `t.X` from the data in the file named `expodata.txt`. We use the function `scan` and wrap it with the function `matrix` to arrange the values in a two-dimensional (2-D) array:

```
> t.X <- matrix(scan("chp3/expodata.txt",skip=1), ncol=2, byrow=T)
Read 22 items
```

The argument `skip = 1` is used to skip reading the first line of the file, which does not contain data but labels. We will talk about labels later. The message "`Read 22 items`" was given by `scan`; it read the 22 data values. Here the `matrix` or 2-D array is configured by reading the file by rows (`byrow = T`) into two columns (`ncol = 2`). Matrices are filled by columns, unless the `byrow = T` argument is used. Note that `byrow` and `ncol` are arguments of the `matrix` function, not of the `scan` function. Double-check that you have the newly created object by using `ls()`.

```
> ls()
[1] "t.X" ...
>
```

Double-check the object contents by typing `t.X` to get

```
> t.X
      [,1]  [,2]
 [1,]    0 1.088
 [2,]    1 1.156
 [3,]    2 1.240
```

```
 [4,]    3 1.383
 [5,]    4 1.500
 [6,]    5 1.709
 [7,]    6 1.851
 [8,]    7 2.075
 [9,]    8 2.250
[10,]    9 2.528
[11,]   10 2.770
```

`t.X` is a matrix object; the `[i,]` denotes a row (across all columns), while the `[,j]` denotes a column (across all rows). Now you can use `dim` to check the dimensions of the matrix:

```
> dim(t.X)
[1] 11 2
>
```

Now assign the first column of `t.X` to object `t` and the second column to object `X`:

```
> t <- t.X[,1]; X <- t.X[,2]
```

Double-check the object contents by typing `t` and `X`:

```
> t
[1] 0 1 2 3 4 5 6 7 8 9 10
> X
[1] 1.088 1.156 1.240 1.383 1.500 1.709 1.851 2.075 2.250 2.528 2.770
```

Set graphics parameters, for example, to axis style "i":

```
>par(xaxs="i",yaxs="i")
```

Plot an $x - y$ line graph of the population data using the argument `type="l"` in the `plot` function. Recall that this is a letter "l" for **line** not number one "1".

```
>plot(t,X,type="l")
```

The result is shown in Figure 3.6. By default, the x- and y-axes are labeled with the variable name. These can be changed using the `xlab=` and `ylab=` arguments. For example, use

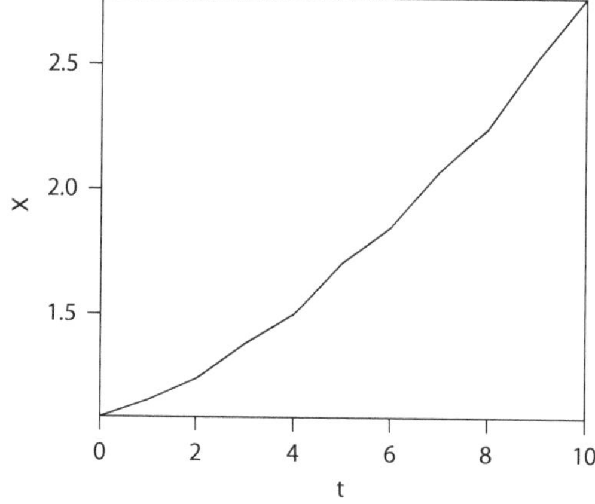

FIGURE 3.6 Plot as x-y using lines.

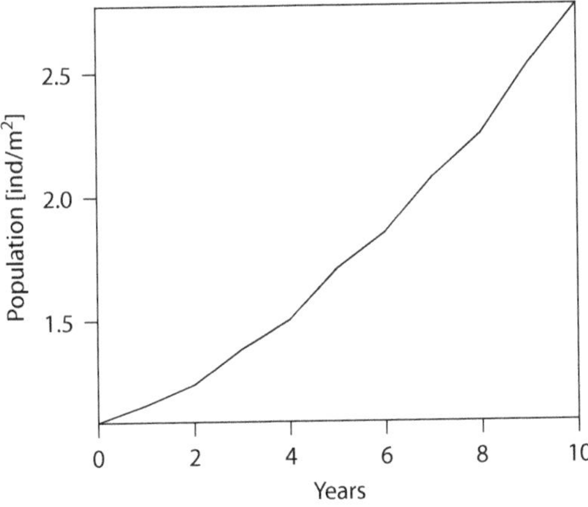

FIGURE 3.7 Plot with customized labels.

```
> plot(t, X, type="l", xlab="Years",ylab="Population [ind]")
```

to obtain Figure 3.7.

3.9.1 Programming Loops and More on Functions

One useful function to write loops is the function `for`, which is a command to execute the succeeding statements controlled by the variation of an index. We will use the simple example of evaluating a function of one variable for different values of a parameter. For example, evaluate an exponential equation for several values of the rate coefficient *r*. Start by creating the sequence of values of t:

```
> t<-seq(0,10,0.1)
```

Then, store the values of *r* in an array:

```
> r <- c(-0.1,0,0.1)
```

Now declare a matrix to store the results; each column will correspond to the values of X for a given value of *r*:

```
> X <- matrix(nrow=length(t), ncol=length(r))
```

Now use the function `for()` to perform a loop:

```
> for(i in 1:3) X[,i]<-exp(r[i]*t)
```

We can ask to see the results; the first few lines are as given in the following:

```
> round(X,2)
      [,1] [,2] [,3]
 [1,] 1.00    1 1.00
 [2,] 0.99    1 1.01
 [3,] 0.98    1 1.02
 [4,] 0.97    1 1.03
 [5,] 0.96    1 1.04
 [6,] 0.95    1 1.05
```

```
 [7,] 0.94    1 1.06
 [8,] 0.93    1 1.07
 [9,] 0.92    1 1.08
[10,] 0.91    1 1.09
...
```

To visualize all the results, use the function `matplot`, which draws several line graphs according to the dimension of the array, and then add a `legend` to identify the curves. The argument `lty` here commands the use of the first three line types available. The `col` argument is the color for the lines. In this case, `col` = 1 indicates a black color for all lines.

```
>matplot(t,X, type="l", lty=c(1:3), col=1)
>legend("topleft", paste("r=",r), lty=c(1:3), col=1)
```

The first argument of `legend` is the location to place the legend by keyword, then the function **paste** is used to compose the label r = with the values of r. See Figure 3.2 earlier in this chapter.

Summarizing, we have the following script:

```
# multiple lines on a plot
t<-seq(0,10,0.1)
r <- c(-0.1,0,0.1)
X <- matrix(nrow=length(t), ncol=length(r))
for(i in 1:3) X[,i]<-exp(r[i]*t)
matplot(t,X, type="l", lty=c(1:3),col=1)
legend("topleft", paste("r=",r), lty=c(1:3),col=1)
```

Note that the pound sign (#) signals a remark (nonexecutable text).

As another example, we can iterate Equation 3.16 using nested programming loops. An outer loop with index i varies parameter lambda and an inner loop varies k. In this case, we plot using symbols and not lines because the time is discrete and not continuous.

```
# difference equation
t<-seq(0,10,1)
lambda <- c(0.9,1,1.1)
X <- matrix(nrow=length(t), ncol=length(lambda))
X[1,] <- 1
for(i in 1:3){
for(k in 2:length(t)) X[k,i]<- lambda[i]*X[k-1,i]
}
matplot(t,X, pch=1:3,col=1)
legend("topleft", paste("lambda=",lambda), pch=1:3,col=1)
```

The result is shown in Figure 3.3 earlier in the chapter.

Exercise 3.11

Generate a linear function $y = ax + b$ with $b = 0.1$ and two values of a, $a = 0.1$ and $a = -0.1$. Plot y for the values of x from 0 to 1. Limit the Y-axis to a minimum of y and a maximum of y. Place a legend.

Exercise 3.12

Write a program in R to produce a figure of $\ln(X/X_0)$ for exponential growth for $r = 0.1$, 0.2, and 0.3. Include a legend.

3.9.2 Exponential from Data

Example: Suppose we have the data set for population growth contained in the file **chp3\expodata.txt** which we read before into object `t.X`.

Let us calculate the growth-rate coefficient, r, by linear regression. First, use columns of `t.X` to create objects for time and population. Then, calculate the natural logarithm of the ratio of each population value to the initial value $\ln(X_n/X_0)$ for all n and assign it to another vector named `lnX`:

```
> t <- t.X[,1]; X <- t.X[,2]
> lnX <- log(X/X[1])
```

Now we will apply the function `lm(y~x)`, which would perform a linear regression of y as a function of x. In this case, we want to run the regression with `zero intercept` since the first position of `lnX` is zero because $\ln(1) = 0$. To apply the `lm( )` function with zero intercept, we write `lm(y~0+x)`. Therefore, we use

```
> lfitpop <- lm(lnX ~ 0 + t)
```

The results of the linear regression have been assigned to an object named `lfitpop`. We can check its contents with

```
> lfitpop
Call:
lm(formula = lnX ~ 0 + t)
Coefficients:
 t
0.091
```

and obtain more details using the **summary** function:

```
> summary(lfitpop)

Call:
lm(formula = lnX ~ 0 + t)
Residuals:
  Min   1Q Median   3Q  Max
-0.051242 -0.031743 -0.003465 0.004288 0.024443

Coefficients:
 Estimate Std. Error t value Pr(>|t|)
t 0.091006 0.001438 63.28 2.36e-14 ***
---
Signif. codes: 0 `***' 0.001 `**' 0.01 `*' 0.05 `.' 0.1 ` ' 1

Residual standard error: 0.02822 on 10 degrees of freedom
Multiple R-Squared: 0.9975, Adjusted R-squared: 0.9973
F-statistic: 4005 on 1 and 10 DF, p-value: 2.363e-14
```

The estimated r coefficient is 0.091. The p value $= 2.36 \times 10^{-14}$ is very low, demonstrating that the slope is different from zero. The $R^2 \sim 0.99$ is high and implies that a large proportion (99%) of the variance is explained by the regression. One can get more information on the results by typing $ and the name of the component, for example, `lfitpop$residuals`.

Now we can plot the *x-y* graph of $\ln(X/X_0)$ versus time to visualize the linearity. Compare this to the line obtained from a linear regression of `lnX` versus `t` to estimate the slope (which is the rate coefficient of the exponential):

```
> plot(t,lnX)
> abline(lfitpop)
```

The result should look like a good fit as shown in Figure 3.8. Note that the line has an intercept equal to 0. The value of the regression coefficient represents the slope of the line, that is to say, the value of r is 0.091 year^{-1}. The R^2 value close to 1 and the low p value indicate a good relationship. A graph of the residual errors would be used to evaluate how well the population growth is represented by the exponential model.

Exercise 3.13

Use the data given above and plot a scatter diagram of $X(t_i)$ versus t_i predicted using the value of $r = 0.091$ calculated by regression. The result should be as shown in Figure 3.9.

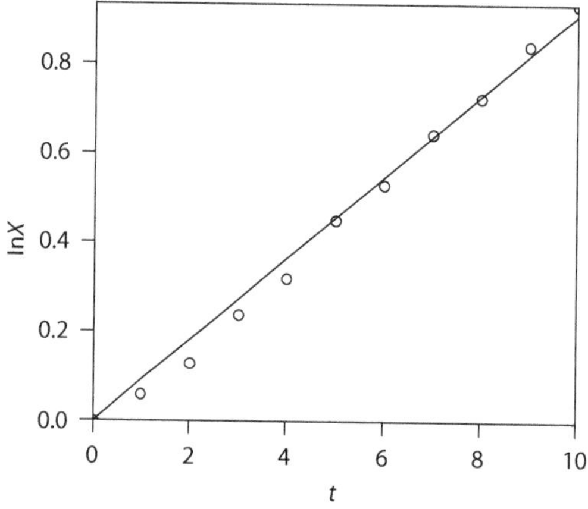

FIGURE 3.8 Estimation of r (rate coefficient) for an exponential population model.

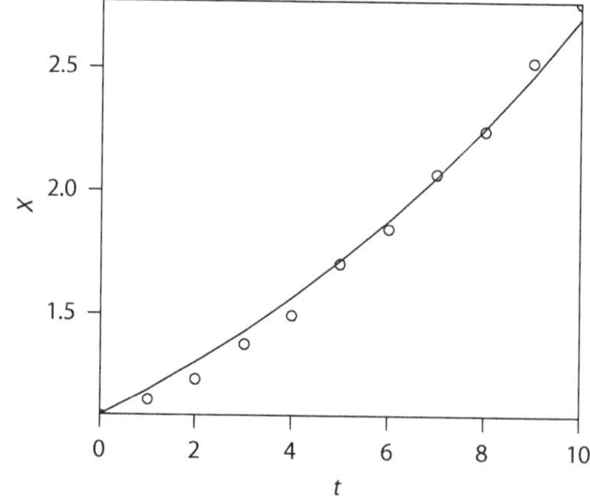

FIGURE 3.9 Data and predicted population.

3.9.3 Nonlinear

The function **nls** performs a nonlinear regression. It requires the functional relationship expressed with operator ~ and a `list` of parameter values to start the iteration. We need to start the optimization algorithm from an initial guess for the coefficient; we will use 0.1 for the initial guess (we know that is around 0.096 based on the linear fit):

```
> nlfitpop <- nls(X~exp(r*t), start=list(r=0.1))
> nlfitpop
Nonlinear regression model
 model: X ~ exp(r * t)
 data: parent.frame()
  r
0.1026
residual sum-of-squares: 0.01384

Number of iterations to convergence: 3
Achieved convergence tolerance: 2.252e-08
```

We have obtained $r = 0.103$. Another way of referring to the contents of the regression object is to use `coefficients(object)`, `residuals(object)`, or `fitted.values(object)`. So, for example, to plot the original data and to add a line for the nonlinear fitted (by fitted we mean predicted or estimated) values, use the following:

```
> plot(t, X)
> lines(t, fitted.values(nlfitpop))
```

The graph in Figure 3.10 shows a good fit.

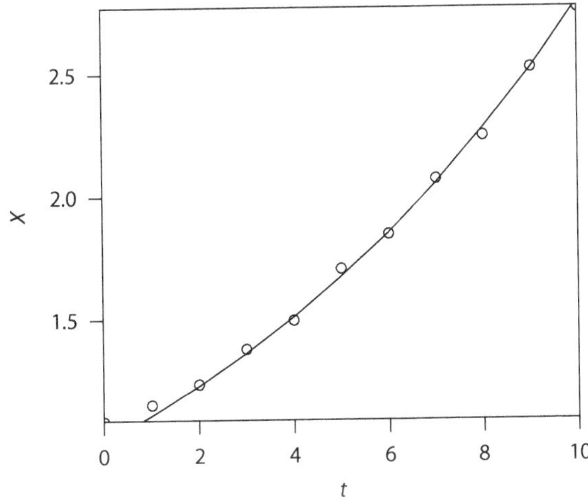

FIGURE 3.10 Data and results using nonlinear regression.

Exercise 3.14

The following data correspond to the decay of concentration (in mg/liter) of a chemical with time (in days). The data are in the file **chp3\expodata2.txt**

```
 t[d] X[mg/l]
 0 1.000
 1 0.905
 2 0.819
 3 0.741
 4 0.670
 5 0.606
 6 0.549
 7 0.496
 8 0.449
 9 0.406
10 0.368
```

Estimate the rate coefficient k by regression (linear and nonlinear) and show the graphs obtained.

3.9.4 More on R Graphics Windows

We can open more graphics device for more plots. This can be done by typing the command `win.graph()` or `windows()`. The size of the graphics windows can be specified with the two numbers for width and height in inches. For example, `windows(4,4)` opens up a new graphics window of width 4 and height 4. The current graph window can be closed with `dev.off()`; all graphics windows can be closed with `graphics.off()`.

You could also direct plots to specific formats using a series of commands, for example, for the *x-y* plot of the population versus years. Use `pdf( )` to generate a PDF file of graph, say the *x-y* plot of the population versus years:

```
> pdf(file="chp3/popul.pdf")
> plot(t,X)
> dev.off()
```

3.9.5 Editing Data in Objects

Example: Assume we want to edit object `t.X`. Use

```
>edit(t.X)
```

to invoke the data editor or use **Data editor ...** from the Edit menu item in the GUI.

3.9.6 Cleanup

Many times, we generate objects that would not be needed after we use them. In this case, it is a good practice to clean up after a session by removing objects with the `rm()` command. A convenient way of doing this is to get a list of the objects with `ls()` and then see what we need to remove.

For example, at this point we may want to keep object `t.X` because it contains the data we may need later. But we may want to delete objects `t` and `X` because they can easily be derived from `t.X`:

```
> ls()
[1] "t.X" "X" "t" ...
> rm(t, X)
```

You can also confirm that objects were removed and get an update of the list to see if there is some more cleanup required. We can reduce clutter by being careful not to generate objects unnecessarily. Some of the objects created could have been avoided using nested commands and by addressing components of arrays using an index. For example, we could have used the nested commands

```
> plot(t.X[,1], t.X[,2])
```

instead of generating new objects `t` and `X`.

4 Model Simulation

In modeling, an important task is calculating the solution of a model. Rarely, we will be able to determine the solution analytically. Therefore, we resort to a numerical calculation performed with a computer. In this chapter, we study various ways of accomplishing this task.

4.1 CONCEPTS OF COMPUTER SIMULATION BASED ON ORDINARY DIFFERENTIAL EQUATIONS

We want to determine the solution $X(t)$ of an ordinary differential equation (ODE) model given by a general function $f(t,p,X(t))$, where p denotes the parameters,

$$\frac{dX(t)}{dt} = f(t,p,X(t)) \tag{4.1}$$

and with a known initial condition, X_0. The generality of this function emphasizes that the rate may depend on the time itself, a set of parameters, and the dependent variable. There are several numerical methods to solve ODEs such as Euler and Runge–Kutta (Swartzman and Kaluzny, 1987).

4.2 EULER METHOD

Because of its simplicity, we will first explain the Euler method. Recall that we can get an approximate value of the derivative dX/dt using the slope as a ratio of small quantities:

$$\frac{dX}{dt} \simeq \frac{X(t+\Delta t) - X(t)}{\Delta t} \tag{4.2}$$

This approximation works better if Δt is very small. Next, substitute this approximation on the left-hand side of the ODE given in Equation 4.1:

$$\frac{X(t+\Delta t) - X(t)}{\Delta t} \simeq f(t,p,X(t)) \tag{4.3}$$

Rewrite this as

$$X(t+\Delta t) \simeq X(t) + \Delta t\, f(t,p,X(t)) \tag{4.4}$$

Now iterate Equation 4.4 with a computer program starting at X_0 with $t = t_0$

$$X_1 = X(\Delta t) = X_0 + \Delta t\, f(t,p,X_0)$$

$$X_2 = X(2\Delta t) = X_1 + \Delta t\, f(t,p,X_1)$$

Or, in general, for the ith time step:

$$X_{i+1} = X\big((i+1)\Delta t\big) = X_i + \Delta t\, f(t,p,X_i) \tag{4.5}$$

until we get to the final value at the final time t_f:

$$X_f = X(t_f) = X(t_f - \Delta t) + \Delta t \, f(t, p, X(t_f - \Delta t))$$

It should be noted that the solution to the ODE has been replaced by a sequence of arithmetic operations (sums and products), yielding a set of values of the solution $X(\Delta t)$, $X(2\Delta t)$, …, $X(i\Delta t)$ separated by the small time step Δt.

Since Δt is small, we do not want to save all the values because there will be too much data if Δt is small and t_f is large; therefore, we usually save or write to an **output file** only every n time steps or every $t_w = n\Delta t$ time units. This will generate $N = \frac{t_f}{n\Delta t}$ data values after the initial condition for a total of $N + 1$ values of X saved or written to the output file.

When executing the program, we need to store in memory data that would have been read from an **input file**: values for the simulation control (which includes the time step, Δt, the initial time, t_0, the final time, t_f, and the time to save or write, t_w). The program saves or writes an output file containing the numerical solution. Each record has two columns, the value of time and the corresponding X value. Thus, the output file will contain this information:

```
Time  X
0     X0
ndt   X(ndt)
2ndt  X(2ndt)
3ndt  X(3ndt)
```

…

and so on until

…

```
Nndt  X(Nndt)
```

It is convenient to write this output file as an **ASCII** text file to facilitate **exporting** it to a variety of software applications, for example, to perform statistical analysis or to generate a graph of the **time series** $X(t)$ versus t using spreadsheets or data analysis programs such as R.

The flowchart given in Figure 4.1 illustrates the simulation process.

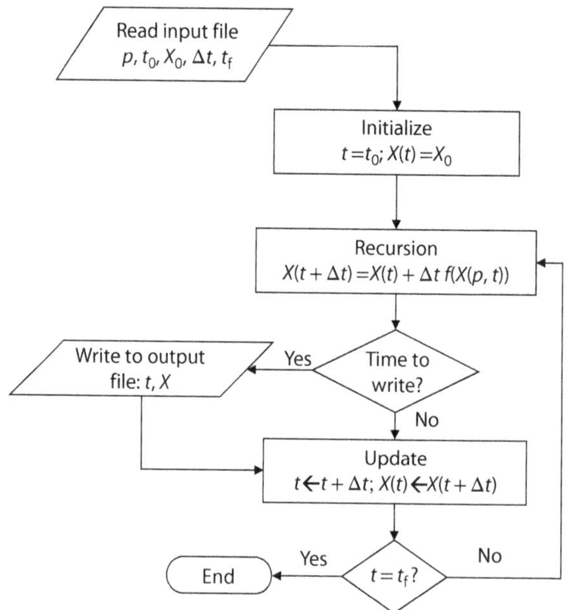

FIGURE 4.1 The simulation flowchart.

Exercise 4.1

What are the files produced by a simulation program? What is the simulation time step? How do you determine how many times the output is written to an output file?

One way to learn a numerical method is to iterate it by hand a few times. So, let us work on the following exercise. We do not need simulation to study the exponential growth or decline because we know the exact mathematical solution of this simple ODE. However, we will use it here because it facilitates the calculations and thus helps explain the main concepts involved in the simulation. Therefore, we will **numerically** solve the exponential model by calculator. The basic relation of Equation 4.4 is simply

$$X(t+\Delta t) = X(t) + \Delta t\, rX(t)$$

where the parameter set p consists of the coefficient r and there is no explicit dependence on t. Thus, the recursive expression Equation 4.5 for any i reduces to

$$X_{i+1} = X((i+1)\Delta t) = X_i + \Delta t\, rX_i \tag{4.6}$$

Exercise 4.2

Use your calculator to simulate the exponential growth using the Euler technique from $X_0 = 1$ ind at $t_0 = 0$ days at the rate of $r = 0.1$ day^{-1}. Use $\Delta t = 0.1$ and $t_f = 1$ day. Write the result for every two time steps and use four significant figures.

Hint:
$X(0) = X_0 = 1$, write to file
$X(0.1) = 1. + 0.01 = 1.01$, skip writing to file
$X(0.2) = 1.01 + 0.01 \times 1.01 = 1.02$, write to file
$X(0.3) = 1.02 + 0.01 \times 1.02 = 1.030$, skip writing to file
$X(0.4) = ?$, write to file
......
$X(1) = ?$, write to file

Complete the calculations. Write (handwrite) the output file as two columns: t and $X(t)$. Plot (just sketch) $X(t)$ versus t using the values in the output file.

In essence, the Euler method is based on including the first-order terms of a Taylor series expansion of function $f(\ )$, that is to say, ignoring terms of second order and above. Thus, the error in each step is proportional to $(\Delta t)^2$. Also, the number of steps is proportional to $1/\Delta t$, and therefore, the accumulated error is proportional to $(\Delta t)^2 \times \frac{1}{\Delta t} = \Delta t$. This is why the Euler method is a first-order method.

4.3 FOURTH-ORDER RUNGE–KUTTA METHOD

The Runge–Kutta methods use values in between each time step and are more precise for the same time step. We will focus on the fourth-order method (RK4). We rewrite Equation 4.1 to explicitly show the dependence on time t in the function $f(\ )$:

$$\frac{dX(t)}{dt} = f(t, p, X(t)) \tag{4.7}$$

The recursive relation is

$$X_{i+1} = X((i+1)\Delta t) = X_i + \frac{\Delta t}{6}(k_1 + 2k_2 + 2k_3 + k_4) \tag{4.8}$$

where

$$\begin{aligned} k_1 &= f(t_i, p, X_i) \\ k_2 &= f\left(t_i + \frac{\Delta t}{2}, p, X_i + \frac{\Delta t}{2}k_1\right) \\ k_3 &= f\left(t_i + \frac{\Delta t}{2}, p, X_i + \frac{\Delta t}{2}k_2\right) \\ k_4 &= f(t_i + \Delta t, p, X_i + \Delta t k_3) \end{aligned} \tag{4.9}$$

Here, note the following:

- k_1 is the slope at the beginning of the time step as in the Euler method.
- k_2 is the slope at the middle of the time step $t_i + \frac{\Delta t}{2}$ with X calculated from k_1 using Euler.
- k_3 is the slope at the middle of the time step $t_i + \frac{\Delta t}{2}$ with X calculated from k_2 using Euler.
- k_4 is the slope at the end of the time step $t_i + \Delta t$ with X calculated from k_3 using Euler.

Compared to Euler's method, we have substituted the slope $\Delta t f(p, X_i)$ in Equation 4.5 for a weighted average of four slopes where more weight is given to the two slopes k_2 and k_3 at the mid-point of the time step. The method is equivalent to dropping the terms of the fifth order and above in a Taylor series; therefore, the error in each time step is proportional to $(\Delta t)^5$. As in the Euler method, the number of steps is proportional to $1/\Delta t$; therefore, the accumulated error is proportional to $(\Delta t)^5 \times \frac{1}{\Delta t} = (\Delta t)^4$. This is why the RK4 method is a fourth-order method.

Exercise 4.3

Use your calculator to simulate one time step of exponential growth using the RK4 method from $X_0 = 1$ ind at $t_0 = 0$ days, at the rate of $r = 0.1$ day^{-1}. Use $\Delta t = 0.1$ and use four significant figures.

Hint:
$X_0 = 1.$
Calculate $k_1 = rX_0 = 0.1 \times 1. = 0.1$
Calculate $k_2 = r(X_0 + (\Delta t/2)k_1) = 0.1 \times (1. + 0.05 \times 0.1) = 0.1 \times 1.005 = 0.1005$
Calculate $k_3 = r(X_0 + (\Delta t/2)k_2) = 0.1 \times (1. + 0.05 \times 0.1005) = 0.1 \times 1.05 = 0.105$
Calculate $k_4 = r(X_0 + (\Delta t)k_3) = 0.1 \times (1.000 + 0.1 \times 0.105) = 0.1 \times 1.010 = 0.101$
Calculate the slope weighted average $(0.1/6) \times (0.1 + 2 \times 0.1005 + 2 \times 0.105 + 0.101) = 0.0102$
Now $X_1 = 1. + 0.0102 = 1.0102$
Compared to $X_1 = 1.01$ of Exercise 4.2, we can appreciate the potential decrease in error.

4.4 RANDOM NUMBER GENERATION

Random variables (RV) are used in modeling to simulate natural processes that include variability or uncertainty. A sequence of samples from a random variable behave as a **realization** of a **stochastic** process. Choosing numbers at random consists of generating a pseudorandom number sequence for the values of the RV.

There are many methods to generate random numbers. Most programming languages provide a basic function to generate numbers that would seem to be drawn from an RV with a uniform probability density function (pdf) between 0 and 1, that is, $U(0,1)$. For example, in R, this is done with **runif(1,0,1)**. Multiple values, say n, can be obtained by **runif(n,0,1)**.

We can transform the RV U with $U(0,1)$ to obtain other distributions, for example, an RV, X, with $U(a,b)$ can be generated from an RV U with $U(0,1)$ by first drawing a random value, u, and then applying the expression

$$x = (b-a)u + a \tag{4.10}$$

When the cumulative density function (cdf) of the desired RV has an explicit formula, a useful method to draw numbers from this RV distribution is to transform via the inverse of the cdf. First, generate u from $U(0,1)$, then apply $x = F^{-1}(u)$. For example, for the exponential pdf,

$$\begin{aligned} u &= F(x) = 1 - \exp(-ax) \\ 1 - u &= \exp(-ax) \\ \ln(1-u) &= -ax \\ x &= -(1/a)\ln(1-u) = F^{-1}(u) \end{aligned} \tag{4.11}$$

For practical purposes, it is the same as using u instead of $1 - u$. So, first generate u and then apply the inverse $x = -\ln(u)/a$.

A value z from a standard normal Z can be generated from two numbers (u_1,u_2) generated from $U(0,1)$ using

$$z = \cos(2\pi u_1)\sqrt{-2\ln(u_2)} \tag{4.12}$$

In R, a uniform RV can be generated in any range [a,b] directly by using **runif(1,a,b)**. In addition, there is a variety of functions for other distributions, such as the normal and exponential distributions. For example, the normal of mean u and standard deviation s can be generated with **rnorm(n,u,s)**, where n is the number of draws. Figure 4.2 provides an example, and we will explain next how to use these functions.

4.5 STOCHASTIC SIMULATION

Assume that the model to be simulated includes one or more RVs each with a known or assumed distribution, say a normal or a uniform pdf. The **Monte Carlo** method to perform the stochastic simulation consists of generating many realizations of the model output. For each realization, the values of the RVs are selected at random from their pdf using a pseudorandom number generator. Then after we complete all the realizations, we aggregate the realizations or compute the statistics such as the mean, median, variance, and confidence intervals. The aggregated output becomes the result of the simulation run. The Monte Carlo method is commonly used for "bootstrap" techniques, one of the "resampling" statistical methods.

For example, suppose that instead of a fixed rate coefficient of increase of a population, we have a coefficient that is distributed with a normal distribution $N(\mu,\sigma)$ with mean μ and standard deviation σ. One way of simulating the population growth would be by applying the equivalent of the Euler or RK4 method, but where at each iteration the function $f(t, p, X)$ is evaluated as rX, where the value of r is selected at random from $N(\mu,\sigma)$. We will experiment with this method later in the computer session.

Another implementation of stochastic population dynamics includes the birth and death process, where the time to an event of increase or decrease of the population is given by a random draw from

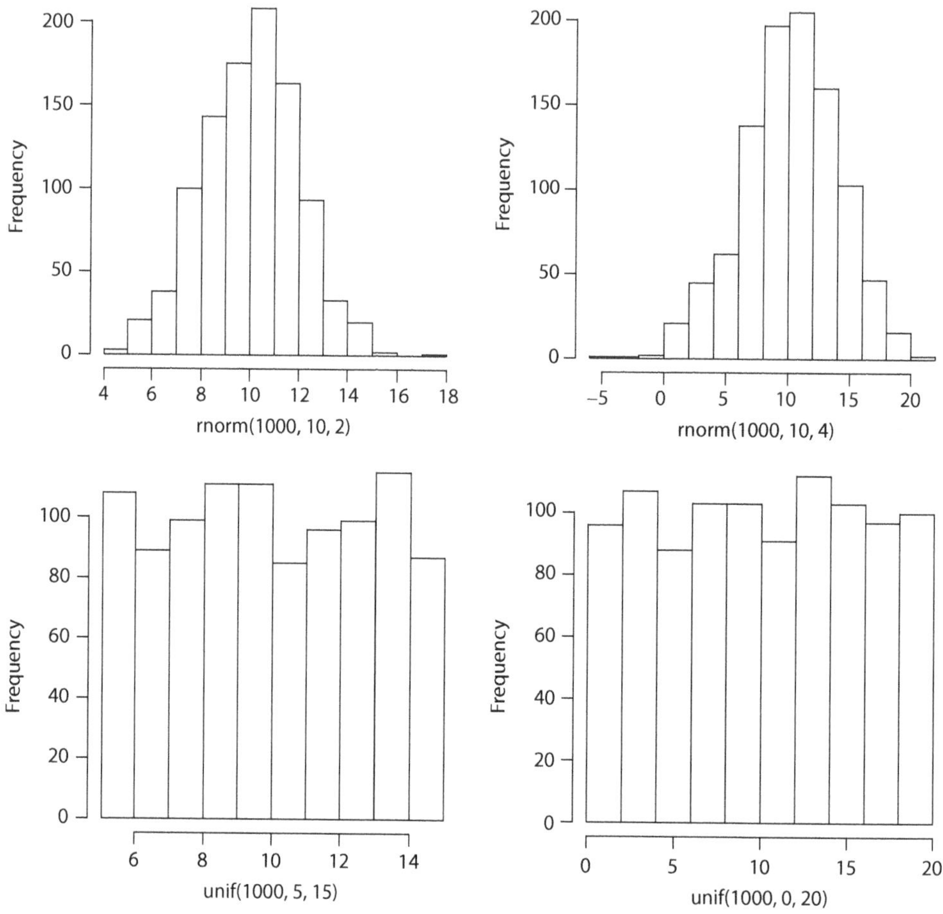

FIGURE 4.2 Normal and uniform distributions obtained by n = 10,000 pseudorandom number generation using R. Top panel: normal with mean 10, standard deviation 2 (left) and 4 (right). Bottom panel: uniform from 5 to 15 (left) and 0 to 20 (right).

an exponential pdf, and then the nature of the event is determined by a discrete probability function determining birth or death.

4.6 COMPUTER SESSION

We assume that you have installed the package **seem** as described in Chapter 1 and thus you are now ready to use some of its functions. In a later section of this chapter, we will explain how they work. To load the seem package, go to **Packages|Load package...**, then browse to **seem**, and select. You do not need to load the package if you have included in the local segment of the **Rprofile.site** file as explained in Chapter 1.

4.6.1 Euler and Runge–Kutta Methods

Let us use the seem functions `expon` and `sim.comp` and the input file **chp4/exp-pop-inp.csv**. These functions have only a didactic purpose; **odesolve**, a package to solve ODE, offers functions that are more efficient, such as **lsoda**. The input file is an **ASCII** text file in **CSV** format (comma-separated values). Open this input file on a text editor as follows or as a spreadsheet (Figure 4.3) to understand its contents:

```
Label,Value,Units,Description
t0,0.00,Day,time zero
tf,100.00,Day,time final
dt,0.10,Day,time step
tw,1.00,Day,time to print
r,0.02,1/Day,intrinsic growth rate
X0,1.00,Indiv,initial population
digX,4,None,significant digits for output file
```

Although a spreadsheet view is convenient, it is important to preserve the CSV ASCII text format when closing and saving.

This file indicates that time runs from zero to 100 in steps of 0.1 and that the results will be written to the file every one unit. The rate coefficient is $r = 0.02$ 1/Day and the initial condition is $X_0 = 1$ indiv. Please recall that we will use [ind] same as [indiv] for population density. There will be four significant digits for X in the output file. To use the `sim.comp` function, we first specify a model function and the input file. The model is a list with component `f` equal to the function `expon`. We explain this function at the end of this chapter.

```
>model<-list(f=expon);file<-"chp4/exp-pop-inp.csv"
```

Although it may seem that a list is unnecessary, we design it this way so that we can add more components later to include forcing terms and discontinuities. Now that we have specified the model and file, type the call statement

```
>out.eu <- sim.comp(model,file,method="euler")
```

Because the filename included the path **chp4/**, the program looks for the input file in this folder. A graphics device opens to show a graph (Figure 4.4), and we can examine the returned object `out.eu` by just typing its name on the console:

	A	B	C	D	E	F	G
1	Label	Value	Units	Description			
2	t0	0	Day	time zero			
3	tf	100	Day	time final			
4	dt	0.1	Day	time step			
5	tw	1	Day	time to write			
6	r	0.02	1/Day	intrinsic growth rate			
7	X0	1	Indiv	initial population			
8	digX	4	None	significant digits for output file			
9							
10							

exp-pop-inp

FIGURE 4.3 The input CSV file also opens as a spreadsheet.

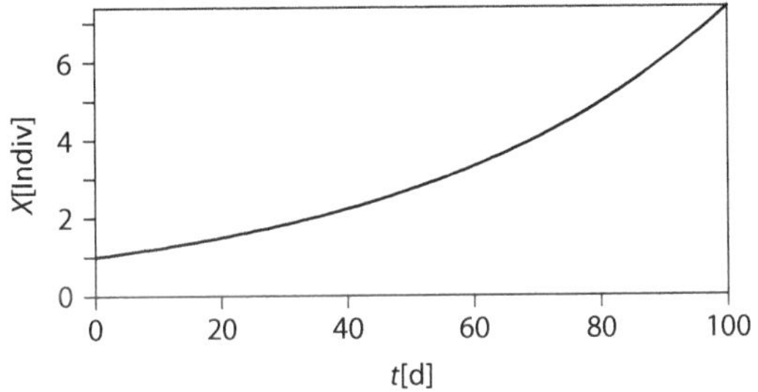

FIGURE 4.4 Simulation of the exponential growth model using the Euler method.

```
> out.eu
$input
  Label Value Units                                  Description
1    t0 0e+00   Day                                    time zero
2    tf 1e+02   Day                                   time final
3    dt 1e-01   Day                                    time step
4    tw 1e+00   Day                                time to write
5     r 2e-02 1/Day                        intrinsic growth rate
6    X0 1e+00 Indiv                           initial population
7  digX 4e+00  None significant digits for output file

$output
    t[Day] X[Indiv]
1        0    1.000
2        1    1.020
3        2    1.041
4        3    1.062
5        4    1.083
6        5    1.105
7        6    1.127
8        7    1.150
9        8    1.173
```

... etc.

This is a `list` with input and output components that can be addressed independently as `out.eu$input` and `out.eu$output`. The input component is just a repeat of the input file, just to make sure that we associate the execution results to the input actually used. The output contains the execution results in two columns with labels that include the units for each variable as extracted from the input file.

In addition, we should have an output CSV file in the folder chp4, named **exp-pop-inp.csv-Euler-out.csv**. We examine this CSV file with a text editor or a spreadsheet. It is convenient to write the output files as an ASCII text file in CSV format to facilitate exporting it to a variety of software applications. The file has a repeated version of the input file (first 8 lines), the labels for t and X (line 9), and two columns of numbers, one for t and another for X, as follows:

```
Label,Value,Units,Description
t0,0,Day,time zero
tf,100,Day,time final
dt,0.1,Day,time step
tw,1,Day,time to print
r,0.02,1/Day,intrinsic growth rate
X0,1,Indiv,initial population
digX,4,None,significant digits for output file
t[Day],X[Indiv]
0,1
1,1.02
2,1.041
3,1.062
4,1.083
5,1.105
6,1.127
7,1.15
8,1.173
9,1.197
```

.. etc.

The function **sim.comp** has an optional argument pdfout that takes the value F by default. When this argument is set to T as pdfout=T, we send the graphs to a PDF file instead of a graphics device in the interactive RGui. This is convenient when we write several statements in batch or when running the program from a terminal. Try

```
> out.eu <- sim.comp(model,file,method="euler",pdfout=T,lab.out="Euler")
```

Once the run is completed, we should have an **exp-pop-inp.csv-Euler-out.pdf** file in addition to the CSV output file. This file contains the same graph shown already in Figure 4.4.

Similarly, we can use RK4 with the same input and obtain the corresponding output:

```
> out.rk <- sim.comp(model,file,pdfout=T,lab.out="RK4")
```

Once this run is completed, we should have two returned objects out.eu and out.rk and four output files in the folder chp4, two CSV text files and two PDF files.

```
Exp-pop-inp.csv-Euler-out.csv
Exp-pop-inp.csv-Euler-out.pdf
Exp-pop-inp.csv-RK4-out.csv
Exp-pop-inp.csv-RK4-out.pdf
```

Examination of the output CSV files for both methods indicates that the results are similar at the beginning of the simulation but then diverge towards the end.

4.6.2 Comparison

Let us compare these Euler and Runge–Kutta results to a reference solution and produce graphs of the error versus time. A reference solution can be calculated with function exp() as in Chapter 3. By using $output, we extract the results from the objects created by the sim.comp function as called in the previous section. Then we calculate the reference with the same number of significant figures:

```
t <- out.eu$output[,1]
X.exp <- signif(exp(0.02*t),4)
X.eu <- out.eu$output[,2]
X.rk <- out.rk$output[,2]
```

Next, we calculate the percent relative error,

$$e(t) = \frac{X_{\mathrm{met}}(t) - X_{\mathrm{ref}}(t)}{X_{\mathrm{ref}}(t)} \times 100 \tag{4.13}$$

where $X_{\mathrm{met}}(t)$ is the value of X for a particular method and $X_{\mathrm{ref}}(t)$ is the value of the reference at the same time t. Using this equation, we calculate the two errors, bind them by column with cbind, and plot them:

```
eu.exp <- 100*(X.eu-X.exp)/X.exp
rk.exp <- 100*(X.rk-X.exp)/X.exp
y <- cbind(eu.exp,rk.exp)

matplot(t,y,type="l",col=1,yaxs="r",xaxs="r",ylab="Percent error")
legend("bottomleft",leg=c("Euler","RK4"),col=1,lty=1:2)
```

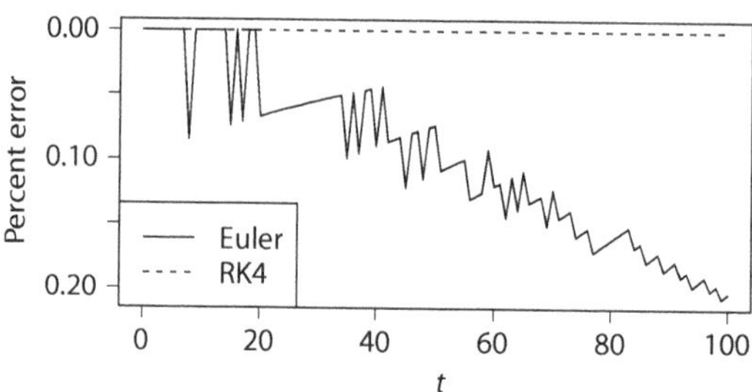

FIGURE 4.5 Numerical method error comparison.

In Figure 4.5, we can appreciate how the error for Euler increases with time, while the one for RK4 is very close to zero.

To perform more runs with different time zero, end time, final time, time step, time to write, parameter *r*, and initial condition, edit the corresponding values in the input file and call the functions again.

Exercise 4.4

Perform Euler runs with decreasing values of $dt = 0.01, 0.001$ and compare the results. Discuss the effect of the time step on the percent relative error. You should note that the execution time increases.

4.6.3 Multiple Runs

The function `sim.mruns` includes an argument to pass information about varying a parameter and perform several runs corresponding to this parameter's values. The argument is `param`; it contains the label of the parameter to be varied and its values. Now call `sim.mruns` with the desired values for the arguments. For example:

```
model<-list(f=expon); file<-"chp4/exp-pop-inp.csv"
param <- list(plab="r", pval = seq(-0.02,0.02,0.01))
out.r <- sim.mruns(model, file, param)
```

Note that the model and file were specified above and we do not really need to repeat, unless these values were modified. The function produces the graphics result (Figure 4.6), returns the object `out.r`, and writes an output file **exp-pop-inp.csv-r-out.csv**. This file contains the following:

```
Label,Value,Units,Description
t0,0,Day,time zero
tf,100,Day,time final
dt,0.1,Day,time step
tw,1,Day,time to write
r,0.02,1/Day,intrinsic growth rate
x0,1,Indiv,initial population
digX,4,None,significant digits for output file
r,-0.02,-0.01,0,0.01,0.02
t[Day],X[Indiv].Run1,X[Indiv].Run2,X[Indiv].Run3,X[Indiv].Run4,X[Indiv].
Run5
```

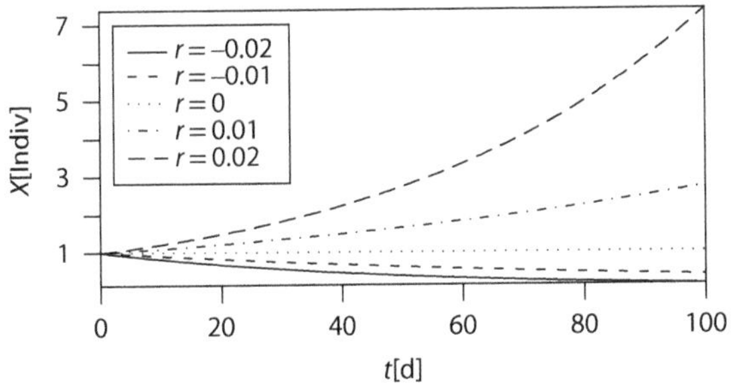

FIGURE 4.6 Multiple runs of the exponential growth by varying the parameter *r*.

```
0,1,1,1,1,1
1,0.9802,0.99,1,1.01,1.02
2,0.9608,0.9802,1,1.02,1.041
3,0.9418,0.9704,1,1.03,1.062
4,0.9231,0.9608,1,1.041,1.083
5,0.9048,0.9512,1,1.051,1.105
6,0.8869,0.9418,1,1.062,1.127
```

... etc.

And examining the R object **out.r** using the console, we see

```
> out.r
$input
  Label Value Units                          Description
1    t0 0e+00   Day                            time zero
2    tf 1e+02   Day                           time final
3    tw 1e+00   Day                        time to write
4     r 2e-02 1/Day                intrinsic growth rate
5    x0 1e+00 Indiv                   initial population
6  digX 4e+00  None significant digits for output file

$param
$param$plab
[1] "r"

$param$pval
[1] -0.02 -0.01  0.00  0.01  0.02


$output
   t[Day] X[Indiv].Run1 X[Indiv].Run2 X[Indiv].Run3 X[Indiv].Run4
X[Indiv].Run5
1       0        1.0000        1.0000      1         1.000         1.000
2       1        0.9802        0.9900      1         1.010         1.020
3       2        0.9608        0.9802      1         1.020         1.041
4       3        0.9418        0.9704      1         1.030         1.062
5       4        0.9231        0.9608      1         1.041         1.083
6       5        0.9048        0.9512      1         1.051         1.105
7       6        0.8869        0.9418      1         1.062         1.127
8       7        0.8694        0.9324      1         1.073         1.150
9       8        0.8521        0.9231      1         1.083         1.174
```

.... etc.

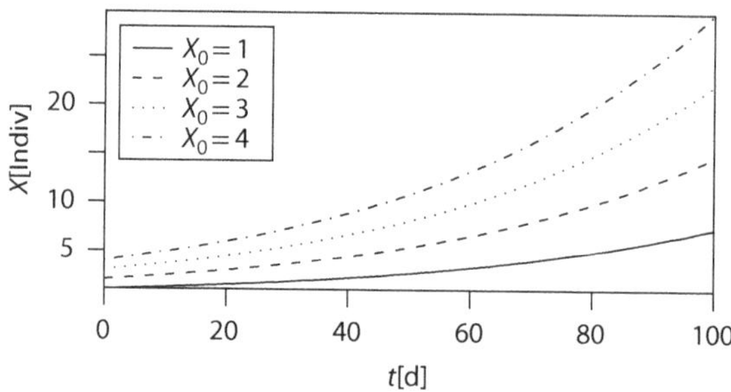

FIGURE 4.7 Multiple runs of the exponential growth by varying the initial condition X_0.

We can vary the initial condition as well by configuring the param definition as follows to obtain the graph shown in Figure 4.7:

```
param <- list(plab="X0", pval = seq(1,4,1))
out.X0 <- sim.mruns(model, file, param)
```

We did not repeat the model and the file statements because we just entered them in the previous run.

To perform another set of runs with a different set of parameter values, *r*, edit the corresponding value in the param definition and run again.

Exercise 4.5

Simulate multiple runs of the exponential growth model with $r = 0, 0.01, 0.02, 0.03$, and 0.04 per day. Use *t* from 0 to 100 days with printed values (time stamps) every day.

The same function can be used to study the exponential decay of a chemical by just changing the input file, for example, **chp4/exp-decay-inp.csv**. As before, open this input file on a text editor as follows or as a spreadsheet to understand its contents:

```
Label,Value,Units,Description
t0,0.00,Day,time zero
tf,10.00,Day,time final
dt,0.10,Day,time step
tw,1.0,Day,time to write
k,-0.2,1/Day,decay rate
X0,100.00,mg/liter,initial concentration
digX,4,None,significant digits for output file
```

Note that now the simulation is only for 10 days, that the parameter label is *k* and takes negative values, and that X_0 is 100 in mg/liter. We can simulate using

```
file <- "chp4/exp-decay-inp.csv"
param <- list(plab="k", pval = seq(-0.2,-0.6,-0.2))
out.r <- sim.mruns(model,file, param)
```

to obtain the result shown in Figure 4.8.

Exercise 4.6

Simulate multiple runs of the exponential decay model with $X_0 = 80, 90, 100, 110$, and 120 mg/liter.

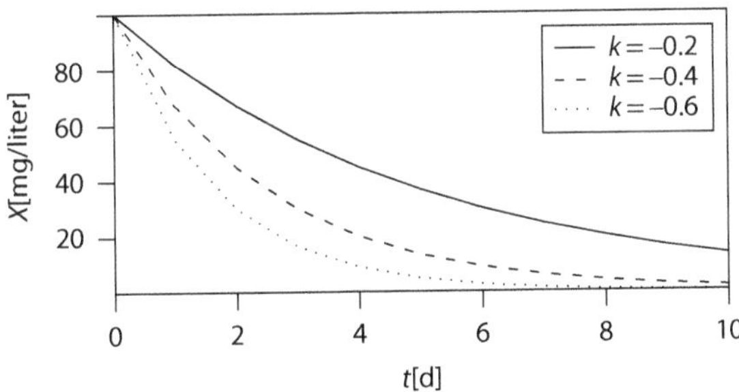

FIGURE 4.8 Multiple runs of the exponential decay by varying the decay rate k.

4.6.4 Random Number Generation

A feature for simulations that include RV is to generate numbers that would seem to be drawn from an RV with uniform pdf between 0 and 1, that is, $U(0,1)$. In R, this is done with `runif(n,0,1)`, where n represents the numbers to be generated and 0,1 represent the interval [0,1].

Let us generate 100 numbers from the uniform probability density function $U(0,1)$:

```
> r <- runif(100,0,1)
```

We can round this to two decimal places:

```
> r <- round(runif(100,0,1),2)
> r
  [1] 0.02 0.64 0.53 0.51 0.69 0.35 0.47 0.09 0.34 0.55 0.84 0.83 0.07 0.89 0.93
 [16] 0.33 0.66 0.03 1.00 0.23 0.81 0.06 0.55 0.63 0.86 0.81 0.71 0.35 0.01 0.24
 [31] 0.10 0.60 0.89 0.81 0.13 0.71 0.57 0.46 0.23 0.97 0.90 0.33 0.62 0.32 0.37
 [46] 0.02 0.73 0.66 0.48 0.05 0.11 0.38 0.42 0.30 0.47 0.32 0.11 0.65 0.13 0.33
 [61] 0.92 0.76 0.54 0.20 0.43 0.30 0.15 0.00 0.89 0.55 0.22 0.83 0.80 0.40 0.56
 [76] 0.62 0.60 0.28 0.66 0.70 0.31 0.40 0.34 0.19 0.69 0.66 0.99 0.45 0.90 0.42
 [91] 0.13 0.04 0.01 0.80 0.98 0.00 0.60 0.32 0.88 0.53
```

We can visualize a histogram as shown in Figure 4.9:

```
> hist(r,freq=F)
```

where the sample seems to be uniformly distributed because the histogram demonstrates a relatively well-spread distribution between 0 and 1. As we already know, the pdf of a uniform distribution between a and b has a height of $1/(b-a)$, in this case 1. We approximate this height in the figure. We can calculate the sample mean of `r` using `mean`, sample variance using `var`, and sample standard deviation using `sd`. So,

```
> round(c(mean(r),var(r),sd(r)),3)
[1] 0.482 0.084 0.290
```

We know that the mean is $(b + a)/2$, in this case 0.5, and the variance $\sigma^2 = \frac{(b-a)^2}{12}$, in this case $1/12 = 0.0833$, and so the standard deviation should be $\sigma = \sqrt{0.0833} = 0.288$. As we can see, the sample results obtained (0.482, 0.084, and 0.290) approximate the mean (expected value, first moment, or population mean) of 0.5, variance of 0.0833, and standard deviation (population) of 0.288.

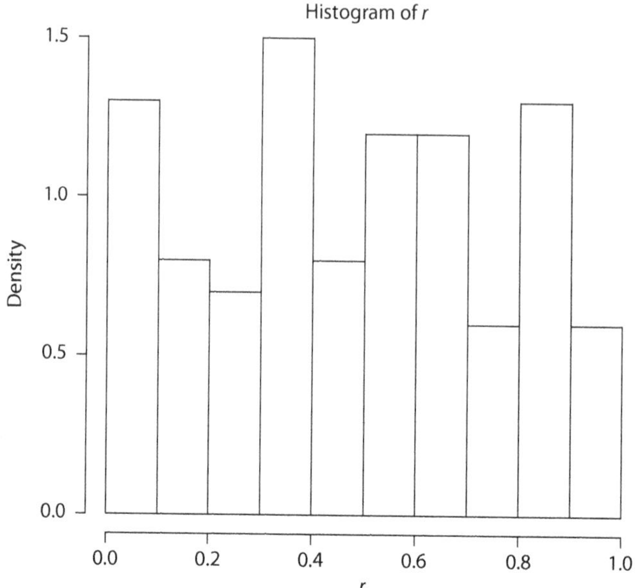

FIGURE 4.9 Histogram of 100 numbers drawn from a uniform distribution (from 0 to 1).

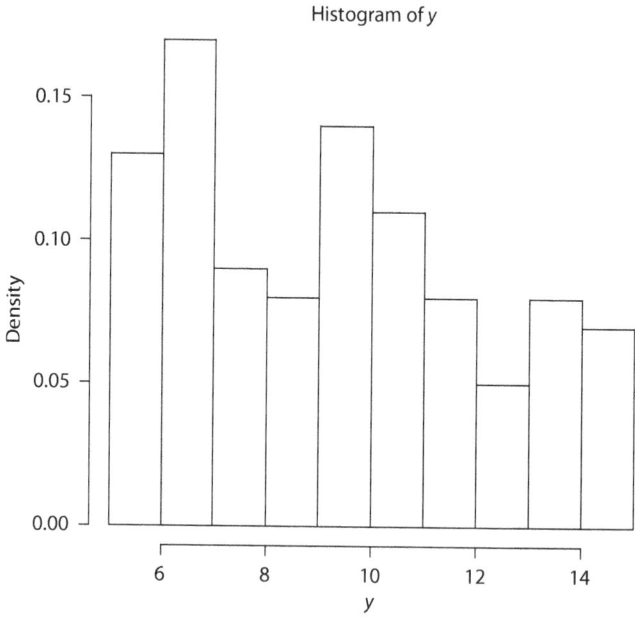

FIGURE 4.10 Histogram of 100 numbers drawn from a uniform distribution (from 5 to 15).

Let us transform the RV X with $U(0,1)$ to obtain an RV Y with $U(a,b)$. First draw a random value x using `runif(1,0,1)` and then apply the expression $y = (b - a)x + a$. Using $a = 5$, $b = 15$,

```
a <- 5;b <- 15
y <- (b-a)*runif(100,0,1)+a
hist(y,prob=T)
```

we obtain the histogram shown in Figure 4.10.

In R, we have a variety of functions already built in for the random number generation of many distributions. These include the uniform, normal, exponential, and many others. These are the functions `rnorm(n,mu,std)`, `runif(n,a,b)`, `rexp(n,rate)`, and so on. Here, for all functions, `n` is the sample size. For the normal, `mu` is μ = mean and `std` is σ = standard deviation; for the uniform, `a` and `b` are the lower (minimum) and the upper (maximum) limits, respectively, of the uniform interval; for the exponential, `rate` is the rate.

Let us generate 10 numbers from a normal distribution of mean $\mu = 0.01$ and standard deviation $\sigma = 0.005$:

```
> r <- rnorm(100,0.01,0.005)
> hist(r)
```

The histogram generated is shown in Figure 4.11 and the sample seems to be normally distributed. We can calculate the sample mean, variance, and standard deviation using

```
> round(c(mean(r),var(r),sd(r)),3)
[1] 0.010 0.000 0.005
```

As we can see, we approximate the mean (expected value, first moment, or population mean) of 0.01 and standard deviation (population) of 0.005. The variance should be 25×10^{-6}.

Exercise 4.7

Generate 1000 random numbers with the uniform probability density function. Use the function `runif(n,a,b)` with a = 0.4 and b = 0.6. What is the expected value or mean (population mean)? What is the value of the sample mean? Compare and discuss. What is the standard deviation? What is the sample standard deviation? Compare. The histogram should demonstrate a relatively well-spread distribution between 0.4 and 0.6.

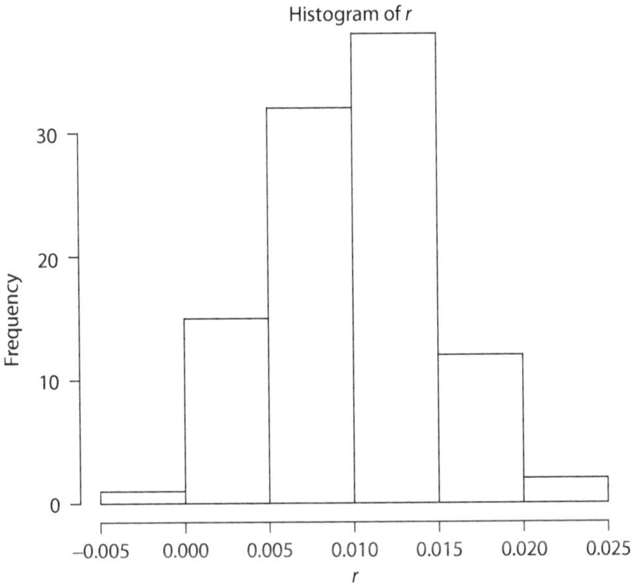

FIGURE 4.11 Histogram of 100 normally distributed numbers.

4.6.5 Custom Random Number Generators

For didactic purposes, in **seem**, we have defined additional random generators `ran.unif`, `ran.exp`, and `ran.norm`, and these are explained at the end of the chapter. These can be called many times using a loop. For example:

```
yunif <- array(); a <- 5;b <- 15
for(i in 1:1000) yunif[i] <- ran.unif(a,b)
hist(yunif,freq=F); title(sub="Uniform in 5,15")
```

The `for` loop function executes the generator 1000 times by changing the value of i from 1 to 1000. Here, `title(sub= )` writes a title under the x-axis; see Figure 4.12.

We can test the other two functions by calling them within a loop to draw samples and then apply the histogram function to obtain the histograms shown in Figure 4.13.

```
# testing using n=1000
par(mfrow=c(2,1))

yexp <- array(); rate <- 1/10
for(i in 1:1000) yexp[i] <- ran.exp(rate)
hist(yexp,freq=F); title(sub="Exponential of rate 0.1")

ynorm <- array();
for(i in 1:1000) ynorm[i] <- ran.norm()
hist(ynorm,freq=F); title(sub="Std Normal")
```

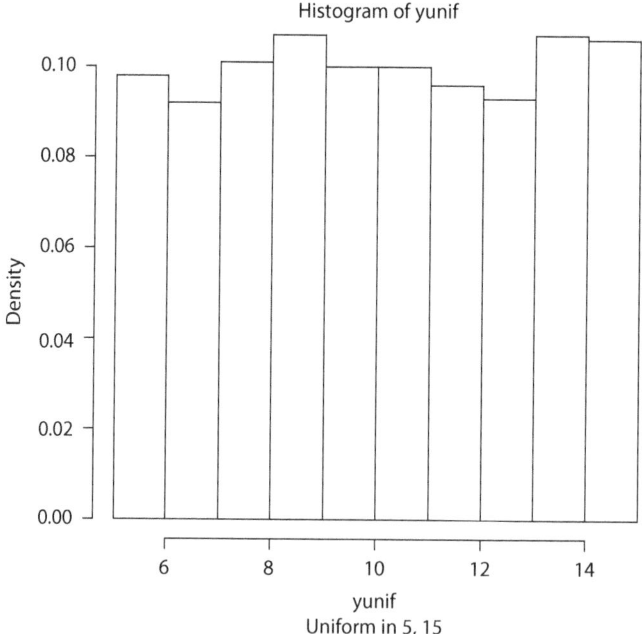

FIGURE 4.12 Histogram of 1000 numbers from a uniform random variable between 5 and 15.

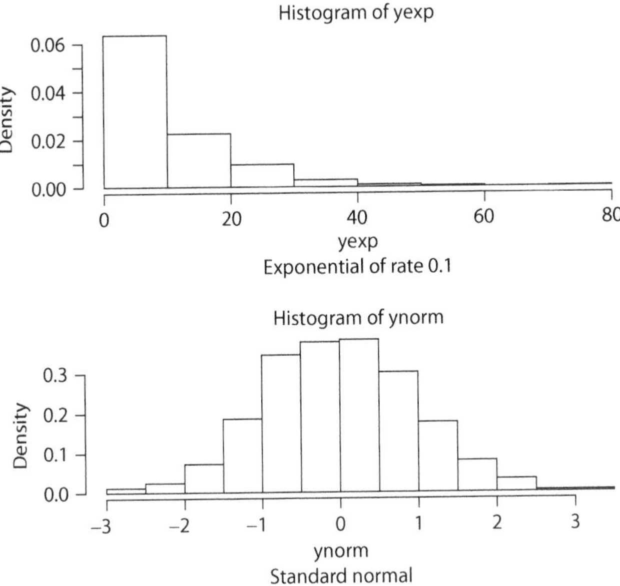

FIGURE 4.13 Histograms of the samples (size 1000) drawn from the exponential and standard normal using transforms from $U(0, 1)$.

The `par(mfrow=c(2,1))` function prepares the graphics device to have two panels arranged vertically (2 rows and 1 column) or one top panel and one bottom panel.

Exercise 4.8

Examine the results of Figures 4.12 and 4.13 with the theoretical pdf for each sample. Use the height, maximum, and spread of the histogram and the pdf for each. Calculate the sample mean, variance, and standard deviation for each sample and compare the results with the theoretical values (population).

4.6.6 Stochastic Simulation

Let us use the seem function `sim.rnum` and the input file **chp4/exp-rnum-inp.csv**. Open this input file on a text editor as shown below.

```
Label,Value,Units,Description
t0,0.00,Day,time zero
tf,100.00,Day,time final
dt,0.10,Day,time step
tw,0.10,Day,time to write
rm,0.01,1/Day,avg intrinsic growth rate
rs,0.10,1/day,sd intrinsic growth rate
n,10,none,number of realizations
X0,10.00,Indiv,initial population
digX,4,None,significant digits for output file
```

This file indicates that the rate coefficient has mean rm = 0.01 and standard deviation rs = 0.1 and that there will be n = 10 realizations. The initial condition is X_0 = 10. Define the model as expon.rand, which is explained at the end of the chapter. Use the `sim.rnum` function by typing the call statement

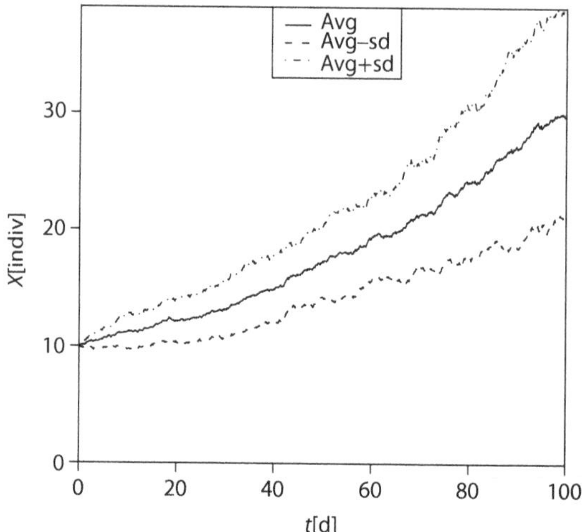

FIGURE 4.14 Result from the Monte Carlo simulation of the exponential growth. Shown are the mean, the mean plus one standard deviation, and the mean minus one standard deviation.

```
model<-list(f=expon.rand); file<-"chp4/exp-rnum-inp.csv"
out.rk <- sim.rnum(model,file)
```

A graphics device opens to show a graph (Figure 4.14) and we can examine the returned resulting object `out.rn` by just typing its name on the console. We can see three lines. The center trace is the average or sample mean of all realizations at each `tw`. The upper trace is the sample mean plus one standard deviation, whereas the lower trace is the sample mean minus one standard deviation.

4.7 SEEM FUNCTIONS EXPLAINED

4.7.1 Function expon

This is a very simple function to define the exponential model:

```
expon <- function(t,p,x) dx <- p*x
```

It is used as part of the argument model to call simulation functions. Although in this case, the model is simple and we could have written it as one line inside the simulation function, it demonstrates the concept of model definition, which we will use later for models that are more complicated.

4.7.2 Functions euler and RK4

For didactic purposes, we will see how to program a simple simulation using the Euler and Runge–Kutta methods using R to numerically solve for X in the model $dX/dt = f(t,p,X)$.

The function `euler` is defined in the following manner:

```
euler <- function(x0, t, f, p, dt){

 # Euler numerical solution of model
 # arguments: x0 initial condition
```

```
#              t times for output
#              f model
#              p parameters
#              dt time step
# arrays to store results
  nt <- length(t); nx <- length(x0)
  ns <- floor((t[2]-t[1])/dt)
  X <- matrix(nrow=nt,ncol=nx)
# first value of X is initial condition
  Xt <- x0; tt <- t[1]
# loops
  for(i in 1:nt){
    X[i,] <- Xt
    for(j in 1:ns){
     tt <- tt + dt;Xt <- Xt + dt*f(tt,p,Xt)
     # X assumed positive
     for(k in 1:nx) if(Xt[k]<0) Xt[k] <- 0
    }
  } # end of run
 return(X)
} # end of function
```

First declare a matrix to store X on the basis of the number of entries of time and the number of entries in variable as given by the initial conditions. In this chapter, this dimension is simply one, but later we will encounter variables with several dimensions. Then we initialize X and time. Then, we write two nested `for` loops. The outer loop controls the times to save the output (as given by `tw`), whereas the inner loop runs the calculation in steps of dt updating X according to Equation 4.6. Note that these loops include more than one line and therefore their scope requires curly braces. For the purposes of this book, all variables X will be positive or zero, so we force X to zero when negative.

Look at the program `RK4` in the console. This is essentially the same function as `euler`, but we have implemented the Runge–Kutta fourth-order method instead of Euler inside the inner loop according to Equations 4.8 and 4.9:

```
# loops
   for(i in 1:nt){
   X[i,] <- Xt
   for(j in 1:ns){
    k1<- f(tt,p,Xt)
    k2<- f(tt+dt/2,p,(Xt+k1*dt/2))
    k3<- f(tt+dt/2,p,(Xt+k2*dt/2))
    k4<- f(tt+dt,p,(Xt+k3*dt))
    kavg <- (dt/6)*(k1+ 2*k2+ 2*k3+ k4)
    tt <- tt + dt; Xt <- Xt + kavg
    # X assumed positive
    for(k in 1:nx) if(Xt[k]<0) Xt[k] <- 0
   }
  } # end of run
```

4.7.3 Function sim.comp

For didactic purposes, we will see how to program a simple simulation using the `euler` and `RK4` functions just described. Open and look at the program `sim.comp` in the console. We will explain the sequence of script lines and their purpose in this program. R does not execute the lines marked with the pound sign (#). These are remarks or comments to document the program. The program is a function with at least two arguments: a model defined as a list and an input filename. Optionally, we specify a numerical method, a logical variable to generate PDF output, and a label to identify output files.

```
sim.comp <- function(model, file, method= "RK4", pdfout=F, lab.out=""){

  # required: model as a list
  #           path/name of input file
  # optional: name of function for num method, default RK4
  #           switch to produce pdf files, default F
  #           label to identify output files, default none
```

We will explain the function in two parts: (1) reading and organizing the input and (2) performing iteration and writing the output. Let us start with the first part.

```
# read input, file format is csv
 dat <- c("character","numeric","character","character")
 input <- read.csv(file,header=T,colClasses=dat)
 # input values and labels
 v <- t(input$Val); vl <- t(input$Lab)

 # t0,tf,dt,tw are in v[1:4]
 t <- seq(v[1],v[2],v[4]); dt <- v[3]
 # p,X0, digX in v[5:7]
 p <- v[5]; X0 <- v[6]; digX <- v[7]
 # units ind and dep vars
 unit <- c(input$Uni[1], input$Uni[6])
```

Here the filename variable `file` is an argument and has an extension `.csv` for comma-separated values; for example, look at the file **chp4/exp-pop-inp.csv**, which we have already examined in the previous section. Instead of `scan` we use `read.csv`. Here, this function assigns the information read from the file to a "data frame" object named **input**. A data frame is an object similar to a matrix but it allows columns of different types, which are specified in the line above as `dat`. We read names like "Label," "Value," and so on from the header (first row) and then read the records in all the rows.

One can address the columns of this data.frame using the dollar sign ($). Thus, we form arrays v and `vl` with the values of the column `input$Val` and `input$Lab`. Here the function `t()` is used twice to transpose from column to row. Do not confuse the R function `t` with the variable t for time. Note that the component names can be abbreviated when there is no ambiguity—for example, `$Lab` for `$Label`. The next line assigns variables from the symbols we recognize. This is just to facilitate writing the next section. Finally, the units for time t and state X are extracted from the input data frame using `$Uni` and stored as an array to facilitate referring to the units.

Next, we look at the second part. First, we call either the function `euler` or `RK4` passing as arguments the initial condition, the time sequence, the function `f` to integrate as a component `model$f` of list `model`, the parameter, and the time step. Then we format the output according to the number of significant figures, organize the results in a data frame, and call another function `onerun.out` to do the heavy work of plotting graphs and writing the files. We will describe this output function in the next section.

```
# call integration
 if(method=="euler")  X <- euler(X0, t, model$f, p, dt)
 else  X <- RK4(X0, t, model$f, p, dt)
 # organize output
 X <- signif(X,digX); output <- data.frame(t,X)
 names(output) <- paste(names(output),"[",unit,"]", sep="")
 # generate output files
 onerun.out(prefix=file, lab.out, input, output, pdfout)

 return(list(input=input,output=output))
} # end of functionv
```

4.7.4 FUNCTION ONERUN.OUT

Now, let us describe the output function. First, it extracts t and X and their labels from the **output** data frame; then it plots X versus t, storing the graph in a PDF file when the pdfout is T; the PDF filename is formed by concatenating the prefix, the output label passed as an argument, the label "-out," and the extension .pdf. Subsequently, it writes the input and output data to a CSV output file named in the same fashion as the PDF file; it uses the function `write.table` with a variety of options and transposes the names of the output to a row before writing them.

```
onerun.out <- function(prefix, lab.out="", input, output, pdfout=F){

 # produces output files for one run and one variable

 # plot the graph; in pdf output file if desired
  t <- output[,1]; X <- output[,-1]
  tlab <- names(output)[1]; Xlab <- names(output)[2]
  if(pdfout==T) pdf(file=paste(prefix,"-",lab.out,"-out.pdf",sep=""))
   par(xaxs="i",yaxs="i")
   plot(t, X, type="l", ylim=c(0,max(X)),xlab=tlab,ylab=Xlab)
   if(pdfout==T) dev.off()

 # writes output file as csv
  fileout.csv <- paste(prefix,"-",lab.out,"-out.csv",sep="")
  write.table(input,fileout.csv, sep=",", row.names=F, quote=F)
  name.out <- t(names(output))
  write.table(name.out,fileout.csv, sep=",", col.names=F,
              row.names=F, append=T, quote=F)
  write.table(output,fileout.csv, sep=",", row.names=F,
              col.names=F, append=T,quote=F)
}
```

It is a good idea to include the input in the output file and in the function's return so that we make sure to know the conditions under which the output was obtained.

4.7.5 FUNCTION SIM.MRUNS

Now we extend the program `sim.comp` from a single simulation run to multiple runs. We need to write an outer loop around the call to the integration. This outer loop varies according to the condition of each run. The new function `sim.mruns` includes an argument to pass the information about the variation in a parameter, for example,

```
>param <- list(plab="r", pval = seq(-0.02,0.02,0.01))
```

It contains the label of the parameter to be varied and its values. These components can be addressed by `param$plab` and `param$pval`, respectively. Recall that the `$` symbol is used to address the components of a list.

```
sim.mruns <- function(model, file, param, pdfout=F, lab.out="") {

 # multiple runs
 # required: model as a list
 #           path/name of input file
 #           param set for multiple runs
 # optional: switch to produce pdf files, default F
 #           label to identify output files, default none
 # read input, file format is csv
 dat <- c("character","numeric","character","character")
 input <- read.csv(file,header=T,colClasses=dat)
 # input values and labels
 v <- t(input$Val); vl <- t(input$Lab)

 # t0,tf,dt,tw are in v[1:4]
 t <- seq(v[1],v[2],v[4]); dt <- v[3]
 digX <- v[7]
 # units ind and dep vars
 unit <- c(input$Uni[1], input$Uni[6])
```

The length of `param$pval` determines the number `np` of the values of the parameter to control the outer loop. The function changes the execution part to include such a loop in the following manner:

```
# number of time stamps and param values
  nt <- length(t); np <- length(param$pval)
 # array to store results
  X <- matrix(ncol=np,nrow=nt)
 # loop for parameter p
  for(i in 1:np) {
   # pick a value of X0 or p from param
    for(j in 5:6) if(param$plab==vl[j]) v[j]<- param$pval[i]
   # p,X0, in v[5:6]
    p <- v[5]; X0 <- v[6]
   # integration
     X[,i] <- RK4(X0, t, model$f, p, dt)
  } # end of parameter p loop
```

The last part of the function prepares the output to identify the units and the run number. Now we use another output function because we have several columns for the output, one for each run. This function is `mruns.out`.

```
# prep and organize output
  X <- signif(X,digX); output <- data.frame(t,X)
  tXlab <- paste(c("t","X"),"[",unit,"]", sep="")
  var.run <- c(tXlab[1], paste(tXlab[2],paste(".
Run",c(1:np),sep=""),sep=""))
  names(output) <- var.run

 # call output function to generate output files
 mruns.out(prefix=file, lab.out=param$plab, input, param, output,
tXlab, pdfout)

 return(list(input=input, param=param, output=output))

} # end of function
```

4.7.6 Function mruns.out

The new output function `mruns.out` is similar to onerun.out but has additional arguments `tXlab` for labeling graphs with units and `param` for the varying parameter. It uses `matplot` instead of `plot` to draw several lines on one graph, one line per run. The output filename includes one more line to provide information about the varying parameter.

```
mruns.out <- function(prefix, lab.out="", input, param, output, tXlab,
pdfout=F){

 # produces output files for multiple runs and one variable

 # produce graph
  t <- output[,1]; X <- output[,-1]
  if(pdfout==T) pdf(file=paste(prefix,"-",lab.out,"-out.pdf",sep=""))
   par(xaxs="i", yaxs="i")
   matplot(t, X, type="l", col=1, xlab=tXlab[1], ylab=tXlab[2])
   legend ("topleft", leg=paste(param$plab,"=",param$pval),
           lty=1:length(param$pval),col=1)
  if(pdfout==T) dev.off()

 # writes output file as csv
  fileout.csv=paste(prefix,"-",lab.out,"-out.csv",sep="")
  write.table(input,fileout.csv, sep=",", row.names=F, quote=F)
  name.param <- t(c(param$plab,param$pval))
  write.table(name.param, fileout.csv, sep=",",
              col.names=F, row.names=F, append=T, quote=F)
  name.out <- t(names(output))
  write.table(name.out,fileout.csv, sep=",", col.names=F,
              row.names=F, append=T, quote=F)
  write.table(output,fileout.csv, sep=",", col.names=F, row.names=F,
              append=T, quote=F)

 }
```

4.7.7 Custom Random Generators

For didactic purposes, we show how to build custom random number generators from the basic uniform distribution between 0 and 1. The following function defines a uniform generator between a and b using the scaling and shifting of $U(0,1)$:

```
ran.unif <- function(a,b){
 u <- runif(1,0,1)  # generate uniform U(0,1)
 x <- (b-a)*u+a     # apply transform
 return(x)    # return the calculated value
}
```

Similarly, the following function generates an exponential using the inverse cumulative method:

```
# random number generators
# obtained from U(0,1) generated by runif(1,0,1)
# exponential
ran.exp <- function(rate){
   u <- runif(1,0,1) # generate uniform U(0,1)
   x <- (1/rate)*(-log(u))
  return(x)
}
```

Furthermore, the following function generates a standard normal from two numbers drawn from a uniform:

```
# standard normal
ran.norm <- function( ){
 # generate two values from uniform
 u1 <- runif(1,0,1); u2 <- runif(1,0,1)
 z <- cos(2*pi*u1)*sqrt(-2*log(u2))
 return(z)
}
```

4.7.8 Function expon.ran

This is a very simple function to define an exponential model with a coefficient drawn from a normal distribution:

```
expon.rand <- function(t,p,x) dx <- rnorm(1,p[1],p[2])*x
```

It is used as a part of the argument model to call simulation functions.

4.7.9 Functions sim.rnum and rnum

The function `sim.rnum` is similar to `sim.comp`. The major differences are that a new function `rnum` is called to perform the integration and a new function `rnum.out` is called to write the output files.

```
# t0,tf,dt,tw are in v[1:4]
t <- seq(v[1],v[2],v[4]); dt <- v[3]
# mu,sd,n in v[5:7]
p <- c(v[5],v[6]); n <- v[7]; X0 <- v[8]; digX<-v[9]
unit <- c(input$Uni[1], rep(input$Uni[8],2))

# call integration
X <- rnum(X0, t, model$f, p, dt,n)
# organize output
X <- signif(X,digX); output <- data.frame(t,X)
names(output) <- paste(names(output),"[",unit,"]", sep="")
tXlab <- paste(c("t","X"),"[",unit,"]", sep="")
# generate output files
rnum.out(prefix=file, lab.out, input, output, tXlab, pdfout)
```

The integration function is based on the `euler` function but includes the use of a three-dimensional (3-D) array X, defined using `structure`, and the calculations of mean and standard deviation of all realizations. The third dimension of array `X` is for the realizations.

```
rnum <- function(x0, t, f, p, dt, n){

 # Numerical solution of random model
 # arguments: x0 init cond
 #            t times for output
 #            f model
 #            p parameters
 #            dt time step
 #            n number of realizations
 # arrays to store results
    nt <- length(t); nx <- length(x0)
    ns <- floor((t[2]-t[1])/dt)
    X <- structure(c(1:(nt*nx*n)),dim=c(nt,nx,n));X[,,]<-NA
    Xm <- matrix(nrow=nt,ncol=nx); Xs <- Xm
 # loops
   # realizations
   for(k in 1:n){
 # first value of X is initial condition
    Xt <- x0; tt <- t[1]
    for(i in 1:nt){
     X[i,,k] <- Xt
     for(j in 1:ns){
      tt <- tt + dt;Xt <- Xt + dt*f(tt,p,Xt)
      # X assumed positive
      if(Xt<0) Xt <- 0
```

```
      }
     } # end of one realization
    } # end of run

  # statistics for the run
       for(i in 1:nt){
        for(j in 1:nx){
         Xm[i,j] <- mean(X[i,j,])
         Xs[i,j] <- sd(X[i,j,])
        }
       }
  return(cbind(Xm,Xs))
 } # end of function
```

4.7.10 Function rnum.out

This function is similar to the `onerun.out` but draws the mean using `plot` and then the upper and lower bounds ($\mu \pm \sigma$) using `lines`:

```
plot(t, Xm, type="l", ylim=c(0,max(X+Xs)),xlab=tlab,ylab=Xlab)
lines(t, Xm-Xs, lty=2)
lines(t, Xm+Xs, lty=4)
legend("top",leg=c("Avg", "Avg-sd", "Avg+sd"),lty=c(1,2,4),col=1)
```

5 Model Evaluation

How good is a model? This is an important question in modeling and entails how to evaluate a model. Usually, the answer to this question depends on the application or the purpose of the model. There are several issues: For example, how good is the computer simulation with respect to the exact model solution? Are there any errors in the simulation code? How good is the assumed model in representing the reality we are trying to model? How good are the parameter estimates? Let us examine some of these questions.

5.1 TESTING THE NUMERICAL METHOD AND THE CODE

We learned from Chapter 4 that to simulate a model, we translate its equations into computer code and then apply some integration numerical method. It is important to make sure that the computer calculation correctly implements the intended mathematical model.

One possibility is to study how good the computer simulation represents the model's exact solution. This is usually done when developing the numerical method or writing the computer code. Nevertheless, being realistic, you do not usually know the exact solution; this is why you are simulating it anyway! Alternatively, we can probably determine what the solution should be for certain special conditions; for example, at **steady state**, or when we isolate a particular component of the model, or use simplified model conditions.

One manner of quantifying the error between the simulated and exact solutions is to calculate the difference between the two values at all times and then aggregate them for the entire simulation run. That is to say, the error, $e(t_i)$, at the simulation time, t_i, is the difference between quantities $X(t_i)$ and $\hat{X}(t_i)$:

$$e(t_i) = X(t_i) - \hat{X}(t_i) \tag{5.1}$$

where $X(t_i)$ and $\hat{X}(t_i)$ are the exact and the simulated solutions at the same time, t_i.

This difference can be expressed relative to the exact solution by calculating its ratio with respect to the exact solution:

$$e_r(t_i) = \frac{e(t_i)}{X(t_i)} = \frac{X(t_i) - \hat{X}(t_i)}{X(t_i)} \tag{5.2}$$

The error could generally vary during the simulation run, that is, $e(t_i)$ varies with t_i for all values of $i = 1, \ldots, n$, where n is the number of values calculated in the simulation run. It can increase or decrease; thus, a simple addition of the errors may lead to the cancellation of positive with negative errors. To avoid this, the error is squared before addition.

$$e^2(t_i) = \left(X(t_i) - \hat{X}(t_i)\right)^2 \tag{5.3}$$

Then, errors for all t_i are added to obtain the total squared error as

$$E = \sum_{i=1}^{n} e^2(t_i) \tag{5.4}$$

TABLE 5.1
Model Evaluation

Days	Data	Model	Error	Relative Error (%)	Squared Error
0	1,000	1,000.00	0.00	0.00	0
1	2,793	2,718.28	74.72	2.68	5,583
2	7,405	7,389.06	15.94	0.22	254
3	20,118	20,085.54	32.46	0.16	1,054
4	54,612	54,598.15	13.85	0.03	192
		Total	136.98	3.08	7,083
		Average	27.40	0.62	1,416.54
		RMS			37.64

and the average or mean squared error as E/n. To express this quantity in the same units as the variable X, we take the square root of the mean squared error to obtain the **root mean square** error (**RMS error** for brevity).

$$\mathrm{RMS}(e) = \sqrt{\frac{E}{n}} = \sqrt{\frac{\sum_{i=1}^{n} e^2(t_i)}{n}} \tag{5.5}$$

Exercise 5.1

Use the results in Chapter 4. Compare simulated X at 0.6 days with exact X at 0.6 days. Hints: What is the difference between the exact and the simulated solutions? What is the relative difference (in %)? What is the total squared error? What is the RMS error?

Table 5.1 shows the details of the calculation for several data values (given in the second column), assuming a population model with $r = 1\ \mathrm{d}^{-1}$. First, calculate all model values using the solution of the exponential model for each time, t. These values are entered in the third column. Second, for each pair of values, calculate the error as data minus model (fourth column), then the relative error as (data – model)/data (mulitplied by 100 to yield %) (fifth column), then the squared error (data – model)^2 as shown in the last column. Third, add all the values to get the totals and divide the total by $n = 5$ to get the averages. Last, take the square root of the average squared error to get the RMS error.

Exercise 5.2

For a population with a rate coefficient of $r = 0.5\ \mathrm{d}^{-1}$, complete Table 5.2. Calculate the modeled values and all error terms (assume that the relative error is with respect to the data value).

5.2 GRAPHICAL EVALUATION

Several types of graphics help to visualize evaluation results (Figure 5.1). We will use, as an example, the results of the evaluation of an exponential decay model as shown in Table 5.3. These results are used to build the plots shown in Figure 5.1. First, there is a graph showing observed and modeled trajectories versus time. This is a very common graphical representation of a model evaluation. Typically, we show the observed values as markers and the model as a line. It is also useful to plot

TABLE 5.2
Model Evaluation $r = 0.5\ d^{-1}$

Days	Data	Model	Error	Relative Error	Squared Error
0	1000	1000			
1	1687				
2	2721				
3	4500				
4	7,474				
		Total			
		Average			
		RMS			

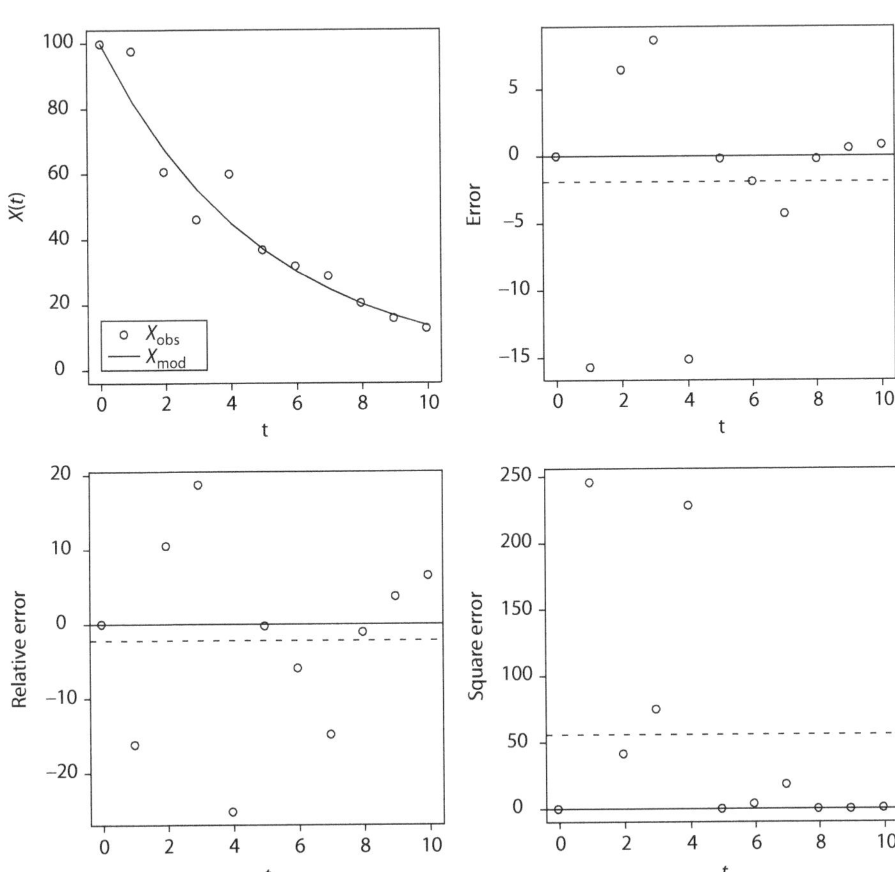

FIGURE 5.1 Plots for model evaluation: modeled versus observed data and errors (absolute, relative, and squared).

the errors versus time: the absolute values, relative values, and square of the absolute errors. We use several panels in the same graph and show horizontal lines for zero (solid) and the averages (dashed).

Another common representation is a scatter plot showing markers for the plot of modeled versus observed together with the unit slope line, which represents what would be the ideal exact fit (Figure 5.2).

TABLE 5.3
Evaluation Results of a Decay Model

t	X_{obs}	X_{mod}	Error	Relative Error	Square Error
0	100	100	0	0	0
1	97.56	81.87	−15.69	−16.08	246.18
2	60.6	67.03	6.43	10.61	41.34
3	46.23	54.88	8.65	18.71	74.82
4	60.04	44.93	−15.11	−25.17	228.31
5	36.97	36.79	−0.18	−0.49	0.03
6	32.01	30.12	−1.89	−5.9	3.57
7	28.93	24.66	−4.27	−14.76	18.23
8	20.41	20.19	−0.22	−1.08	0.05
9	15.94	16.53	0.59	3.7	0.35
10	12.71	13.53	0.82	6.45	0.67
		Total	20.87	−24	613.56
		Avg	−1.9	−2.18	55.78
		RMS			7.47

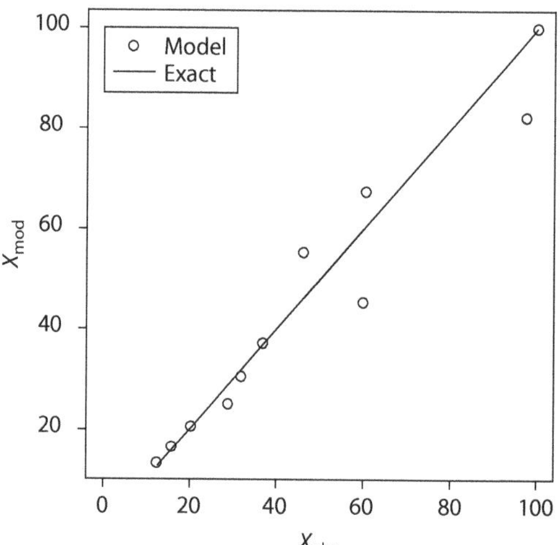

FIGURE 5.2 Plot of observed versus modeled results together with the ideal unit slope line.

5.3 TESTING THE MODEL

How good is the model? Are the assumptions valid for the problem we are trying to solve? To answer these questions, the modeled output needs to be compared to available data or data to be collected to test the model. The comparison can be graphical, using data and simulated values on the same plot, but statistical analysis is needed for quantitative results. We can calculate the relative difference or the RMS difference between model output and real data using expressions such as Equations 5.2 and 5.5. In this case, the subscript i corresponds to all data points, which could be only a subset of all simulated values. To avoid circularity, the data points used for evaluation should not be the same as those used for calibration.

Example: Assume that you have data on changes of a population over time. Let us suppose we want to use the simple exponential model to forecast population changes. We can estimate the value of r or calibrate the model using the data for one period of time. Run or calculate the solution

TABLE 5.4
Data on Population Growth

t[years]	0	0.5	1	1.5	2	2.5	3	3.5	4	4.5	5	5.5	6	6.5	7	7.5	8	8.5	9	9.5	10
X[indiv]	1	1	1.1	1.1	1.2	1.4	1.6	1.9	2.2	2.8	3.5	4.5	6.1	8.3	11.6	16.7	24.5	37.1	57.4	91.2	148

and produce output for another period. Then, compare the real data for this period with the simulated data. This comparison will serve as a test of how good the simple exponential represents the population.

Exercise 5.3

Suppose we have the data on population growth given in Table 5.4. Use the simple exponential model and estimate the value of *r*, or calibrate the model using the entire data set using linear regression. Plot the data and the simulated values in the same graph. Use both the data and the log-transformed data. Is the exponential model a good model for this data set? Estimate the value of *r* or calibrate the model using data for the period 0–5.0 years. Now, run or calculate the model and produce output for the period 5.5–10.0 years. Then, compare real data for this period with the simulated data. Do the simulated values correspond to the data values? Calculate the value of *r* using the data for the period 5.5–10 years. Further calculations done by breaking down the data set show that *r* increases with time. We conclude that a simple exponential model is not a very good model for this population since a constant does not represent well the coefficient *r* throughout the entire study period. We could still apply an exponential model, but a more complicated one in which *r* varies with time.

5.4 TESTING THE PARAMETER VALUES: SENSITIVITY ANALYSIS

This is one of several types of parametric analysis and a very important component of modeling. We are interested in evaluating the changes in the solution of a model due to changes in the values of the parameters. We evaluate the variation of the solution $X(t, p)$ of a model due to the variation of parameter p. For relatively small variations of p, or local analysis, we can evaluate the gradient of a target function of the solution $g(X(t, p))$ with respect to p evaluated at a **nominal** value of the parameter, which is the selected point for analysis.

For example, for one parameter, the gradient is the derivative of $g(X(t, p))$ with respect to p evaluated at a nominal value, p_n:

$$S = \left.\frac{\mathrm{d}\, g(X(p,t))}{\mathrm{d}\, p}\right|_{p=p_n} \tag{5.6}$$

For several parameters, we use the partial derivatives ∂/∂ with respect to each one of the parameters:

$$S_k = \left.\frac{\partial g(X(p,t))}{\partial p_k}\right|_{p_k=p_{k\,n}} \tag{5.7}$$

where p_k represents one of several parameters $k = 1, 2, \ldots$. However, this case can be more complicated because of the possible interactions among the effects of the parameters. We consider this more complicated case in Chapter 8.

We need a target function, $g(\,)$, to convert the solution or output of a dynamic model, which is a trajectory or a function of time, into a single number. This function can be defined using a characteristic of the dynamics (if we have the analytical solution), and then, the derivative could also be calculated in a closed form. In practice, however, all we have is the simulated solution for a set of values of p, and we are limited to calculating the function $g(\,)$ of a set of values of the simulation run for each one of the values of the parameter. Function $g(\,)$ can be, for example, the maximum, minimum, average, RMS, or ending value (at $t = t_f$) of the run. In addition, we could use the maximum, minimum, average, RMS, or ending value (at $t = t_f$) of the difference between the run and the nominal run. Thus, a target function summarizes the entire run in just one value. We will refer to these as **metrics** of the run. Then, Equation 5.7 is approximated by the slope of $g(\,)$ at the nominal p_n as obtained for a discrete set of values of p. The word metric as used here does not mean that the function obeys the properties of a metric in the mathematical sense.

In practice, we are limited to an approximation of the derivative given by the slope or the ratio of discrete and small changes of $g(\,)$ with p. When we use the **difference of metrics**, we write

$$\frac{\Delta g(X(t,p_i))}{\Delta p_i} = \frac{g(X(t,p_i) - g(X(t,p_n))}{p_i - p_n} \tag{5.8}$$

which, for simplicity, we write as

$$\frac{\Delta g(X_i)}{\Delta p_i} = \frac{g(X_i) - g(X_n)}{p_i - p_n} \tag{5.9}$$

This slope is calculated with respect to the nominal value around which we perform the analysis.

Often, we express the ratio $\Delta g(X_i) / \Delta p_i$ in terms of the percent relative change of the target or the metric with respect to the percent relative change of the parameter nominal value:

$$S(X_i) = \frac{\Delta g(X_i) / g(X_n)}{\Delta p_i / p_n} \tag{5.10}$$

In this case, sensitivity is expressed as %/%, that is, what percent of $g(X_i)$ changes with respect to $g(X_n)$ when we change p_i by 1% with respect to p_n.

When we use **metrics of the difference** between the run and nominal, we can write

$$\frac{\Delta g(X(t,p_i))}{\Delta p_i} = \frac{g(X(t,p_i) - X(t,p_n))}{p_i - p_n} \tag{5.11}$$

which, for simplicity, we write as

$$\frac{\Delta g(X_i)}{\Delta p_i} = \frac{g(X_i - X_n)}{p_i - p_n} \tag{5.12}$$

In relative terms,

$$S(X_i) = \frac{g\left((X(t,p_i) - X(t,p_n)) / X(t,p_n)\right)}{\Delta p_i / p_n} \tag{5.13}$$

In this book, we select six metrics, $g_j(\,)$, defined as follows:

$$\begin{aligned}
g_1(X_i) &= \max_t (X_i(t)) \\
g_2(X_i) &= \min_t (X_i(t)) \\
g_3(X_i) &= X_i(t_f) \\
g_4(X_i) &= \frac{1}{t_f} \sum_t (X_i(t) - X_n(t)) \\
g_5(X_i) &= \sqrt{\frac{1}{t_f} \sum_t (X_i(t) - X_n(t))^2} \\
g_6(X_i) &= \max_t \left|(X_i(t) - X_n(t))\right|
\end{aligned} \tag{5.14}$$

where $g_1(\,)$ is the maximum value of X_i for run i; $g_2(\,)$ is the minimum value of X_i for run i; $g_3(\,)$ is the value of X_i for run i; $g_4(\,)$ is the average of the difference $X_i - X_n$ for run i; $g_5(\,)$ is the RMS of the difference $X_i - X_n$ for run i; and $g_6(\,)$ is the maximum of the absolute value of the difference $X_i - X_n$ for run i. Note that the last two metrics always yield positive values. Their sign is inverted for values corresponding to the parameter values below the nominal.

Calculations proceed as follows:

1. Select a set of m values of p.
2. Execute m simulation runs, one for each value of p, say, run i using $p = p_i$, yielding results X_i.
3. Calculate all target functions or metrics, $g_j(X_i)$, for each run i.
4. Now organize metrics $g_j(X)$ versus p_i and calculate sensitivity.
   - $g_j(X)$ versus p_i, absolute response of the metric versus absolute values of p.
   - $\Delta g_j(X_i)$ versus Δp_i, relative response of the metric to relative changes in p.
   - $S_j(X_i)$ versus Δp_i, sensitivity with respect to p.
5. For visualization, we can plot one curve, $X(t)$, versus t for each run (family of curves), metrics $g_j(X)$ versus p, $\Delta g_j(X)$ versus Δp, and S_j versus Δp.

For example, assume an exponential model with $X_0 = 1$, $t_f = 100$, and nominal $r_n = 0.02$. Let us do the calculations for $r_i = 0.03$ using the end value metric $g_3(\,)$:

$$g_3(X(tf, r_i)) = X_0 \times \exp(t_f \times r_i) = 1 \times \exp(100 \times 0.03) = 20.08$$

$$g_3(X(t_f, r_n)) = X_0 \times \exp(t_f \times r_n) = 1 \times \exp(100 \times 0.02) = 7.38$$

$$\frac{\Delta g_3(X_i)}{\Delta p_i} = \frac{g_3(X(t_f, r_i)) - g_3(X(t_f, r_n))}{r_i - r_n} = \frac{20.08 - 7.38}{0.03 - 0.02} = \frac{12.70}{0.01}$$

$$S_3(X_i) = \frac{\Delta g_3(X_i) / g_3(X(t_f, r_n)}{\Delta p_i / p_n} = \frac{12.70 / 7.38}{0.01 / 0.02} = \frac{1.72}{0.5} = 3.44$$

This means that a 50% increase in r at nominal $r = 0.02$ produces a 172% change in the X value at the end of the run ($t = 100$). The sensitivity ratio is 3.44%/%.

Exercise 5.4

Consider the exponential decay model. For a fixed simulation time $t_f = 1$ and initial condition $X_0 = 100$, how does the metric $g(X) = \min(X)$ change with k at nominal $k_n = -0.1$ for a 10% decrease in k? Hint: Note that for the exponential model with a negative rate coefficient, the minimum value is equal to the end value. Be careful with signs.

5.4.1 Selecting Values of the Parameter

The range of parameter variation determines the type of sensitivity analysis. A small range implies a limited change in parameter values or local analysis, whereas a large range implies a global analysis. In practical terms, the **nominal** is the best guess or the best estimate value for the parameter; then, the parameter is **perturbed** or varied above and below the nominal value to obtain a total of

m levels or values for the parameter. For example, denoting the percent change of the parameter value by v, we could have

- $m = 2$, two levels, nominal and nominal + v
- $m = 3$, three levels, nominal − v, nominal, and nominal + v
- $m = 5$, five levels, nominal − $2v$, nominal − v, nominal, nominal + v, and nominal + $2v$

and more levels by assuming increasing multiples of v.

Determining v requires some knowledge of the range of variation of the parameter or of its **variability** (Swartzman and Kaluzny, 1987, pp. 17–24). In such cases, when we have descriptive statistics of the parameter value, we can select the average as the nominal and v proportional to the standard deviation, for example, make v equal to double the standard deviation. For example, assume that in an exponential model, we have mean $r = 0.02$ and $\sigma = 0.005$, then $v = 2\sigma = 0.01$. Then, for $m = 3$, the levels are $(0.02 - 0.01)$, $(0.02 - 0.00)$, and $(0.02 + 0.01)$ or $r = 0.01, 0.02, 0.03$. In this case, $v = 50\%$ of the nominal value.

This regular scheme to generate values of the parameter could bias some of the results for more complicated models. Alternatively, we can select values by random sampling. There are two different ways of performing random sampling: **simple** and **stratified**. In simple, for each run, we pick a number from the assumed or known distribution of the parameter where the mean is the nominal value. We can select, for example, $N(\mu, \sigma)$, a normal distribution with mean μ and standard deviation σ or a uniform distribution, $U(a, b)$, with min = a, max = b.

In stratified, we select subranges of the distribution and then sample within each subrange to reduce the number of runs and force exploration of the whole range. One special type of stratified sampling is that in which each subrange is sampled once, that is, the subrange is selected randomly without replacement. This method will be discussed in Chapter 8.

Note that performing sensitivity analysis by random sampling is not necessarily the same as a Monte Carlo simulation because we keep the selected parameter value fixed for each realization. When the model is deterministic, one run suffices for each parameter value. If the model were stochastic, then sensitivity analysis would imply multiple Monte Carlo simulation runs.

5.4.2 Sensitivity Plots

As explained above, in practice, sensitivity analysis is performed by making several simulation runs defined by varying a parameter and examining the change of one or several metrics of the output. A metric summarizes the entire run in just one value. We can visualize the effect of changing the parameter on the value of the metric by means of x-y graphs. We can perform an odd number of runs where we assume that the nominal value corresponds to the center value in the set of parameter values.

The $x - y$ graph of the value of the metrics for each parameter value $g_j(X)$ versus p illustrates the absolute changes in metrics and parameters (Figure 5.3, upper panel). In addition, relative values in percent for both metric and parameter, $\Delta g_j(X)$ versus Δp, can be plotted to visualize the impact of relative changes (Figure 5.3, lower panel).

The slope of each curve of the lower panel of the figure is the sensitivity, S_j, of a metric to the parameter at the nominal value. This slope varies depending on the parameter value. At each parameter value, we can calculate the slope in relative units. The ratio at the nominal value is not defined because it yields a division by zero, and in this case, the point is interpolated. The slopes of the curves can be plotted against the percent change in the parameter value, S_j, versus Δp (Figure 5.4).

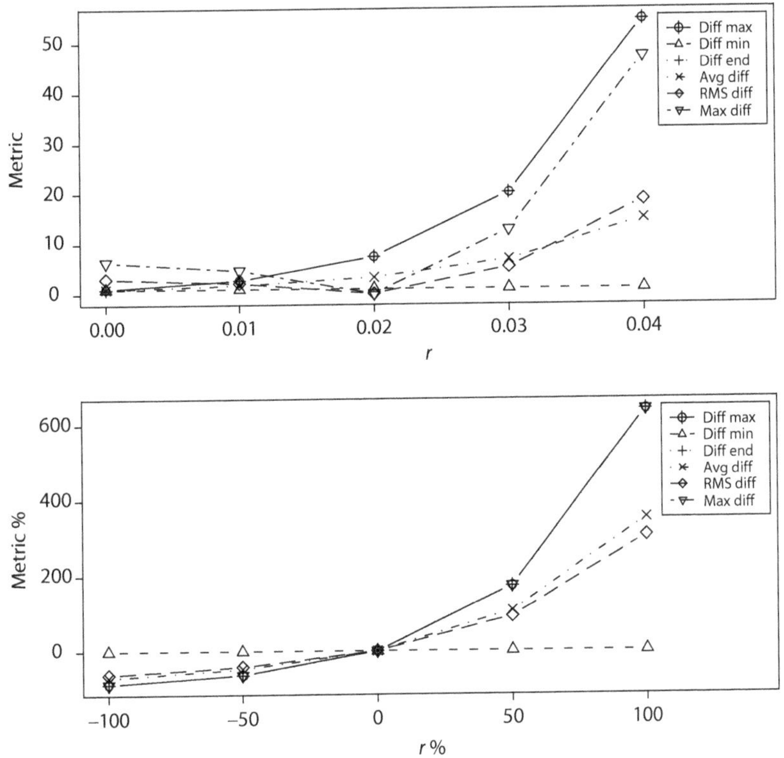

FIGURE 5.3 Sensitivity plots: absolute and relative values.

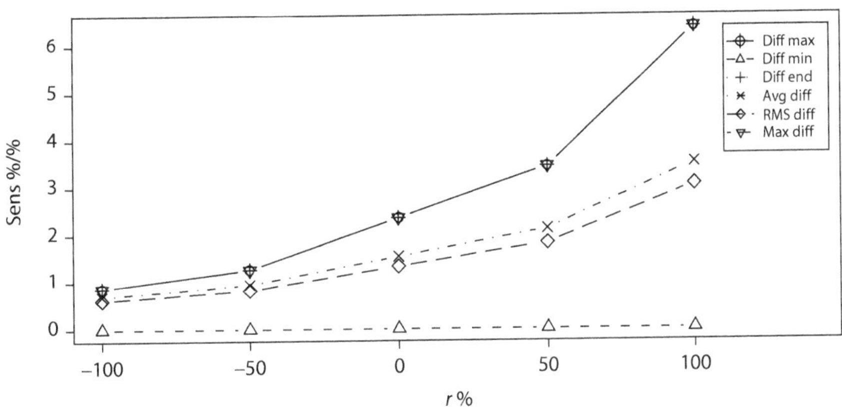

FIGURE 5.4 Sensitivity plots: slopes.

Similar figures are obtained when parameter values are selected by sampling (Figure 5.5), but an additional figure can be obtained by running a linear regression of the relative changes in metrics $\Delta g_j(X)$ versus the relative parameter change, Δp. In this case, the regression is performed with zero intercept so that its line crosses the *y*-axis zero at the nominal value (Figure 5.6, lower panel). In such a case, the high values at the right and the low values at the left make the slopes at the nominal value higher than those obtained in the upper panel. Limiting the excursions in parameter value would make both approaches yield similar results.

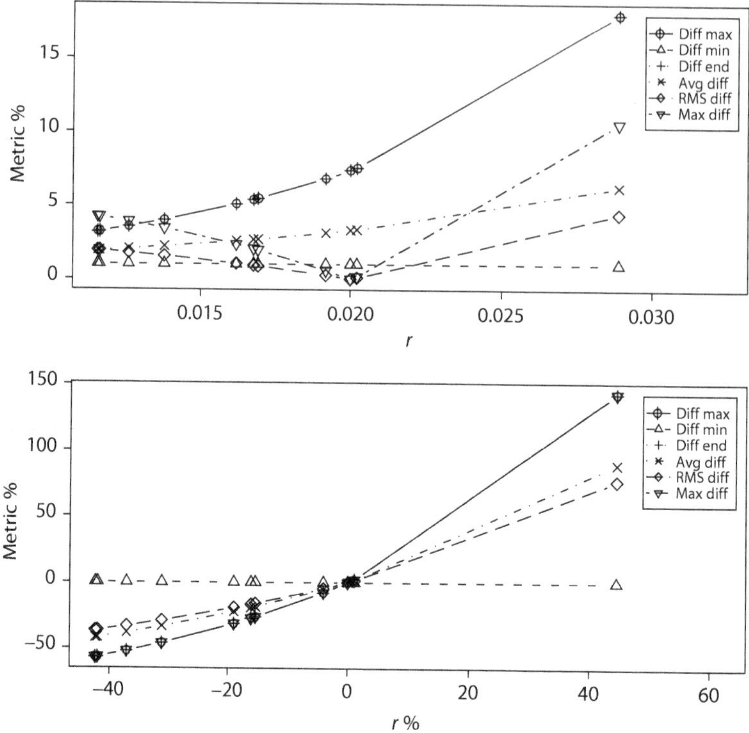

FIGURE 5.5 Sensitivity plots: absolute and relative values. Parameter values are selected by sampling.

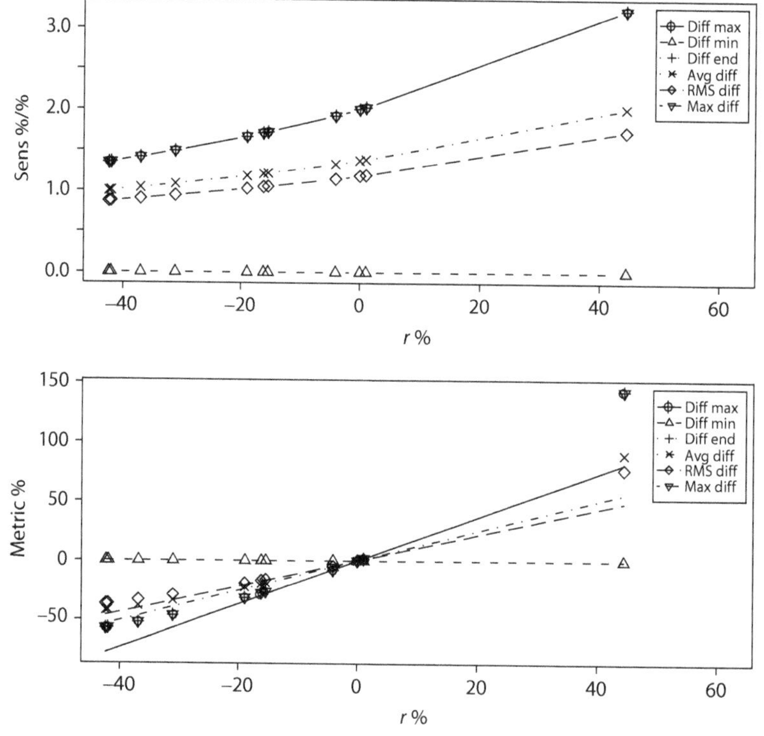

FIGURE 5.6 Sensitivity plots: slopes and regression.

5.5 COMPUTER SESSION

5.5.1 Model Evaluation

Let us calculate the error of a model with respect to the observed. Suppose that after calibrating an exponential model of chemical decay and obtaining $k = -0.2\ d^{-1}$, we gather more data in an attempt to validate the model and obtain the following data starting with a concentration of 100 mg/liter. The data file is **chp5/expodecay-data.txt**.

```
t[d]  X[mg/l]
0  100.00
1  97.56
2  60.60
3  46.23
4  60.04
5  36.97
6  32.01
7  28.93
8  20.41
9  15.94
10 12.71
```

We want to evaluate how good the model is by means of the relative and RMS errors. We start by reading the data file using a scan and converting it into a two-column matrix.

```
> t.Xobs <- matrix(scan("chp5/expodecay-data.txt", skip=1), ncol=2,
byrow=T)
> t.Xobs
      [,1]   [,2]
 [1,]    0 100.00
 [2,]    1  97.56
 [3,]    2  60.60
 [4,]    3  46.23
 [5,]    4  60.04
 [6,]    5  36.97
 [7,]    6  32.01
 [8,]    7  28.93
 [9,]    8  20.41
[10,]    9  15.94
[11,]   10  12.71
>
```

We want to run the model with $k = -0.2$ as given in the **chp5/exp-decay-inp.csv** input file which contains

```
Label,Value,Units,Description
t0,0.00,Day,time zero
tf,10.00,Day,time final
dt,0.1,Day,time step
tw,1.00,Day,time to write
k,-0.2,1/Day,decay rate
X0,100.00,mg/l,initial concentration
digX,4,None,significant digits for output file
```

Let us make a run by calling the function sim.comp, as discussed in Chapter 4.

```
model <- list(f=expon)
t.Xmod <- sim.comp(model,"chp5/exp-decay-inp.csv")
```

As a result, we now have two objects: one for the observed data t.Xobs, and another for the simulated data t.Xmod. Now, we work with both the observed and modeled data to calculate the errors. We have built a function model.eval in the **seem** package, which will be explained at the end of the chapter. This function produces graphics output and writes an output file that contains the evaluation results.

```
# assign values for calculating errors
t <- t.Xobs[,1]; Xobs <- t.Xobs[,2]; Xmod <- t.Xmod$output[,2]
# run model evaluation
exp.eval <- model.eval(t, Xobs, Xmod, "chp5/eval-decay")
```

We can display the results in the console by typing the object exp.eval.

```
> exp.eval
$result
       t   Xobs   Xmod abs.err rel.err sq.err
 [1,]  0 100.00 100.00    0.00    0.00   0.00
 [2,]  1  97.56  81.87  -15.69  -16.08 246.07
 [3,]  2  60.60  67.03    6.43   10.61  41.38
 [4,]  3  46.23  54.88    8.65   18.71  74.85
 [5,]  4  60.04  44.93  -15.11  -25.16 228.21
 [6,]  5  36.97  36.79   -0.18   -0.49   0.03
 [7,]  6  32.01  30.12   -1.89   -5.90   3.57
 [8,]  7  28.93  24.66   -4.27  -14.76  18.23
 [9,]  8  20.41  20.19   -0.22   -1.08   0.05
[10,]  9  15.94  16.53    0.59    3.70   0.35
[11,] 10  12.71  13.53    0.82    6.48   0.68

$label.result.tot.avg
[1] "abs.tot" "rel.tot" "sq.tot"  "abs.avg" "rel.avg" "sq.avg"  "rms"

$result.tot.avg
[1] -20.86 -23.96 613.42  -1.90  -2.18  55.77   7.47
```

The last value of the last record is the RMS error.

Use an editor to look at the result file **eval-decay.txt**.

```
time    Xobs    Xmod abs.err  rel.err   sq.err
0.00  100.00  100.00    0.00     0.00     0.00
1.00   97.56   81.87  -15.69   -16.08   246.07
2.00   60.60   67.03    6.43    10.61    41.38
3.00   46.23   54.88    8.65    18.71    74.85
4.00   60.04   44.93  -15.11   -25.16   228.21
5.00   36.97   36.79   -0.18    -0.49     0.03
6.00   32.01   30.12   -1.89    -5.90     3.57
```

```
  7.00    28.93    24.66    -4.27    -14.76     18.23
  8.00    20.41    20.19    -0.22     -1.08      0.05
  9.00    15.94    16.53     0.59      3.70      0.35
 10.00    12.71    13.53     0.82      6.48      0.68
abs.tot rel.tot sq.tot abs.avg rel.avg sq.avg rms
-20.86   -23.96   613.42    -1.90     -2.18     55.77    7.47
```

Graphics of evaluation results are in the file **eval-decay.pdf**. The graphics are the same as in Figures 5.1 and 5.2, which were presented as examples earlier. The function `model.eval` could have been used with argument `pdfout=T` to store the graphics in a PDF file.

Exercise 5.5

Use the function `model.eval` to perform the evaluation of the exponential model with $k = -0.2$ with respect to the data found in the file **chp5/expodecay-exercise.txt**. Note that you need to rerun the model for $t = 0$ to $t = 20$. Discuss the evaluation results.

5.5.2 Sensitivity Analysis

We will make multiple runs by varying a parameter and examining the change of one or several metrics of the output. These all apply to $X(t)$ for the entire simulation run; that is, a metric summarizes the entire run in just one value. Therefore, we can analyze the effect of changing the parameter on the value of the metric and plot x–y graphs for visualization.

We calculate metrics from the simulation output of the program `sim.mruns` used in Chapter 4 by writing a new **seem** function to calculate the metrics, perform sensitivity analysis, and produce the plots. The function is `sens` and takes as arguments the simulation output, a list of parameter labels and values, the nominal value of the parameter, and the prefix of the filename to store the graphs. The `sens` function is explained at the end of the chapter. Often, we perform an odd number of runs where the nominal value corresponds to the center value in the set of parameter values, and we vary by a percent. However, the nominal's place can be anywhere as dictated by the argument `pnom`.

First, select the parameter values. Let us make $m = 5$ simulation runs, with nominal $r = 0.02$ and $\sigma = 0.005$, which gives $v = 50\%$ or 0.01.

```
# give nom and sd
pnom=0.02; psd =0.005; plab = c("r"); runs=5
# factorial
pval <- seq((pnom -(runs-1)*psd),(pnom+ (runs-1)*psd),2*psd)
param <- list(plab=plab, pval=pval, pnom=pnom, fact=T)
```

Therefore, the set of values for parameter r is $r \pm 0.01$ and $r \pm 0.02$, which yields 0.00, 0.01, 0.02, 0.03, 0.04. The `fact` logical argument is set to T for factorial analysis.

Next, produce output by a call to `sim.mruns` with the input file **exp-sens-inp.csv**. Finally, perform sensitivity analysis by a call to `sens` with the output just generated, the `param` set, and the filename to store results.

```
t.X <- sim.mruns(model,"chp5/exp-sens-inp.csv", param)
s.y <- sens(t.X$output, param, "chp5/exp-sens-plots-fact-out")
```

The output `s.y` is a list with the following components: metric values corresponding to param values (stored in the component `$val.val`), percent change in the metric to percent change in the parameter value (stored in the component `$perc.perc`), and percent ratio change in the metric to percent change in the parameter (stored in the component `$perc.ratio`).

```
> s.y
$val.val
  Param Val Diff Max Diff Min Diff End Avg Diff RMS Diff Max Diff
1      0.00    1.000        1    1.000    1.000    3.046    6.389
2      0.01    2.718        1    2.718    1.732    2.144    4.671
3      0.02    7.389        1    7.389    3.295    0.000    0.000
4      0.03   20.090        1   20.090    6.785    5.268   12.701
5      0.04   54.600        1   54.600   14.870   18.562   47.211
$perc.perc
  Param Perc Diff Max Diff Min Diff End Avg Diff RMS Diff Max Diff
1       -100   -86.466        0  -86.466  -69.652  -61.682  -86.466
2        -50   -63.216        0  -63.216  -47.423  -41.389  -63.216
3          0     0.000        0    0.000    0.000    0.000    0.000
4         50   171.891        0  171.891  105.926   91.127  171.891
5        100   638.936        0  638.936  351.277  304.622  638.936

$perc.ratio
  Param Perc Diff Max Diff Min Diff End Avg Diff RMS Diff Max Diff
1       -100     0.865        0    0.865    0.697    0.617    0.865
2        -50     1.264        0    1.264    0.948    0.828    1.264
3          0     2.351        0    2.351    1.533    1.325    2.351
4         50     3.438        0    3.438    2.119    1.823    3.438
5        100     6.389        0    6.389    3.513    3.046    6.389
>
```

The `sens` function also produced plots using these values. These can be stored in a PDF output file using the option pdfout = T. The contents of this file are illustrated in Figures 5.3 and 5.4.

A couple of detailed issues are taken care of by `sens`. First, the ratio at the nominal is estimated by interpolation because the ratio is undefined (it would yield a division by zero). Second, the `sens` function takes the absolute value of the nominal in the denominator, thus making sure that the sign is inverted for negative parameter values. For example, the difference of one of the values with respect to a negative nominal (–0.06 – (–0.08))/(–0.08) should be positive because it represents an increase in the parameter value.

Exercise 5.6

Perform sensitivity analysis for five levels of the chemical decay coefficient, k. The nominal is –0.08 and σ = 0.005. Use the file **chp5/exp-decay-inp.csv**, which has t from 0 to 10 with time stamps every day and initial condition 100 mg/liter.

As discussed in Chapter 4, random numbers can be generated by many methods; for example, recall that you can use simple commands to generate normal and uniformly distributed numbers. We can generate a normal with `rnorm(n, u, s)`, where n is the number of samples, and u and s are the mean and standard deviations, respectively; also, for a uniform we can use `runif(n, a, b)`, where a and b are the min and max of the uniform. For example, let us use `rnorm` for simple random sampling to generate 10 values of r from the normal with mean 0.02 and standard deviation 0.005, which would yield an equivalent $v = 2\sigma = 0.01$ or 50% of the nominal as in the previous example.

We include the mean 0.02 in the set to make sure that it can be used as nominal as directed in the call to `sens`.

```
pnom=0.02; psd =0.005; plab = c("r"); runs=10
pval <- round(c(pnom,rnorm(runs,pnom,psd)),5)
param <- list(plab=plab, pval=pval, pnom=pnom, fact=F)
t.X <- sim.mruns(model,"chp5/exp-sens-inp.csv", param)
s.y <- sens(t.X$output, param, "chp5/exp-sens-plots-samp-out")
```

Therefore, we execute 11 runs, the nominal plus 10 random values of r. The logical argument `fact` is set to `F` to negate factorial analysis and perform sampling. The graphical results are depicted in Figures 5.5 and 5.6. Examination of the output on the console indicates an additional component `reg.slope.R2`, which contains regression results. The `$perc.ratio` and the `reg.slope.R2` components are as follows:

```
$perc.ratio
   Param Perc Diff Max Diff Min Diff End Avg Diff RMS Diff Max Diff
1      -42.35    1.349        0    1.349    0.999    0.871    1.349
2      -42.00    1.353        0    1.353    1.002    0.873    1.353
3      -37.05    1.413        0    1.413    1.037    0.902    1.413
4      -31.05    1.490        0    1.490    1.082    0.940    1.490
5      -18.95    1.665        0    1.665    1.183    1.025    1.665
6      -16.10    1.710        0    1.710    1.209    1.046    1.710
7      -15.35    1.722        0    1.722    1.216    1.053    1.722
8       -4.10    1.921        0    1.921    1.326    1.145    1.921
9        0.00    2.003        0    2.003    1.371    1.183    2.003
10       1.05    2.024        0    2.024    1.382    1.192    2.024
11      44.45    3.225        0    3.225    2.012    1.731    3.225

$reg.slope.R2
      Diff Max Diff Min Diff End Avg Diff RMS Diff Max Diff
slope    1.834        0    1.834    1.264    1.095    1.834
r2       0.854      NaN    0.854    0.905    0.909    0.854
```

We see that at the nominal run (position 9 in this case), the ratios in percent 2.003, 0, 2.003, 1.371, 1.182, 2.003 have a similar pattern to the regression line slopes 1.834, 0, 1.834, 1.264, 1.095, 1.834 but have a higher value for all metrics. The R^2 values are all above 0.8, indicating a reasonable fit. However, the diff max, diff end, and max diff have lower R^2 due to high values at the right as can be seen in Figure 5.6.

Note that there is a reduced range of p with respect to the previous example. This occurs because the $\pm 2\sigma$ interval in a normal has high probability. The range could be increased by assuming higher variance, using a uniform distribution with a broad range (as in Exercise 5.7) and stratified sampling (which will be discussed in Chapter 8).

Exercise 5.7

Generate a sample of parameter values for 10 runs of an exponential growth model with nominal 0.02 and $\sigma = 0.005$ but explore a wider range from 0 to 0.04. Perform sensitivity analysis. Hint: Assume a uniform distribution and the mean equal to the nominal.

5.6 SEEM FUNCTIONS EXPLAINED

5.6.1 FUNCTION MODEL.EVAL

This function is too lengthy to reproduce here in its entirety. It can be studied from the console. Some highlights are the calculation of the errors,

```
# absolute error: individual, total and average
abs.err <- Xmod-Xobs
abs.err.tot <- sum (abs.err)
abs.err.avg <- abs.err.tot/length(Xobs)

# relative error in percent: individual, total and average
rel.err <- 100*(Xmod-Xobs)/Xobs
rel.err.tot <- sum(rel.err)
rel.err.avg <- rel.err.tot/length(Xobs)

# squared error: individual, total and average
sq.err <- (Xmod-Xobs)^2
sq.err.tot <- sum (sq.err)
sq.err.avg <- sq.err.tot/length(Xobs)
# rms error
rms.err <- sqrt(sq.err.avg)
```

and how to build an image with four panels that use available space more efficiently,

```
# graph to visualize use four panels
mat <- matrix(1:4,2,2,byrow=T)
layout(mat, widths=rep(7/2,4), heights=rep(7/2,4), TRUE)
par(mar=c(4,4,1,.5),xaxs="r", xaxs="r")
```

5.6.2 FUNCTION SENS

This function is also too lengthy to reproduce here in its entirety. We only highlight the arguments,

```
# function to calculate sens metrics for multiple runs

sens <- function(t.X, param, fileout, pdfout=F) {

# sensitivity analysis
# arguments:
# t.X time seq and model results
# param list of plabel, param, and factor values
# fileout prefix to store file and plots
```

the calculation of metrics,

```
# for each parameter value calculate the metrics
 for(j in 1:np){
   d.max[j] <- max(X[,j])
   d.min[j] <- min(X[,j])
```

```
    d.end[j] <- X[nt,j]
    avg.d[j] <- mean(X[,j])
    err.m[,j] <- X[,j] - X[,nnom]
    rms.d[j] <- sqrt(mean((err.m[,j])^2))
    max.d[j] <- max(abs(err.m[,j]))
  }
  y <- cbind(d.max, d.min, d.end, avg.d, rms.d, max.d)
  val.val <- data.frame(round(cbind(p,y),3))
  names(val.val) <- c("Param Val",label.m)

  err.r <- err.m
  for(j in 1:np){
    for(i in 1:nt){
     if (X[i,nnom] ==0) err.r[i,j] <- 0 else
       err.r[i,j] <- (err.m[i,j])/X[i,nnom]
    }
    rms.d[j] <- sqrt(mean((err.r[,j])^2))
    max.d[j] <- max(abs(err.r[,j]))
  }
```

and the regression,

```
# regression for random sampling
if(param$fact==F){
ycoeff <- matrix(nrow=ndex,ncol=2); yp.est <- yp;r2 <- array()
for(i in 1:4){
ycoeff[i,] <- lm(yp[,i]~0+pp)$coeff
r2[i] <- summary(lm(yp[,i]~0+pp))$r.square
yp.est[,i] <- ycoeff[i,1]*pp
}
for(i in 5:6){
  #ppa <- abs(pp)
  ppa <- pp
  ycoeff[i,] <- lm(yp[,i]~0+ppa)$coeff
  r2[i] <- summary(lm(yp[,i]~0+ppa))$r.square
  yp.est[,i] <- ycoeff[i,1]*ppa
}

slope <- ycoeff[,1]
reg.slope.R2 <- data.frame(round(rbind(slope, r2),3))
names(reg.slope.R2) <- label.m
 }
```

6 Nonlinear Models

Let us return to the linear ordinary differential equation (ODE) model given by

$$\frac{dX(t)}{dt} = kX(t) \tag{6.1}$$

We are now interested in cases where the per capita rate is not a constant, k, but varies with X. For example, the per capita rate itself may be a function $f(X)$, as occurs when you consider feedback of the population density to control the per capita rate:

$$\frac{dX(t)}{dt} = f(X(t))X(t) \tag{6.2}$$

The solution is no longer a simple exponential and the model is no longer linear; it becomes **nonlinear**.

There are multiple environmental applications of nonlinear models. In fact, most environmental models include complex feedback loops and nonlinear behavior. Two simple models are **logistic** growth in population dynamics and **Michaelis–Menten–Monod** kinetics in the fate of chemicals.

6.1 POPULATIONS: LOGISTIC GROWTH

Logistic growth is the simplest model that incorporates density dependence. Its simplicity is due to assuming a nonstructured and spatially homogeneous population. Density dependence is what makes the model nonlinear; the per capita rate is dependent on population density. It considers the effect of crowding to decrease the rate of change (Hallam, 1986b; Keen and Spain, 1992).

The per capita rate exhibits a linear decrease of the per capita rate with respect to density modeled by

$$f(X) = r\left(1 - \frac{X}{K}\right) \tag{6.3}$$

where K is a new parameter characterizing the potential maximum value of X; it is defined as the **carrying capacity** of the environment and has the same units as X. The net rate of change is quadratic or **parabolic** because the linear decrease of the per capita rate is multiplied by the proportional increase with density.

In Figure 6.1, we see the linear decrease of the per capita rate and the parabolic form of the net rate for three values of r and for the same value of $K = 5$. Clearly, the slope of the per capita rate increases with parameter r. In this figure, note that the peak or maximum net rate occurs at $X = K/2$. For example, when $K = 5$, the peak occurs at $X = 5/2 = 2.5$. The value of the peak can be determined by taking the derivative and making it equal to zero, as explained in Chapter 2. We get a peak value of $rK/4$, for example, for $r = 0.2$, the peak value is given as $0.2 \times 5/4 = 0.25$.

In Figure 6.2, we see the linear decrease of the per capita rate and the parabolic form of the net rate for three values of K and the same value of $r = 0.4$. The slope of the per capita rate increases with parameter K. We confirm that the peak net rate occurs at $X = K/2$ and the peak increases with K. For example, when $K = 4$, the peak occurs at $X = 4/2 = 2$ and the peak value is $rK/4 = 0.4 \times 4/4 = 0.4$.

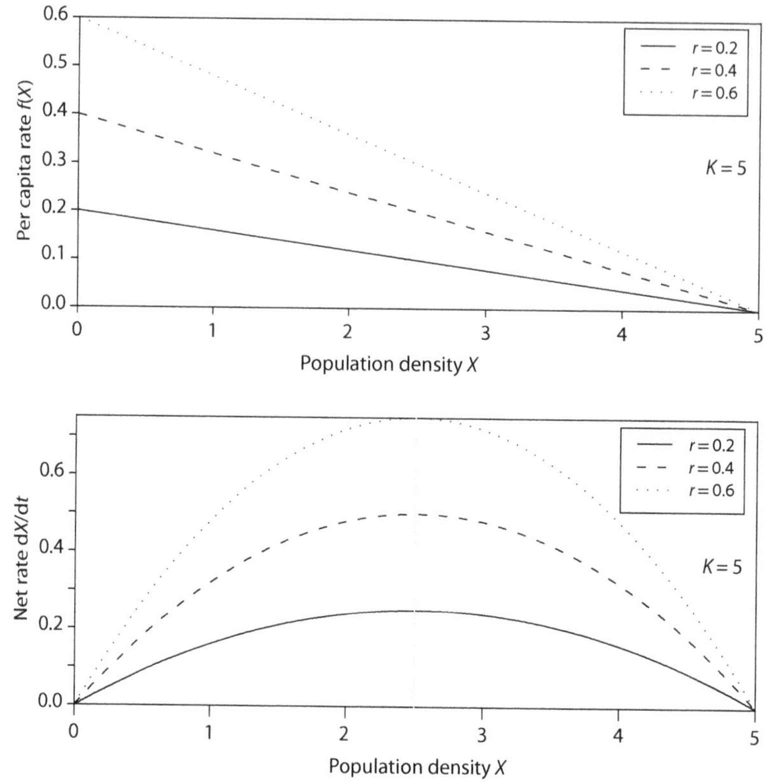

FIGURE 6.1 The per capita rate decreases with density (top panel) and the net rate is quadratic or parabolic as a function of density (bottom panel). Several lines are drawn according to r while $K = 5$ for all lines.

The complete model is then

$$\frac{dX}{dt} = rX\left(1 - \frac{X}{K}\right) \tag{6.4}$$

(Hallam, 1986b). The units vary according to how quickly a population changes. For example, for a fast growing population we may select days as time unit and therefore the units are t [d], r [1/d], X [ind], dX/dt [ind/d], K [ind], and $r(1-X/K)$ [1/d].

When X is low, the logistic model behaves like an exponential model with r as the rate coefficient. When X is very high, the rate decreases and tends to reach zero as X reaches a maximum value of K, that is, the saturating population (carrying capacity of the environment). There are two important cases as follows:

- $r(1-X/K)<0$: net rate of change is negative and X decreases.
- $r(1-X/K)>0$: net rate of change is positive and X increases.

The solution of the logistic is

$$X(t) = \frac{KX_0}{\left(X_0 - (X_0 - K)\exp(rt)\right)} \tag{6.5}$$

where $X_0 = X(0)$ is the initial density or density at time $t = 0$. This solution displays different behaviors according to the initial condition when plotted against time (top panel of Figure 6.3). First, two behaviors can occur depending on whether or not the initial condition exceeds the carrying capacity.

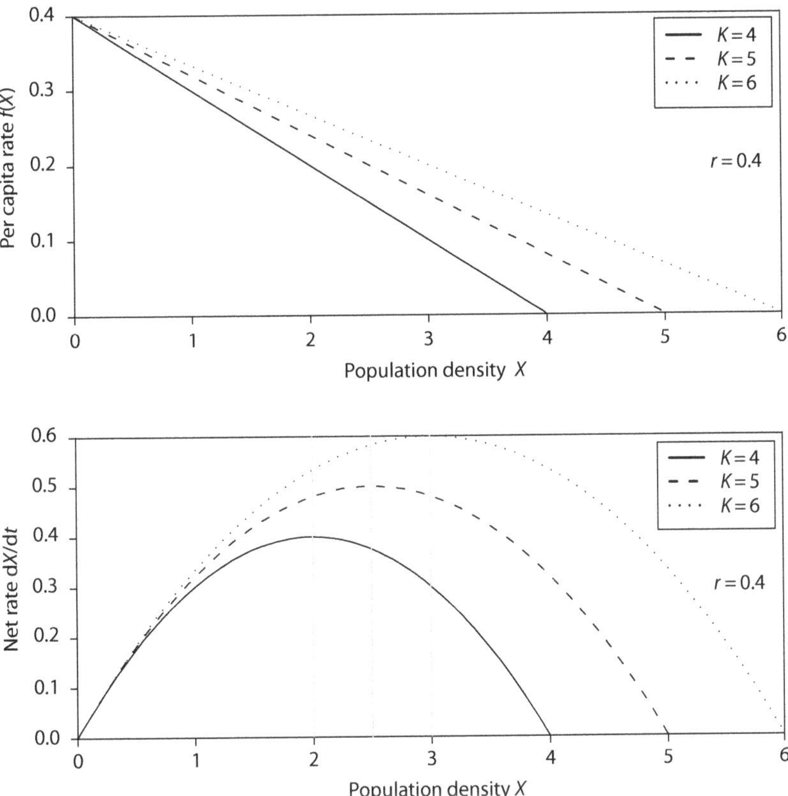

FIGURE 6.2 The per capita rate decreases with density (top panel) and the net rate is quadratic or parabolic as a function of density (bottom panel). Several lines are drawn according to K while $r = 0.4$ for all lines.

One tends to K from above (X declining) if $X > K$ and the other tends to K from below (X growing) if $X < K$. This occurs because for $X > K$, the net rate becomes negative, as can be seen in the bottom panel. Second, for the growing trajectory, there are two different behaviors according to whether or not the initial condition exceeds half of the carrying capacity. One is sigmoid or S-shaped and happens when $X_0 < K/2$, and the other is monotonic and happens when $X_0 > K/2$. This is understandable because the net rate changes from increasing to decreasing at the peak of the parabola, which occurs at $X = K/2$. In the example of Figure 6.3, the peak occurs at $X = 2.5$ and it is reached at $t \sim 4$ for the trajectory starting at $X_0 = 1$.

The logistic model has one more parameter, K, compared to the exponential model. Model calibration and sensitivity also involve K. The logistic is a **nonlinear** ODE that has a simple closed-form solution, which means that we do not really need a computer simulation.

However, to understand how to do a simulation, just note that the **algorithm** works the same way as it does in Chapter 4. When using Euler, the new net rate enters into the right-hand side of the equation:

$$X(t) = X(t-\Delta t) + \Delta t\, X(t-\Delta t)\big(r(1 - X(t-\Delta t)/K)\big) \tag{6.6}$$

A logistic model can be expanded and varied by making the per capita rate a more complicated function of X, for example, a nonlinear function of X leading to negative or positive compensation. In addition, logistic-like models are the basis of several other models, like tree diameter growth.

Exercise 6.1

What is the maximum rate of growth for K = 100 ind and r = 0.1 year^{-1}?

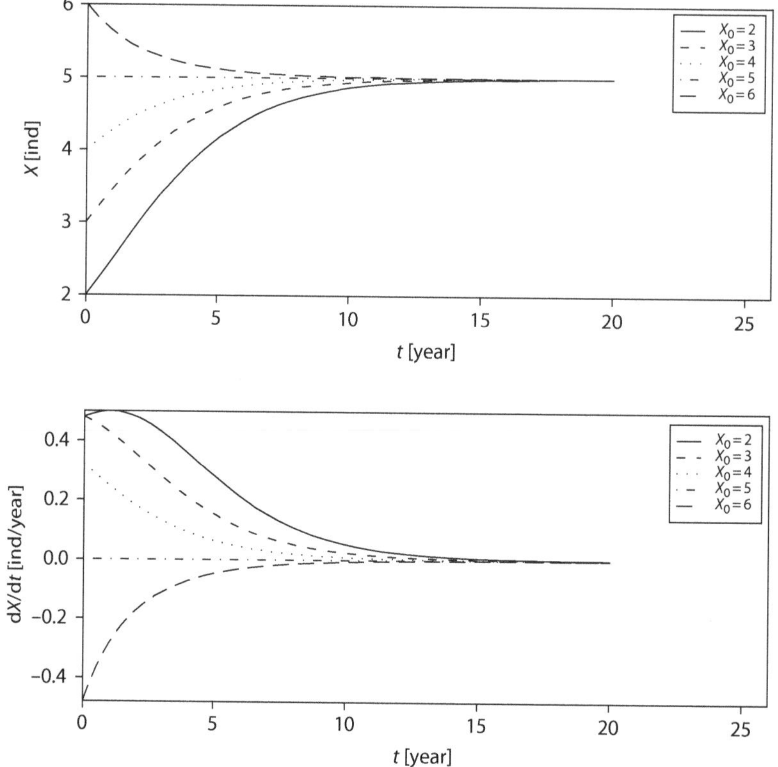

FIGURE 6.3 The behavior of the logistic model depends on the initial condition. Top panel is the population and bottom panel is the net rate. $K = 5$ and $r = 0.4$ are used for all lines.

Exercise 6.2

In a logistic model, consider two cases according to the values of density X: (a) $X = K/4$ or (b) $X = 3K/4$. Which case yields the largest rate of change, (a) or (b)? Sketch the solutions with respect to time when the initial value corresponds to (a) and (b).

6.2 FATE OF CHEMICALS: MICHAELIS–MENTEN–MONOD

When the reaction rate is not a constant but a function of the concentration, we have a nonlinear model; for example, in a bimolecular reaction with equal reactant concentrations, the per-unit rate increases with concentration X in units of M·L^{-3} and can be modeled as

$$\frac{\mathrm{d}X(t)}{\mathrm{d}t} = -(kX(t))\,X(t) = -kX^2(t) \tag{6.7}$$

and the net rate is parabolic (Hemond and Fechner, 1994; Keen and Spain, 1992). Likewise, for two reactants with concentrations X_1 and X_2 in the units of M·L^{-3}, the model is given by

$$\frac{\mathrm{d}X_1(t)}{\mathrm{d}t} = -(kX_2(t))\,X_1(t) \tag{6.8}$$

When $X_2(t)$ is nearly constant with time, the second-order process can be modeled linearly by using $k' = kX_2$ as an approximation for the rate constant. The pseudo-first-order kinetics can then be modeled exponentially.

Exercise 6.3

What are the units for k?

We focus on the M3 kinetics here. Biodegradation by microorganisms is often modeled in a similar manner to enzyme kinetics. A common assumption beyond simple exponential decay is that the rate coefficient increases with concentration but eventually saturates or reaches a maximum at high concentrations.

The rate coefficient changes with concentration in the following form:

$$-f(X) = -K_{max}\left(\frac{1}{K_h + X}\right) \tag{6.9}$$

where K_{max} is the maximum rate or the rate that would be achieved when X is very high and K_h is the half-rate or the value of concentration at which the rate is half of the maximum. Parameter K_{max} has the units of rate, that is to say, concentration per unit time ($M{\cdot}L^{-3}{\cdot}T^{-1}$), for example, $mg{\cdot}liter^{-1}{\cdot}d^{-1}$. The parameter K_h has the unit of concentration ($M{\cdot}L^{-3}$), for example, $mg{\cdot}liter^{-1}$. Alternatively, K_{max} is expressed in $M{\cdot}L^{-3}{\cdot}T^{-1}$ per cell to explicitly consider the bacterial population degrading the chemical.

Per-unit rate can be scaled with respect to K_{max} (that is to say, expressed as a fraction of K_{max}) as $-(X/(K_h + X))$. As such, it is unitless and this is a hyperbolic function; it shows an asymptotic or limiting behavior with X. Using the per-unit rate to formulate the model, the net rate becomes

$$\frac{dX}{dt} = -f(X)X = -K_{max}\left(\frac{X}{K_h + X}\right) \tag{6.10}$$

Checking the units again, we see that the left-hand side is the net rate given in $M{\cdot}L^{-3}{\cdot}T^{-1}$ and K_{max} also has the unit $M{\cdot}L^{-3}{\cdot}T^{-1}$; thus, the per-unit rate on the left-hand side is unitless. The net reaction rate exhibits the hyperbolic pattern shown in Figure 6.4 as it varies with the parameters K_{max} and K_h. Note that the half-rate always occurs at the corresponding $K_{max}/2$ (top panel) and that $K_{max}/2$ always occurs for the various K_h (bottom panel).

Exercise 6.4

Assume that K_{max} is 100 $mg{\cdot}liter^{-1}{\cdot}d^{-1}$ and $K_h = 0.3$ $mg{\cdot}liter^{-1}$. What is the net rate at $X = K_h$? What is the rate at $X = 2K_h$? What about when $X = 0.1K_h$?

In addition, it is important to realize that the slope of $f(X)X$ at the origin ($X = 0$) is K_{max}/K_h. We can see this because the derivative is given as $\frac{K_h K_{max}}{(K_h + X)^2}$, which at $X = 0$ evaluates to K_{max}/K_h. It is easy to see this fact geometrically (Figure 6.5). Denoting α as the slope at $X = 0$, we can write M3 as $f(X) = \frac{K_{max}}{\frac{K_{max}}{\alpha} + X}$.

What value of chemical concentration will produce a reaction rate of three-quarters of the maximum value? The value of X can be solved from

$$-\frac{3}{4}K_{max} = -K_{max}\left[\frac{X}{K_h + X}\right]$$

that is, after canceling $-K_{max}$ on both sides, we have

$$\frac{3}{4} = \left[\frac{X}{K_h + X}\right]$$

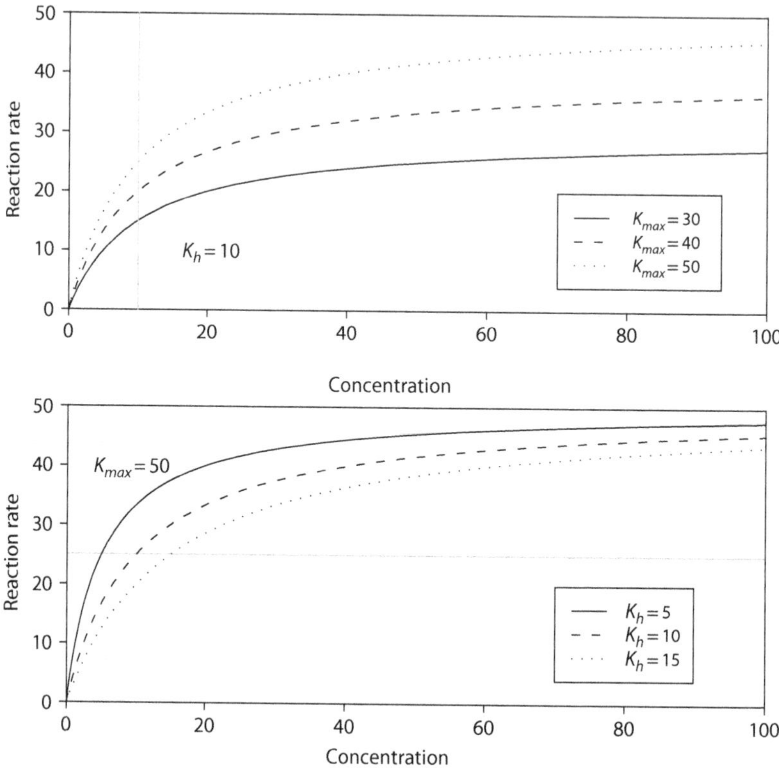

FIGURE 6.4 Rate versus concentration for varying K_{max} and fixed K_h (top) and for varying K_h with fixed K_{max} (bottom).

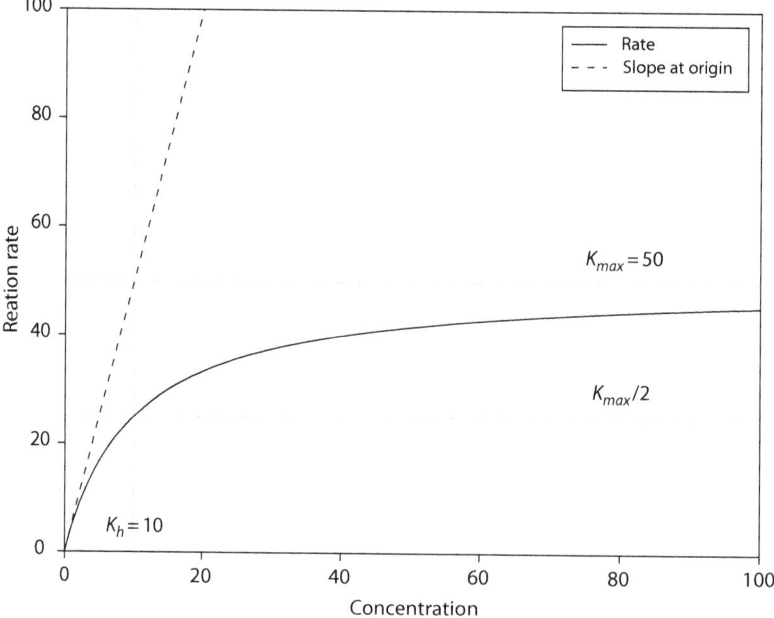

FIGURE 6.5 Rate versus concentration and slope α at $X = 0$, illustrating that this slope is the ratio K_{max}/K_h.

and solving, we get $3K_h + 3X = 4X$ and finally $X = 3K_h$. Therefore, a concentration of three times the half-rate will yield three-quarters of the maximum rate.

6.3 MONOD KINETICS

This model can be used to describe bacterial population growth by linking biodegradation rate to growth rate by means of a yield coefficient, y. The assumption is that microbe growth is fueled by the energy released during biodegradation. Thus, the chemical being biodegraded can be considered food to the microbe. The cell yield (y) is the number of new microbes per concentration of the chemical degraded in units of cells/mg. Denote by X_2 the bacterial population density in units of cells/liter. A simple linear model of this population includes growth when the resource is present and a constant death rate:

$$\frac{dX_2}{dt} = \mu X_2 - DX_2 \tag{6.11}$$

where the rate parameter μ is the specific growth rate coefficient and can be further modeled using M3 enzyme-kinetics:

$$\frac{dX_2}{dt} = yK_{max}\left[\frac{X_1}{K_h + X_1}\right]X_2 = \mu_{\max}\left[\frac{X_1}{K_h + X_1}\right]X_2 \tag{6.12}$$

where the new parameter maximal specific growth rate ($\mu_{\max}$) lumps the effect of cell yield (y) and the maximum reaction rate (K_{max}). That is to say, $\mu_{\max} = yK_{max}$. In effect, the maximal growth rate is that achieved when the food is unlimited.

The equation for the resource X_1 is given by

$$\frac{dX_1}{dt} = -K_{max}\left[\frac{X_1}{K_h + X_1}\right]X_2 \tag{6.13}$$

In this case, K_{max} is expressed in $M{\cdot}T^{-1}$ per cell to explicitly consider the bacterial population. Therefore, the unit of $\mu_{\max}$ is (cells/M) $M{\cdot}T^{-1}{\cdot}cell^{-1} = T^{-1}$. Equation 6.12 is the **Monod** model. It can be used to model a batch culture (chemical and microbe are put in a closed vessel), a continuous culture, or a chemostat (chemical is continuously added to the vessel).

Equations 6.10 and 6.12 are also referred to as M3 kinetics because the Monod model contains the same hyperbolic form as in Equation 6.10. This type of model is an important component of more complicated models that we will encounter many times in this book. We will find this in predator–prey models, nutrient uptake by phytoplankton models, and photosynthesis models.

Exercise 6.5

What is the net bacterial growth rate when $X_1 = 0.1$ mg·liter^{-1}, $y = 10^3$ cell·mg^{-1}, $K_{max} = 2 \times 10^{-7}$ mg·s^{-1}·cell^{-1}, $K_h = 0.1$ mg·liter^{-1}, and the bacterial population $X_2 = 10^3$ cells·liter^{-1}?

6.4 COMPUTER SESSION

We use seem's `sim` function to simulate the logistic, M3 decay, and Monod batch models. The function `sim` can make a single run or multiple runs according to a `list` named `param` as defined in the previous chapters when we used the function `sim.mruns`. The `sim` function is explained at the end of this chapter. As new features, it includes the following: (1) you can vary the initial condition or any one of the parameters; (2) it calculates the rate dX/dt graph versus t and X versus t; and (3) it graphs rate dX/dt versus X to understand the nonlinear behavior.

6.4.1 Logistic

We use seem's `sim` function with the `logistic` model and input file **logistic-inp.csv**. The `logistic` function is straightforward:

```
logistic <- function(t,p,x){
r<-p[1]; K<-p[2]
dx <- r*x*(1-x/K)
return(dx)
}
```

The input file **logistic-inp.csv** contains

```
Label,Value,Units,Description
t0,0.00,Year,time zero
tf,20.00,Year,time final
dt,0.1,Year,time step
tw,1.0,Year,time to write
r,0.4,1/Year,intrinsic growth rate
K,5.00,Ind,carrying capacity
X0,1.00,Ind,initial population
digX,4,None,significant digits for output file
```

We use the logistic function and the input file in sim in the following manner to make one run:

```
logis <- list(f=logistic)
t.X <- sim(logis,"chp6/logistic-inp.csv")
```

Only the model and the input file are the required arguments to `sim`. In this mode, the simulation conducted by `sim` is one run with nominal parameter values given in the input file. You will see two graphs displayed in the first graphics device (Figure 6.6). One is the behavior of X versus time and the other is the change of the rate dX/dt versus time. The second graphics device has the graph of rate versus X (Figure 6.7).

Next, use the logistic function to make several runs by varying the parameter r.

```
param <- list(plab="r", pval = seq(0.2,0.6,0.2))
t.X <- sim(logis,"chp6/logistic-inp.csv",param)
```

Two graphs are displayed in the first graphics device (Figure 6.8) as shown earlier. One is X versus time and the other is dX/dt versus time. The second graphics device has the graph of rate versus X (Figure 6.9). As seen before, setting `pdfout = T` will produce a PDF file for the graphs instead of using the R GUI. The text file **logistic-inp.csv-r-out.csv** contains the results.

Exercise 6.6

Discuss the family of curves obtained for each graph. Recall that for exponential behavior, the rate versus X graph would be a straight line, but for logistic behavior, it should be parabolic.

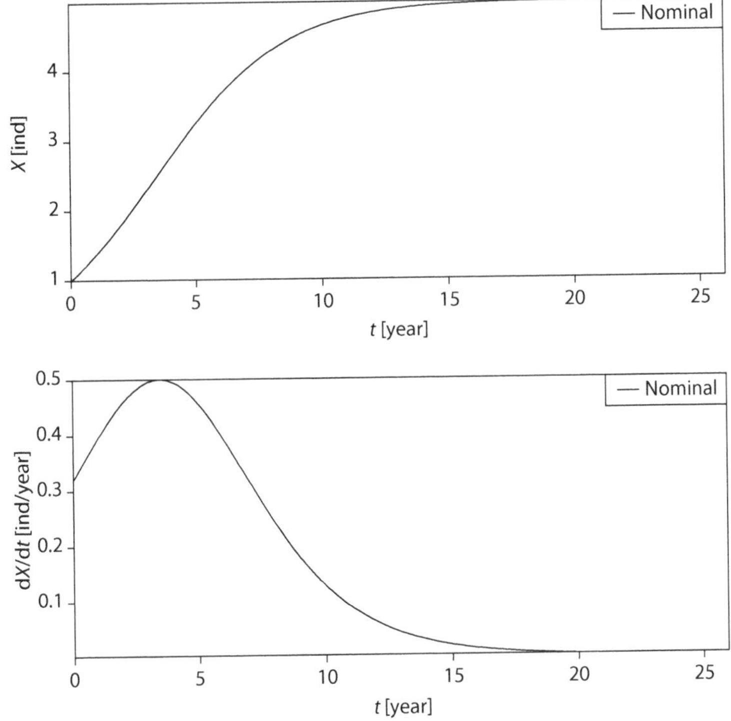

FIGURE 6.6 Logistic density (top panel) and rate (bottom panel) for the nominal parameter values.

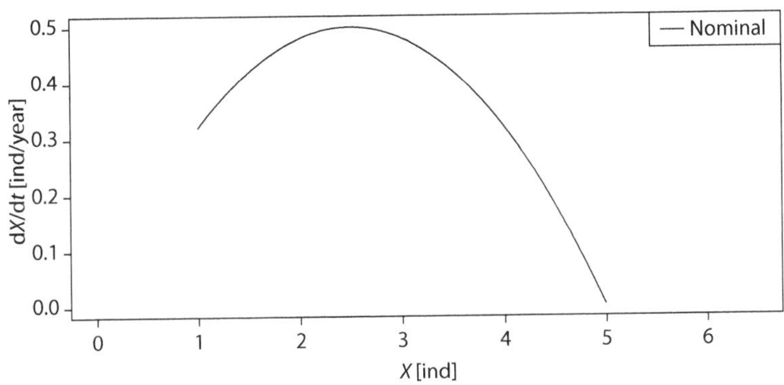

FIGURE 6.7 Logistic rate versus density plotted for the nominal parameter values.

We can explore the effect of varying the carrying capacity and initial conditions:

```
param <- list(plab="K", pval = seq(4,6,1))
t.X <- sim(logis,"chp6/logistic-inp.csv",param,pdfout=T)
param <- list(plab="X0", pval = c(1,3,6))
t.X <- sim(logis,"chp6/logistic-inp.csv",param,pdfout=T)
```

Here we have set `pdfout` = `T` to obtain four more figures. Two figures are in **logistic-inp.csv-K-out.pdf** for the carrying capacity and two are in **logistic-inp.csv-X0-out.pdf** for the initial conditions. We show the first page of the figures for carrying capacity in Figure 6.10. The first page

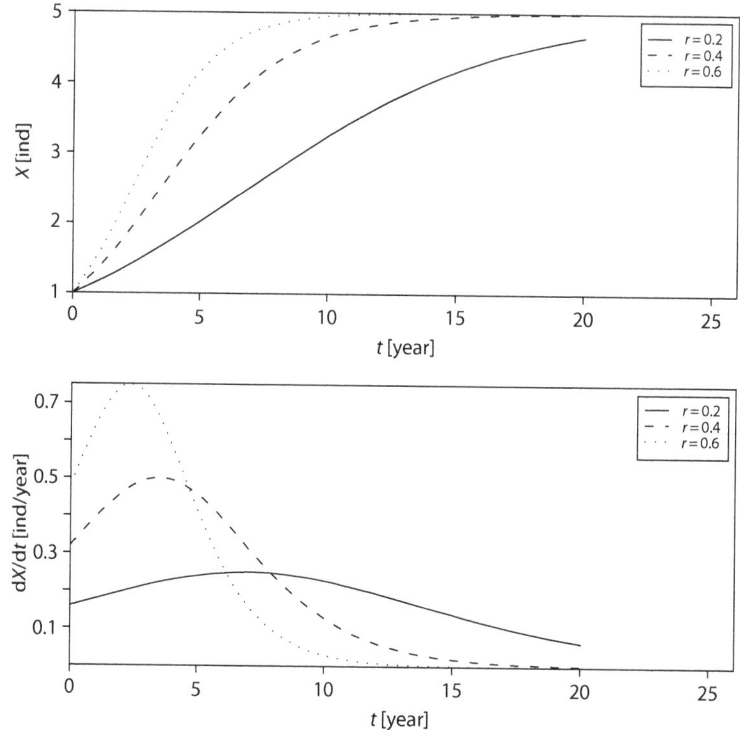

FIGURE 6.8 Logistic density (top) and rate (bottom) for several values of the rate coefficient.

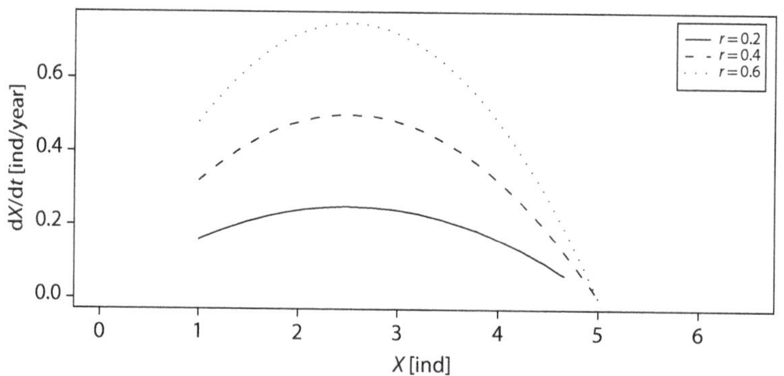

FIGURE 6.9 Logistic rate versus density plotted for several values of the rate coefficient.

of the figures for initial conditions is similar to the one already shown in Figure 6.3, which we have already discussed. It demonstrates the stability of the equilibrium corresponding to the carrying capacity, K, by using the initial conditions X_0 in three different ranges: when $X_0 < K/2$, X should increase in sigmoid form and go toward K; when $> K/2$, X should increase in nonsigmoid form and go toward K; when $X_0 > K$, X should decrease and go toward K. The plots for all the initial conditions converge to the same equilibrium, the carrying capacity.

Exercise 6.7

You apply a pesticide to control an insect pest by killing 70% of the pest population. Assume a logistic growth at a rate $r = 0.2$ [1/d] and $K = 1000$. How long does it take before the population builds back up to 80% of K? Hint: Use a logistic model, with $r = 0.2$ and $X_0 = 0.3 \times K$, and look for the time when $X = 0.8 \times K$.

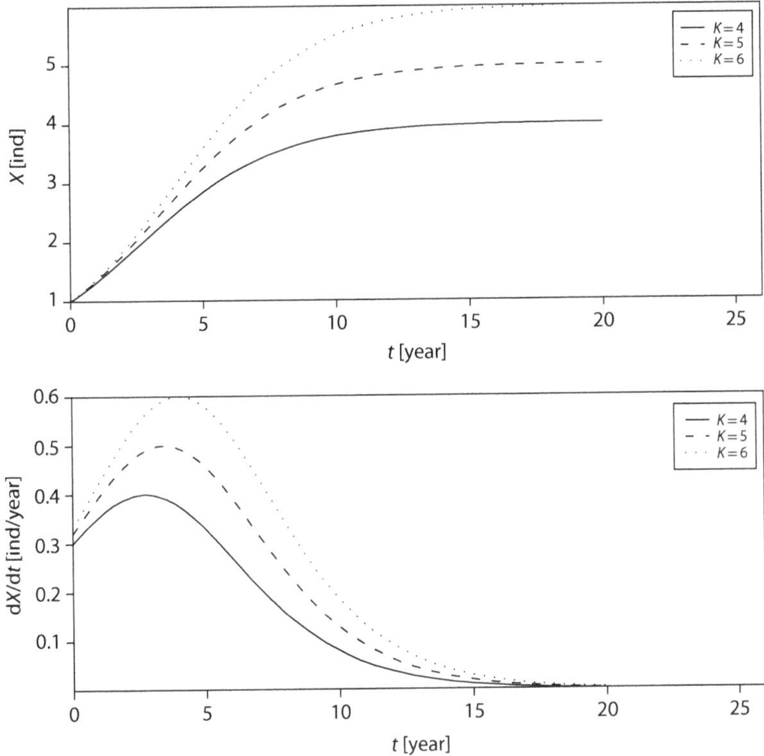

FIGURE 6.10 Logistic numerical solution for several values of carrying capacity. Density (top) and rate (bottom).

In addition, we study the sensitivity of the logistic model with respect to the initial conditions using the `sens` function from the last chapter:

```
pnom = 4; psd=0.5; plab = c("X0"); runs=5
pval <- seq((pnom -(runs-1)*psd),(pnom+ (runs-1)*psd),2*psd)
param <- list(plab=plab, pval=pval, pnom=pnom, fact=T)
t.X <- sim(logis,"chp6/logistic-inp.csv",param,pdfout=T)
s.y <- sens(t.X$output, sens.par, fileout="chp6/logis-sens-out",pdfout=T)
```

We obtain the plots in a PDF output file **logis-sens-out.pdf** as we have selected `pdfout = T`; see Figures 6.11 and 6.12. We can see that the `diff end` value is not sensitive to changes in X_0. Of course, `diff min` follows X_0 when this is less than K and stays at K whenever X_0 is larger than K. `Diff max` is K and is not sensitive to X_0 whenever X_0 is less than K, but follows X_0 when this is larger than K.

Exercise 6.8

Discuss the sensitivity of Avg diff, RMS diff, and Max diff to X_0.

Exercise 6.9

Perform sensitivity analysis of the logistic with respect to r and discuss.

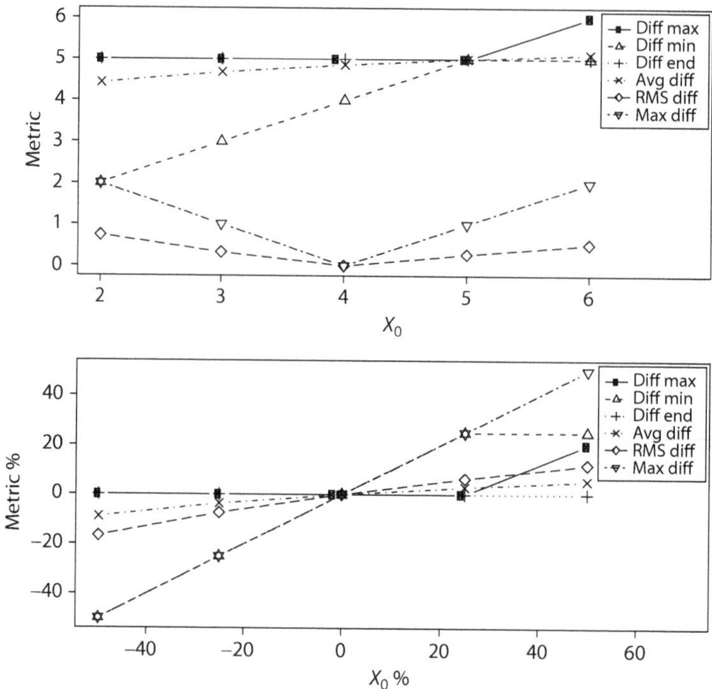

FIGURE 6.11 Sensitivity plots: absolute (top panel) and relative (bottom panel) values for the logistic with respect to the initial conditions.

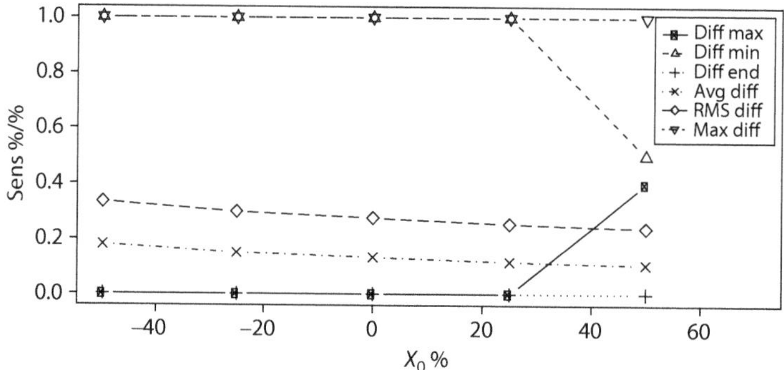

FIGURE 6.12 Sensitivity plots: slopes for the logistic with respect to the initial conditions.

6.4.2 M3 Chemical Decay

Now we will use the **M3** function explained at the end of this chapter. This function is very similar to the logistic function but uses Equation 6.10 as a model equation.

```
M3 <- function(t,p,x){
# p[1] negative for decay
Kmax<-p[1]; Kh<-p[2]
dx <- Kmax*x/(x+Kh)
return(dx)
}
```

This is M3 chemical kinetics. Recall that Kmax is the maximum net rate and K_h is the half-rate concentration. Variable x represents concentration in mg·liter^{-1}.

As an example, we will use the following nominal values K_{max} = 40 (mg·liter^{-1})d^{-1}, K_h = 30 mg·liter^{-1}, and initial condition 100 mg·liter^{-1}. The input file is **m3decay-inp.csv**:

```
Label,Value,Units,Description
t0,0.00,d,time zero
tf,10.00,d,time final
dt,0.01,d,time step
tw,0.10,d,time to write
Kmax,40.00,mg/l/d,max rate
Kh,30.00,mg/l,half-rate concentration
X0,100.00,mg/l,concentration
digX,4,None,significant digits for output file
```

Now we use M3 to form a list m3decay for the function call and vary K_{max}, then K_h, and finally X_0.

```
m3decay<- list(f=M3)
param <- list(plab="Kmax", pval = seq(-20,-60,-10))
t.X <- sim(m3decay,"chp6/m3decay-inp.csv", param)
t.X <- sim(m3decay,"chp6/m3decay-inp.csv", param, pdfout=T)

param <- list(plab="Kh", pval = seq(20,60,10))
t.X <- sim(m3decay,"chp6/m3decay-inp.csv", param)
t.X <- sim(m3decay,"chp6/m3decay-inp.csv", param, pdfout=T)

param <- list(plab="X0", pval = seq(60,140,20))
t.X <- sim(m3decay,"chp6/m3decay-inp.csv", param)
t.X <- sim(m3decay,"chp6/m3decay-inp.csv", param, pdfout=T)
```

We have produced graphics in both interactive graphics and three PDF files. The first pages of each file are reproduced here from Figures 6.13 through 6.15. The concentration curves resemble the exponential decay, but it is not exactly exponential. Inspection of the dX/dt curves on the lower panel of each figure reveals that the rate decreases as the chemical concentration drops, but the decrease in the rate is not proportional to the decrease in the concentration. Just as an example, we also reproduce here the rate versus concentration graph for varying K_h (Figure 6.16).

We can also perform sensitivity analysis with respect to one of the parameters, say K_h, in the same manner as we did for the logistic, yielding figures in the **m3decay-sens-out.pdf** file (Figures 6.17 and 6.18):

```
pnom=40;psd=5; plab = c("Kh"); runs=5
pval <- seq((pnom -(runs-1)*psd),(pnom+ (runs-1)*psd),2*psd)
param <- list(plab=plab, pval=pval, pnom=pnom, fact=T)
t.X <- sim(m3decay,"chp6/m3decay-inp.csv", param, pdfout=T)
#s.y <- sens(t.X$output, param, fileout="chp6/m3decay-sens-out")
s.y <- sens(t.X$output, param, fileout="chp6/m3decay-sens-out",pdfout=T)
```

Exercise 6.10

Discuss the sensitivity analysis of M3 decay with respect to K_h.

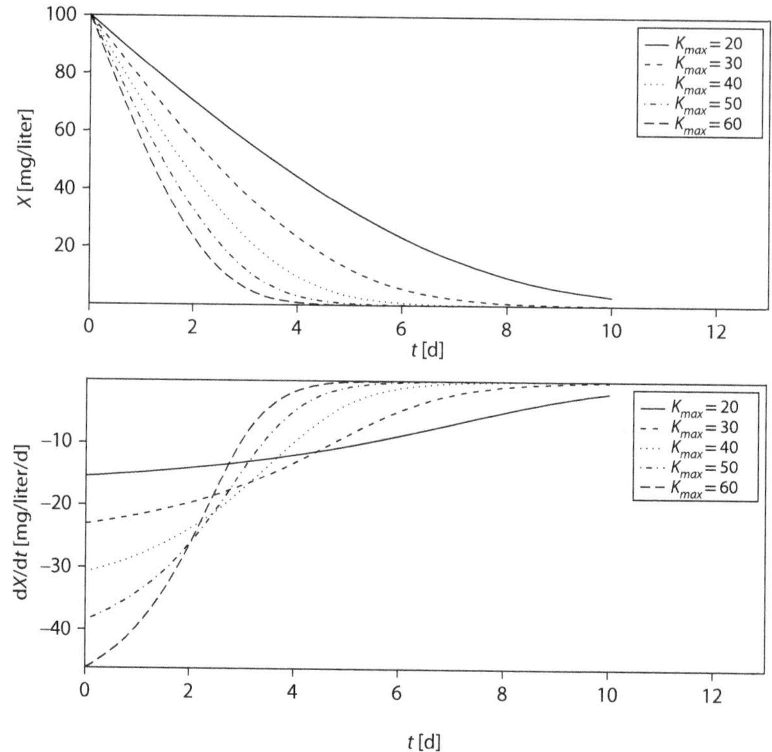

FIGURE 6.13 Multiple runs of M3 kinetics by varying max rate (K_{max}). Density (top panel) and rate (bottom panel).

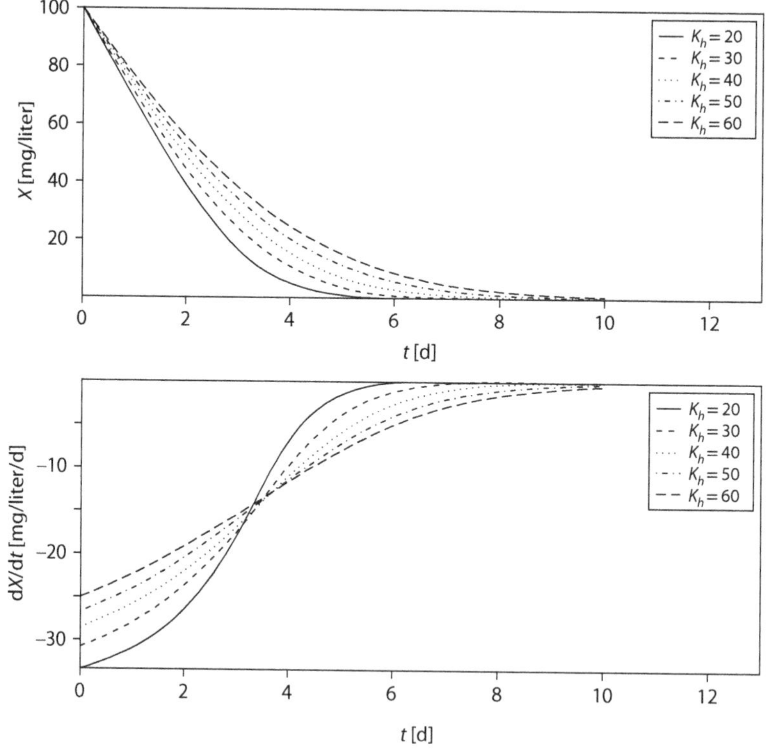

FIGURE 6.14 Multiple runs of M3 model by varying the half-rate (K_h). Density (top panel) and rate (bottom panel).

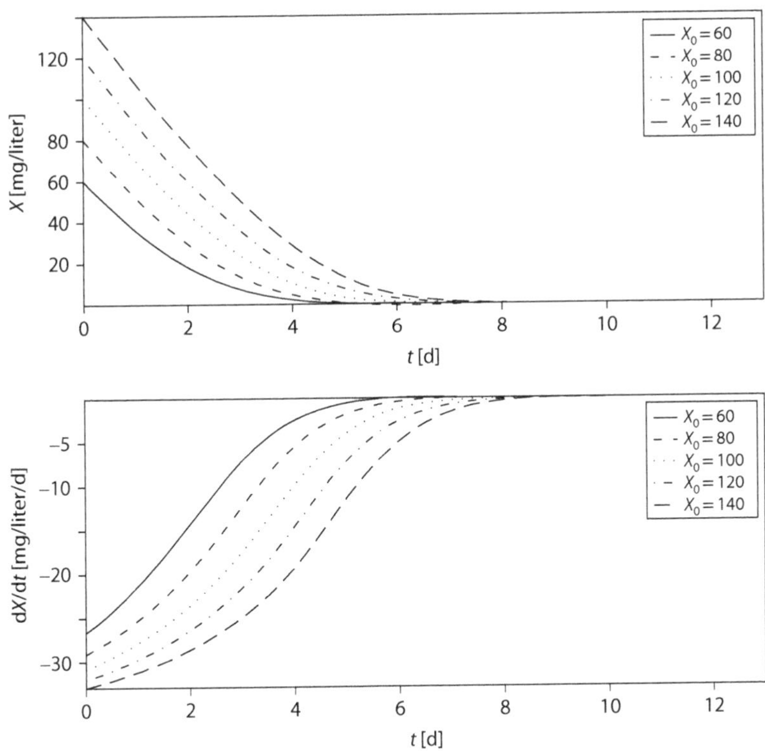

FIGURE 6.15 Multiple runs of M3 kinetics by varying the initial condition (X_0). Density (top panel) and rate (bottom panel).

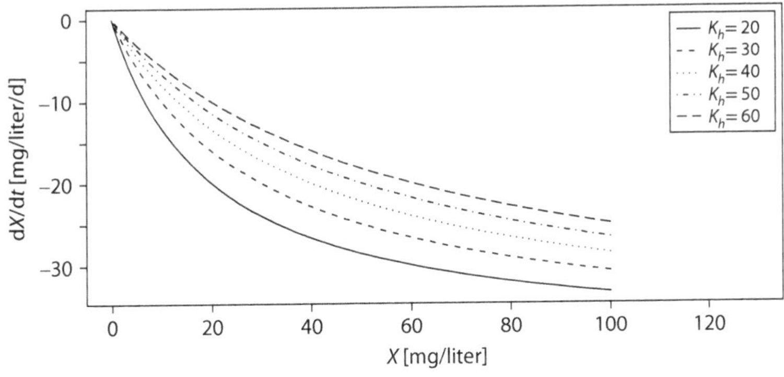

FIGURE 6.16 Reaction rate versus concentration plotted for several values of K_h.

6.4.3 Monod Batch

We will use the model `monod.batch`. The function is very similar to the logistic and M3 functions but we have varied the model equation to account for two dependent variables, as explained at the end of the chapter.

As an example, we will use the following nominal values given in the input file **monod-batch-inp.csv**:

```
Label,Value,Units,Description
t0,0.00,d,time zero
tf,10.00,d,time final
dt,0.01,d,time step
```

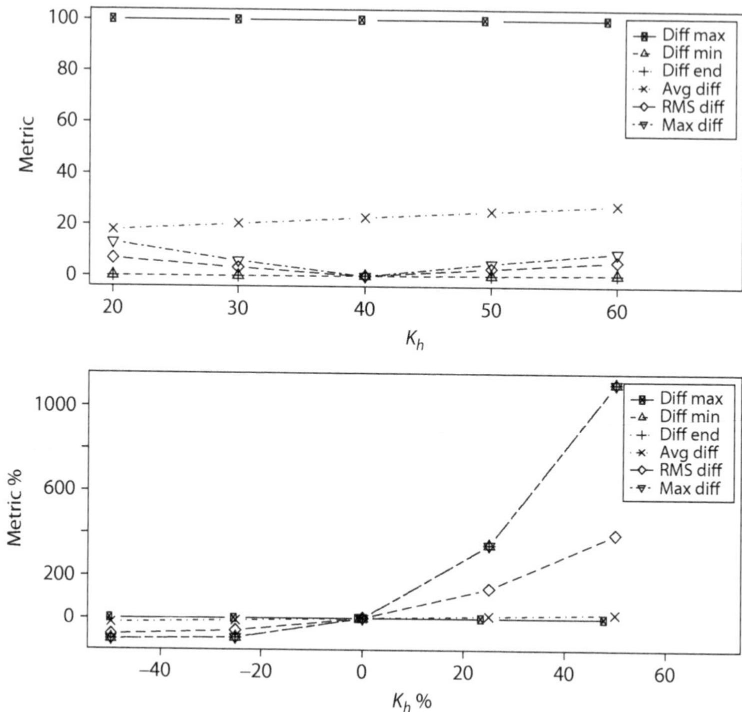

FIGURE 6.17 Sensitivity plots: absolute (top panel) and relative (bottom panel) values for m3decay with respect to K_h.

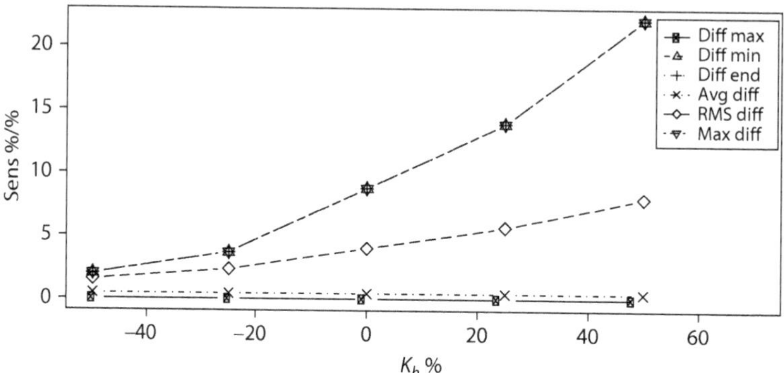

FIGURE 6.18 Sensitivity plots: slopes for m3decay with respect to K_h.

```
tw,0.10,d,time to write
Kmax,1,mg/d/cell,max rate
Kh,1,mg/l,half-rate concentration
y,10.00,cell/mg,Yield
D,0.15,cell/l/d,death rate
X1.0,100.00,mg/l,substrate initial condition
X2.0,10.00,cell/l,population initial condition
digX,4,None,significant digits for output file
```

Now we run the function with the parametric variation. We will vary K_{max}, K_h, y, D, and then the initial conditions. We send all the graphics output to a PDF.

```
monod <- list(f=monod.batch)
param <- list(plab="Kmax", pval = seq(0.5,1.5,0.5))
t.X <- sim(monod,"chp6/monod-batch-inp.csv", param,pdfout=T)

param <- list(plab="Kh", pval = seq(0.5,1.5,0.5))
t.X <- sim(monod,"chp6/monod-batch-inp.csv", param,pdfout=T)

param <- list(plab="y", pval = seq(5,15,5))
t.X <- sim(monod,"chp6/monod-batch-inp.csv", param,pdfout=T)

param <- list(plab="D", pval = seq(0.10,0.20,0.05))
t.X <- sim(monod,"chp6/monod-batch-inp.csv", param,pdfout=T)

param <- list(plab="X1.0", pval = seq(90,110,10))
t.X <- sim(monod,"chp6/monod-batch-inp.csv", param,pdfout=T)

param <- list(plab="X2.0", pval = seq(9,11,1))
t.X <- sim(monod,"chp6/monod-batch-inp.csv", param,pdfout=T)
```

Examine the graphical output in the **monod-batch-inp.csv-X1.0-out.pdf** and **monod-batch-inp.csv-Kmax-out.pdf** files. The first page of each file is illustrated in Figure 6.19 for X1.0 and in Figure 6.20 for K_{max}. The second page has a graph called the **phase-plane** portrait. It shows the trajectory of $X2$ versus $X1$ with time implicit (Figure 6.21). We see how the culture population has grown fast and used up all the resource, and it is in decline after that due to the death rate.

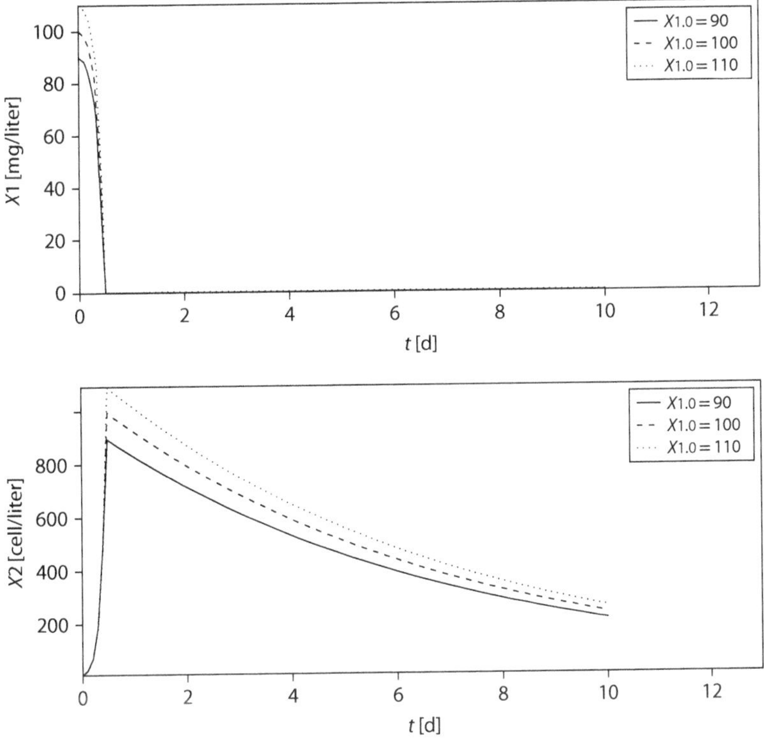

FIGURE 6.19 Monod batch culture: effect of changing the initial condition of the substrate. Top panel: dynamics of the concentration $X1$. Bottom panel: dynamics of the population $X2$.

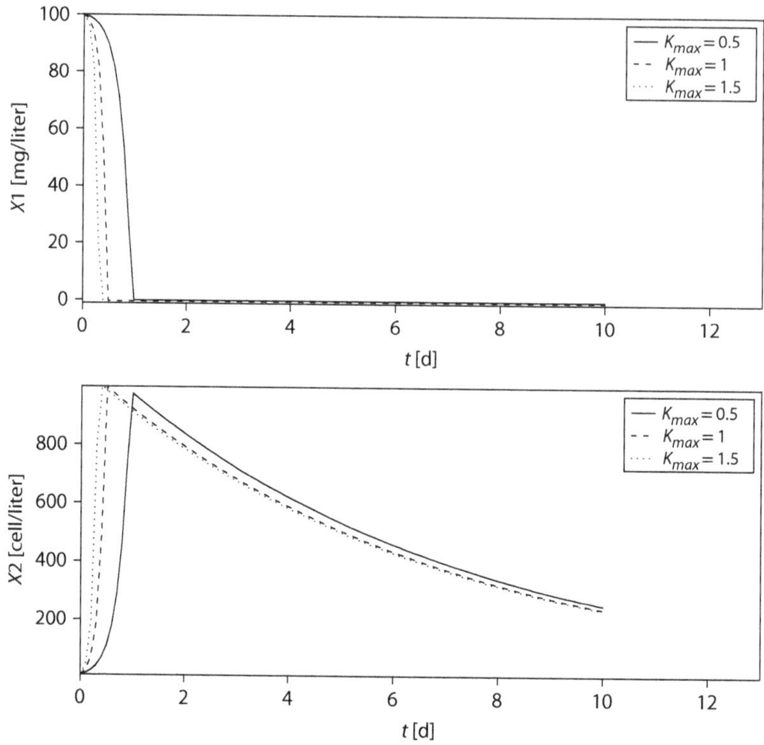

FIGURE 6.20 Monod batch culture: effect of changing the maximum rate (K_{max}). Top panel: dynamics of the concentration $X1$. Bottom panel: dynamics of the population $X2$.

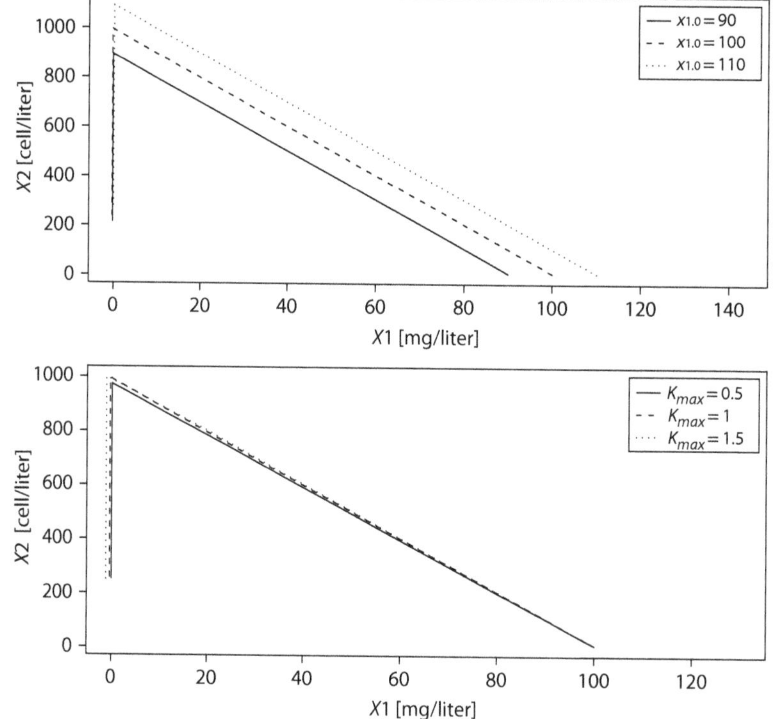

FIGURE 6.21 Monod batch culture: phase-change portrait showing the effect of changing the initial conditions of the substrate (top panel) and the maximum rate K_{max} (bottom panel).

Exercise 6.11

Produce graphs and analyze the effect of changing yield, *y*, and changing death rate, *D*. Discuss.

We can also perform sensitivity analysis with respect to one of the parameters, say K_h, in the same manner as we did for the `logistic` and `m3decay` functions.

```
# give nom and sd
pnom=2;psd=0.25; plab = c("Kh"); runs=5
pval <- seq((pnom -(runs-1)*psd),(pnom+ (runs-1)*psd),2*psd)
param <- list(plab=plab, pval=pval, pnom=pnom, fact=T)
# to call a run
t.X <- sim(monod,"chp6/monod-batch-inp.csv", param,pdfout=T)
```

However, in this case, we have two state variables and have to decide which one of the two variables to analyze before calling the sens function. In the following, we call sens one variable at a time by selecting the first columns 1 and 2–6 for time and the variable *X*1 and then the columns 1 and 7–11 for the time and the variable *X*2:

```
# use variable 1
v2=1;i =1; v1<- v2+1;v2 <- runs+(v1-1); v <-c(1,v1:v2)
s.y <- sens(t.X$output[,v], param, fileout="chp6/monod-sens-
X1Khout",pdfout=T)
# use variable 2
i =2; v1<- v2+1;v2 <- runs+(v1-1); v <-c(1,v1:v2)
s.y <- sens(t.X$output[,v], param, fileout="chp6/monod-sens-X2Khout",pdfout=T)
```

Figure 6.22 shows the graphics result for *X*1 on the top panel and for *X*2 at the bottom panel. The patterns are relatively similar, but we see from the values on the *y*-axes that *X*1 is much less sensitive to K_h than *X*2.

6.4.4 Build Your Own Model Simulation

Now that you have seen examples of three model functions (`logistic, M3, and monod.batch`) and their corresponding input files, you can build your own simulation using `sim` in four easy steps.

First, let me guide you through an example. Suppose we want to simulate a model on the basis of the logistic function but with the term *X* as a power function (X^a):

$$dX/dt = rX^a\left(1-\frac{X}{K}\right) \tag{6.14}$$

where *a* is a new parameter affecting how quickly *X* grows with *X* when away from the equilibrium, *K*.

Step 1: Edit the logistic function logistic to have a new name `logistic.compen` with a new parameter `p[3]` assigned to a and raise x to power a, that is, x^a:

```
logistic.compen <- function(t,p,x){
r<-p[1]; K<-p[2]; a <- p[3]
dx <- r*x^a*(1-x/K)
return(dx)
}
```

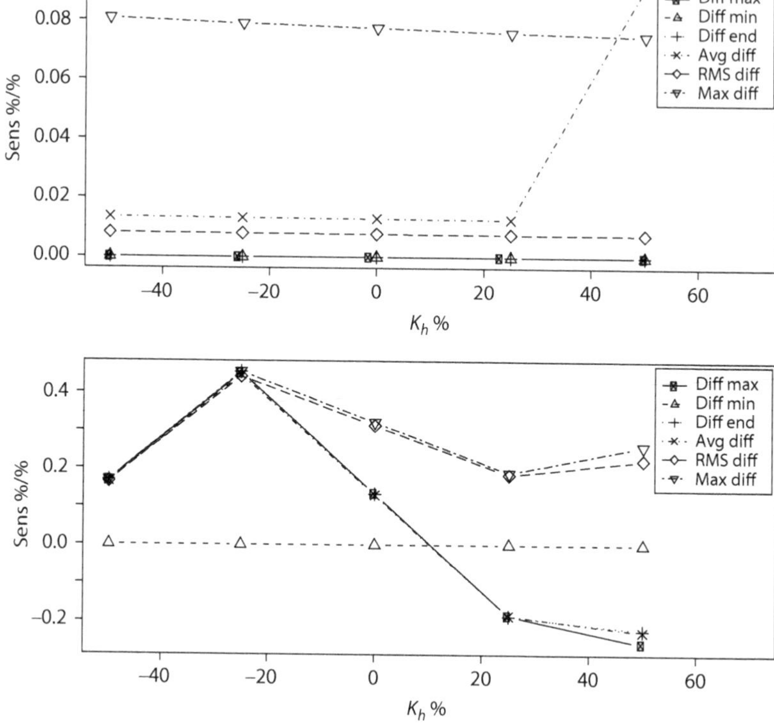

FIGURE 6.22 Sensitivity analysis of Monod batch model. Top: variable X_1 (concentration). Bottom: variable X_2 (culture).

Step 2: Run this script or save it in a file **myown.R** and source it.

Step 3: Edit the input file **logistic-inp.csv** to insert a new line for parameter a:

```
a,1.5,none,shape coefficient
```

Insert it after the line for parameter K and before the line for X_0:

```
Label,Value,Units,Description
t0,0.00,Year,time zero
tf,20.00,Year,time final
dt,0.1,Year,time step
tw,1.0,Year,time to write
r,0.4,1/Year,intrinsic growth rate
K,5.00,Ind,carrying capacity
a,1.5,none,shape coefficient
X0,1.00,Ind,initial population
digX,4,None,significant digits for output file
```

Save this file as `logistic-compen-inp.csv`.

Step 4: We use the new model and the input file in sim in the following manner to make one run:

```
logis.a <- list(f=logistic.compen)
t.X <- sim(logis.a,"chp6/logistic-compen-inp.csv")
```

With only the model and the input file specified into sim, we get the nominal run. Now, vary the parameter a:

```
param <- list(plabel="a",pval=c(0.5,1,1.5))
t.X <- sim(logis.a,"chp6/logistic-compen-inp.csv",param,pdfout=T)
```

You should have obtained a set of graphs including those shown in Figure 6.23. You should note that the rise to carrying capacity is faster for values of a greater than 1 and slower for values of a lower than 1. The border case $a = 1$ is the logistic response.

Exercise 6.12

Implement a simulation of decay following allosteric-type kinetics $\frac{dX}{dt} = -K_{max}\left(\frac{X^m}{K_h + X^m}\right)$ and explore the effect of the parameter m.

You can now appreciate the reason for developing `sim` as a generic function. Even though `sim` is more complicated than a specific function, we recover the investment because it facilitates building new models.

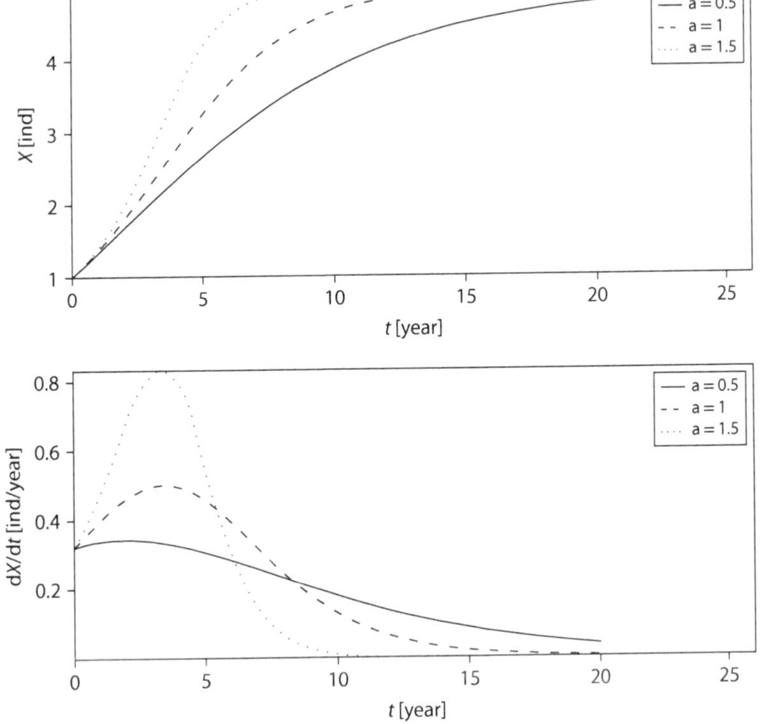

FIGURE 6.23 Modified logistic dynamics.

6.5 SEEM FUNCTIONS EXPLAINED

6.5.1 FUNCTIONS LOGISTIC, M3, AND MONOD.BATCH

The three functions `logistic`, `M3`, and `monod.batch` define the dynamics of the logistic, M3, and Monod models. The functions are implementations of the ODE corresponding to these models. The functions are used to define the `f` component of the model list object to call the `sim` function.

```
logistic <- function(t,p,x){
r<-p[1]; K<-p[2]
dx <- r*x*(1-x/K)
return(dx)
}
```

```
M3 <- function(t,p,x){
# p[1] negative for decay
Kmax<-p[1]; Kh<-p[2]
dx <- Kmax*x/(x+Kh)
return(dx)
}
```

```
monod.batch <- function(t,p,x){
# x[1] substrate, x[2] microbe
Kmax<-p[1]; Kh<-p[2]; y<-p[3]; D<-p[4]
dx1 <- -x[2]*Kmax*x[1]/(x[1]+Kh)
dx2 <- -dx1*y - D*x[2]
dx<-c(dx1, dx2)
return(dx)
}
```

6.5.2 FUNCTION SIM

The first part of the `sim` function is like the sim.mruns function of Chapter 4, but we include the rates and generalize using positions for the parameters and initial conditions, since we can have more than one. This function will be a workhorse for the remaining chapters.

```
sim <- function(model, file, param=NULL, pdfout=F, lab.out="") {

# multiple runs
# required: model as a list
#   path/name of input file
# optional: param set for multiple runs
#   switch to produce pdf files, default F
#   label to identify output files, deafult none

# read input, file format is csv
dat <- c("character","numeric","character","character")
input <- read.csv(file,header=T,colClasses=dat)
```

```
# input values and labels
v <- t(input$Val); vl <- t(input$Lab)
# t0,tf,dt,tw are in v[1:4]
t <- seq(v[1],v[2],v[4]); dt <- v[3]
# position of first initial condition
iNS <- max(which((vl=="X1.0")),which((vl=="X0")))
last <- length(vl); NS <- last-iNS # number of states
mp <- c(5:(iNS-1)); mX0 <- c(iNS:(iNS+NS-1)); digX <- v[last]
# units ind and dep vars
if(NS>1){
unit <- c(input$Uni[1], rep(input$Uni[iNS],NS))
tXlab <- paste(c("t",paste("X",seq(1:NS),sep="")),"[",unit,"]", sep="")
} else {
unit <- c(input$Uni[1], input$Uni[iNS],
  paste("(",input$Uni[iNS],")/",input$Uni[1],sep=""))
tXlab <- paste(c("t","X","dX/dt"),"[",unit,"]", sep="")
}
```

In this case, we collect the two parameters in an array, p, which will be an argument to the model function f. A key to the functionality of sim is detecting the position iNS of the initial condition in the input file. For this purpose, it assumes that the initial condition is labeled X_0 for one-dimensional systems or starts with X1.0 for multidimensional systems. In addition, it is assumed that the parameter starts listing in position 5 of the file. Now we know that the parameters are listed from 5 to iNS-1 and the initial conditions are from iNS to the next to last position. This information is used in the parametric loop when making a decision on whether to vary the initial condition or a parameter and which parameter to vary.

For the sake of generality, the value of NS is tested to decide whether to include the rate dX.dt for one-dimensional systems using the model function or ignore it for the multidimensional systems. Also for the sake of generality, we have made the param argument optional with default value NULL. Therefore, the null condition of param is tested with is.null(param) to decide on making one run or many according to the values of param.

```
# number of time stamps and runs
  nt <- length(t)
  if(is.null(param)) np <- 1 else {
  if(length(param$plab)==1) param$pval<- matrix(param$pval)
  np <- dim(param$pval)[1]
  }
# arrays to store results
  X <- structure(1:(np*nt*NS),dim=c(nt,np,NS))
  if(NS==1) dX.dt <- matrix(nrow=nt,ncol=np)
# loop for parameter p that varies
  for(i in 1:np) {
     # pick a value of X0 or p from the param set
     if(is.null(param)==F){
      for(k in 1:length(param$plab)){
      for(j in mp[1]:mX0[length(mX0)])
        if(param$plab[k]==vl[j]) v[j]<- param$pval[i,k]
      }
     }
```

```
      # parameters
      p <- v[mp[1]:mp[length(mp)]]
      # X0 initial condition
      X0 <- v[mX0[1]:mX0[length(mX0)]]
      # integration
      out <- RK4(X0, t, model$f, p, dt)
      # rename x part of out for simpler use
      if(NS>1) for(k in 1:NS) X[,i,k] <- out[,k]
      # calculate rate for one-state systems
      else{
       X[,i,1] <- out
       for(j in 1:nt) dX.dt[j,i] <- model$f(t[j],p,X[j,i,1])
      }
   } # end of parameter p loop
```

The last part of this is also more demanding because it should form labels with different variables. In addition, it calls a new function `vars.out` to produce graphs and output files.

```
 # prep and organize output as data frame
   if(NS==1) X <- cbind(X[,,1],dX.dt)
   X <- signif(X,digX)
   output <- data.frame(t,X)
 # forms names for output vars and runs
   nv <- length(tXlab)
   var.run <- c(tXlab[1],

 paste(tXlab[2],paste(".Run",c(1:np),sep=""),sep=""))
   for(i in 3:nv) var.run <- c(var.run, paste(tXlab[i],paste(".
 Run",c(1:np),sep=""),sep=""))
   names(output) <- var.run

 # call output function to generate output files
   vars.out(file, lab.out, input, param, output, tXlab, pdfout)

   return(list(input=input, param=param, output=output))
 } # end of function
```

6.5.3 Function vars.out

This function is long but that is the price we pay so that it can be generic and reusable for future models with two or more variables and simulations of two or more runs. In addition, we plot $X2$ versus $X1$ to describe a phase–plane portrait of the system. To make the legend more readable, we have expanded the horizontal axes to 50% more than the maximum value of t and start the legend at the top right. For the sake of space, we do not explain this function with more detail. The interested reader can examine the function on the console.

7 Stability and Disturbances

7.1 STABILITY

Stability is a model concept related to differential equations. An **equilibrium point** is a value of X where the rate of change, dX/dt, is zero. Stability is a property of an equilibrium point: the state returns to that equilibrium after a **perturbation**. A model can have several equilibrium points and each point can have different stability properties; some points can be stable and others unstable. Stability has been a topic of great interest in ecological models as a way of understanding when populations, communities, and ecosystems can remain unaltered or be sustained long term (Hallam, 1986b; Swartzman and Kaluzny, 1987).

Stability analysis is the process of determining the stability properties of all equilibrium points of the model. First, we find the equilibrium points by looking for the values of X that make the net rate equal to zero, that is, what is the value of X^* such that

$$\frac{dX}{dt} = 0 \tag{7.1}$$

is satisfied when evaluated at this point, X^*. In other words, the equilibrium points are the **roots** of Equation 7.1.

For illustration, let us consider only one state variable, X, and the plot shown in Figure 7.1. The vertical axis is the net rate of change and the horizontal axis is the state variable. There are two equilibrium points, $X^* = 20$ and $X^* = 80$. These are the values, where the rate goes through zero.

To determine stability, test for the sign of the rate of change around the equilibrium point; see the arrows in Figure 7.1. Around $X^* = 20$, values of X larger than the equilibrium have a negative rate, whereas values of X lower than the equilibrium have a positive rate. Trajectories return to this equilibrium, and therefore, it is **stable**. The opposite result is displayed for the case of $X^* = 80$, trajectories run away from this point. Values of X larger than 80 lead to further increases in X, whereas values of X lower than 80 lead to decreases in X. Therefore, this equilibrium is **unstable**.

The equilibrium $X^* = 20$ is only **locally** stable, meaning that trajectories return to this point in a neighborhood around the equilibrium or excursions produced for relatively small perturbations. In fact, if the perturbation carries the trajectory above $X = 80$, then it will not return to $X = 20$.

7.1.1 Stability of Linear Dynamics

There is only one point that satisfies the equilibrium conditions for a linear model. That is, there is only one point, X^*, that satisfies Equation 7.1. If this point is **locally** stable, then it is also **globally** stable, meaning for the entire range or domain of X or excursions produced by large perturbations.

Consider the exponential model. To find the equilibrium points, look for the values of X^* that make the net rate equal to zero. That is, what is X^* such that $dX/dt = kX^* = 0$. In this case, the only equilibrium is at zero, that is, $X^* = 0$. Now examine the rate around this point. The rate is positive near X^* if k is positive, and therefore, X increases away from $X^* = 0$. That is, the equilibrium is unstable if k is positive or X grows without limits. On the contrary, the rate is negative near X^* if k is negative and X decreases back to $X^* = 0$. That is, the equilibrium is stable if k is negative or X decreases to zero (Figure 7.2).

We can consider examples of two environmental interpretations: (1) a population modeled exponentially can either grow ($r > 0$) or decline ($r < 0$) to extinction and (2) chemical decay implies negative rate coefficient, $k < 0$, then $X^* = 0$ is always stable.

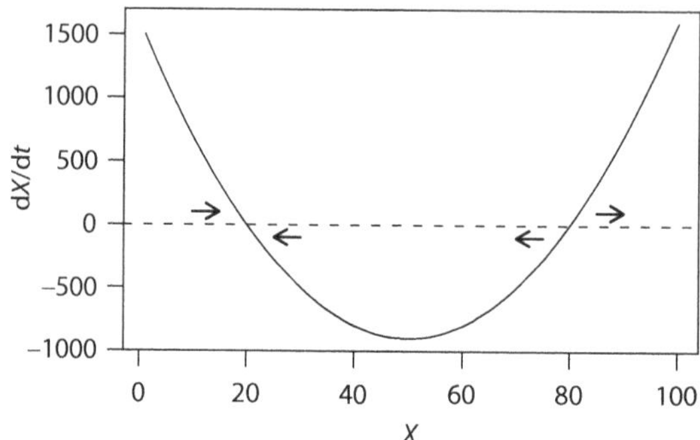

FIGURE 7.1 Equilibrium points and stability.

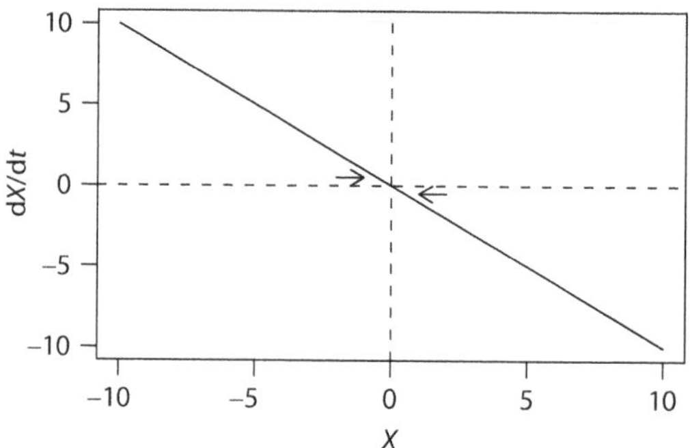

FIGURE 7.2 Stable equilibrium for the exponential model with a negative coefficient.

7.1.2 Stability of Nonlinear Dynamics

Because a nonlinear model may have several equilibrium points, local stability of an equilibrium point does not imply global stability. That is, an equilibrium point may be stable for small disturbances but become unstable for large disturbances.

Take the logistic model, for example. To find the equilibrium points, look for the values of X that make the net rate equal to zero. That is, find X^* such that $\mathrm{d}X/\mathrm{d}t = rX^*\left(1 - X^*/K\right) = 0$. There are two roots since this occurs for $X^*_1 = 0$ and $X^*_2 = K$. The stability of X^*_2 can be examined by looking at the rate around this point. For $X > K$, the net rate is negative and the population declines to K, but for $X < K$, the net rate is positive and the population grows to K. Therefore, X^*_2 is stable and X^*_1 is unstable.

As another example, the equilibrium points of the Michaelis–Menten model are the roots of

$$\frac{\mathrm{d}X}{\mathrm{d}t} = -K_{max}\left[\frac{X^*}{K_h + X^*}\right] = 0 \tag{7.2}$$

The equilibrium is $X^* = 0$. The net rate is negative, the concentration decreases, and $X^* = 0$ is stable.

7.2 DISTURBANCES

Disturbances could represent two different situations: those produced by **natural causes** (naturally occurring wildfires and extreme weather events) and **human-induced** (management and resource utilization).

The stability discussion presented in Section 7.1 implicitly assumes that a disturbance moves the state away from the equilibrium at time zero. To explicitly consider an **external** factor or an

exogenous disturbance, we can add a term u to the net rate to represent this externally controlled factor.

$$\frac{dX}{dt} = f[(t, p, X(t)] + u \tag{7.3}$$

Thus, the forcing term u, must have units of net rate (e.g., ind/day). Being a rate, this term can be positive or negative. When u is positive, we have an inflow rate (e.g., immigration, pollutant discharge). When u is negative, we have an outflow rate (e.g., harvest, pollutant cleanup).

The solution of the **forced** model (Equation 7.3) has two parts: the **autonomous** response or unforced behavior and the **forced** response. The first one represents how the system would respond in the absence of disturbance ($u = 0$). The second one is that part of the response that is determined by the disturbance. For linear systems, this is the **convolution** of the exponential with the forcing term. **Convolution** is a mathematical operator performed on two functions of time; it basically means integrating up one function multiplied by the other one that is shifted at different time lags.

As examples, we will include forcing terms in the population exponential model

$$\frac{dX}{dt} = rX + u$$

and in the logistic model

$$\frac{dX}{dt} = rX\left(1 - \frac{X}{K}\right) + u$$

Now we model the forcing term, u, with more detail. First, denote by I the intensity of the disturbance. There are at least two alternatives: a **proportional** and an **absolute** disturbance.

First, let us consider the case when u is proportional to the state:

$$u = IX \tag{7.4}$$

For populations, this means that the disturbance is **relative** or proportional to the density; it affects a fraction of the individuals, and therefore, I has units of per capita rate (e.g., 1/day).

Second, consider u to be a constant regardless of the state:

$$u = I \tag{7.5}$$

In this case, the disturbance is absolute or independent of density. It affects I individuals regardless of how many there are, and therefore, the units of I are net rate units (ind/day). Let us examine more details on both alternatives considering the exponential and the logistic models.

7.2.1 Example: Exponential

The autonomous response shows an increase in population if r is positive, that is, unlimited exponential growth. The disturbance will impose a **bound** to this growth whenever the impact of the disturbance is negative, for example, when harvesting a population. Let us consider absolute mode harvesting with intensity, H_a. The sign of the net rate

$$\frac{dX}{dt} = rX + H_a$$

will determine whether the population grows or declines. Find the equilibrium of the forced system by solving

$$\frac{dX}{dt} = rX + H_a = 0$$

to find X^* at which $rX^* + H_a = 0$:

$$X^* = -\frac{H_a}{r}$$

This equilibrium is such that the intrinsic growth rate is balanced with the harvest rate. This equilibrium is unstable because the trajectories diverge from this point. In other words, the trajectories are **repelled** from the equilibrium. See the arrows in Figure 7.3, where $H_a = -0.02$ and $r = 0.02$, giving $X^* = 1$. If X is less than X^*, then the harvest rate is larger and the population declines to extinction. But, if X is larger than X^*, the harvest rate cannot keep up with the growth rate and the population grows without restraint. That is, X never settles at X^* (see Figure 7.4, where $H_a = -0.02$ and $r = 0.02$, giving $X^* = 1$).

Now consider a relative or proportional harvest with intensity H_p. The sign of the net rate $rX + H_pX$ determines the stability. The equilibrium is the root of

$$(r + H_p)X^* = 0 \tag{7.6}$$

that is, $X^* = 0$. Its stability depends on the sign of $r + H_p$. For example, making H_p negative and larger than r could control the growth of a pest population. This is equivalent to increasing the per capita death rate.

When the exponential model has a negative r (declining population) or when it represents the degradation of a chemical, adding a positive disturbance makes the response stable. The equilibrium is a steady-state value. For example, if we consider a concentration-independent uptake, U_a, and depuration at the rate coefficient, k, of a toxicant by an organism, we can write the model

$$\frac{dX}{dt} = -kX + U_a \tag{7.7}$$

where variable X is in units of mass concentration, say mg/kg, that is, milligram of toxicant per kilogram of body weight. The depuration rate coefficient is in units of T^{-1} and U_a is in units of M M^{-1} T^{-1} (say mg kg^{-1} d^{-1}). This is one of the simplest models in ecotoxicology. The equilibrium point is $X^* = U_a/k$. The stability is determined from the signs of $U_a - kX$ around this value. If X is above $-U_a/k$, the stability will decrease, but if X is below $-U_a/k$, it will tend to increase. Thus, trajectories

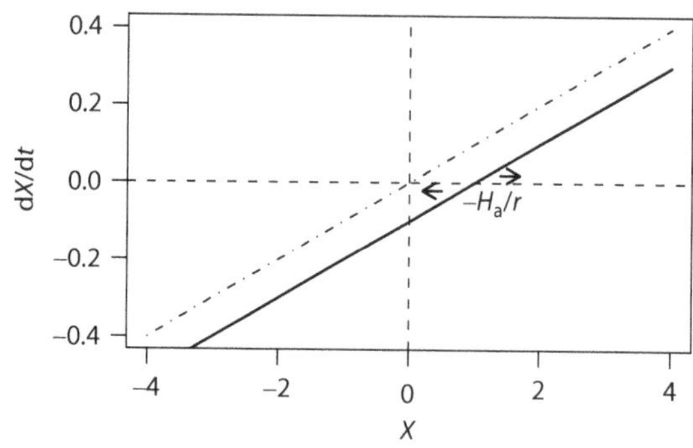

FIGURE 7.3 Unstable equilibrium for the exponential growth harvested in absolute mode.

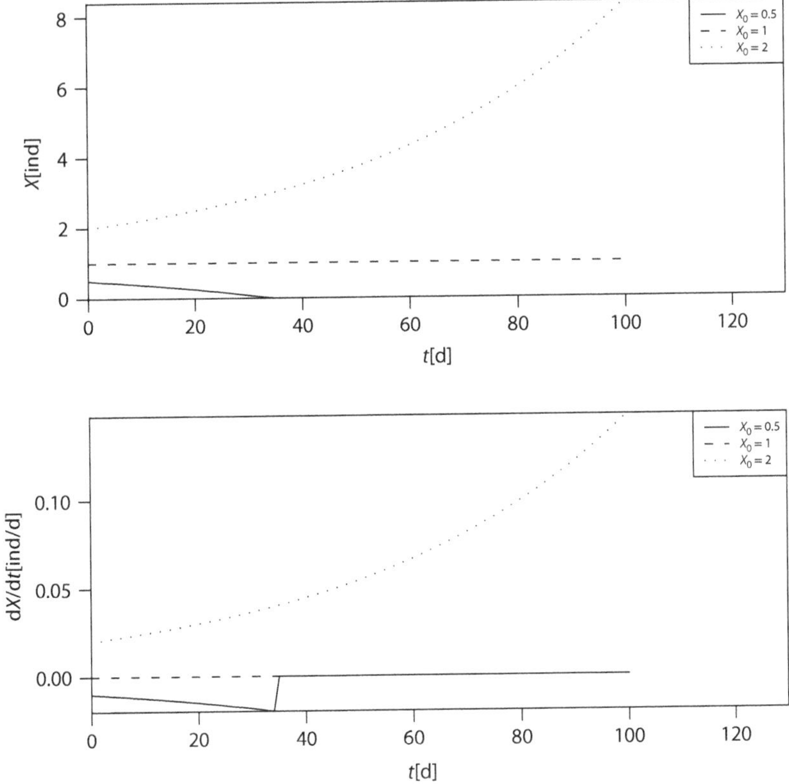

FIGURE 7.4 Forced exponential: growth with harvesting. Density (top panel) and rate (bottom panel).

converge to U_a/k. We will treat ecotoxicology with more details in Chapter 11. Note that the end value is a stable equilibrium (steady state), and it represents bioaccumulation (see Figure 7.5, where $k = 0.2$).

Exercise 7.1

An organism is able to get rid of (depurates) a toxic substance at a constant rate coefficient of $k = 0.5\text{d}^{-1}$. However, the ambient toxic concentration stays constant, forcing the organism to take in the toxic substance at a rate of 1 mg $\text{kg}^{-1}\,\text{d}^{-1}$. What is the equilibrium value of the chemical concentration in the organism? Note that we use concentration in units of toxicant mass in milligrams per unit of body weight in kilograms. Is this equilibrium stable? Draw a graph of the concentration as a function of time.

7.2.2 Example: Logistic

The disturbance term will reduce the equilibrium when the impact of the perturbation on the population is negative, for example, when harvesting and clear cutting (Hallam, 1986b).

Consider the absolute mode harvesting case. The sign of

$$rX\left(1+\frac{X}{K}\right)+H_{\text{a}}$$

will determine whether the population grows or declines. Use $\text{d}X/\text{d}t = 0$ to find the equilibrium or X^* at which

$$rX^*\left(1-\frac{X^*}{K}\right)+H_{\text{a}}=0 \tag{7.8}$$

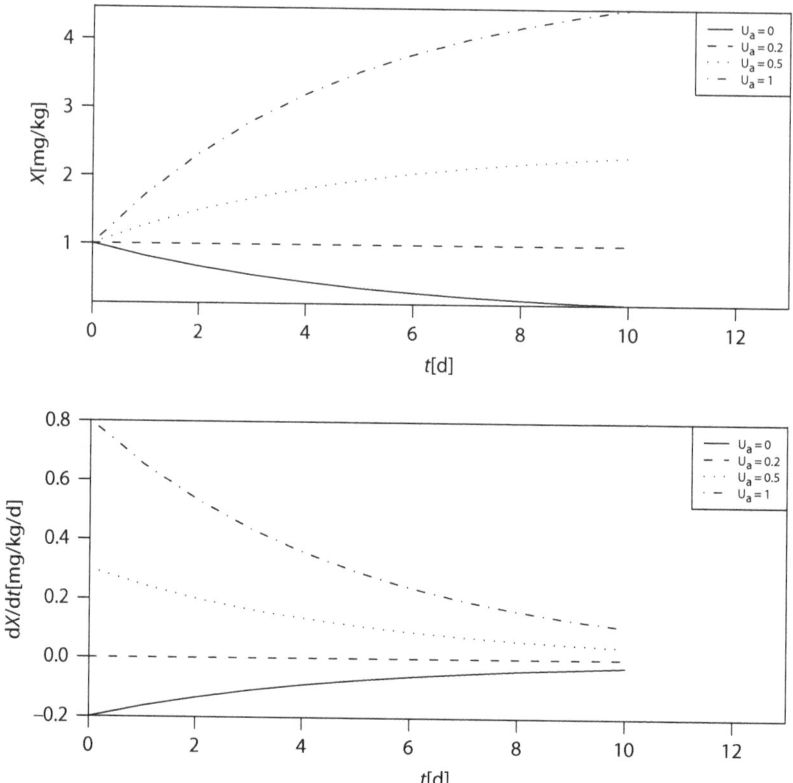

FIGURE 7.5 Forced exponential: bioaccumulation.

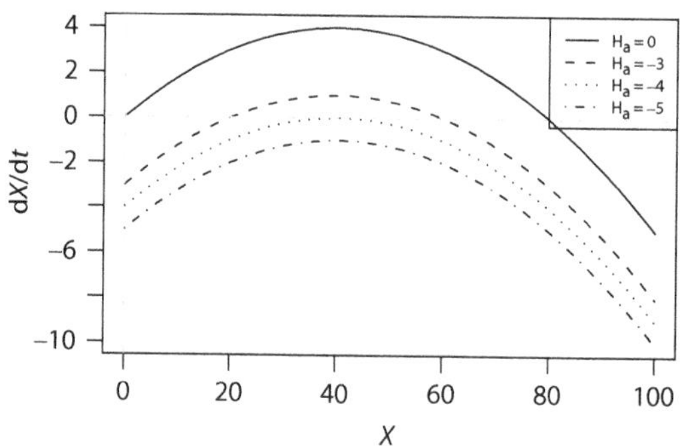

FIGURE 7.6 Equilibrium points for harvested logistic model in absolute mode.

This equation is quadratic:

$$rX^* - \frac{rX^{*2}}{K} + H_a = 0 \tag{7.9}$$

Here, H_a is negative (because it is a harvesting term), and the results vary according to the relative value of the harvest intensity with respect to the maximum rate, as shown in Figure 7.6, where $r = 0.2$ and $K = 80$.

The equilibrium points are real numbers if the harvesting does not exceed a "tolerable" rate as occurs for $H_a = -3$, as shown in Figure 7.6. The equilibrium with the lowest value is unstable, but the other equilibrium points are stable. This last one will be the new reduced "carrying capacity." In addition, perturbing the population to values below the lower equilibrium value will make the population extinct. That is, we now have a threshold. See Figure 7.7 for the trajectories starting at various X_0.

However, if H_a equals or exceeds the tolerable rate, the results are very different. If H_a is equal to the tolerable rate, the equilibrium is critically stable ($H_a = -4$ in Figure 7.6). Slight perturbations that bring X below this point will make the population extinct (see Figure 7.8). If H_a exceeds the tolerable rate (e.g., $H_a = -5$ in Figure 7.6), then the equilibrium points are not real numbers (no intersection of growth rate with harvest) and the population will always become extinct (Figure 7.9).

Consider the relative or proportional harvest case with intensity H_pX. The sign of $rX(1 - X/K) - H_pX$ will determine the stability. Find the equilibrium using $dX/dt = 0$:

$$\left[r\left(1-\frac{X^*}{K}\right)-H_p\right]X^*=0 \tag{7.10}$$

Also a quadratic equation yielding two equilibria $X^*_1 = 0$ and $X^*_2 = K(1 - H_p/r)$. The harvested logistic in the proportional mode is just like a logistic but with $r - H_p$ for the intrinsic rate instead of just r. One of the new equilibria, X^*_2, is stable, whereas X^*_1 is unstable. See Figure 7.10, which was produced with $r = 0.2$ and $K = 80$.

Exercise 7.2

When H_p is equal to $r/2$, the stable equilibrium is $K/2$ and referred to as the maximum sustainable yield. Assume $r = 0.2$ and $K = 80$. What is the maximum sustainable yield? For which value of H_p does it occur? Identify this curve in Figure 7.10.

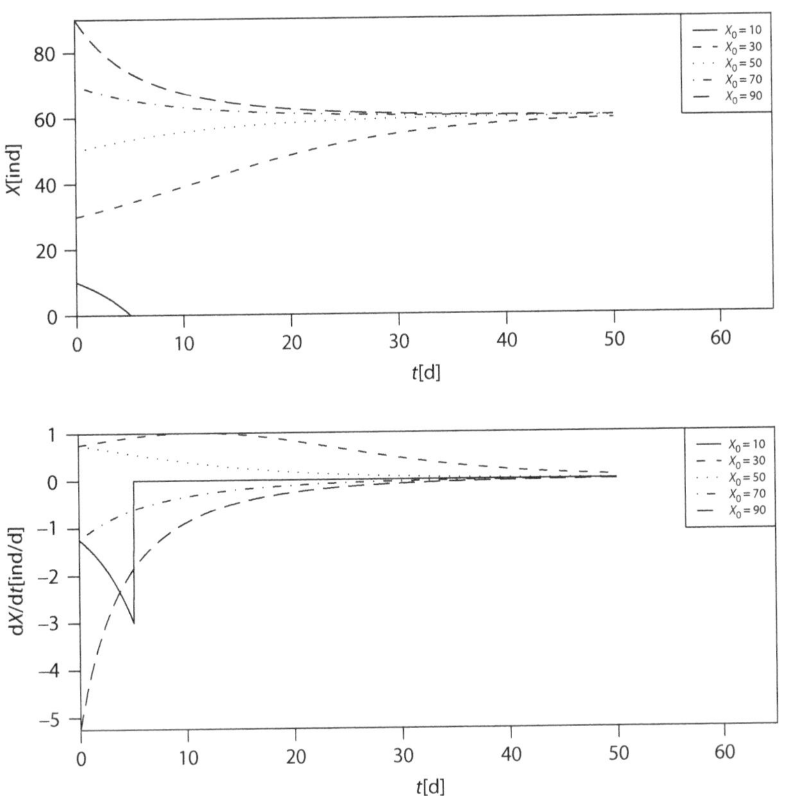

FIGURE 7.7 Harvesting the logistic below the tolerable rate.

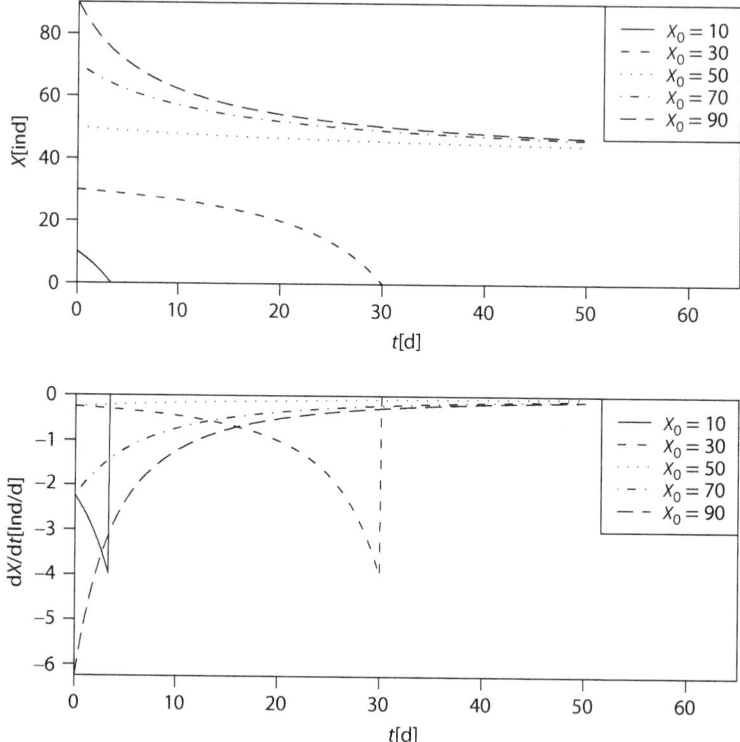

FIGURE 7.8 Harvesting the logistic at the tolerable rate.

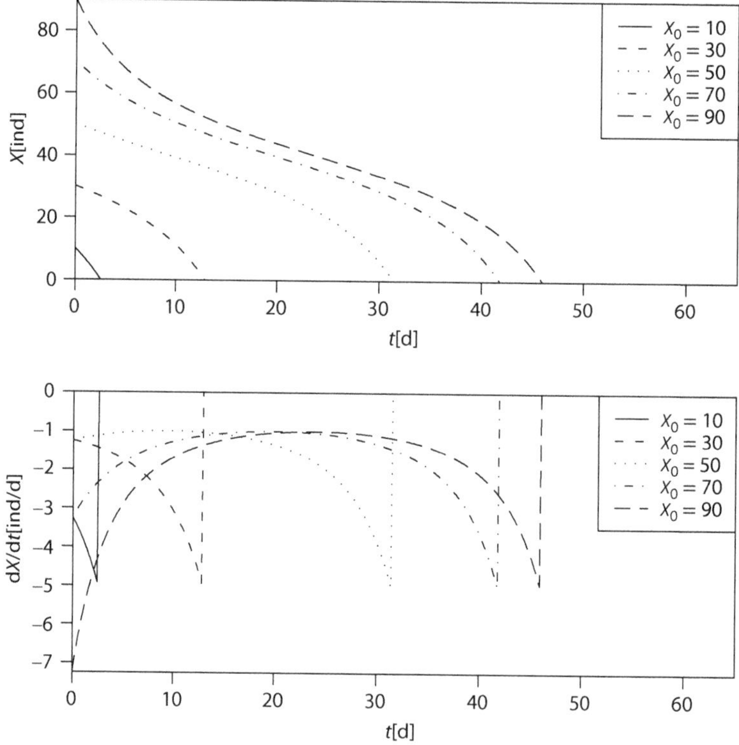

FIGURE 7.9 Harvesting the logistic above the tolerable rate.

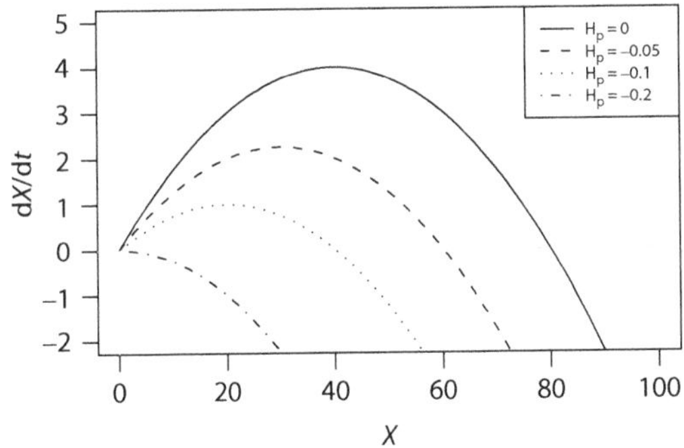

FIGURE 7.10 Harvested logistic in proportional mode.

7.3 VARIABILITY AND STRESS

Environmental variability and stress can be modeled by time varying intrinsic rate and carrying capacity (Hallam, 1986b; Swartzman and Kaluzny, 1987). Conceptually, we make the external forcing affect the model parameters, instead of adding a new term. As examples, we will consider the exponential and the logistic models.

7.3.1 Review of Sine and Cosine Functions

A convenient building block of periodic functions is the **sine** wave function

$$x(t) = a\sin(\omega t) \tag{7.11}$$

where the angular frequency is

$$\omega = 2\pi f = 2\pi/T \tag{7.12}$$

which is proportional to the frequency, f, and is given in units of radians. The frequency is the inverse of the period, T, of oscillations. One full cycle is for $\omega t = 2\pi$. The sine is equal to 0 at 0, π, and 2π, equal to a at $\pi/2$, and equal to $-a$ at $3\pi/2$. This repeats for every cycle. In Figure 7.11, we illustrate two full cycles of the period $T = 5$. The **cosine** function is the sine function but shifted by 90 degrees, or $\pi/2$ radians (Figure 7.12):

$$x(t) = a\cos(\omega t) = a\sin(\omega t - \pi/2)$$

Therefore, the cosine is equal to 0 at $\pi/2$ and $3\pi/2$, equal to a at 0 and 2π, and equal to $-a$ at π. This repeats for every cycle.

7.3.2 Periodic Variations in the Exponential Model

Variations in the intrinsic rate coefficient r could be due to (1) an increase in stress leading to gradual sublethal changes in reproductive factors and (2) seasonal changes producing periodic variability of reproductive output. That is, r could have two components, one gradually varying r_m and another periodically varying r_p as

$$r(t) = r_m(t) + r_p(t) \tag{7.13}$$

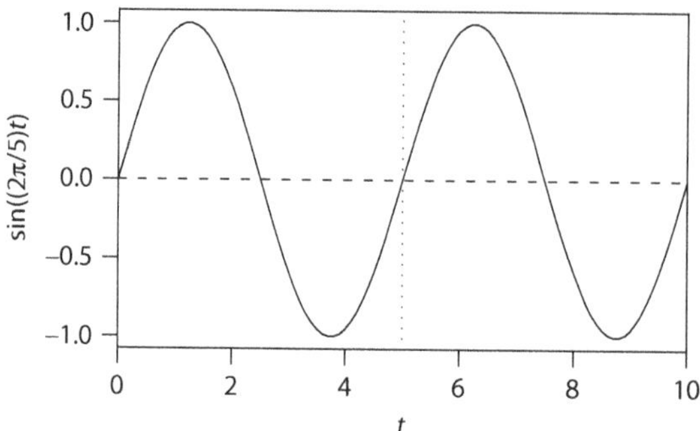

FIGURE 7.11 Sine wave function of amplitude $a = 1$ and period $T = 5$.

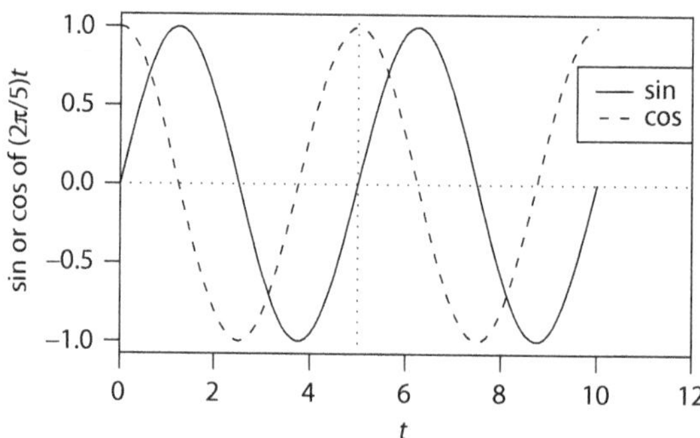

FIGURE 7.12 Cosine and sine functions.

The gradual part could be modeled as a linear decrease from r_o (the nonstressed value of r) with time, as

$$r_m(t) = r_o - r_d t \tag{7.14}$$

and the periodic part could be modeled using a sine function:

$$r(t) = r_m(t) + r_a \sin\left[(2\pi/T)t\right] \tag{7.15}$$

where r_m is the mean value of oscillations, r_a is the amplitude of the oscillations, and T is the period of oscillations. In Figure 7.13, we see the effect of changing r_d. For $r_d = 0$, the oscillations are sustained and the population grows exponentially but modulated periodically according to $r_a = 0.2\ \text{d}^{-1}$ and $T = 2$ d. But as r_d increases, the population declines because of the gradual decrease in r. For this figure, we have used $r_a = 0.06$ and a period $T = 1$ with three different values of r_m. Alternatively, variations in r could also be random, r taking values from a distribution with mean r_m and variance r_v.

7.3.3 Periodic Variations in the Logistic Model

Variations in K could be due to a deteriorating environment, such as a decrease in food supply or scarcity of water. This is modeled as a drift in K with time. Variations in K may be due to seasonal effects inducing periodic changes in carrying capacity, whether deterministic or random (Keen and Spain, 1992, pp. 109–112). We can use similar formulations as in Equations 7.13 and 7.15, but applied to the logistic model. In this case, we vary parameter K:

$$K(t) = K_0 + K_d t + K_a \cos(2\pi t/T)$$

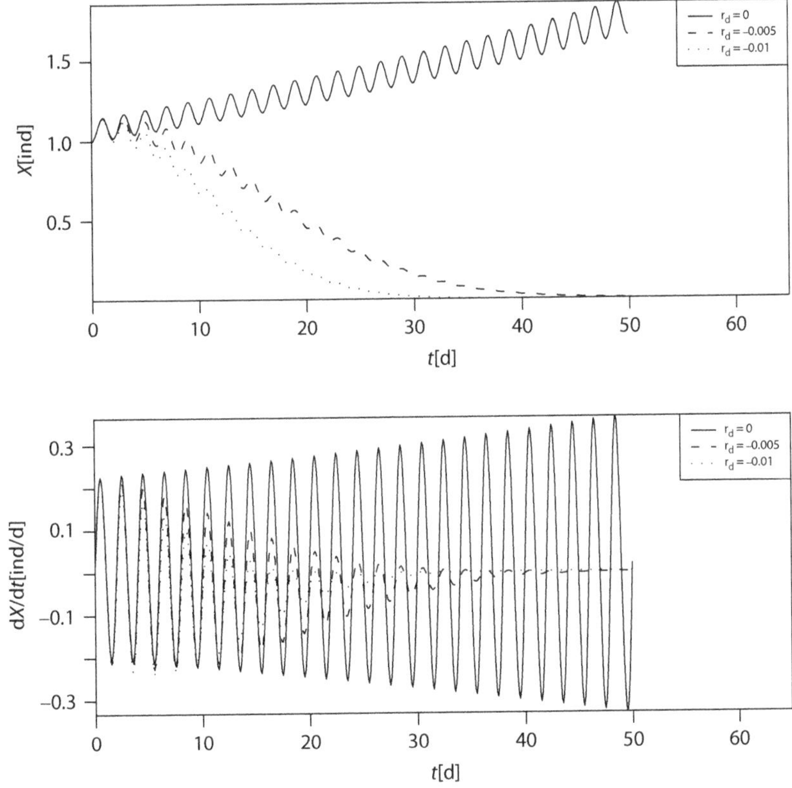

FIGURE 7.13 Exponential growth with gradual and sinusoidal variations of intrinsic growth rate.

In Figure 7.14, we illustrate how the ability to track this environmental fluctuation depends on the value of the intrinsic growth rate, r. For this figure, we have used $K_0 = 80$, $K_d = -0.1$, $K_a = 10$, and $T = 10$. The relative value of r compared to the frequency of oscillations, T, determines the behavior. For low values of r (thereby producing low values of the product rT), the population cannot track the environmental fluctuation, and therefore, the intensity of its fluctuation is damped or diminished. For higher values of r, the population can track the fluctuations and degradation in carrying capacity. This corresponds to higher values of the product, rT.

7.4 SUDDEN AND IMPULSIVE DISTURBANCES

In many cases, the disturbance, u, is a sequence of sudden events, that is, the disturbance does not occur all the time but at certain intervals. To model this situation, two parameters can be used to characterize the disturbance. These two parameters relate to how often the system is disturbed and by how much, that is, the frequency of the disturbance and intensity of each event.

The results are similar to a continuous disturbance, but now the trajectories exhibit a continuous curve punctuated by discontinuities at the times of occurrence of the disturbance events. For example, when the population is modeled with an exponential, it shows steady increases but sudden decrements by episodes of disturbance (e.g., harvesting, kills). We will cover two cases: exponential and logistic population growth (Acevedo, 1980b, 1981; Acevedo and Raventós, 2003).

7.4.1 Example: Exponential Growth

Suppose that every T time units there is a instantaneous reduction of the population by a fraction of its value as shown in Figure 7.15. The population graph looks like a saw-tooth curve. Each drop

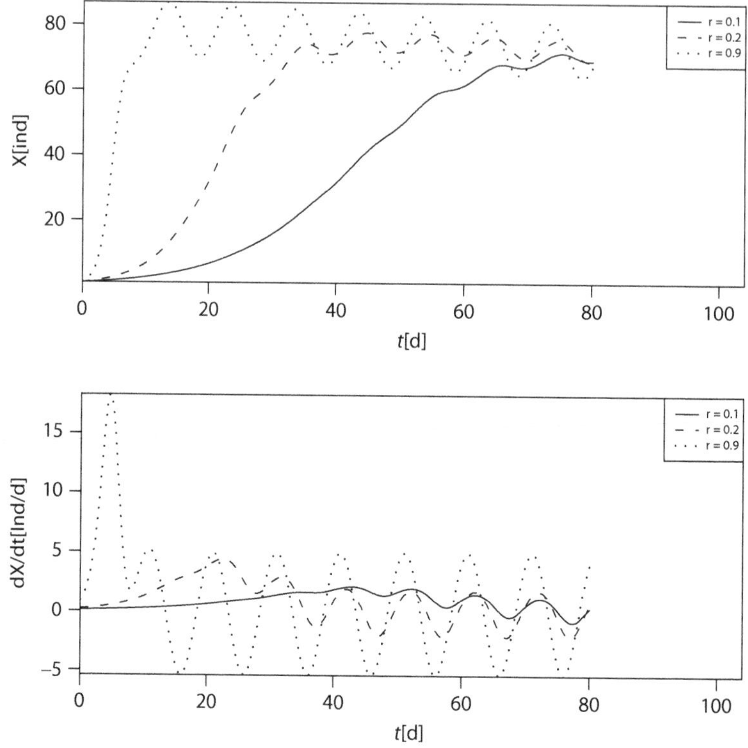

FIGURE 7.14 Logistic growth with periodic variation of carrying capacity, K, and for different values of parameter r.

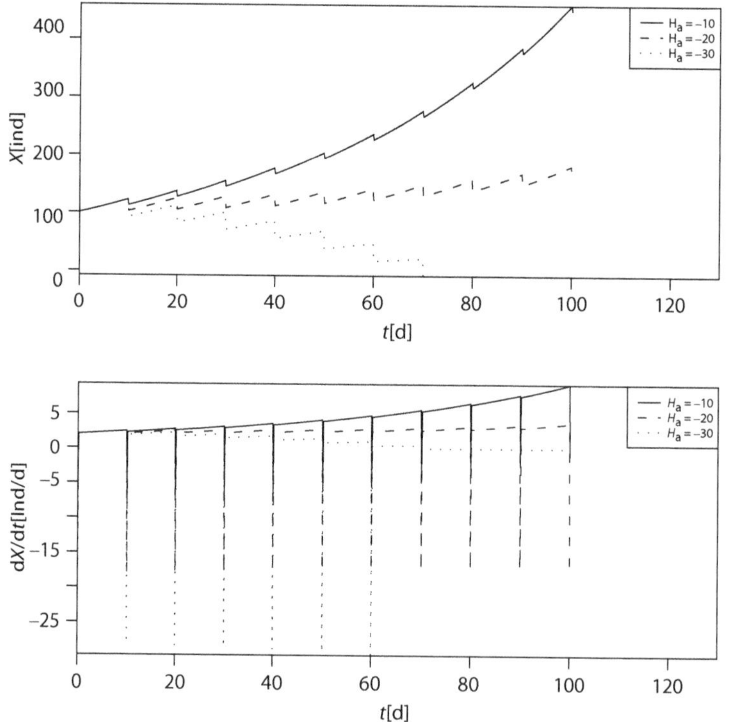

FIGURE 7.15 Exponential growth punctuated by sudden and repetitive disturbances.

corresponds to the occurrence of a disturbance event, which repeats every 10 days as can be appreciated from the values in the X axes.

Immediately before the event number $k + 1$, the population is determined from the growth before the earlier event, k,

$$X\left[(k+1)T\right]^{-} = \exp(rT)X\left[kT\right]^{+} \tag{7.16}$$

where the superscript "+" indicates the value immediately after the event and "–" indicates the value just before the event.

Right after event $k + 1$, the population is reduced by H_p with respect to the value given in Equation 7.16 and then

$$X\left[(k+1)T\right]^{+} = \exp(rT)X\left[kT\right]^{+}(1-H_p) \tag{7.17}$$

which can be abbreviated as

$$X(k+1)^{+} = \alpha X(k)^{+} \tag{7.18}$$

where the new parameter $\alpha = \exp(rT)(1-H_p)$ combines the period, T, and the intensity, H_p, of the disturbance. This is now a finite **difference equation**. The solution is found by the iteration for $k = 1, 2, 3, \ldots,$

$$\begin{aligned} X(1)^{+} &= X(0)^{+}\alpha \\ X(2)^{+} &= X(1)^{+}\alpha = X(0)^{+}\alpha^{2} \\ X(3)^{+} &= X(2)^{+}\alpha = X(0)^{+}\alpha^{3} \end{aligned} \tag{7.19}$$

which can be generalized as

$$X(k)^{+} = X(0)^{+}\alpha^{k} \tag{7.20}$$

The sequence of population values, $X(k)^{+}$, grows without bound when $\alpha > 1$, declines to 0 when $\alpha < 1$, and remains constant when $\alpha = 1$. In these last two situations, the population is limited in growth.

The product rT represents the ratio of the intrinsic growth rate and the frequency $1/T$ of the disturbance. When $rT >> 1$, the population grows rapidly in comparison with the rate of occurrence of the disturbance events, but for $rT << 1$, the disturbance is more frequent than the rate of growth of the population.

For example, if a pest population grows with $r = 0.3\ d^{-1}$, we can calculate how often you need to apply a control application that covers 90% of the population so that it will eventually lead to a population decline. The coverage indicates that $H_p = 0.9$, and therefore, $\alpha = 0.1\exp(0.3T)$. For decline, we need $\alpha < 1$ and T should be less than $T = \ln(10)/0.3 = 7.67$ d, and therefore, you need at least a weekly application.

Exercise 7.3

What coverage level is needed to control a pest population with $r = 0.3\ d^{-1}$ if the control can be applied only monthly?

To describe the trajectory in the continuous time interval in between events, we can use a fraction m, $0 < m < 1$, such that any time instant, t, between kT and $(k + 1)T$ is given by $t = (k + m)T$ (see Figure 7.15). Thus, the population value at that time is

$$X\left[(k+m)T\right] = \exp(rmT)X\left[kT\right]^{+} \tag{7.21}$$

For example, if we want to harvest a population that has $r = 0.1$ per month, allowing an annual harvest such that the population persists in the long term (sustainable), what should be the maximum harvest intensity? What is the population in the middle of the year in between harvest events? In this case, for sustainability, we want $\alpha > 1$; therefore, $\alpha = \exp(1.2)(1 - H_p)$ must be larger than 1 and then H_p must be less than $1 - \exp(-1.2) = 0.69$. Midyear means $m = 0.5$, and then, the population is $\exp(0.6) = 1.82$ times larger than the population just after the harvest event.

Exercise 7.4

For the values given in the example, if the harvest was 80%, how often can you allow harvest so that the population does not collapse or go extinct? What is the population at one-third of the time in between the events?

When the harvest is not a fraction of X, that is, if the reduction is of intensity, H_a, independent of X, the recurrence equation is

$$X\left[(k+1)T\right]^{+} = \exp(rT)X\left[k\right]^{+} - H_{a} \tag{7.22}$$

and the solution is

$$X(k)^{+} = X(0)^{+}\exp(krT) - H_{a}\sum_{i=1}^{k}\exp\left[(i-1)rT\right] \tag{7.23}$$

Note that the disturbed population is unstable because the exponentials grow without bound.

Even though α could be made less than 1, when we have immigration, f, immediately after the event, we write

$$X(k+1)^{+} = \alpha X(k)^{+} + f \tag{7.24}$$

and its solution is similar to the one in Equation 7.23:

$$X(k)^{+} = X(0)^{+}\alpha^{k} + f\sum_{i=1}^{k}\alpha^{i} \tag{7.25}$$

This last sum converges to a value, S, since $\alpha < 1$, and therefore, the population does not decline but it achieves a stationary value, fS, and this means that the arrival of new individuals compensates for the increased mortality due to the disturbance.

If immigration does not occur immediately after the event and is delayed by a time, t_i, shorter than the time to the next disturbance episode, that is, $t_i = mT$, where $0 < m < 1$, then

$$X(k+1)^{+} = \alpha X(k)^{+} + f\exp(-rmT) \tag{7.26}$$

and its steady-state value is $f\exp(-rmT)S$, that is, $\exp(rmT)$ times less than the previous one when there was no delay in immigration.

7.4.2 Example: Logistic Growth

Using the solution to the logistic model, the population just before event $k + 1$ is given as a function of the value just after the prior event by

$$X[(k+1)T]^{-} = \frac{K}{1+\dfrac{\left(K - X[kT]^{+}\right)\exp(-rt)}{X[kT]^{+}}} \tag{7.27}$$

Immediately after the event $k + 1$, the population is reduced to an amount, H_p, with respect to the value given in Equation 7.27 and then

$$X[(k+1)T]^{+} = X[(k+1)T]^{-}(1-H_p) \tag{7.28}$$

Substituting Equation 7.28 in Equation 7.27,

$$X(k+1)^{+} = \frac{\alpha X(k)^{+}}{1+\dfrac{[\exp(rt)-1]X(k)^{+}}{K}} \tag{7.29}$$

where the parameter $\alpha = \exp(rT)(1-H_p)$, as before, combines the period and intensity of disturbance. This new difference equation $X(k+1)+ = f[X(k)^{+}]$ is defined by a nonlinear function $f(X) = \alpha X/(1 + qX)$, where $q = [\exp(rT)-1]/K$. The equilibrium or fixed points $X^* = f(X^*)$, that is, $X^* = (\alpha - 1)/q$. The asymptotic value is then a fraction, b, of the carrying capacity, K, where

$$b = \frac{\alpha - 1}{\dfrac{\alpha}{1-H_p} - 1}$$

The solution converges to a nonzero value as long as $\alpha > 1$. When $\alpha < 1$, the population goes asymptotically to extinction. Rewriting condition $\alpha > 1$ as $H_p < 1 - \exp(-rT)$, we can see that when the population grows slowly compared with the frequency of disturbance, $rT \ll 1$, we may require a strict bound for H_p in order to achieve the persistence of the population. On the other hand, if the population grows rapidly compared to the frequency of disturbance, $rT \gg 1$, then the intensity, H_p, can be large and still achieve persistence. Figure 7.16 illustrates two different cases of parameter rT dictated by two different values of r.

For example, a pest grows logistically with $r = 0.3\ \text{d}^{-1}$ and $K = 1000$. How often should we apply a control measure that covers 90% of the population in such a way that we have 10 individuals left after each application? Coverage indicates $H_p = 0.9$, while 10 individuals means 1% of K and $b = 0.01$, which requires that $\alpha = (1 - b)/[1 - b/(1 - H_p)]$ must be $\alpha = 0.99/0.9 = 1.1$ and then using $\alpha = \exp(0.3T)0.1$, we conclude that T must be $T = \ln(11)/0.3 = 7.99$ d.

Exercise 7.5

Using the data from the previous example, what coverage is needed to control the population to 1% of *K*, if we are limited to monthly applications?

Using m, $0 \le m \le 1$ to denote any instant in the interdisturbance interval, we can write

$$X[(k+m)T] = \frac{K}{1+\dfrac{(K - X[kT]^{+})\exp(-rmt)}{X[kT]^{+}}} \tag{7.30}$$

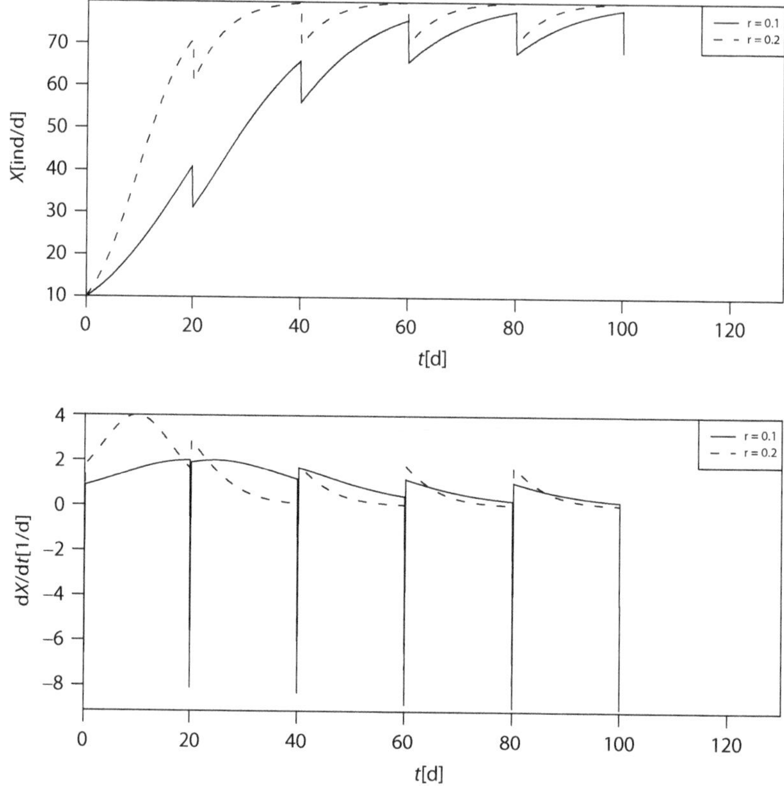

FIGURE 7.16 Logistic growth punctuated by sudden and repetitive disturbances for two different values of r.

Note that when $m = 1$, we recover Equation 7.27. The asymptotic value is a fraction, c, of K:

$$c = \frac{\alpha - 1}{\alpha - 1 + \exp[rT(1-m)]\,H_p} \tag{7.31}$$

We can think of an index of persistence as the average of the population in the interval between disturbances,

$$K\left[1 + (rT)^{-1}\ln(1 - H_p)\right]$$

which summarizes the relation between the population parameters or strategy, r, K, and the disturbance regime parameterized by H_p, T.

For example, we want to harvest a population with $r = 0.1$ month^{-1} and $K = 1000$ by means of a yearly harvest such that in a long run 900 individuals remain after each application. What is the maximum intensity allowable? What would be the population in the middle of the year between the harvests? In this case, $T = 12$ months and $b = 0.9$, and we want $\alpha > 1$, with a value such that $\alpha = \exp(rT)(1 - H_p)$ and also $\alpha = (1 - b)/[1 - b/(1 - H_p)]$. Solving for H_p, we get

$$H_p = 1 - \left\{\frac{[(1-b) + \exp(rT)b]}{\exp(rT)}\right\}$$

Substitute $rT = 1.2$ and $H_p = 0.07$ and evaluate $\alpha = \exp(1.2)(1 - 0.07) = 3.08$. Midyear means $m = 0.5$, and therefore, the population is $1000c$, where $c = 2.08/[2.08 + \exp(0.6)0.0698] = 0.942$, that is, 942 individuals.

Exercise 7.6

Assume that the harvest is 80%. How often can you harvest while avoiding population collapse? What is the population at one-third of the interval?

We could plan an optimal application of this type of impulse control to achieve desired objectives (Brewer, 1979). In many cases, the disturbance does not occur either at fixed intervals or as we have modeled it here. An alternative is to model the disturbance as a Poisson process (Hanson and Tuckwell, 1981), and then, the interval time is a random variable distributed exponentially with rate $p = 1/T$.

7.5 COMPUTER SESSION

We will study the exponential and logistic models subject to continuous as well as sudden disturbances (absolute and relative) and with drift and periodic variation of the per capita growth rate and carrying capacity. In **seem**, we have functions to model the exponential and logistic with disturbance (that can be used to model growth of a population subject to harvest or pollutant decay subject to uptake) and functions to model periodic variation of a parameter.

7.5.1 Forced Exponential and Logistic: Population Growth with Harvest

First, let us look at the input file **exp-harvest-inp.csv**. It has two new parameters that act as disturbance modes: **absolute** or fixed (H_a) and **relative** or proportional (H_p), as explained earlier in this chapter.

```
Label,Value,Units,Description
t0,0.00,d,time zero
tf,10.00,d,time final
dt,0.010,d,time step
tw,0.10,d,time to write
r,0.2,1/d,intrinsic growth rate
Hp,-0.3,1/d,proportional harvest
Ha,0,Ind/d,fixed harvest
X0,1.00,Ind,initial population
digX,4,None,significant digits for output file
```

Here, the fixed harvest is set to zero and the relative is equal to −0.3, which exceeds the growth rate. Define a new model function `exp.f` using `expon.forced` (explained at the end of this chapter) and run `sim` in the following manner:

```
exp.f <- list(f=expon.forced)
param <- list(plab="Ha",pval=c(0,-0.1,-0.2))
t.X <- sim(exp.f,"chp7/exp-harvest-inp.csv",param)
```

which would vary the absolute rate. We obtain the results shown in Figure 7.17. Even when $H_a = 0$, the population declines because H_p exceeds r. Both absolute and relative modes produce a population decline. Adding $H_a > 0$ drives the population quickly to extinction.

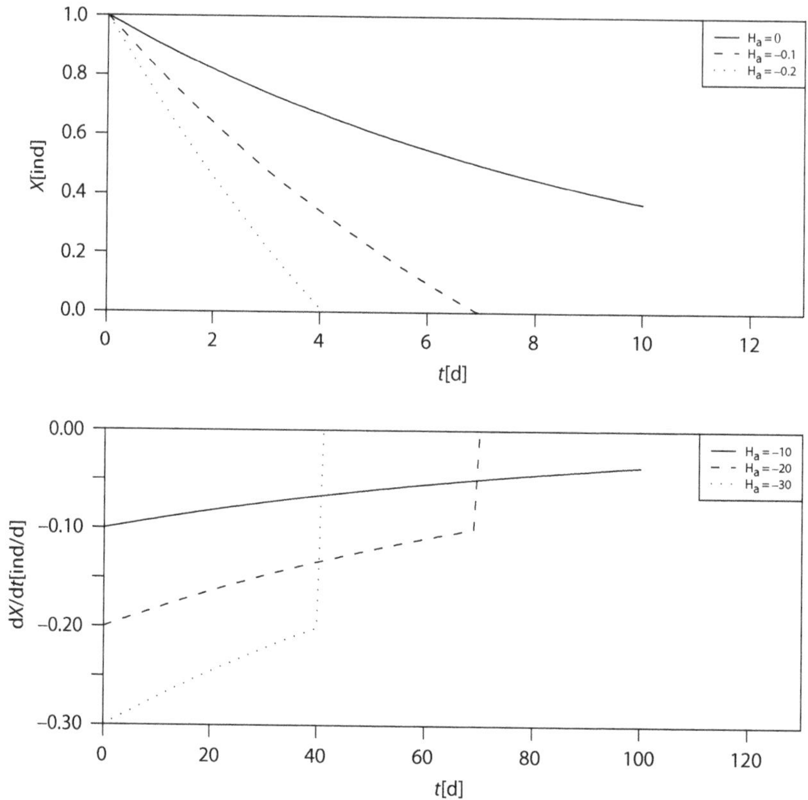

FIGURE 7.17 Exponential growth model leads to a decline if harvest rate exceeds growth rate.

Exercise 7.7

Assume that there is no absolute harvesting. Calculate the value of the relative harvest rate at which the population switches from growth to decline. Generate a graph with two curves, one for growth and one for decline, to support your answer.

Exercise 7.8

Generate a plot of X versus t for several values of absolute harvest rate and a plot of the `Diff end` metric as a function of intensity. Discuss sensitivity analysis results in %/%.

Now, let us look at the input file **logis-harvest-inp.csv**.

```
Label,Value,Units,Description
t0,0.00,Year,time zero
tf,50.00,Year,time final
dt,0.01,Year,time step
tw,0.1,Year,time to write
r,0.2,1/Year,intrinsic growth rate
K,80.0,Ind,carrying capacity
Hp,0,1/d,proportional harvest
Ha,-3,Ind/d,fixed harvest
X0,1.00,Ind,initial population
digX,4,None,significant digits for output file
```

Here, the absolute harvest is set to –3, and the relative is equal to –0, which produces two equilibria. Define the model and the parameter set, and run sim in the following manner:

```
logis.f <- list(f=logistic.forced)
param <- list(plab="X0",pval=seq(10,90,20))
t.X <- sim(logis.f,"chp7/logis-harvest-inp.csv",param)
```

which would vary the initial conditions to obtain the results shown in Figure 7.7.

7.5.2 Forced Exponential: Bioaccumulation

Let us consider a forced exponential to model uptake, U, and depuration, k, of a toxicant in an organism. Recall that the end value is the stable equilibrium (steady state) and represents bioaccumulation in units of mg/kg. Let us use the input file **exp-bioaccu-inp.csv**. It has a parameter, U_a, set at 0.5 that acts as the absolute disturbance mode and has a parameter, U_p, set at 0 for relative or proportional. We will use only U_a and leave U_p at zero to assume that uptake is independent of concentration.

```
Label,Value,Units,Description
t0,0.00,d,time zero
tf,10.00,d,time final
dt,0.10,d,time step
tw,1.00,d,time to write
k,-0.2,1/d,depuration rate coeff
Up,0.0,1/d,proportional uptake
Ua,0.5,(mg/Kg)/d,fixed uptake
X0,1.0,mg/Kg,initial concentration
digX,4,None,significant digits for output file
```

Run sim in the following manner:

```
param <- list(plab="Ua",pval=c(0,0.2,0.5,1))
t.X <- sim(exp.f,"chp7/exp-bioaccu-inp.csv",param)
```

which would vary the uptake rate. We obtain the results shown in Figure 7.5. For $U_a = 0.5$, the bioaccumulation is near 2.5 as expected because steady state is $U_a/k = 0.5/0.2 = 2.5$ mg/kg.

Exercise 7.9

What is the effect of having $U_a = 0$? Why does the concentration flatline at $U_a = 0.2$? Why does an uptake rate above 0.2 cause bioaccumulation?

7.5.3 Environmental Variability and Stress

We will simulate a population modeled exponentially with parameter r that varies periodically. For this, we use the input file **exp-var-inp.csv** that contains the following:

```
Label,Value,Units,Description
t0,0.00,d,time zero
tf,50.00,d,time final
```

```
dt,0.010,d,time step
tw,0.10,d,time to write
r0,0.01,1/d,mean intrinsic growth rate
rd,-0.005,(1/d)/d,drift intrinsic growth rate
ra,0.2,1/d,amplitude variation of intrinsic growth rate
T,2,d,period of variation of intrinsic growth rate
X0,1.00,Ind,initial population
digX,4,None,significant digits for output file
```

Define a model `exp.v` using f = expon.var, specify the values of `rd` in the manner indicated in Figure 7.13, and apply the `sim` function:

```
exp.v <- list(f=expon.var)
param <- list(plab="rd",pval=c(0,-0.005,-0.01))
t.X <- sim(exp.v,"chp7/exp-var-inp.csv",param)
```

To obtain the same result shown in that figure, a synthetic view of the dynamics is given in Figure 7.18 depicting rate versus population.

Exercise 7.10

Change the amplitude (0.1, 0.2, 0.3) and period (2, 4, 6) of the oscillations of *r*. Generate a plot of *X* versus *t* for these values of amplitude and period. Discuss the results.

Exercise 7.11

Compare a gradually deteriorating intrinsic rate (drift) to the periodic changes of the intrinsic rate. (1) Force the periodic variation to be the dominant factor; to do this, make the amplitude of the periodic variation much larger than the gradual deterioration. (2) Force the drift to be the dominant factor; to do this, make the amplitude of the periodic variation much smaller than the drift. (3) Compare the results. Note: Use the root mean square (RMS) (analog to the variance) as a measure for the comparisons. (4) Consider the effect of the period by asking what happens if the oscillatory stress changes faster than the intrinsic rate of population growth (stress has a short period compared to doubling time). What if the oscillatory stress is slow compared to the intrinsic rate of the population growth? (Stress has a long period compared to doubling time.) Hint: Prepare

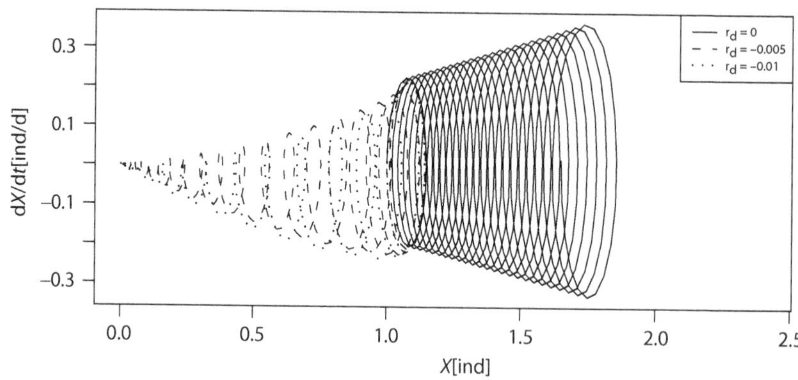

FIGURE 7.18 Rate versus population for the exponential growth model with periodic variation of the intrinsic growth rate.

a table comparing the RMS values for dominance of drift or periodic stress, with a sentence on the meaning of the table; use a graph comparing fast and slow periodic stress, and include a sentence on the meaning of the graph.

Similarly, we can simulate a population modeled logistically with parameter K that varies periodically. For this, we use the function `logistic.var` and the input file `logis-var-inp.csv`. Apply the function by varying r in the manner indicated in Figure 7.14 and obtain the same result shown in that figure:

```
logis.v <- list(f=logistic.var)
param <- list(plab="Kd",pval=c(0,-0.5,-0.6))
t.X <- sim(logis.v,"chp7/logis-var-inp.csv",param)
```

A synthetic view of the dynamics is given in Figure 7.19, depicting the rate versus population.

7.5.4 Sudden and Repetitive Disturbances

We will simulate a population modeled exponentially subject to a sudden disturbance. For this purpose, we use the input file **exp-sud-inp.csv**, which contains:

```
Label,Value,Units,Description
t0,0.00,d,time zero
tf,100.00,d,time final
dt,0.01,d,time step
tw,0.1,d,time to write
r,0.02,1/d,intrinsic growth rate
T,10.0,d,period forcing function
Hp,0.0,1/d,relative intensity forcing function
Ha,-10.0,Ind/d,absolute intensity forcing function
x0,100.00,Ind,initial population
digX,4,None,significant digits for output file
```

Build two functions, `expon.z` and `expon.g`, to handle discontinuities (these functions are explained at the end of the chapter), Now use them to define a model `exp.sud` with three components in the list: `f, z, and g`. The remaining two lines are the same; build a parameter set by varying H_a in the manner indicated in Figure 7.15 and apply the sim function.

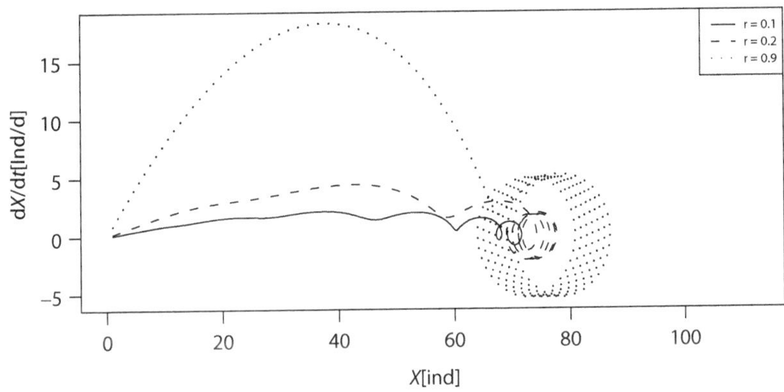

FIGURE 7.19 Rate versus population for the logistic growth model with periodic variation of the carrying capacity.

```
exp.sud <- list(f=expon,z=expon.z,g=expon.g)
param <- list(plab="Ha",pval=c(-10,-20,-30))
t.X <- simd(exp.sud,file="chp7/exp-sud-inp.csv",param)
```

We obtain the same result as shown in that figure.

Exercise 7.12

You want to apply a pesticide to control an insect pest only at discrete points in time, that is, every *n* days. Each pesticide application will kill 70% d^{-1} of the population. What is the maximum period between the applications, which will "control" (i.e., achieve zero or negative long-term growth) the pest population growing at $r = 0.2$ d^{-1} in between applications? Generate graphs and a brief sentence discussing your answer. Hint: Use proportional (relative) mode, with $r = 0.2$ and intensity $H_p = -0.7$. Start with 2 days for the application period, then try extending this period to more days, for example, 3, 4, 5, ..., to see how large you can make it before the population has long-term positive growth. Tip: As you extend the application period, you may need to extend the simulation time tf.

Finally, we simulate a population modeled logistically subject to a sudden disturbance. For this, we use the list `logis.sud` built from `logistic`, `logistic.z`, and `logistic.g` (the latter two are explained at the end of the chapter) and the input file **logis-sud-inp.csv**, and apply the function by varying *r* in the manner indicated in Figure 7.16:

```
logis.sud <- list(f=logistic,z=logistic.z,g=logistic.g)
param <- list(plab="r",pval=c(0.1,0.2))
out.f <- simd(logis.sud,file="chp7/logis-sud-inp.csv",param)
```

We obtain the same result as shown in that figure.

7.5.5 Build Your Own

Simulate a logistic subject to episodic disturbances. Instead of a fixed period, *T*, make it random according to the exponential distribution with rate $p = 1/T$. That is, model the disturbance as a Poisson process (Hanson and Tuckwell, 1981). Proceed in four steps as discussed in Chapter 6.

Step 1: Edit the function `logistic.z` to have a new name, **logistic.rand.z**, with parameter p[3] used to calculate the Poisson rate. Define `tz` as an array, then initialize a counter, `i`, and find the possible next `tz`. In order to The term "possible" is used we need to check that it does not exceed the final time. Only if it does not, we assign it to `tz[i]` and repeat incrementing the counter every time.

```
logistic.rand.z <- function(t,p,x){
 # random poisson impulse occurrence
 rate <- 1/p[3];tz <- array()
 i=1;tnext <- t[1]+rexp(1,rate)
 while(tnext < t[length(t)]) {
  tz[i] <- tnext
  tnext <- tz[i] + rexp(1,rate)
  i<-i+1
 }
 return(round(tz,0))
}
```

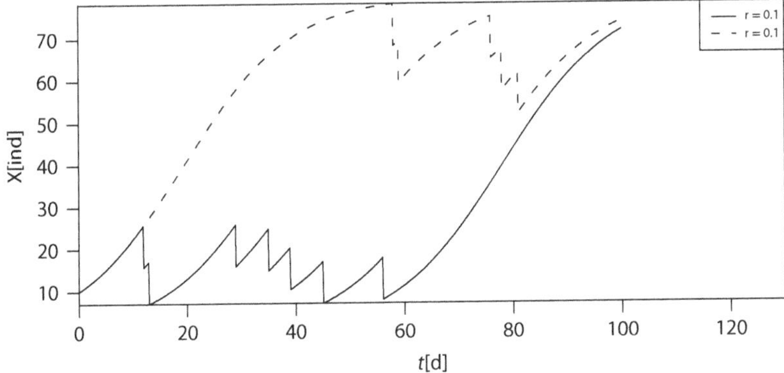

FIGURE 7.20 Two realizations of the logistic growth model subject to a disturbance modeled as a Poisson process.

Step 2: Run this script or save it in a file **myown.R** and source it.
Step 3: There is no need to modify the input file since `p[3]` will be used to calculate the rate.
Step 4: We use the new model and the input file in `sim` in the following manner to make one run:

```
logis.rand <- list(f=logistic,z=logistic.rand.z,g=logistic.g)
t.X <- simd(logis.rand,file="chp7/logis-sud-inp.csv")
```

Only with the model and input file called by sim do we get the nominal run. Now, make two runs for parameter r but keep it fixed at $r = 0.1$ so that the result is two realizations of the random process:

```
param <- list(plab="r",pval=c(0.1,0.1))
out.f <- simd(logis.rand,file="chp7/logis-sud-inp.csv",param)
```

You should have obtained a set of graphs including the one shown in Figure 7.20.

Exercise 7.13

Implement a simulation of a modified logistic from Chapter 6, $dX/dt = rX^a\left(1-\frac{X}{K}\right)$, subject to random Poisson sudden disturbances as above.

You can now appreciate the advantage of defining a model as a list. We can have several modules for f, z, and g and mix and match them to form a diversity of models.

7.6 SEEM FUNCTIONS EXPLAINED

7.6.1 Model Functions: Forced Expon and Logistic

The `expon.forced` and `logistic.forced` functions are simple modifications of the `expon` and `logistic` functions. Parameters H_p and H_a are passed as components of parameter p.

```
expon.forced <- function(t,p,x){
  r <- p[1]; Hp <- p[2]; Ha <- p[3]
  dx <- r*x + Hp*x + Ha
  return(dx)
  }

logistic.forced <- function(t,p,x){
  r<-p[1]; K<-p[2]; Hp <- p[3]; Ha <- p[4]
  dx <- r*x*(1-x/K) + Hp*x + Ha
  return(dx)
  }
```

7.6.2 Model Functions: Parameter Variability

The `expon.var` and `logistic.var` functions are also direct modifications of `expon` and `logistic`, and these are the first cases we have encountered where we include the time variation in the model definition.

```
expon.var <- function(t,p,x){
  # dx = (rm + rd*t + ra*sin(2pit/T))*x
  r <- p[1] + p[2]*t + p[3]*sin(2*pi*t/p[4])
  dx <- r*x
  return(dx)
  }

logistic.var <- function(t,p,x){
  # K(t) = K0 + Kd*t + Ka*sin(2pi t/T)
  r<- p[1];K <- p[2] + p[3]*t + p[4]*sin(2*pi*t/p[5])
  dx <- r*x*(1-x/K)
  return(dx)
  }
```

7.6.3 Model Functions: Sudden Disturbances

The `expon.z` and `expon.g` functions accommodate for discontinuities. As described later, in simd, we have opted to define two separate functions: one continuous (`f`) and one discontinuous (`.g`). In addition, the discontinuous functions receive as an argument the set of discontinuities defined in the function `.z`. This strategy will allow more generality when we work with different sets of discontinuities.

First `expon.z` is simply a set of fixed times with period T given by p[2] and `expon.g` detects these times and applies the disturbance intensities:

```
expon.z <- function(t,p,x){
  # periodic impulse train
  tz <- seq(t[1],t[length(t)],p[2])
  return(tz)
}
```

```
expon.g <- function(t,p,x,tz){
  # linear disturbance at impulse times
  u <- 0
  for(j in 1:length(tz))
  if(x>0) if(abs(t - tz[j])<0.00001) u <- p[3]*x + p[4]
  return(u)
}
```

Similarly, we define `logistic.z` and `logistic.g` taking into account that the position in array p of period and disturbance intensities is different in the logistic model.

```
logistic.z <- function(t,p,x){
  # periodic impulse train
  tz <- seq(t[1],t[length(t)],p[3])
  return(tz)
}

logistic.g <- function(t,p,x,tz){
  # linear disturbance at impulse times
  u <- 0
  for(j in 1:length(tz))
 if(x>0) if(abs(t - tz[j])<0.00001) u <- p[4]*x + p[5]
 return(u)
}
```

7.6.4 Function simd

This function is a modification of `sim` that allows for discontinuities. The only changes with respect to `sim` are in the parameter loop, where just before the call to integration, we calculate `tz`, invoking function `z`; we call a new Runge–Kutta function RK4D (see next section), which has new arguments `g` and `tz`; and the rate calculation for one-dimensional (1-D) systems uses function `g`.

```
  # discontinuity times set
 tz <- model$z(t,p,x)
 # integration
 out <- RK4D(X0, t, model$f, p, dt, model$g, tz)
 # rename x part of out for simpler use
 if(NS>1) for(k in 1:NS) X[,i,k] <- out[,k]
 # calculate rate for one-state systems
 else{
  X[,i,1] <- out
  for(j in 1:nt)
  dX.dt[j,i] <- model$f(t[j],p,X[j,i,1]) +
                model$g(t[j],p,X[j,i,1],tz)
 }
```

7.6.5 Function RK4D

We have opted to rewrite the RK4 function as a new RK4D function that explicitly includes g in the calculation in the following manner:

```
for(i 3in 1:nt){
 X[i,] <- Xt
 for(j in 1:ns){
   k1<- f(tt,p,Xt)
   k2<- f(tt+dt/2,p,(Xt+k1*dt/2))
   k3<- f(tt+dt/2,p,(Xt+k2*dt/2))
   k4<- f(tt+dt,p,(Xt+k3*dt))
   kavg <- (dt/6)*(k1+ 2*k2+ 2*k3+ k4)
   tt <- tt + dt; Xt <- Xt + kavg
   Xt <- Xt + g(tt,p,Xt,tz)
 }
}# end of run
```

It is possible to define the functions differently. This strategy separates the change of rate due to discontinuities from the continuous part, thus allowing more generality when we work with different sets of discontinuities.

8 Sensitivity Analysis, Response Surfaces, and Scenarios

8.1 MULTIPARAMETER SENSITIVITY ANALYSIS

As explained in Chapter 5, we are interested in evaluating the changes in the solution of a model because of the changes in the values of the parameters. Although the concept of sensitivity analysis is relatively simple when we vary only one parameter at a time, some difficulties arise when analyzing the effect of various parameters jointly. Among these difficulties, we can list the following: the need for numerous runs, interaction among the parameters, different amount of variability among parameters, and difficulty in defining the appropriate sensitivity metric. In this chapter, we address some of these issues.

Sensitivity analysis allows us to understand the effect of uncertain parameters. Statistical tools are useful in sensitivity analysis, particularly the analysis of variance (ANOVA) for factorial selection of parameter values and multiple regression analysis for random sampling of parameter values (see Ford, 1999, pp. 373–380; Swartzman and Kaluzny, 1987, pp. 217–251; Kirchner, 1992, pp. 37–41).

8.1.1 Brief Review of Multiple Regression

We will extend the simple linear regression model to more than one independent variable, X. That is, we now have several (m) independent or factor variables, X_i, that influence one dependent or response variable, Y.

As before, we develop a linear least square estimator of Y:

$$\hat{Y} = b_0 + b_1X_1 + b_2X_2 + \cdots + b_mX_m \tag{8.1}$$

This is the equation of a plane with **intercept** b_0 and **coefficients** b_k, $k = 1,\ldots,m$. Note that we now have $m + 1$ regression parameters to estimate; these are m coefficients and one intercept. For each observation, i, we have a set of data points, y_i, x_{ki}, and we have the **estimated** value of Y at the specific points, x_{ki}:

$$\hat{y}_i = b_0 + b_1x_{1i} + b_2x_{2i} + \cdots + b_mx_{mi} \tag{8.2}$$

This is an extension of the simple regression problem reviewed in Chapter 3. We need to find a set of b_k, $k = 1,\ldots,m$, that minimizes the square error, that is, find the set b_k such that

$$\min_{b_k} q = \min_{b_k} \sum_{i=1}^{n} e_i^2 = \min_{b_k} \sum_{i=1}^{n} (y_i - \hat{y}_i)^2 \tag{8.3}$$

The solution is found in a similar way to that explained in Chapter 3. We can also use the matrix method as will be explained in Chapter 9. We will give the solution for two variables, X_k, where $k = 1,2$, for simplicity. These expressions extend to more than two variables. Without going into all the details, the solution is given by the following means and covariances:

$$b_0 = \bar{Y} - b_1\bar{X}_1 - b_2\bar{X}_2$$
$$b_1 = \frac{(s_{X_2})^2 s_{cov(X_1,Y)} - s_{cov(X_2,Y)} s_{cov(X_1,X_2)}}{(s_{X_1})^2(s_{X_2})^2 - (s_{cov(X_1,X_2)})^2} \tag{8.4}$$
$$b_2 = \frac{(s_{X_1})^2 s_{cov(X_2,Y)} - s_{cov(X_1,Y)} s_{cov(X_1,X_2)}}{(s_{X_1})^2(s_{X_2})^2 - (s_{cov(X_1,X_2)})^2}$$

Note that the covariance between X_1 and X_2 plays an important role here. If this covariance were to be zero, that is, if X_1 and X_2 are uncorrelated, then the solution simplifies to

$$b_0 = \bar{Y} - b_1\bar{X}_1 - b_2\bar{X}_2$$
$$b_1 = \frac{s_{cov(X_1,Y)}}{s^2_{X_1}} \tag{8.5}$$
$$b_2 = \frac{s_{cov(X_2,Y)}}{s^2_{X_2}}$$

where b_k are the **partial** or **marginal** coefficients, that is, the rate of change of Y with one of the X_k while holding all the other X_j constants. We can see that, in the special case of uncorrelated X_1 and X_2, this marginal change of Y with X_1 depends only on the covariance of X_1 and Y and on the variance of X_1. However, when X_1 and X_2 are correlated, this marginal coefficient is affected by the variance of the other variable, the covariance of the other variable with Y, and the covariance of X_1 and X_2.

Using the definition of correlation coefficient, we have

$$r_{(X,Y)} = \frac{s_{cov(X,Y)}}{s_X s_Y} \tag{8.6}$$

We can rewrite the solution of the two uncorrelated variables as

$$b_0 = \bar{Y} - b_1\bar{X}_1 - b_2\bar{X}_2$$
$$b_1 = \frac{r_{(X_1,Y)} s_Y}{s_{X_1}} \tag{8.7}$$
$$b_2 = \frac{r_{(X_2,Y)} s_Y}{s_{X_2}}$$

We will not review the details of ANOVA methods here because of the lack of space. For the purpose of this chapter, the main concept we need to understand is that we can statistically detect significant differences of a response variable with respect to several factors by examining the p value of the statistical test conducted, for example, an F-test in parametric analysis or a chi-square test in nonparametric analysis.

8.1.2 Factorial Design

Assume that we have a nominal or best estimate value for each parameter. Then, each parameter is perturbed or varied above and below this nominal value to obtain a total of m levels or values for the parameter (Kirchner, 1992; Swartzman and Kaluzny, 1987). As we explained in Chapter 5 by denoting v as the percent change of the parameter value, we can have the following: $m = 2$, two levels, nominal and nominal $+ v$; $m = 3$, three levels, nominal $- v$, nominal, and nominal $+v$; $m = 5$, five levels, nominal $-2v$, nominal $-v$, nominal, nominal $+v$, and nominal $+2v$; and more levels by assuming increasing multiples of v.

In analyzing the effect of several parameters, using the same levels and same v for each parameter standardizes the design to facilitate comparison across parameters. The differences in variability among parameters imply using v according to various multiples of the standard deviation.

For example, assume that in a logistical model we have K with mean 100 and r with mean 0.2 and we do a factorial design with $v = 50\%$ for both parameters. So, $v = 50$ for K and $v = 0.1$ for r. This is equivalent to assuming a standard deviation $\sigma = 25\%$ of the mean for both parameters. The three levels for K are 50, 100, and 150, and those for r are 0.1, 0.2, and 0.3. Therefore, we have the combinations as given in Table 8.1.

Once we define values of the parameters, the next step in factorial analysis is combining the values of all the parameters. There are two different factorial approaches: **full** or **complete** and **fractional**. In full factorial, we combine all values of all parameters, whereas in fractional, we use a subset of all the combinations.

We will focus on full factorial and combine the values of all parameters by holding one value constant for one parameter and varying the other parameters across all the other values. Repeat this for all parameters and values and assume p parameters and m levels for each parameter. The number of runs is denoted by n. For example, for $p = 2$, the number of runs is $n = m \times m = m^2$; for $p = 3$, the number of runs is $n = m \times m \times m = m^3$. In general, then, the number of runs is simply given as

$$n = m^p \tag{8.8}$$

As an example, let us use the logistic and values given in Table 8.1. In this case, $p = 2$ parameters and $m = 3$ levels as given above, so the number of runs required is $n = 3^2 = 9$, as shown in Table 8.2. The last row lists the parameters that have been perturbed in each run. The runs include the following interactions:

- One baseline or nominal run, all nominal values (run 5)
- Four runs for the individual effect (one-way) of each parameter (runs 2 and 8 for K and runs 4 and 6 for r)
- Four runs for the pairwise combinations (two-way) of the two parameters (runs 1, 3, 7, and 9)

Simulating the logistic with the values given in Table 8.2, we obtain the results shown in Figure 8.1.

TABLE 8.1
Three Levels for Two Parameters

Parameter	Nominal	σ	v	v (%)	Nominal $-v$	Nominal	Nominal $+v$
r	0.2	0.05	0.1	50	0.1	0.2	0.3
K	100	25	50	50	50	100	150

TABLE 8.2
Full Factorial for Two Parameters, Three Levels

Run	1	2	3	4	5	6	7	8	9
r	0.1	0.2	0.3	0.1	0.2	0.3	0.1	0.2	0.3
K	50	50	50	100	100	100	150	150	150
Perturbed	*r, K*	*K*	*r, K*	*r*	*nom*	*r*	*r, K*	*K*	*r, K*

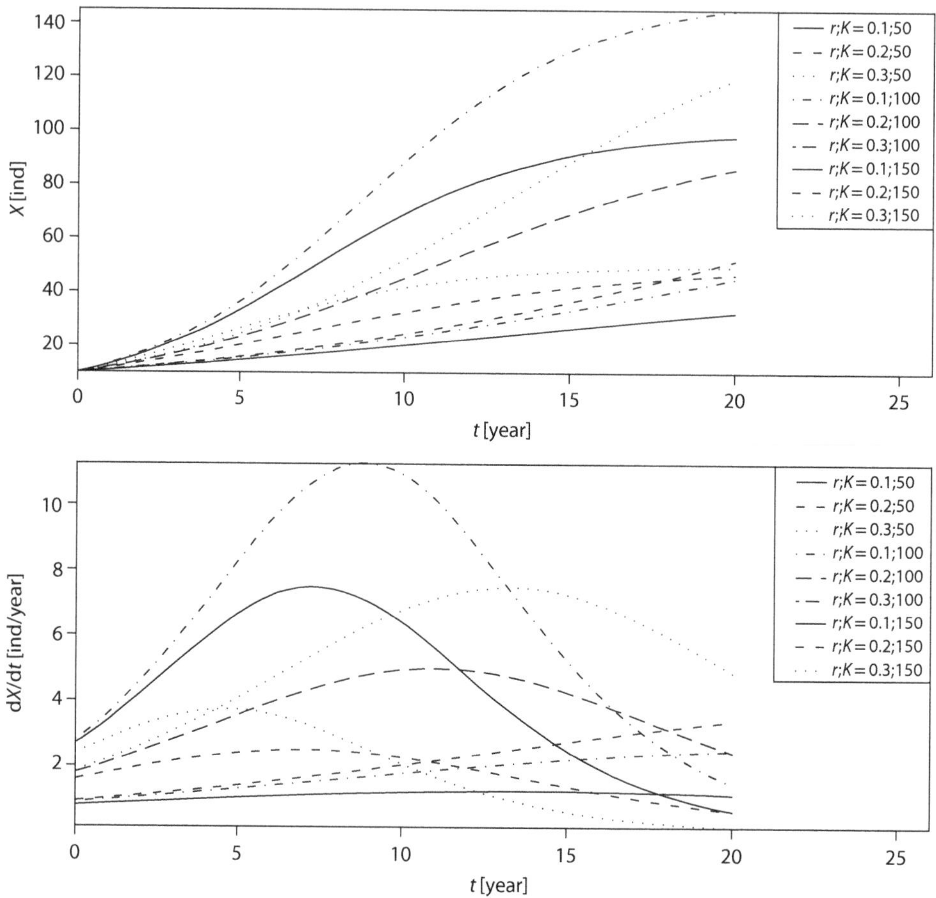

FIGURE 8.1 Logistic simulated with the combination of parameters r and K.

Exercise 8.1

Assume that K and r have the same means as given above and that the variances of K and r are 16 and 0.0016, respectively. Write a table of simulation runs for the two parameters (r, K) and the two levels (nominal and nominal + v).

Our next task is to calculate the effect of the parameter variation by using the results from the simulation runs. When we calculate the metric using the methods described in Chapter 5, we obtain the results for the 9 runs and 6 metrics as shown in Table 8.3. This information allows us to draw the interaction graphs as shown in Figure 8.2 (where the metrics are in absolute numbers) and Figure 8.3 (where the metrics and the parameters are relative to the nominal or baseline and converted to percent).

The upper two rows of the panels indicate that there is no interaction for the first four metrics, since the lines do not cross; however, metrics 5 and 6 (last row) are related to the difference with respect to the nominal and show interaction.

There is a set of runs for which a parameter is varied and another set for which it is not. The average effect of varying a parameter is the average of the output of the first set minus the average of the output of the second set. Some metric of the model output is denoted by $g_j(\,)$ as given in Chapter 5. For example, the average effect of the parameter variation on this metric is calculated by averaging

TABLE 8.3
Metrics from the Runs

Run	r	K	Diff Max	Diff Min	Diff End	Avg Diff	RMS Diff	Max Diff
1	0.1	50	32.44	10	32.44	20.58	31.18	53.41
2	0.2	50	46.59	10	46.59	30.75	20.23	39.26
3	0.3	50	49.51	10	49.51	36.64	16.35	36.34
4	0.1	100	45.09	10	45.09	24.72	25.77	40.77
5	0.2	100	85.85	10	85.85	46.26	0	0
6	0.3	100	97.82	10	97.82	61.95	17.58	25.24
7	0.1	150	51.82	10	51.82	26.63	23.23	35
8	0.2	150	119.4	10	119.4	57.05	15.17	33.55
9	0.3	150	145	10	145	83.12	43.4	61.42

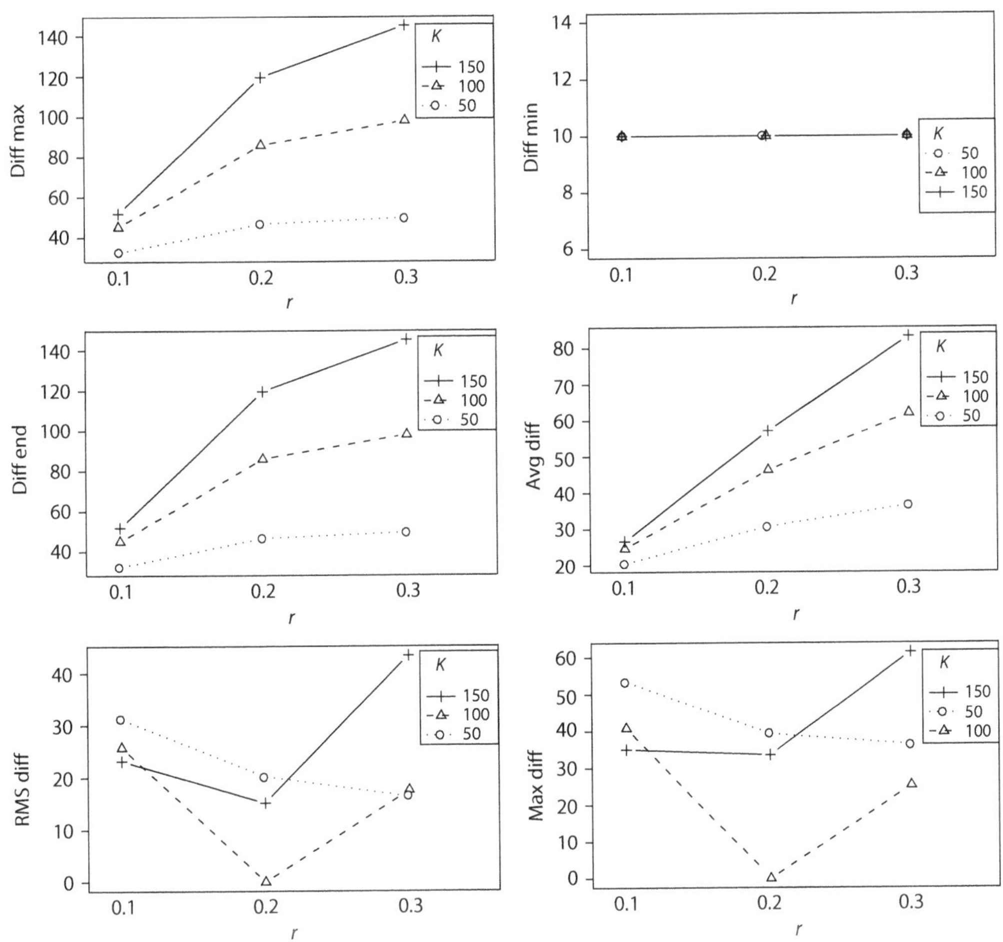

FIGURE 8.2 Interaction graphs of the variation of metrics as we change r and K.

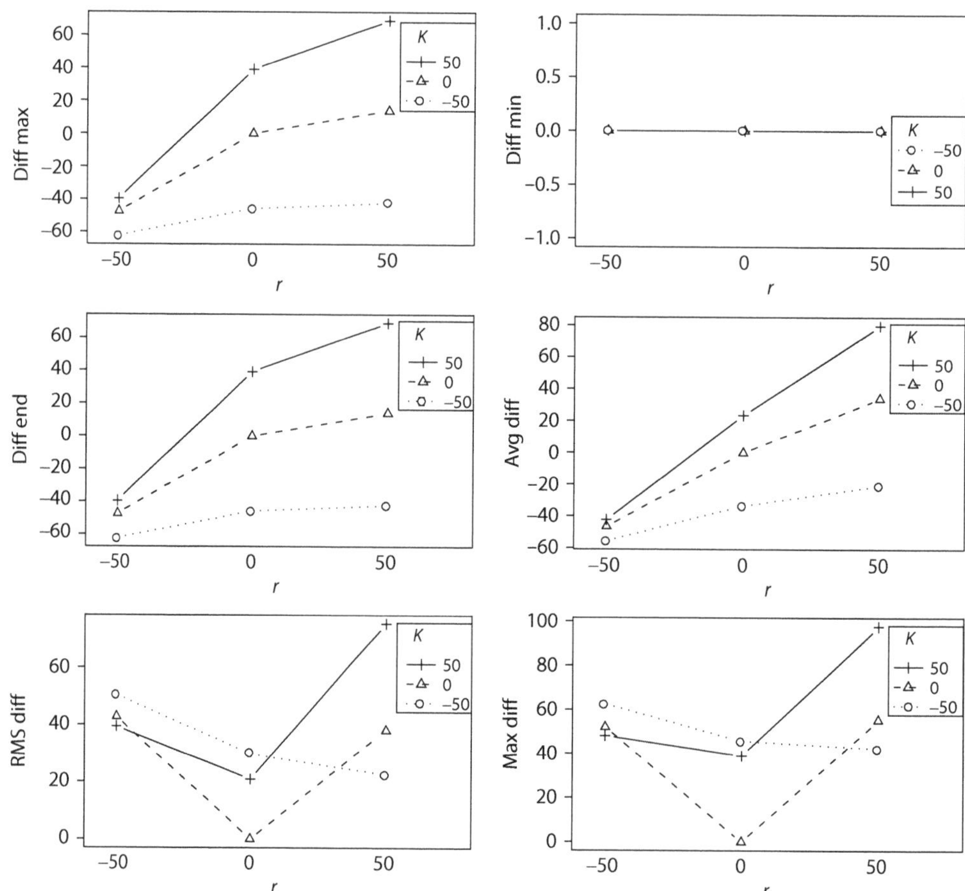

FIGURE 8.3 Interaction graphs of the variation of metrics as we change r and K using relative values in percent.

the metric values obtained in all runs involved in the perturbation of parameter r (runs 1, 4, 7, 3, 6, and 9) minus the ones not involved in the perturbation (runs 2, 5, and 8). Using a simplified notation, we have

$$\overline{g_j}(r) = \text{avg}\left(\left|g_j(r)\right|, \left|g_j(r,K)\right|\right) - \text{avg}\left(\left|g_j(\text{nom})\right|, \left|g_j(K)\right|\right) \tag{8.9}$$

where avg denotes the average of the quantities in parentheses. Here, the pair $g_j(r), g_j(r,K)$ denotes the value of the metric for runs with r perturbed and the value of the metric for runs with both r, K perturbed, respectively; the pair $g_j(\text{nom}), g_j(K)$ denotes the value of the metric for both r, K nominal and K perturbed, respectively. We take the absolute value (converting negative to positive) of the metrics, so that we can average the effects.

Using the information from Table 8.3, we obtain

$$\overline{g_j}(r) = \frac{184.4}{6} - \frac{16.85}{3} = 30.73 - 5.62 = 25.11 \tag{8.10}$$

The interaction between two parameters, say r and K, can be calculated by the difference of the effect of one parameter, r, under two different conditions of the other parameter, K, perturbed or nominal. For example, let us calculate the effect of r in interaction with K. For K perturbed, the effect of r is obtained by the effects of runs 1, 3, 7, and 9 (i.e., when both r and K are perturbed), whereas when K is nominal, the effect of r is obtained by the effects of runs 4 and 6 (i.e., when r is perturbed and K is not). Then, the effect of r on the interaction with K is given by

$$\overline{g_j}(r,K)=\frac{\sum_{\text{runs}1,3,7,9} g_j(r,K)}{4}-\frac{\sum_{\text{runs}4,6} g_j(r)}{2}=\frac{123}{4}-\frac{61.42}{2}=0.04 \tag{8.11}$$

Exercise 8.2

Using the results for the diff max metric given above, calculate the effect of K and the effect of K under interaction with r.

Exercise 8.3

Using tables resulting from Exercise 8.1, determine the formulas for the effect of r and the effect of the interaction r, K.

Exercise 8.4

Consider two levels ($m = 2$, nominal and nominal $+v$) and three parameters ($p = 3$) denoted by P_1, P_2, and P_3. Determine the formulas for effects and interaction.

Exercise 8.5

Consider the Michaelis–Menten–Monod model. How many parameters does it include? Write down the parameter names. How many runs are required for a full-factorial sensitivity analysis at two levels? Write a table of these simulation runs. Write a formula for the average effect of perturbing K_h.

Note that the number of runs, n, can increase dramatically with the number of parameters, even for a few levels. For three levels and three parameters, $n = 3^3 = 27$; for four levels and four parameters, $n = 3^4 = 81$; and for five levels and five parameters, $n = 3^5 = 243$. Therefore, it is interesting to find ways of reducing the number of runs and still examine all the parameters. This is accomplished by fractional factorial design; but the downside of reducing the runs is that you also confound the effect of some of the pairwise or two-way interactions. We do not discuss this topic further in this book and assume that we can perform a full factorial. This is made more and more feasible now days by increased computer performance.

8.1.3 Latin Hypercube Sampling

Assume that you have p parameters and that they have different distributions and ranges (Kirchner, 1992; Swartzman and Kaluzny, 1987). The range of the parameter R is denoted by $[R_{min}, R_{max}]$. For n simulation runs, subdivide the range into n quantiles. Each quantile is from R_i to R_{i+1} in such a way that the probability is the same for all intervals.

$\text{Prob}[R_i, R_{i+1}] = p$ for all intervals and the sum of the probabilities for all intervals is 1. This requires making the intervals larger in areas of low probability density. This allows us to sample the entire range. An example is shown in Figure 8.4, where a normal variable with mean 10 and $\sigma = 5$ has been divided into 10 intervals of equal probability of 0.1. The quantiles are $-\infty$, 3.60, 5.79, 7.38,

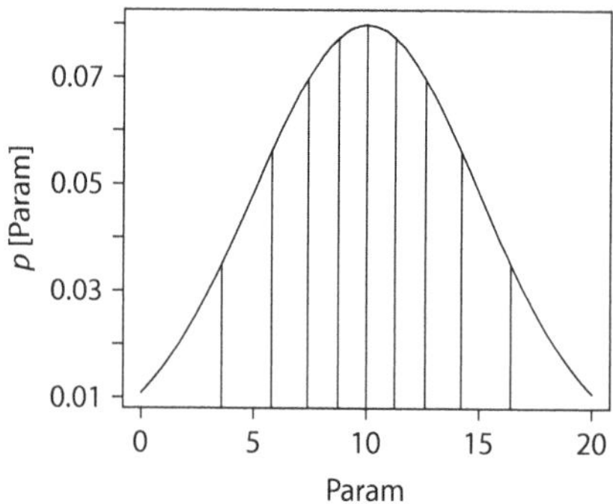

FIGURE 8.4 Dividing a normal distribution in 10 intervals for stratified sampling.

8.73, 10.00, 11.26, 12.62, 14.21, 16.41, and ∞. For a uniform distribution over $[R_{min}, R_{max}]$, the size of each interval is the same.

To select the runs, pick a quantile for each parameter (randomly without replacement and assume uniform distribution) and then pick a parameter value from the quantile (randomly according to the parameter distribution).

For example, consider two parameters, P1 and P2. Assume uniform distribution for all the parameters with [0,1] range. To produce 10 runs by Latin hypercubes, consider the following:

- Divide [0,1] into 10 intervals of size 1/10.
- For P1: Create a **discrete** or **mass** uniform distribution for 10 intervals with values 1,…,10 all with prob = 1/10. This is variable *I*.
- For P2: Create a **discrete or mass** uniform distribution for 10 intervals with values 1,…,10 all with prob = 1/10. This is variable *J*.
- Run 1:
  - Sample variable *I* randomly and use the selected value *i* (from 1 to 10) for selecting the interval of P1.
  - Sample the selected interval from the uniform density between $[R_i, R_{(i+1)}]$.
  - Repeat for P2, with variable *J*, to select interval *j* and the values within this interval.
- Run 2:
  - Adjust the pdf of *I* to remove *i* and rescale the probabilities to 1/9, and sample for the new interval *i*.
  - Repeat for the pdf of *J* and get a new interval *j*.
  - Sample the selected interval *i* for a value of P1.
  - Repeat for *j* and P2.
- Run 3:
  - Repeat the above by adjusting the pdf to 1/8.
- And so on for runs 4, 5, …, 9.
- Run 10:
  - Select the remaining intervals.
  - Sample P1 and P2 from this interval.

This analysis is conducted using multiple regression in such a way that the metric of the output is a linear function of the parameters:

$$g_j(X) = b_0 + b_1 P_1 + b_2 P_2 \tag{8.12}$$

Now the values of the marginal coefficients b_1 and b_2 quantify the sensitivity of the metric to the corresponding parameter. To compare one parameter with another, it is necessary to standardize the values of the parameters, that is to say, for each value subtract the mean and divide by the standard deviation. In this way all the parameters have mean 0 and $\sigma = 1$.

8.2 RESPONSE SURFACE AND SCENARIOS

In principle, sensitivity analysis can be performed for any type of parameter, either endogenous (intrinsic or internal) or exogenous (driving variables or disturbances). In practice, it is convenient to recognize that sensitivity analysis to the driving variables is a **response surface** for the model. This allows us to ask the question: How does the model respond to variations in the external forcing?

A common application of models is **scenario–consequence** analysis, that is, **what-if** analysis or **what** will happen to the model output **if** the external forcing was to correspond to this set of external conditions, **what if** the external conditions change to this other set of values, and so on.

Scenarios are usually selected as a limited set (you do not want to execute too many runs) of **plausible** combinations of values of driving or external variables. It is convenient to select scenarios based on factorial or sampling methods, but it is also common to select the values of interest based on the questions asked of the model as well as expert judgment.

If multiple runs are stored as a surface response, then exploring scenarios is just a lookup procedure, unless the response surface is not very smooth.

As an example, consider an exponential model (parameterized by r) with sudden and repetitive disturbances of intensity, U, and period, T; we would proceed in the following manner:

- Deal with uncertainties in the intrinsic parameters (only one in this case, r) by sensitivity analysis.
- Deal with uncertainties in the disturbance regime by response surface (which is a sensitivity analysis on the two external parameters, U and T).
- Explore a set of scenarios of disturbance regime and look up the metric values that correspond to the selected scenarios of disturbance.
- Draw conclusions.

Being realistic, for complex models it is difficult to have a smooth and complete response surface, so you may proceed directly to perform scenario–consequence analysis and make only the runs required by the setup of this analysis, which should be fewer than the number required for a response surface.

8.3 COMPUTER SESSION

8.3.1 Factorial Design

Several `seem` functions are useful in sensitivity and response function analysis. The function `mway-fact`, explained at the end of this chapter, helps determine a full factorial design. For example, using values of r and K from Table 8.1, we can design the factorial of three levels at 50% for both parameters:

```
# gives nom lab, v in % and levels
pnom= c(0.2,100); plab= c("r","K"); perc.v= 50; levels=3
param <- mway.fact(pnom,plab,perc.v,levels)
```

We obtain nine runs or pairs of values of r and K corresponding to the combinations given in Table 8.2. Some of the components of the resulting `param` are as follows:

```
> param
$plab
[1] "r" "K"
$pval
      [,1] [,2]
 [1,]  0.1   50
 [2,]  0.2   50
 [3,]  0.3   50
 [4,]  0.1  100
 [5,]  0.2  100
 [6,]  0.3  100
 [7,]  0.1  150
 [8,]  0.2  150
 [9,]  0.3  150
$pnom
[1]   0.2 100.0
$fact
[1] TRUE
$pv
     [,1] [,2]
[1,]  0.1   50
[2,]  0.2  100
[3,]  0.3  150
>
```

Note that `$pval` contains a set of combined parameter values. We use the list `param` to call the logistic function as we have done before in Chapter 6 and apply the function `msens`, which will be explained at the end of this chapter.

```
logis <- list(f=logistic)
t.X <- sim(logis,"chp8/logistic-100-inp csv",param,pdfout=T,lab
out="sens-fact-")
s.y <- msens(t.X$output, param, "chp8/logis-100-sens-plots-fact-
out",pdfout=T,resp=F)
```

In similar fashion to the function `sens` in Chapter 5, arguments to `msens` are the simulation output, the `param` list obtained earlier, and the name of the file. Optionally, here we will set `pdfout = T` to obtain a PDF file and use the character string `lab.out = "sens-fact-"` to identify the output files.

From the simulation, we get the results already displayed in Figure 8.1. Look at the PDF file **chp8/logistic-100-inp.csv-sens-fact-rK-out.pdf**. Its first page corresponds to the interaction plots of Figure 8.2, whereas the second page corresponds to the graphs of Figure 8.3.

Look at the output `s.y` on the console. The first component `$Ma` has the table of values already shown in Table 8.3, whereas the component `$Mp` is the same as in the table but converted to its relative values in percent.

```
> s.y
$Ma
    r   K  Diff Max  Diff Min  Diff End  Avg Diff  RMS Diff  Max Diff
1 0.1  50     32.44        10     32.44     20.58     31.18     53.41
2 0.2  50     46.59        10     46.59     30.75     20.23     39.26
3 0.3  50     49.51        10     49.51     36.64     16.35     36.34
```

```
4 0.1 100      45.09        10       45.09       24.72       25.77       40.77
5 0.2 100      85.85        10       85.85       46.26        0.00        0.00
6 0.3 100      97.82        10       97.82       61.95       17.58       25.24
7 0.1 150      51.82        10       51.82       26.63       23.23       35.00
8 0.2 150     119.40        10      119.40       57.05       15.17       33.55
9 0.3 150     145.00        10      145.00       83.12       43.40       61.42

$Mp
    r    K  Diff Max  Diff Min  Diff End  Avg Diff  RMS Diff  Max Diff
1 -50  -50    -62.21         0    -62.21    -55.50     50.44     62.50
2   0  -50    -45.73         0    -45.73    -33.52     29.74     45.73
3  50  -50    -42.33         0    -42.33    -20.80     22.62     42.33
4 -50    0    -47.48         0    -47.48    -46.57     42.81     51.96
5   0    0      0.00         0      0.00      0.00      0.00      0.00
6  50    0     13.94         0     13.94     33.92     38.08     55.29
7 -50   50    -39.64         0    -39.64    -42.43     39.43     48.03
8   0   50     39.08         0     39.08     23.32     20.89     39.08
9  50   50     68.90         0     68.90     79.69     75.27     97.62
```

Included in the output are the p values of the ANOVA and nonparametric test to examine whether there is a significant difference in the response of the metrics. We do not discuss this component in this book. It suffices to say that all metrics except Diff min vary with both parameters r and K.

The last components of `s.y` contain the average effect. The component `$Eff` has the average effect of all the metrics of each parameter and the component `$Eff.Int` has the effect of the parameter in interaction with the other.

```
$Eff
    Diff Max  Diff Min  Diff End  Avg Diff  RMS Diff  Max Diff
r     17.480         0    17.480    27.538    27.894    31.350
K     29.175         0    29.175    15.713    12.769    20.133

$Eff.Int
    Diff Max  Diff Min  Diff End  Avg Diff  RMS Diff  Max Diff
r     22.560         0    22.560     9.358     6.494     8.997
K     10.865         0    10.865    21.184    21.620    20.214
```

Exercise 8.6

Consider the forced exponential with an uptake rate. The depuration rate k has mean 0.1 and the uptake rate U has mean 20. Run a full factorial design of the three levels for each parameter using a 50% variation. Obtain a plot for the runs, perform sensitivity analysis, obtain the plots, and discuss the results.

8.3.2 Stratified Random Sampling: Latin Hypercube Sampling

We now perform sensitivity analysis by sampling in the more complicated case of more than one parameter that varies at a time. Examine the function `mway.samp`. This **seem** function performs Latin hypercubes to obtain n numbers by sampling using a uniform distribution from a min to a max value determined from the desired percent variation. The function varies several parameters simultaneously. In the following, we use 10 runs, r with mean 0.2 and K with mean 100, and a percent variation of 50%.

```
pnom= c(0.2,100); plab= c("r","K"); perc.v= 50; runs=10
param <- mway.samp(pnom,plab,perc.v,runs)
```

Please note that the random selection could lead to a variety of combinations of r and K, for example, low r and high K, low r and low K, and so on. Some components of `param` are as follows:

```
$plab
[1] "r" "K"
$pval
       [,1]    [,2]
 [1,] 0.20 100.00
 [2,] 0.18  57.30
 [3,] 0.18 133.94
 [4,] 0.14  91.19
 [5,] 0.30 122.69
 [6,] 0.26 101.48
 [7,] 0.12  76.41
 [8,] 0.24  82.16
 [9,] 0.12 117.55
[10,] 0.22 147.54
[11,] 0.25  65.99
$pnom
[1]   0.2 100.0
$fact
[1] FALSE
>
```

Note that the values of r and K cover the range of variation uniformly. Use these values to make the runs and then call the function `msens` to calculate the metrics and produce the plots in PDF files that will be named **logistic-100-inp.csv-sens-samp-rK-out.pdf** and **logis-100-sens-plots-samp-out.pdf**. Both are in folder **chp8**.

```
t.X <- sim(logis,"chp8/logistic-100-inp.csv",param,pdfout=T,lab.out="sens-
samp-")
s.y <- msens(t.X$output, param, "chp8/logis-100-sens-plots-samp-
out",pdfout=T)
```

We reproduce here some of the contents of the output PDF files. In Figure 8.5 from **logistic-100-inp.csv-sens-samp-rK-out.pdf**, we see the results of the simulation runs. In Figure 8.6 from **logis-100-sens-plots-samp-out.pdf**, we see the scatter plots of the calculated versus the estimated metrics together with the line of slope 1 corresponding to the conditions where both values are the same. The estimated metrics were obtained with the coefficients of multiple regression, one for each parameter, given below in component `reg.slope.R2` of `s.y`.

```
> s.y
$Ma
     r      K  Diff Max  Diff Min  Diff End  Avg Diff  RMS Diff  Max Diff
1 0.20 100.00     85.85        10     85.85     46.26      0.00      0.00
2 0.18  57.30     50.74        10     50.74     31.44     18.81     35.11
3 0.18 133.94    100.10        10    100.10     48.28      4.88     14.25
4 0.14  91.19     61.05        10     61.05     32.29     16.61     25.13
5 0.30 122.69    119.40        10     19.40     71.95     29.39     40.37
6 0.26 101.48     96.60        10     96.60     57.20     12.36     17.45
7 0.12  76.41     47.68        10     47.68     26.69     23.55     38.17
8 0.24  82.16     77.55        10     77.55     47.08      3.65      8.30
9 0.12 117.55     59.50        10     59.50     30.22     18.88     27.91
```

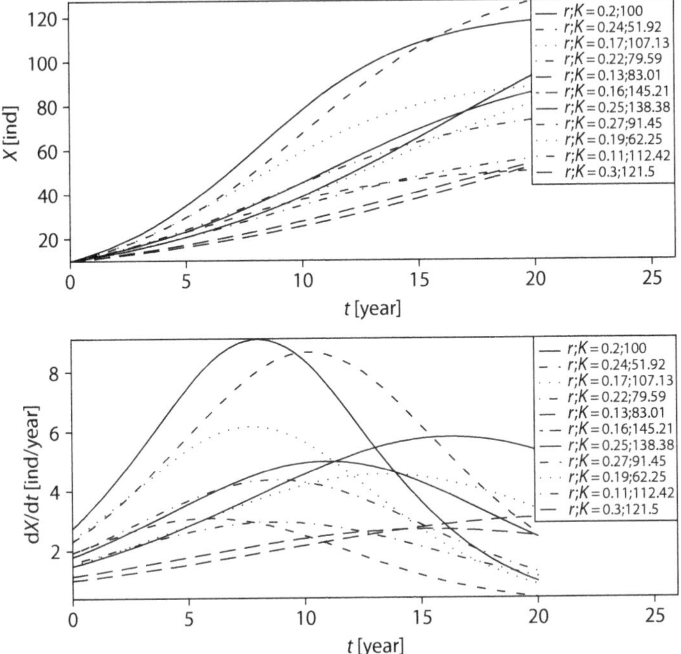

FIGURE 8.5 Simulation runs using stratified sampling.

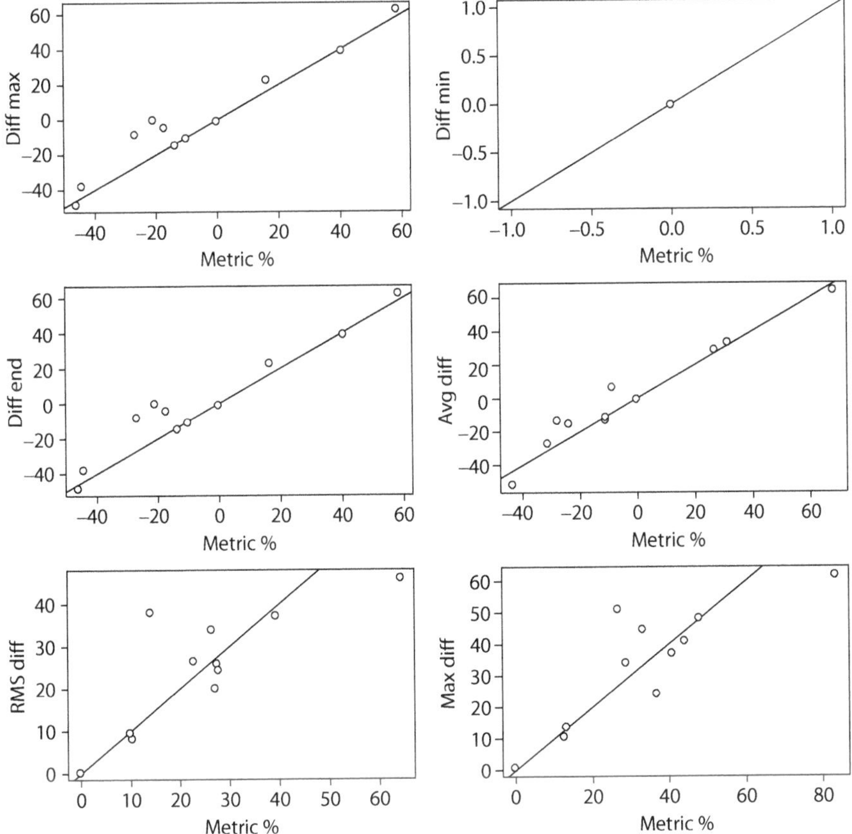

FIGURE 8.6 Sensitivity to r and K for a logistic. Scatter plots and identity lines.

```
10   0.22   147.54    126.20   10   126.20    62.55    21.15    40.35
11   0.25    65.99     63.59   10    63.59    41.55     9.17    22.26

$Mp
      r        K  Diff Max  Diff Min  Diff End  Avg Diff  RMS Diff  Max Diff
1     0     0.00      0.00         0      0.00      0.00      0.00      0.00
2   -10   -42.70    -40.90         0    -40.90    -32.03     28.53     40.90
3   -10    33.94     16.60         0     16.60      4.38      6.53     16.60
4   -30    -8.81    -28.89         0    -28.89    -30.20     27.96     34.30
5    50    22.69     39.08         0     39.08     55.54     55.29     74.36
6    30     1.48     12.52         0     12.52     23.65     24.78     34.26
7   -40   -23.59    -44.46         0    -44.46    -42.30     38.70     47.26
8    20   -17.84     -9.67         0     -9.67      1.78      8.27     12.81
9   -40    17.55    -30.69         0    -30.69    -34.66     32.44     39.89
10   10    47.54     47.00         0     47.00     35.22     31.25     47.00
11   25   -34.01    -25.93         0    -25.93    -10.18     12.62     25.93

$reg.slope.R2
     Diff Max   Diff Min   Diff End   Avg Diff   RMS Diff   Max Diff
r       0.177          0      0.177      0.246      0.229      0.278
K       0.232          0      0.232      0.160      0.079      0.152
R2      0.918        NaN      0.918      0.965      0.909      0.931
>
```

The relative values of the parameter are standardized as long as we use the same percent variation for all the parameters. However, the `msens` function standardizes the relative values of the parameters by subtracting the mean and dividing the result by the standard deviation.

Exercise 8.7

Perform Latin hypercubes sampling using nominal $k = 0.01$ and $u = 20$. Use 50% variation. Perform sensitivity analysis based on this sampling. Discuss the coefficients of the regression and R^2. Recall that the end value is bioaccumulation.

We can use `msens` for more than two parameters. For example, take the logistic and include the initial condition, X_0. Then, we have

```
# give nom lab, v in % and runs
pnom= c(0.2,100,1); plab= c("r","K","X0"); perc.v= 50; runs=30
param <- mway.samp(pnom,plab,perc.v,runs)
t.X <- sim(logis,"chp8/logistic-100-inp csv",param,pdfout=T,lab. out="sens-
samp-3param-")
s.y <- msens(t.X$output, param, "chp8/logis-100-sens-plots-samp-3param-
out",pdfout=T)
```

This yields results for regression that includes all three parameters,

```
$reg.slope.R2
     Diff Max   Diff Min   Diff End   Avg Diff   RMS Diff   Max Diff
r       0.667       0.00      0.667      0.663      0.497      0.691
K       0.102       0.00      0.102      0.068     -0.109     -0.141
X0      0.206       0.29      0.206      0.291      0.222      0.291
R2      0.943       1.00      0.943      0.910      0.789      0.821
```

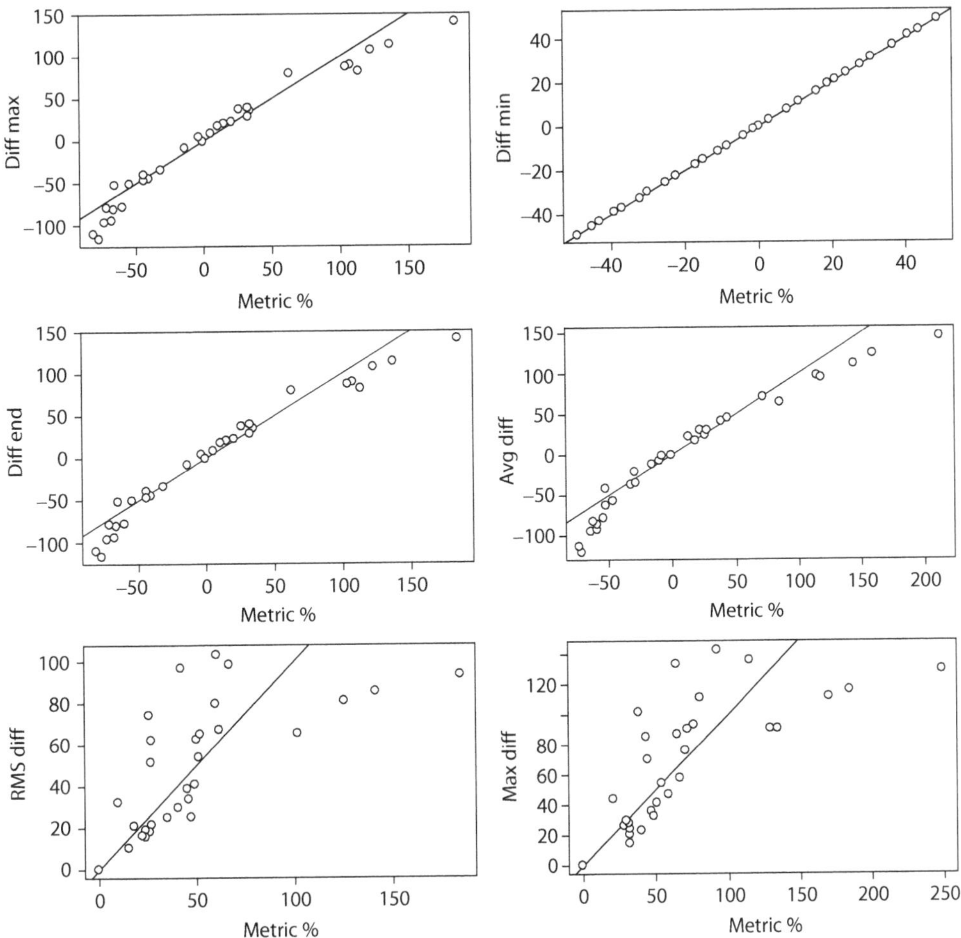

FIGURE 8.7 Scatter plots of the variation of metrics and corresponding regression lines as we change r and K.

and the graphical output of the file **chp8/logis-100-sens-plots-samp-3param-out.pdf** as shown in Figure 8.7.

8.3.3 Response Surface

We now expand the range of parameter variation to look at the response surface to the parameters. For various illustrations, we now base the examples on the function m3decay given in Chapter 6. First, we fix the value of K_{max} and combine K_h and X_0 in a **full factorial** mode using nine levels:

```
m3decay<- list(f=M3)
pnom= c(30,100); plab= c("Kh","X0"); perc.v= 10; levels=9
param <- mway.fact(pnom,plab,perc.v,levels))
```

Therefore, for a pair of parameters, there will be $9 \times 9 = 81$ simulation runs. If we were to explore all combinations of the three parameters, the simulation consists of $9 \times 9 \times 9 = 729$ runs.

```
t.X <- sim(m3decay,"chp8/m3decay-inp.csv",param,pdfout=T,lab.out="resp-
fact-")
s.y <- msens(t.X$output, param, "chp8/m3decay-sens-plots-resp-
fact-out",pdfout=T,resp=T)
```

We have used metrics of 5-day runs to evaluate the response functions as we change the values of the parameters. In this case, we select only four metrics since we are not comparing with the nominal. These are max, min, end, and avg. To visualize the results, we can plot all four metrics as a function of two of the parameters as contained in the file **chp8/m3decay-sens-plots-resp-fact-out.pdf** (Figure 8.8). The function produces a grid of cells with varying grey shades according to the metric's value, and overlapping contour maps for all the metrics for a pair of parameters. Since we have three parameters, we would have to do two other plots for each metric: one for the pair K_h, K_{max} (Figure 8.9) and another for the pair K_{max}, X_0 (Figure 8.10).

These can be generated using

```
pnom= c(40,30); plab= c("Kmax","Kh"); perc.v= 10; levels=9
param <- mway.fact(pnom,plab,perc.v,levels)
```

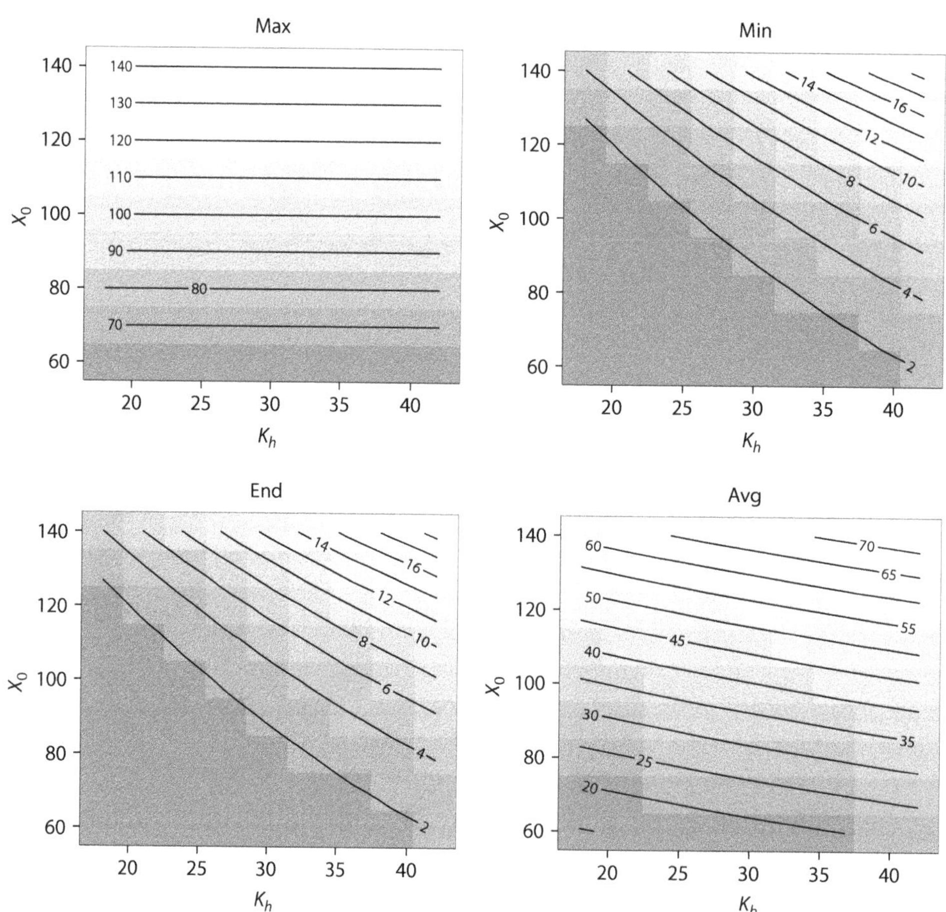

FIGURE 8.8 Response surfaces of M3 decay model for pair K_h, X_0.

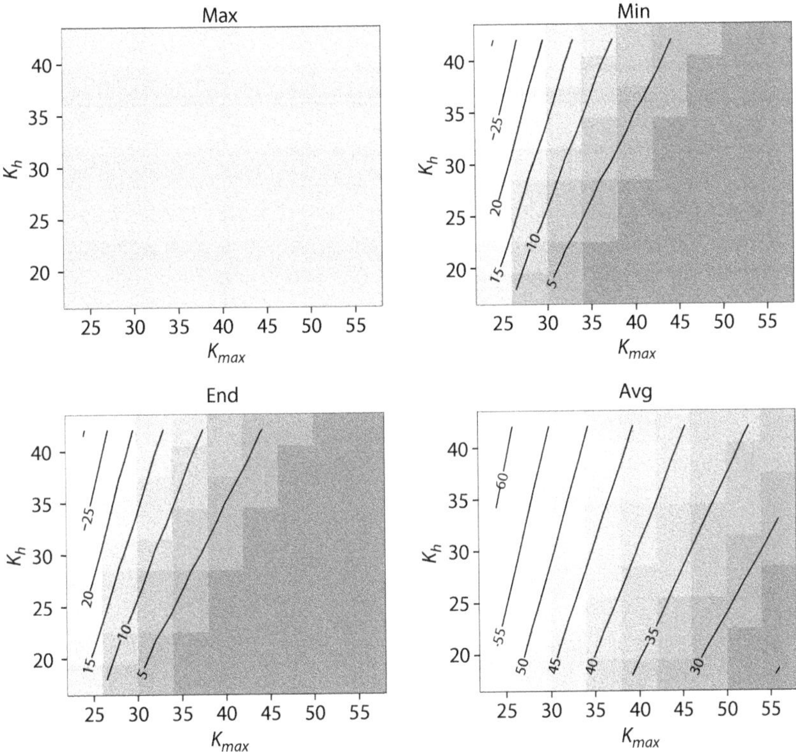

FIGURE 8.9 Response surfaces of M3 decay model for parameter pair K_{max}, K_h.

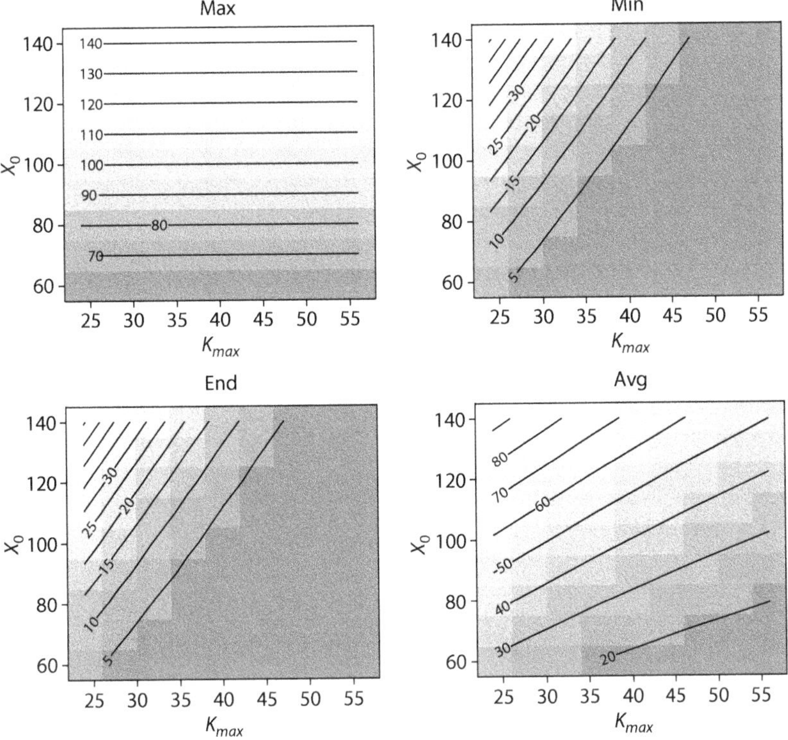

FIGURE 8.10 Response surfaces of M3 decay model for parameter pair K_{max}, X_0.

and

```
pnom= c(40,100); plab= c("Kmax","X0"); perc.v= 10; levels=9
param <- mway.fact(pnom,plab,perc.v,levels)
```

Selecting the set of values for each parameter is not trivial, since we need to avoid a combination of values that would produce invalid results, for example, a very high K_{max} with a very low K_h and X_0 would tend to produce an excessive value for the rate at low concentrations.

These figures provide the general behavior of response of the model to the three parameters. Now we can think of sensitivity as a local gradient with respect to the parameter evaluated at a nominal point on these surfaces.

To examine the values obtained to make these maps, we can look at the list `$x` where these results are stored. This list has six elements, one for each metric. The first four of each set of six correspond to the metrics visualized in the figures. The other two refer to the difference with respect to nominal. For example, for the pair K_h, X_0

```
> s.y$x
[[1]]
   60 70 80 90 100 110 120 130 140
18 60 70 80 90 100 110 120 130 140
21 60 70 80 90 100 110 120 130 140
24 60 70 80 90 100 110 120 130 140
27 60 70 80 90 100 110 120 130 140
30 60 70 80 90 100 110 120 130 140
33 60 70 80 90 100 110 120 130 140
36 60 70 80 90 100 110 120 130 140
39 60 70 80 90 100 110 120 130 140
42 60 70 80 90 100 110 120 130 140

[[2]]
     60    70    80    90   100   110   120   130   140
18 0.000 0.000 0.000 0.000 0.000 0.000 0.000 0.000 0.000
21 0.000 0.000 0.000 0.000 0.000 0.000 0.000 0.000 0.001
24 0.000 0.000 0.000 0.000 0.000 0.001 0.001 0.002 0.003
27 0.000 0.000 0.001 0.001 0.001 0.002 0.004 0.006 0.009
30 0.001 0.001 0.002 0.003 0.005 0.007 0.011 0.016 0.024
33 0.002 0.003 0.005 0.007 0.011 0.017 0.025 0.036 0.053
36 0.005 0.007 0.011 0.016 0.024 0.035 0.050 0.072 0.102
39 0.010 0.015 0.022 0.032 0.046 0.065 0.091 0.128 0.177
42 0.018 0.027 0.039 0.056 0.079 0.110 0.152 0.209 0.285
```

This is a fairly large object; therefore, we can address specific entries of `$x`. For example, for metric 4, Avg, this is `$x[[4]]`.

```
[[4]]
      60     70     80     90    100    110    120    130    140
18 7.428  9.532 11.884 14.483 17.329 20.424 23.765 27.353 31.191
21 7.874 10.052 12.478 15.151 18.072 21.240 24.657 28.319 32.232
24 8.319 10.572 13.072 15.820 18.814 22.057 25.548 29.284 33.269
27 8.764 11.092 13.666 16.488 19.557 22.874 26.438 30.249 34.309
30 9.210 11.611 14.260 17.156 20.299 23.691 27.328 31.214 35.346
```

```
33  9.656 12.131 14.854 17.824 21.041 24.507 28.219 32.179 36.386
36 10.100 12.650 15.447 18.491 21.782 25.322 29.108 33.140 37.420
39 10.546 13.169 16.040 19.158 22.523 26.135 29.994 34.100 38.452
42 10.990 13.688 16.632 19.823 23.262 26.947 30.878 35.056 39.477
```

Exercise 8.8

We have a container with 2 kg (recall 1 kg = 1000 g = 10^6 mg) of a chemical on board a small boat floating in a 10,000 liter pond. We are worried that the container is about to suddenly leak and spill a fraction of its content into the pond before we can stop it. We want a model to tell us the concentration of chemical 5 days after the potential spill.

Assumptions: There is no water circulation in the pond (we do not want to do hydrodynamical modeling just yet), the compound dissolves completely and instantaneously in water (we do not want to do complex fate modeling just yet), and the concentration is the same everywhere in the pond, that is, homogeneous (we do not want to do transport–dispersion–mixing modeling just yet).

Therefore, the external forcing could be represented by the amount of chemical at $t = 0$ or the initial condition of X because we are trying to model the recovery after a spill of this chemical. Using the response surfaces from the last example, if the bottle is about half full right after the spill, how much dissolved chemical do you expect to remain in the pond 10 days after the spill? Assume $K_{max} = 40$ and $K_h = 30$. How would it be different from the result for $K_{max} = 30$ and $K_h = 40$?

8.3.4 Scenarios

Consider the situation of Exercise 8.8. Assume that we are uncertain of the amount spilled and of the K_{max} and K_h values at which the chemical will degrade in the pond. We want to run the scenarios with three factors: (1) amount spilled, (2) K_{max} of kinetics for the decay, and (3) K_h of kinetics.

Using a variation of 20% and nominal values of half the bottle, $K_{max} = 40$, and $K_h = 30$, we obtain $X_0 = 80$, 100, and 120; $K_{max} = 32$, 40, and 48; and $K_h = 24$,30, and 36. So, the full factorial implies $3 \times 3 \times 3 = 27$ runs.

```
pnom= c(40,30,100); plab= c("Kmax","Kh","X0"); perc.v= 20; levels=3
param <- mway.fact(pnom,plab,perc.v,levels)
```

The following set of scenarios is obtained by combining these values of the three factors:

```
> param
$plab
[1] "Kmax" "Kh""X0"
$pval
      [,1] [,2] [,3]
 [1,]  32    24   80
 [2,]  40    24   80
 [3,]  48    24   80
 [4,]  32    30   80
 [5,]  40    30   80
 [6,]  48    30   80
 [7,]  32    36   80
 [8,]  40    36   80
 [9,]  48    36   80
[10,]  32    24  100
[11,]  40    24  100
```

```
[12,] 48   24  100
[13,] 32   30  100
[14,] 40   30  100
[15,] 48   30  100
[16,] 32   36  100
[17,] 40   36  100
[18,] 48   36  100
[19,] 32   24 120
[20,] 40   24 120
[21,] 48   24 120
[22,] 32   30 120
[23,] 40   30 120
[24,] 48   30 120
[25,] 32   36 120
[26,] 40   36 120
[27,] 48   36 120
```

We run simulations and the msens function using this set of values:

```
t.X <- sim(m3decay,"chp8/m3decay-inp.csv",param,pdfout=T,lab.out="resp-fact-")
s.y <- msens(t.X$output, param, "chp8/m3decay-sens-plots-scen-fact-Kmax-
Kh-X0-out",pdfout=T,resp=T)
```

Examine the numerical results:

```
> s.y
$Ma
   Kmax  Kh   X0  Diff Max  Diff Min  Diff End  Avg Diff  RMS Diff  Max Diff
1    32  24   80        80      2.56      2.56     31.79      9.94     20.00
2    40  24   80        80      0.53      0.53     25.83     15.11     20.00
3    48  24   80        80      0.10      0.10     21.70     19.63     26.10
4    32  30   80        80      4.75      4.75     34.21      8.86     20.00
5    40  30   80        80      1.40      1.40     28.04     13.12     20.00
6    48  30   80        80      0.38      0.38     23.62     17.56     23.14
7    32  36   80        80      7.12      7.12     36.39      8.53     20.00
8    40  36   80        80      2.65      2.65     30.13     11.50     20.00
9    48  36   80        80      0.92      0.92     25.50     15.71     21.13
10   32  24  100       100      6.31      6.31     45.34      6.05      8.22
11   40  24  100       100      1.46      1.46     37.10      2.91      4.12
12   48  24  100       100      0.29      0.29     31.19      9.46     13.77
13   32  30  100       100      9.77      9.77     48.01      8.91     11.80
14   40  30  100       100      3.21      3.21     39.74      0.00      0.00
15   48  30  100       100      0.91      0.91     33.56      6.77      9.66
16   32  36  100       100     13.12     13.12     50.39     11.50     15.08
17   40  36  100       100      5.36      5.36     42.18      2.67      3.64
18   48  36  100       100      1.94      1.94     35.84      4.28      6.11
19   32  24  120       120     13.12     13.12     60.61     21.30     24.84
20   40  24  120       120      3.67      3.67     50.16     12.21     20.00
21   48  24  120       120      0.78      0.78     42.29      8.35     20.00
22   32  30  120       120     17.60     17.60     63.34     23.87     27.41
23   40  30  120       120      6.67      6.67     53.09     14.32     20.00
24   48  30  120       120      2.05      2.05     45.06      8.74     20.00
25   32  36  120       120     21.65     21.65     65.77     26.26     30.03
```

```
26    40    36 120    120    9.88    9.88    55.77    16.56    20.03
27    48    36 120    120    3.85    3.85    47.68     9.94    20.00
```

The sens graphics output is in the file **chp8/m3decay-sens-plots-scen-fact-*Kmax*-*Kh*-X0-out.pdf**. It has three pages, one for each pair of parameters. As an example, Figure 8.11 shows one of these pages. We have used `sens` to plot the response maps for each pair calculated for the average of the third parameter. Therefore, we have only three maps for each metric. A more complete set of maps would be obtained by plotting the response of each pair for each value of the third parameter. This would yield 27 maps for each metric.

We can explore the answers to questions like what is the worst situation (largest concentration) after 5 days? We can simply find the maximum of the sixth column of `s.y$Ma` because this column corresponds to the end value

```
> max(s.y$Ma[,6])
[1] 21.65
> which(s.y$Ma[,6]==max(s.y$Ma[,6]))
[1] 25
```

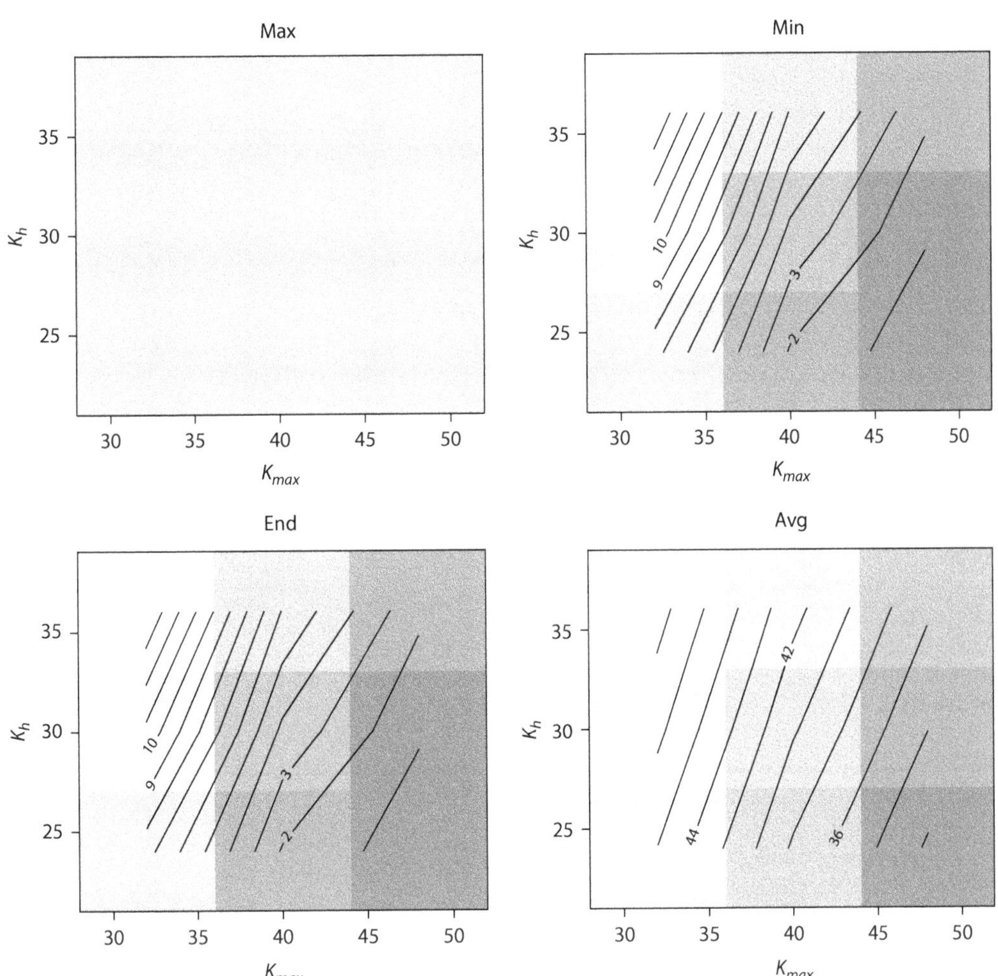

FIGURE 8.11 Results for 27 scenarios of M3 decay model selected by factorial of three factors, K_{max}, K_h, and X_0. Plots shown are for the pair K_{max}, K_h averaged over all values of X_0.

Therefore, the worst situation is to have 21.65 mg/liter after 5 days, and this occurs for scenario 25, consisting of $K_{max} = 32$, $K_h = 36$, and $X_0 = 120$.

Exercise 8.9

Using the results in `s.y$Ma`, if we know that half of the bottle contents are gone, and we're still uncertain about the chemistry of the compound, what would be the worst situation or largest concentration of the chemical you would expect to remain after 5 days? What is the best situation or lowest concentration after 5 days?

8.4 SEEM FUNCTIONS EXPLAINED

8.4.1 Function to Perform Factorial Analysis

This function `mway.fact` uses two R functions, `gl` and `factor`.

```
mway.fact <- function(pnom,plab,perc.v,levels){

npar <- length(pnom)
pv <- matrix(nrow=levels,ncol=npar); pf <- list(); pfr <- list()
n <- which(seq(3,17,2)==levels)
for(i in 1:npar){
  v <- perc.v*pnom[i]/100
  pv[,i] <- seq(pnom[i]-n*v, pnom[i]+n*v,v)
  pf[[i]] <- factor(pv[,i])
}
pval <- matrix(nrow=levels^npar,ncol=npar)
for(i in 1:npar){
  pfr[[i]] <- gl(levels,levels^(i-1),levels^npar,labels=pv[,i])
  pval[,i] <- pv[c(pfr[[i]]),i]
}
return(list(plab=plab, pval=pval, pnom=pnom, fact=T, pv=pv, pf=pf,
pfr=pfr))
}
```

8.4.2 Function to Perform Latin Hypercubes

Function `mway.samp` is limited to the uniform distribution but can be extended to sample other distributions.

```
mway.samp <- function(pnom,plab,perc.v,runs){

# latin hypercubes from a uniform distribution
# how many parameters
np <- length(pnom)

# intervals
n <- runs
# param.range has min, max for each parameter
```

```
min <- pnom-perc.v*pnom/100; max <- pnom+perc.v*pnom/100

# nominal are means
# divide interval in n equal parts
 pequal <- (max - min)/n
 interval <- matrix(ncol=np, nrow=n)
# sample without replacement
 for(i in 1:np) interval[,i] <- sample(n, replace=F)
# sample from each interval
 pval <- interval
 for (i in 1:n) {
    for(k in 1:np){
        j <- interval[i,k]
        minj <- pequal[k]*(j-1) + min[k]
        maxj <- pequal[k]*j + min[k]
        pval[i,k] <- runif(1, minj, maxj)
    }
 }
 pval=rbind(pnom, pval,deparse.level=0)

 return(list(plab=plab, pval=round(pval,2), pnom=pnom, fact=F))
```

8.4.3 Function to Perform Multiparameter Sensitivity Analysis

Function msens is much more complicated than the previous two, and it is not reproduced or discussed fully for the sake of space. Only some highlights are given here.

The pairwise combinations for more than two parameters employ the function **factorial** and the function tapply. The latter applies the mean to the response using the parameters as factors and produces tables x and xp, which are the basis to calculate the effects and maps of the response functions.

```
# pairwise selection for more than 2 param
 nc <- factorial(npar)/(factorial(2)*factorial(npar-2))
 combo <- matrix(nrow=nc,ncol=2);k=1
 for(i in 1:(npar-1)){
  for(j in (i+1):npar) {combo[k,]<- c(i,j);k=k+1}
 }
 # pairwise response matrices
 nlev <- dim(param$pv)[1]
 x <- list(); xp <- list()
 j=0
 for(k in 1:nc){
 for(i in 1:ndex){
  j=j+1
  x[[j]] <-  round(tapply(y[,i],list(p[,combo[k,1]],p[,combo[k,2]]),
  mean),3)
  xp[[j]] <- round(tapply(yp[,i],list(p[,combo[k,1]],p[,combo[k,2]]),
  mean),3)
 }
 }
```

The average effects are computed from the tables xp. As implemented in the next few lines, it computes the effects for a pair of parameters only.

```
Eff <- matrix(NA,2,ndex); Eff.Int <- Eff
for(i in 1:ndex){
Y <- abs(xp[[i]])
Eff[1,i] <- mean(Y[-2,])- mean(Y[2,])
Eff[2,i] <- mean(Y[,-2])- mean(Y[,2])
Eff.Int[1,i] <-mean(Y[-2,-2]) - mean(Y[-2,2])
Eff.Int[2,i] <-mean(Y[-2,-2]) - mean(Y[2,-2])
}
Eff <- data.frame(round(Eff,3)); names(Eff) <- label.m; row.names(Eff)<-
param$plab
Eff.Int <- data.frame(round(Eff.Int,3)); names(Eff.Int) <- label.m; row.
names(Eff.Int)<- param$plab
```

Multiple regression is performed using the `lm` of each metric versus the relative parameter values, `ps`, which had been previously standardized so that we can compare coefficients across the parameters. We extract the coefficients and the R2 from the summary of `lm`, which is a list. Then, we use the coefficients to calculate the predicted metrics in order to employ these later to graph the scatter plots of metric calculated versus metric predicted. Finally, all information is packed in a data.frame to be returned by the function.

```
ycoeff <- matrix(nrow=ndex,ncol=npar); yp.est <- yp;r2 <- array()
for(i in 1:rdex){
  ycoeff[i,] <- lm(yp[,i]~0+ps)$coeff
  r2[i] <- summary(lm(yp[,i]~0+ps))$r.square
}
for(i in (rdex+1):ndex){
  ppa <- abs(ps)
  ycoeff[i,] <- lm(yp[,i]~0+ppa)$coeff
  r2[i] <- summary(lm(yp[,i]~0+ppa))$r.square
}
for(k in 1:nruns){
  for(i in 1:4) yp.est[k,i] <- sum(ycoeff[i,]*ps[k,])
  for(i in 5:6) yp.est[k,i] <- sum(ycoeff[i,]*ppa[k,])
}
slope <- t(ycoeff)
reg.slope.R2 <- data.frame(round(rbind(slope, t(r2)),3))
names(reg.slope.R2) <- label.m
row.names(reg.slope.R2) <- c(param$plab,"R2")
```

Interaction plots are produced using the function `interaction.plots`, for example:

```
xt <- length(param$pv[,2])
x1 <- p[,1]; x2 <- p[,2]
for(j in 1:ndex){
  interaction.plot(x1,x2,y[,j],ylab=label.m[j],xlab=param$plab[1],
             type="b", pch=1:xt, trace.label=param$plab[2])
}
```

Response maps are generated by means of the function `image` and the overlapping isolines are generated using the function `contour` with the option `add = T`.

```
label.r.nc <-rep(label.r,nc)
for(k in 1:nc){
x1 <- param$pv[,combo[k,1]]; x2 <- param$pv[,combo[k,2]]
for(j in (rdex*(k-1)+1):(rdex*k)) {

image(x1,x2,x[[j]],xlab=param$plab[combo[k,1]],ylab=param$plab[combo
o[k, 2]],col=grey(10:20/20),pty="s")
  contour(x1,x2,x[[j]],add=T)
  title(label.r.nc[j])
 } # j loop
} # k loop
```

Finally, scatter plots are simpler:

```
for(j in 1:ndex){
 plot(yp[,j],yp.est[,j],ylab=label.m[j],xlab="Metric %")
 abline(a=0,b=1)
}
```

You can study more details of this function by viewing it on the console.

Part III

Multidimensional Models

Structured Populations, Communities, and Ecosystems

9 Linear Dynamical Systems

So far in this book, we have dealt with one dependent variable, the state variable, X, and one independent variable, t, except for the Monod model in Chapter 6 that considers two dependent variables. In the remainder of the book, we treat models with more than one dependent variable to be able to track a population by age or size class, the movement of a material in multiple compartments, the interactions among many species in an ecological community, and the interactions among ecosystem variables.

Therefore, the models to analyze and simulate become **systems** of differential equations with state variables $X_1(t)$, $X_2(t)$, ..., $X_n(t)$, where n is the dimension of the system. Since we need to solve the equations simultaneously, we use mathematical strategies that take advantage of the relationships among variables to find a solution. A very useful strategy is to arrange the interactions among the variables in an array called a **matrix** such that the interaction between X_i and X_j becomes an entry in row i and column j of the matrix.

More specifically, when the interactions among the variables X_i are linear, then we can use **linear algebra** to analyze the system and calculate the dynamics resulting from the interaction. Therefore, matrices and linear algebra underlie the methods of **linear dynamical systems**. Moreover, we can approximate nonlinear interactions by linearization around the equilibrium points of the state variables to provide results in the neighborhood of those points. Therefore, a good grasp of linear algebra is of great importance to understand the remaining chapters of this book. Sections 9.1 through 9.7 of this chapter provide a review of linear algebra, and then in the remaining sections we summarize some major aspects of linear dynamical systems.

9.1 MATRICES

A matrix is an array of numbers organized into rows and columns. The position of each element in the array is identified by its position in row i and column j. In the following matrix **A**, element a is in row 1 and column 1; element b is in row 1 and column 2; element c is in row 1 and column 3. Note that we use bold font to denote matrices.

$$\mathbf{A} = \begin{bmatrix} a & b & c \\ d & e & f \\ g & h & i \end{bmatrix} \tag{9.1}$$

Exercise 9.1

Identify the row and column number of the remaining elements in matrix **A**.

A convenient notation is to use the column and row numbers as subindices of the entries and, therefore, write matrix **A** as

$$\mathbf{A} = \begin{bmatrix} a_{11} & a_{12} & a_{13} \\ a_{21} & a_{22} & a_{23} \\ a_{31} & a_{32} & a_{33} \end{bmatrix} \tag{9.2}$$

9.2 DIMENSIONS OF A MATRIX

The number of rows and columns of a matrix represent the **dimensions** of the matrix. If the number of rows in a matrix is different from the number of columns, it is a **rectangular matrix**. An $n \times m$ matrix **B** has n rows and m columns:

$$\mathbf{B} = \begin{bmatrix} b_{11} & b_{12} & . & . & . & b_{1m} \\ b_{21} & b_{22} & . & . & . & b_{2m} \\ . & . & & & & . \\ . & . & & & & . \\ . & . & & & & . \\ b_{n1} & b_{n2} & . & . & . & b_{nm} \end{bmatrix} \tag{9.3}$$

Exercise 9.2

Write a 2×3 matrix with entries b_{ij} as matrix **B**.

9.3 VECTORS

A matrix with only one column is a **column vector** and a matrix with only one row is a **row vector**. For example, a column vector with three entries has dimensions of 3×1; a row vector with four entries has dimensions of 1×4. The column vector $\boldsymbol{x}$ and the row vector $\boldsymbol{y}$ shown in the following equations have three rows and four columns, respectively.

$$\boldsymbol{x} = \begin{bmatrix} x_{11} \\ x_{21} \\ x_{31} \end{bmatrix} \tag{9.4}$$

$$\boldsymbol{y} = \begin{bmatrix} y_{11} & y_{12} & y_{13} & y_{14} \end{bmatrix} \tag{9.5}$$

The dimensions of column vector $\boldsymbol{x}$ and row vector $\boldsymbol{y}$ are 3×1 and 1×4, respectively.

Exercise 9.3

What would be the dimensions of a row vector with n entries? What would be the dimensions of a column vector with n entries?

We can think of a vector geometrically as an $n \times 1$ matrix in such a way that the entries define the coordinates of a point in n-dimensional space. The length of the vector will be the distance from the origin of the coordinates to the point, whereas an arrow pointing from the origin toward the point indicates the direction of the vector. We can easily visualize vectors of dimension 2×1 on a plane by drawing an arrow from the origin of the coordinates to the point, with coordinates given by the elements of the vector. For example, Figure 9.1 shows vectors $\boldsymbol{x}_1 = \begin{bmatrix} 2 \\ 1 \end{bmatrix}$ and $\boldsymbol{x}_2 = \begin{bmatrix} -1 \\ 1 \end{bmatrix}$ in a two-dimensional (2-D) space. Their lengths are $\sqrt{2^2 + 1^2} = \sqrt{5} = 2.24$ for $\boldsymbol{x}_1$ and $\sqrt{(-1)^2 + 1^2} = \sqrt{2} = 1.41$ for $\boldsymbol{x}_2$.

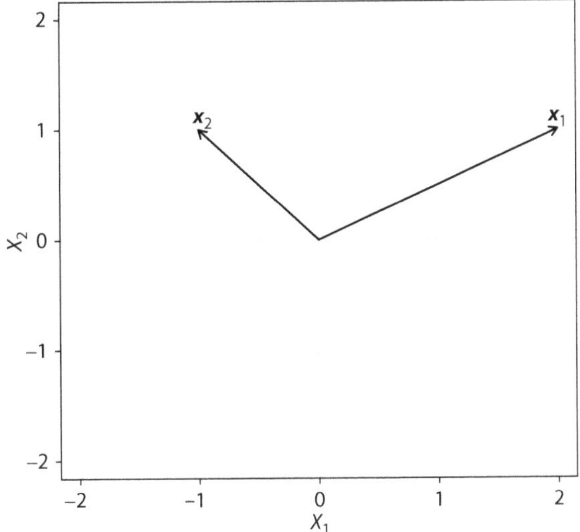

FIGURE 9.1 Geometric interpretation of vectors in two dimensions.

9.4 SQUARE MATRICES

Matrix **A** in Equation 9.2 is a square matrix because it has the same number of columns as rows, three rows and three columns, that is, a 3×3 matrix. Square matrices represent linear systems of equations and linear dynamical systems that have the same number of unknown variables as linearly independent equations.

The **main diagonal** of a square matrix is composed of all of the entries for which the column number i is equal to the row number j. The main diagonal of matrix **A** would consist of the elements a_{11}, a_{22}, and a_{33}. That is, all elements a_{ij} such that $i = j$. All entries above the diagonal would have $j > i$; for example, in matrix **A**, these elements are a_{23}, a_{12}, a_{13}. Entries below the diagonal would have i greater than j. For example, in matrix **A**, these are a_{21}, a_{31}, a_{32}.

Exercise 9.4

Write a 2×2 matrix. Determine the elements above and below the diagonal.

9.4.1 TRACE

The **trace** of a square matrix is the sum of all of the elements in the main diagonal. That is,

$$\mathbf{tr(A)} = \sum_{i=1}^{n} a_{ii} \tag{9.6}$$

For example, consider matrix **D**:

$$\mathbf{D} = \begin{bmatrix} 1 & 4 \\ 5 & 2 \end{bmatrix}$$

The trace of **D** is $1 + 2 = 3$.

9.4.2 SYMMETRIC MATRICES

A square matrix is said to be **symmetric** if all the entries above the main diagonal are equal to the corresponding entries below the main diagonal, $a_{ij} = a_{ji}$. The following matrix **C** is an example of a symmetric matrix.

$$\mathbf{C} = \begin{bmatrix} 2 & 1 & 4 \\ 1 & 2 & 3 \\ 4 & 3 & 2 \end{bmatrix} \tag{9.7}$$

9.4.3 Identity Matrices

An **identity** matrix is a symmetric matrix with all diagonal entries equal to 1, that is, $a_{ii} = 1$, and all off-diagonal entries equal to 0, that is, $a_{ij} = 0$, when $i \neq j$. The identity matrix works with matrix multiplication (which we will cover soon) as the multiplicative identity, one, works with multiplication. That is, to say, $\mathbf{AI} = \mathbf{A}$ and $\mathbf{IA} = \mathbf{A}$. The following matrix $\mathbf{I}$ is a 3×3 identity matrix:

$$\mathbf{I} = \begin{bmatrix} 1 & 0 & 0 \\ 0 & 1 & 0 \\ 0 & 0 & 1 \end{bmatrix} \tag{9.8}$$

9.5 MATRIX OPERATIONS

9.5.1 Addition and Subtraction

Matrix addition (or subtraction) is the most intuitive operation in matrices. If the dimensions of each of the summand matrices match, then the matrices can be added (or subtracted) together entry by entry. $\mathbf{C} = \mathbf{A} + \mathbf{B}$ is obtained by $c_{ij} = a_{ij} + b_{ij}$ for all entries ij. The following example shows the addition of two 3×2 matrices:

$$\begin{bmatrix} 1 & 2 & 3 \\ -1 & -2 & -3 \end{bmatrix} + \begin{bmatrix} -4 & -5 & -6 \\ 4 & 5 & 6 \end{bmatrix} = \begin{bmatrix} 1-4 & 2-5 & 3-6 \\ -1+4 & -2+5 & -3+6 \end{bmatrix} = \begin{bmatrix} -3 & -3 & -3 \\ 3 & 3 & 3 \end{bmatrix}$$

9.5.2 Scalar Multiplication

By **scalar**, we mean a single number or a 1×1 matrix. A scalar can multiply a matrix to produce a new matrix "proportional" to the previous one. We multiply the scalar by each entry within the matrix. If $\mathbf{D} = k\mathbf{B}$, then $d_{ij} = k \times b_{ij}$ for each i and j. For example, when matrix $\mathbf{A} = \begin{bmatrix} a_{11} & a_{12} & a_{13} \\ a_{21} & a_{22} & a_{23} \\ a_{31} & a_{32} & a_{33} \end{bmatrix}$ is multiplied by a scalar with a value equal to 3, we would get $3\mathbf{A} = \begin{bmatrix} 3a_{11} & 3a_{12} & 3a_{13} \\ 3a_{21} & 3a_{22} & 3a_{23} \\ 3a_{31} & 3a_{32} & 3a_{33} \end{bmatrix}$.

Also, for example, $2\mathbf{C}$ will be $2 \times \mathbf{C} = 2 \times \begin{bmatrix} 2 & 1 & 4 \\ 1 & 2 & 3 \\ 4 & 3 & 2 \end{bmatrix} = \begin{bmatrix} 4 & 2 & 8 \\ 2 & 4 & 6 \\ 8 & 6 & 4 \end{bmatrix}$.

9.5.3 Linear Combination

A linear combination of vectors is the sum of the scalar multiplication of a set of vectors by a set of coefficients. That is, vector $\boldsymbol{y}$ of dimensions $n \times 1$ is

$$\boldsymbol{y} = c_1\boldsymbol{x}_1 + c_2\boldsymbol{x}_2 + \cdots + c_m\boldsymbol{x}_m = \sum_{i=1}^{m} c_i\boldsymbol{x}_i \tag{9.9}$$

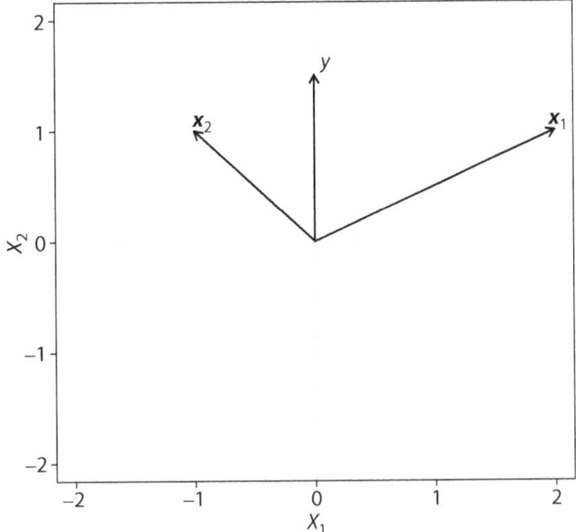

FIGURE 9.2 Linear combination of two vectors in two dimensions.

where scalars c_i are coefficients and $\boldsymbol{x}_i$ are vectors, all of dimensions $n \times 1$. For example, take vectors $\boldsymbol{x}_1 = \begin{bmatrix} 2 \\ 1 \end{bmatrix}$ and $\boldsymbol{x}_2 = \begin{bmatrix} -1 \\ 1 \end{bmatrix}$ that we plotted before and coefficients $c_1 = 0.5$, $c_2 = 1$:

$$\boldsymbol{y} = 0.5 \times \begin{bmatrix} 2 \\ 1 \end{bmatrix} + 1 \times \begin{bmatrix} -1 \\ 1 \end{bmatrix} = \begin{bmatrix} 0 \\ 1.5 \end{bmatrix}$$

Figure 9.2 shows vector $\boldsymbol{y}$ that results from this linear combination.

9.5.4 Matrix Multiplication

Matrix multiplication is more complex than addition and scalar multiplication. The simplest multiplication is an entry-wise product or the Hadamard product of two matrices $\mathbf{C} = \mathbf{A} \circ \mathbf{B}$, where $c_{ij} = a_{ij} \times b_{ij}$, and the two multiplier matrices **A**, **B** should have the same dimensions. This product has an application in image and map processing; however, for linear dynamical systems, we require a more complicated matrix multiplication because the matrix elements correspond to the coefficients of equations. Therefore, in the remainder of the text, we will refer to **matrix multiplication** of two matrices as the operation of obtaining the product by summing the element-wise product of each of the rows of the first matrix with each of the columns of the second matrix. This operation is **not commutative**. Therefore, we need to distinguish between **premultiplication** and **postmultiplication**. To multiply two matrices, we must match the number of columns in the first multiplier to the number of rows in the second:

$$\mathbf{A}_{n \times r} \times \mathbf{B}_{r \times m} = \mathbf{C}_{n \times m}$$

where $n \times r$ are the dimensions of matrix **A** and $r \times m$ are the dimensions of matrix **B**. Here, matrix **A** premultiplies matrix **B**. The resulting matrix **C** has dimensions $n \times m$. A nonrigorous representation of this match is

$$n \times \underbrace{r \times r}_{\text{must match}} \times m \rightarrow n \times m$$

where the two values in the center must be equal.

When the dimensions of the matrices match, the two matrices are multiplied by summing the entry-wise product of each of the rows of the first matrix with each of the columns of the second matrix. For example, in the following, a 2 × 3 matrix premultiplies a 3 × 1 matrix:

$$\begin{bmatrix} 1 & 2 & 3 \\ 4 & 5 & 6 \end{bmatrix} \times \begin{bmatrix} 7 \\ 8 \\ 9 \end{bmatrix} = \begin{bmatrix} 1\times 7+2\times 8+3\times 9 \\ 4\times 7+5\times 8+6\times 9 \end{bmatrix} = \begin{bmatrix} 50 \\ 131 \end{bmatrix} \tag{9.10}$$

For example, take matrix **C** from Equation 9.7:

$$\mathbf{CI} = \begin{bmatrix} 2 & 1 & 4 \\ 1 & 2 & 3 \\ 4 & 3 & 2 \end{bmatrix} \begin{bmatrix} 1 & 0 & 0 \\ 0 & 1 & 0 \\ 0 & 0 & 1 \end{bmatrix} = \begin{bmatrix} 2 & 1 & 4 \\ 1 & 2 & 3 \\ 4 & 3 & 2 \end{bmatrix} = \mathbf{C}$$

Exercise 9.5

What would the dimensions of the resulting matrix be if a 3 × 4 matrix is postmultiplied with a 4 × 5 matrix?

Exercise 9.6

Perform the following multiplication:

$$\begin{bmatrix} 1 & 4 & -7 \\ 2 & -5 & 8 \end{bmatrix} \times \begin{bmatrix} 1 & 2 \\ 3 & 4 \\ 5 & 6 \end{bmatrix}$$

A special case of multiplication is very important to mention here: when we postmultiply a square matrix $n \times n$ by a column vector $n \times 1$, we obtain another vector $n \times 1$. Thus, we can see this matrix multiplication as a transformation of the vector into another, usually with different lengths and directions. For example, $\mathbf{A}\boldsymbol{x}_1 = \begin{bmatrix} 1 & 1 \\ -2 & 4 \end{bmatrix}\begin{bmatrix} 2 \\ 1 \end{bmatrix} = \begin{bmatrix} 3 \\ 0 \end{bmatrix}$ transforms vector $\boldsymbol{x}_1 = \begin{bmatrix} 2 \\ 1 \end{bmatrix}$ by rotating it clockwise 45° and lengthening it from $\sqrt{2^2+1^2} = \sqrt{5} = 2.24$ to $\sqrt{3^2+0^2} = \sqrt{9} = 3$ (Figure 9.3).

9.5.5 Determinant of a Matrix

The **determinant** of a matrix is an operation that **assigns a scalar to a square matrix**. Matrix inversion and other complex matrix processes require this operation. Determinants are only relevant for square matrices. The determinant of matrix **A** is usually denoted as $|\mathbf{A}|$ or det(**A**). There is a simple formula for evaluating the determinant of a 2 × 2 or 3 × 3 matrix. The determinant is the sum of the products of the elements in the upper left to lower right diagonals minus the sums of the products of the lower left to upper right diagonals. For a 2 × 2 matrix, this is

$$\begin{vmatrix} a_{11} & a_{12} \\ a_{21} & a_{22} \end{vmatrix} = a_{11}a_{22} - a_{21}a_{12} \tag{9.11}$$

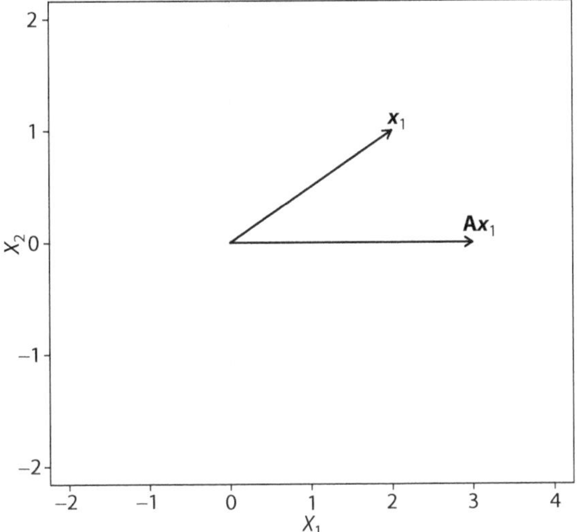

FIGURE 9.3 A vector $\boldsymbol{x}_1$ transformed to another vector by matrix multiplication.

and, for a 3 × 3 matrix, this is

$$\begin{vmatrix} a_{11} & a_{12} & a_{13} \\ a_{21} & a_{22} & a_{23} \\ a_{31} & a_{32} & a_{33} \end{vmatrix} = (a_{11}a_{22}a_{33} + a_{21}a_{32}a_{13} + a_{31}a_{12}a_{23}) - (a_{31}a_{22}a_{13} + a_{21}a_{12}a_{33} + a_{11}a_{32}a_{23}) \tag{9.12}$$

For example, the determinant of the following matrix **C** would be 5:

$$|\mathbf{C}| = \begin{vmatrix} 2 & 1 \\ 1 & 3 \end{vmatrix} = 2 \times 3 - 1 \times 1 = 5$$

The determinant of matrix **C** from Equation 9.7 is 2:

$$\begin{aligned} |\mathbf{C}| &= \left| \begin{bmatrix} 2 & 1 & 4 \\ 1 & 2 & 3 \\ 4 & 3 & 2 \end{bmatrix} \right| \\ &= 2 \times 2 \times 2 + 4 \times 1 \times 3 + 4 \times 1 \times 3 - 4 \times 2 \times 4 - 1 \times 1 \times 2 - 2 \times 3 \times 3 \\ &= 8 + 12 + 12 - 16 - 2 - 12 = 2 \end{aligned}$$

As you can see, increasing the dimensions from 2 × 2 to 3 × 3 increased the number of calculations significantly. Matrices represent transformations in n space, and evaluating the determinant of an $n \times n$ matrix requires more calculations. Practically, we use computers to calculate determinants for matrices of large dimensions.

Exercise 9.7

Find the determinant for the 2 × 2 identity matrix. On the basis of this calculation, what do you think the determinant of a 5 × 5 identity matrix would be? Why?

9.5.6 Matrix Transposition

Transposition consists of exchanging the rows and columns of a matrix. The transpose of matrix **A** would be denoted as $[\mathbf{A}]'$ or $[\mathbf{A}]^T$. This would result in a row vector becoming a column vector. For example, a matrix with dimensions 2 × 3 would become a 3 × 2 matrix as shown in the following:

$$\begin{bmatrix} 6 & 5 & 4 \\ 3 & 2 & 1 \end{bmatrix}^T = \begin{bmatrix} 6 & 3 \\ 5 & 2 \\ 4 & 1 \end{bmatrix} \tag{9.13}$$

Exercise 9.8

Find the transpose of the square symmetric matrix **C** in Equation 9.7. Do you think the transpose of a square symmetric matrix is always the same matrix?

The **major product** of a matrix is a square symmetric matrix formed by the product of the matrix and its transpose.

Exercise 9.9

Multiply the matrix and its transpose given in Equation 9.13. What are the dimensions of the product? Is it symmetric?

Exercise 9.10

Consider the following matrices:

$$\mathbf{A} = \begin{bmatrix} 5 & 0 \\ 0 & 5 \end{bmatrix} \qquad \mathbf{B} = \begin{bmatrix} 4 & 1 \\ 3 & 0 \\ 2 & -1 \end{bmatrix}$$

a. Can you add these? If yes, find **A** + **B**; if not, explain why.
b. Can you find **AB**? **BA**? If yes, do it; if not, why not?
c. What is the transpose of **B**?
d. Find the det($\mathbf{B}^T\mathbf{B}$).

The **major product** of a matrix is useful in linear regression methods. For example, we form a two-column matrix with the data values or observations x_i of a variable X in a sample of size n in column 2 and all 1s in column 1 in the following manner:

$$\mathbf{X} = \begin{bmatrix} 1 & x_1 \\ 1 & x_2 \\ 1 & x_3 \\ \cdots & \cdots \\ 1 & x_n \end{bmatrix} \tag{9.14}$$

Now, if we transpose and premultiply **X** by its transpose $\mathbf{X}^T$, we will obtain a square matrix

$$\mathbf{X}^{\mathrm{T}}\mathbf{X} = \begin{bmatrix} 1 & 1 & 1 & \dots & 1 \\ x_1 & x_2 & x_3 & \dots & x_n \end{bmatrix} \begin{bmatrix} 1 & x_1 \\ 1 & x_2 \\ 1 & x_3 \\ \dots & \dots \\ 1 & x_n \end{bmatrix} = \begin{bmatrix} n & \sum x_i \\ \sum x_i & \sum x_i^2 \end{bmatrix} \tag{9.15}$$

The entries are the sum of the squares of x_i in the diagonal and the sum of x_i in the off-diagonal entries. The matrix $\mathbf{S_x} = \mathbf{X}^{\mathrm{T}}\mathbf{X}$ plays an important role in regression. For example, suppose we have five values for X, $x_i = 1, 2, 2, 1, 0$:

$$\mathbf{S_x} = \mathbf{X}^{\mathrm{T}}\mathbf{X} = \begin{bmatrix} 1 & 1 & 1 & 1 & 1 \\ 1 & 2 & 2 & 1 & 0 \end{bmatrix} \begin{bmatrix} 1 & 1 \\ 1 & 2 \\ 1 & 2 \\ 1 & 1 \\ 1 & 0 \end{bmatrix} = \begin{bmatrix} 5 & 1+2+2+1+0 \\ 1+2+2+1+0 & 1+4+4+1+0 \end{bmatrix} = \begin{bmatrix} 5 & 6 \\ 6 & 10 \end{bmatrix}$$

Exercise 9.11

Suppose we have six values for X, $x_i = 2, 1, 0, 0, 1, 2$. Calculate $\mathbf{S_x} = \mathbf{X}^{\mathrm{T}}\mathbf{X}$.

9.5.7 Matrix Inversion

Division is not strictly defined for matrices. However, recall that the division of two numbers a/b is defined by multiplying a by the inverse $1/b = b^{-1}$. For two matrices $\mathbf{A}$ and $\mathbf{B}$, we multiply $\mathbf{A}$ by the inverse of $\mathbf{B}$, denoted as $\mathbf{B}^{-1}$, instead of dividing $\mathbf{A}$ by $\mathbf{B}$. The inverse of matrix $\mathbf{A}$ is another matrix defined as a matrix that when multiplied by $\mathbf{A}$ yields the identity matrix $\mathbf{I}$. That is, $\mathbf{AA}^{-1} = \mathbf{I}$. The inverse only exists for square matrices with linearly independent rows. The determinant of a matrix can be used to determine if the rows of the matrix are linearly independent. Linearly independent square matrices, that is, nonsingular matrices, have nonzero determinants; others, that is, singular matrices, have determinants equal to zero. In other words, for the inverse of $\mathbf{A}$ to exist, $\mathbf{A}$ must be nonsingular:

$$|\mathbf{A}| \neq 0$$

To calculate the inverse of $\mathbf{A}$ form the **adjoint** matrix adj $\mathbf{A}$ and divide by the determinant:

$$\mathbf{A}^{-1} = \frac{\text{adj}\,\mathbf{A}}{|\mathbf{A}|} \tag{9.16}$$

where adj $\mathbf{A}$ is formed by the cofactors of the entries of the transpose of $\mathbf{A}$ or equivalently by the transpose of the matrix of cofactors of $\mathbf{A}$:

$$\text{adj}\,\mathbf{A} = \left(\mathbf{A}^{\mathrm{T}}\right)^{\mathrm{c}} = \left(\mathbf{A}^{\mathrm{c}}\right)^{\mathrm{T}} \tag{9.17}$$

The cofactor of an entry a_{ij} is the determinant of the matrix that remains after deleting the row i and the column j that correspond to this entry a_{ij}. The sign depends on whether the sum of row and column numbers (i and j) for the entry position is even (assign a positive sign) or odd (assign a negative sign).

To illustrate how to calculate the adjoint and the inverse, we will apply Equations 9.16 and 9.17 to a 2×2 matrix.

$$\mathbf{A} = \begin{bmatrix} 4 & 10 \\ 10 & 30 \end{bmatrix} \tag{9.18}$$

Step 1: Calculate the transpose. In this case, it is just the same matrix because it is symmetric.

Step 2: Calculate the cofactors of $\mathbf{A}^{\mathrm{T}}$. In this simple case, the matrix remaining after deleting a row and a column is simply a scalar. The determinant of a scalar is just the same scalar. Take entry a_{11}, delete row 1 and column 1, and the matrix remaining is just scalar 30; the sign is positive because $1 + 1 = 2$ is even. Take entry a_{12}, delete row 1 and column 2, and the matrix remaining is scalar 10; the sign is negative because $1 + 2 = 3$ is odd. Take entry a_{21}, delete row 2 and column 1, and the matrix remaining is just scalar 10; the sign is negative because $2 + 1 = 3$ is odd. Take entry a_{22}, delete row 2 and column 2, and the matrix remaining is just scalar 4; the sign is positive because $2 + 2 = 4$ is even. Using these cofactors, the adjoint matrix is

$$\operatorname{adj}\mathbf{A} = \left(\mathbf{A}^{\mathrm{T}}\right)^{\mathrm{c}} = \begin{bmatrix} 30 & -10 \\ -10 & 4 \end{bmatrix}$$

Step 3: Calculate the determinant:

$$|\mathbf{A}| = \left|\begin{bmatrix} 4 & 10 \\ 10 & 30 \end{bmatrix}\right| = 4 \times 30 - 10 \times 10 = 20 \tag{9.19}$$

Step 4: Divide the matrix of cofactors by the determinant:

$$\mathbf{A}^{-1} = \frac{\operatorname{adj}\mathbf{A}}{|\mathbf{A}|} = \frac{\left(\mathbf{A}^{\mathrm{T}}\right)^{\mathrm{c}}}{|\mathbf{A}|} = \frac{1}{20}\begin{bmatrix} 30 & -10 \\ -10 & 4 \end{bmatrix} = \begin{bmatrix} 1.5 & -0.5 \\ -0.5 & 0.2 \end{bmatrix} \tag{9.20}$$

A practical way of inverting a matrix of low dimensions, without remembering cofactors, is simultaneously performing operations on the matrix and the identity matrix so as to convert all the diagonal elements to 1s and the off-diagonal elements to 0s. For example,

Step 1: Place matrix **A** besides an identity matrix:

$$\begin{bmatrix} 4 & 10 \\ 10 & 30 \end{bmatrix} \begin{bmatrix} 1 & 0 \\ 0 & 1 \end{bmatrix}$$

Step 2: Row 1 of both matrices is divided by the element in the first row, first column to produce 1 at the a_{11} entry:

$$\begin{bmatrix} 1 & 2.5 \\ 10 & 30 \end{bmatrix} \begin{bmatrix} 0.25 & 0 \\ 0 & 1 \end{bmatrix}$$

Step 3: The first row scaled by the first element in row 2 is subtracted from row 2 to reduce a_{21} to 0:

$$\begin{bmatrix} 1 & 2.5 \\ 0 & 5 \end{bmatrix} \begin{bmatrix} 0.25 & 0 \\ -2.5 & 1 \end{bmatrix}$$

Step 4: Row 2 is divided by an element in the second column to give $a_{22} = 1$:

$$\begin{bmatrix} 1 & 2.5 \\ 0 & 1 \end{bmatrix} \begin{bmatrix} 0.25 & 0 \\ -0.5 & 0.2 \end{bmatrix}$$

Step 5: Row 2 is scaled by the remaining nonzero off-diagonal terms and subtracted from row 1 to reduce the remaining off-diagonal a_{12} term to 0:

$$\begin{bmatrix} 1 & 0 \\ 0 & 1 \end{bmatrix} \begin{bmatrix} 1.5 & -0.5 \\ -0.5 & 0.2 \end{bmatrix}$$

The final matrix on the right-hand side is the inverse of the matrix on the left-hand side in the first step. Multiply them to verify that their product is an identity matrix. We showed this procedure for pedagogical purposes; we typically use computers to calculate the inverse of a large matrix.

9.6 SOLVING SYSTEMS OF LINEAR ALGEBRAIC EQUATIONS

Based upon the elements of matrices reviewed so far, we can solve systems of equations. For example,

$$\begin{aligned} 4x_1 + 10x_2 &= 38 \\ 10x_1 + 30x_2 &= 110 \end{aligned} \tag{9.21}$$

By using the coefficients of each of the equations as entries in a matrix, we can rewrite Equation 9.21 as follows:

$$\begin{bmatrix} 4 & 10 \\ 10 & 30 \end{bmatrix} \begin{bmatrix} x_1 \\ x_2 \end{bmatrix} = \begin{bmatrix} 38 \\ 110 \end{bmatrix} \tag{9.22}$$

You can check your matrix multiplication skills by multiplying the first two matrices. It should yield a matrix containing the left-hand side of the expressions in Equation 9.21. Denoting the vector of unknowns x_1 and x_2 as $\boldsymbol{x} = \begin{bmatrix} x_1 \\ x_2 \end{bmatrix}$, the matrix equation is now in the form $\mathbf{A}\boldsymbol{x} = \boldsymbol{b}$, where $\mathbf{A} = \begin{bmatrix} 4 & 10 \\ 10 & 30 \end{bmatrix}$ and $\boldsymbol{b} = \begin{bmatrix} 38 \\ 110 \end{bmatrix}$.

To solve for $\boldsymbol{x}$ in the equation

$$\mathbf{A}\boldsymbol{x} = \boldsymbol{b} \tag{9.23}$$

multiply both sides of the equation by the inverse of $\mathbf{A}$. That is,

$$\mathbf{A}^{-1}\mathbf{A}\boldsymbol{x} = \mathbf{I}\boldsymbol{x} = \boldsymbol{x} = \mathbf{A}^{-1}\boldsymbol{b} \tag{9.24}$$

gives

$$\boldsymbol{x} = \mathbf{A}^{-1}\boldsymbol{b} \tag{9.25}$$

We already calculated the inverse of **A** in Section 9.5.7 and it is given in Equation 9.20:

$$\mathbf{A}^{-1} = \begin{bmatrix} 1.5 & -0.5 \\ -0.5 & 0.2 \end{bmatrix}$$

Apply Equation 9.24, which, after multiplication, is simplified to

$$\begin{bmatrix} 1 & 0 \\ 0 & 1 \end{bmatrix}\begin{bmatrix} x_1 \\ x_2 \end{bmatrix} = \begin{bmatrix} 2 \\ 3 \end{bmatrix} \tag{9.26}$$

Therefore, the unique solutions to the system of equations in Equation 9.21 is $x_1 = 2$ and $x_2 = 3$. If you graph both equations, the lines should intersect at the point (2, 3).

A typical use of this method is to perform **regression analysis** based on solving for the slope b_1 and the intercept b_0 of a line of best fit between X and Y using the observations y_i and x_i. We need to solve the matrix equation

$$\begin{bmatrix} n & \sum_{i=1}^{n} x_i \\ \sum_{i=1}^{n} x_i & \sum_{i=1}^{n} x_i^2 \end{bmatrix}\begin{bmatrix} b_0 \\ b_0 \end{bmatrix} = \begin{bmatrix} \sum_{i=1}^{n} y_i \\ \sum_{i=1}^{n} x_i y_i \end{bmatrix} \tag{9.27}$$

which can be abbreviated by using the name $\mathbf{S_x}$ for the major product matrix in the left, the name $\boldsymbol{b}$ for the vector of slope and intercept, and the name $\mathbf{S_y}$ for the right-hand side as follows:

$$\underbrace{\begin{bmatrix} n & \sum_{i=1}^{n} x_i \\ \sum_{i=1}^{n} x_i & \sum_{i=1}^{n} x_i^2 \end{bmatrix}}_{\mathbf{S_x}}\underbrace{\begin{bmatrix} b_0 \\ b_1 \end{bmatrix}}_{\mathbf{b}} = \underbrace{\begin{bmatrix} \sum_{i=1}^{n} y_i \\ \sum_{i=1}^{n} x_i y_i \end{bmatrix}}_{\mathbf{S_y}}$$

Written in matrix form, the equation is

$$\mathbf{S_x}\boldsymbol{b} = \mathbf{S_y} \tag{9.28}$$

Note that $\mathbf{S_x}$ is a symmetrical matrix involving only the observations of X. Solving the matrix equation for unknown $\boldsymbol{b}$ allows for the calculation of intercept b_0 and slope b_1 of the regression line. The solution is

$$\boldsymbol{b} = \mathbf{S_x}^{-1}\mathbf{S_y} \tag{9.29}$$

9.7 EIGENVALUES AND EIGENVECTORS

As we discussed before, usually when we postmultiply a square matrix by a column vector, say $\mathbf{A}\boldsymbol{x}$, the result is vector $\boldsymbol{y}$ with a different length and direction than the original vector (recall Figure 9.3). However, there is a particular class of vector $\boldsymbol{v}$ for each square matrix that when premultiplied by the matrix $\mathbf{A}\boldsymbol{v}$ the resulting vector **preserves the direction** and only changes the length of the original vector by a scalar factor λ, that is, $\lambda\boldsymbol{v}$. This class of vector is called an **eigenvector**. The scale factor associated with the transformation of an eigenvector is called an **eigenvalue**. Formally, an eigenvalue–eigenvector pair of $\mathbf{A}$ is any real or complex scalar–vector pair denoted as $(\lambda, \boldsymbol{v})$, such that $\mathbf{A}\boldsymbol{v} = \lambda\boldsymbol{v}$. The equality will hold even if we multiply the equation by any scalar k. Therefore, there are an infinite number of eigenvectors associated with a particular eigenvalue.

9.7.1 Finding Eigenvalues

The equation in the definition provided above states

$$\mathbf{A}\boldsymbol{v} = \lambda\boldsymbol{v} \tag{9.30}$$

This is equivalent to

$$\mathbf{A}\boldsymbol{v} - \lambda\boldsymbol{v} = \mathbf{0} \tag{9.31}$$

or by just changing the sign to

$$\lambda\boldsymbol{v} - \mathbf{A}\boldsymbol{v} = \mathbf{0} \tag{9.32}$$

Then, by factoring out the vector $\boldsymbol{v}$ and inserting the identity matrix $\mathbf{I}$ (remember that the identity matrix does not change the value of the vector), we get

$$(\lambda\mathbf{I} - \mathbf{A})\boldsymbol{v} = 0 \tag{9.33}$$

If the inverse of $\lambda\mathbf{I} - \mathbf{A}$ exists, then the solution will be $\boldsymbol{v} = 0$, the null eigenvector, which is not very interesting. Therefore, nonnull vectors of $\boldsymbol{v}$ exist only if $\lambda\mathbf{I} - \mathbf{A}$ does not have an inverse. The inverse does not exist when the matrix $\lambda\mathbf{I} - \mathbf{A}$ is singular, that is, when

$$|\lambda\mathbf{I} - \mathbf{A}| = 0 \tag{9.34}$$

This condition is the **characteristic equation** for $\mathbf{A}$ and is an nth-degree polynomial in λ for an $n \times n$ matrix. Therefore, there are at most n distinct values of λ that satisfy the characteristic equation in Equation 9.34.

As an example, let us consider the following matrix:

$$\mathrm{A} = \begin{bmatrix} 1 & 1 \\ -2 & 4 \end{bmatrix} \tag{9.35}$$

The characteristic equation of this matrix is

$$\left| \begin{bmatrix} \lambda & 0 \\ 0 & \lambda \end{bmatrix} - \begin{bmatrix} 1 & 1 \\ -2 & 4 \end{bmatrix} \right| = \begin{vmatrix} \lambda - 1 & -1 \\ 2 & \lambda - 4 \end{vmatrix} = 0 \tag{9.36}$$

Calculating the determinant gives

$$(\lambda - 1)(\lambda - 4) + 2 = \lambda^2 - 5\lambda + 6 = 0 \tag{9.37}$$

and factoring gives

$$(\lambda - 3)(\lambda - 2) = 0 \tag{9.38}$$

From this, we can conclude that $\lambda_1 = 3$ and $\lambda_2 = 2$ are two distinct solutions to the quadratic equation (Equation 9.37). Each one of these eigenvalues is associated with an eigenvector.

9.7.2 Finding Eigenvectors

The eigenvectors can be found by solving Equation 9.31, $\mathbf{A}\boldsymbol{v} = \lambda\boldsymbol{v}$, for the eigenvalues above. Take the $\lambda_1 = 3$ eigenvalue, and denote v_1 and v_2 as the entries of the eigenvector $\boldsymbol{v}_1$ associated with this eigenvalue; the equation $\mathbf{A}\boldsymbol{v}_1 = \lambda\boldsymbol{v}_1$ would be as follows:

$$\begin{bmatrix} 1 & 1 \\ -2 & 4 \end{bmatrix}\begin{bmatrix} v_1 \\ v_2 \end{bmatrix} = 3\begin{bmatrix} v_1 \\ v_2 \end{bmatrix} \tag{9.39}$$

yielding

$$v_1 + v_2 = 3v_1 \quad \text{and} \quad -2v_1 + 4v_2 = 3v_2 \tag{9.40}$$

We can see that $v_2 = 2v_1$ will satisfy both equations in Equation 9.40.

This result emphasizes the point that each eigenvector is a member of a class of eigenvectors that satisfy the condition $v_2 = 2v_1$; any eigenvector that satisfies this condition will work. Arbitrarily choose

$$\boldsymbol{v}_1 = \begin{bmatrix} v_1 \\ v_2 \end{bmatrix} = \begin{bmatrix} 1 \\ 2 \end{bmatrix} \tag{9.41}$$

as the representative eigenvector. Geometrically, we can see that if we transform premultiplying by $\mathbf{A}$, we get

$$3\begin{bmatrix} v_1 \\ v_2 \end{bmatrix} = 3\begin{bmatrix} 1 \\ 2 \end{bmatrix} = \begin{bmatrix} 3 \\ 6 \end{bmatrix}$$

which has the same direction but is three times longer than $\boldsymbol{v}_1$ (Figure 9.4). The length is $\sqrt{6^2 + 3^2} = \sqrt{3^2(2^2 + 1^2)} = 3 \times 2.27 = 6.71$. The eigenvector that associates with $\lambda_2 = 2$ is $\boldsymbol{v}_2 = \begin{bmatrix} 1 \\ 1 \end{bmatrix}$; when multiplied by A, it is scaled by 2 (Figure 9.4), that is, $2\begin{bmatrix} 1 \\ 1 \end{bmatrix} = \begin{bmatrix} 2 \\ 2 \end{bmatrix}$ with length $\sqrt{2^2 + 2^2} = \sqrt{4 \times 2} = 2 \times 1.41 = 2.82$.

Exercise 9.12

Show that $\boldsymbol{v}_2 = \begin{bmatrix} 1 \\ 1 \end{bmatrix}$ is an eigenvector associated with the other eigenvalue $\lambda = 2$ in the example above. Multiply by A and demonstrate that it is scaled by 2.

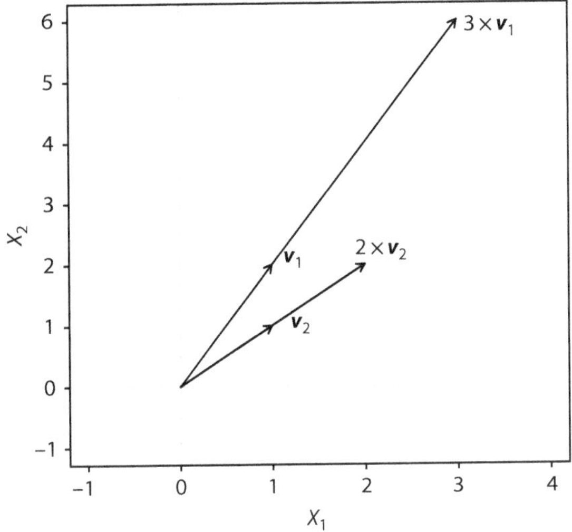

FIGURE 9.4 A transformed eigenvector preserves direction and scales in length.

9.7.3 Complex Eigenvalues

When solving a characteristic equation, sometimes we get complex eigenvalues. Let us briefly review the concept of complex numbers before going further. The square root of a negative number yields an **imaginary** number. To denote these, we use the notation

$$i = \sqrt{-1} \tag{9.42}$$

For example:

$$x = \sqrt{-4} = 2i$$

A complex number has a real part and an imaginary part. For example, $x = 2 \pm 4i$ has real part equal to 2 and two values for the imaginary part $4i$ and $-4i$. Complex numbers occur in **conjugate** pairs like this, where the imaginary parts have the same value but opposite in sign.

9.7.4 Quadratic Equation

In general, for a 2×2 system, the characteristic equation $|\lambda\mathbf{I} - \mathbf{A}| = 0$ is

$$\lambda^2 - (a_{11} + a_{22})\lambda + (a_{11}a_{22} - a_{12}a_{21}) = 0 \tag{9.43}$$

Here, we identify the trace of $\mathbf{A}$, $T = a_{11} + a_{22}$, and the determinant of $\mathbf{A}$, $\Delta = a_{11}a_{22} - a_{12}a_{21}$, and rewrite the characteristic equation as $\lambda^2 - T\lambda + \Delta = 0$ with roots

$$\lambda = \frac{T \pm \sqrt{T^2 - 4\Delta}}{2} = \frac{T \pm \sqrt{D}}{2} \tag{9.44}$$

where $D = T^2 - 4\Delta$ is called the discriminant of $\mathbf{A}$. For positive discriminant $D > 0$, the eigenvalues are real, but when the discriminant is negative $D < 0$, we will get complex λ_1, λ_2, with real part $T/2$

and imaginary parts $\pm\frac{\sqrt{D}}{2}i$. We can see that the determinant and trace determine the properties of the eigenvalues λ_1, λ_2. By the way, it is easy to see that the sum of the eigenvalues is equal to the trace T regardless of the value of D. This is true for a square matrix of higher dime nsion, n:

$$\text{Tr}(\mathbf{A}) = \sum_{i=1}^{n} \lambda_i \tag{9.45}$$

For example, in the matrix in Equation 9.35, the trace is 5 and the sum of the eigenvalues is $3+2=5$. Also, for example, consider the matrix $\mathbf{A} = \begin{pmatrix} 1 & 2 \\ -3 & 1.5 \end{pmatrix}$; it has $T = \text{Tr}(\mathbf{A}) = 2.5$ and $\Delta = \det(\text{A}) = 7.5$; therefore, $D = T^2 - 4\Delta = -30$, and the eigenvalues are complex, $\lambda = 1.25 \pm 2.44i$; we show them in Figure 9.5, where we plot the eigenvalue in the complex plane (imaginary vs. real parts). It is of great interest to see what happens to the eigenvalues as we change some element of the matrix. It is a type of sensitivity analysis.

A nice tool is the **root locus** graph, which is just a plot of the eigenvalues as we sweep one parameter, for example, one element of the matrix. As an example in the matrix **A** above, as we decrease a_{11}, the eigenvalues move from their original position decreasing their imaginary part until they become real (Figure 9.6). When the eigenvalues reach the Im = 0 axis, they become real and negative. At the intersection, the eigenvalues are repeated, then they split into two opposite directions, one becoming more negative and the other less negative until it crosses the Re = 0 axis and becomes positive (Figure 9.6).

Let us catalog the various eigenvalue types as a function of T and Δ while visualizing the **root loci** (Figures 9.8 and 9.9). In these figures, we plot the eigenvalue in a complex plane while sweeping it by the trace, T, in these figures. In each graph, we plot an eigenvalue for three different values of the determinant.

- Case 1: **Positive nonzero determinant**. The discriminant may be positive or negative depending on the sign of $T^2 - 4$; therefore, we can have real or complex eigenvalues. The sign of the real part depends on the sign of T.
  - Case 1.1: **Positive nonzero determinant and negative nonzero trace**. When $\Delta > 0$ and $T < 0$, we can have a positive or negative discriminant, D (Figure 9.7, top left panel). This situation will produce eigenvalues with negative real parts (root loci in Figure 9.8, left side); then, if $D > 0$, or $T^2 > 4$, the eigenvalues will be real (no imaginary part)

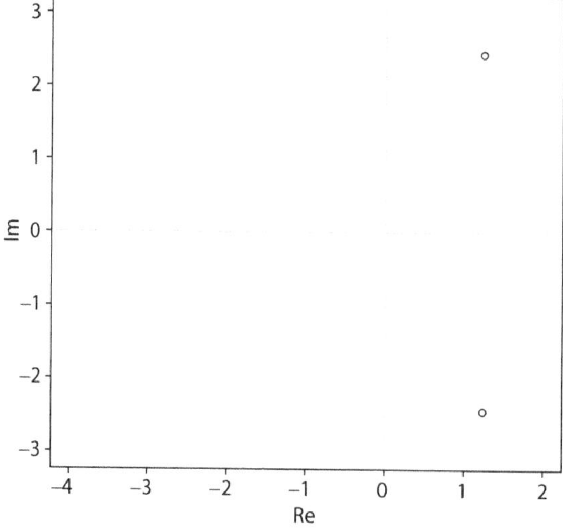

FIGURE 9.5 A complex eigenvalue in the complex plane (imaginary vs. real).

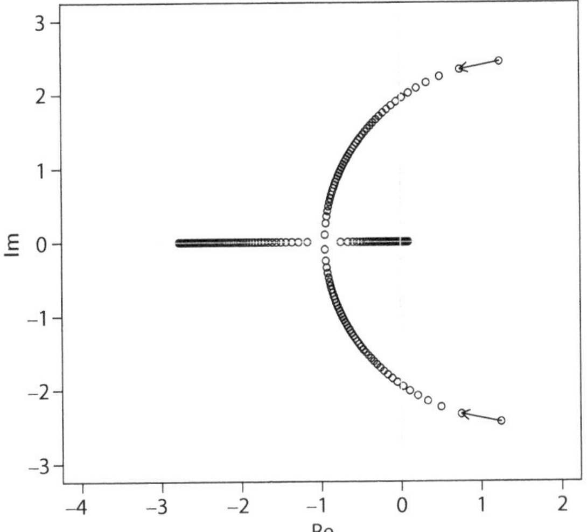

FIGURE 9.6 A root locus for a complex eigenvalue pair as we change element a_{11} of matrix **A**.

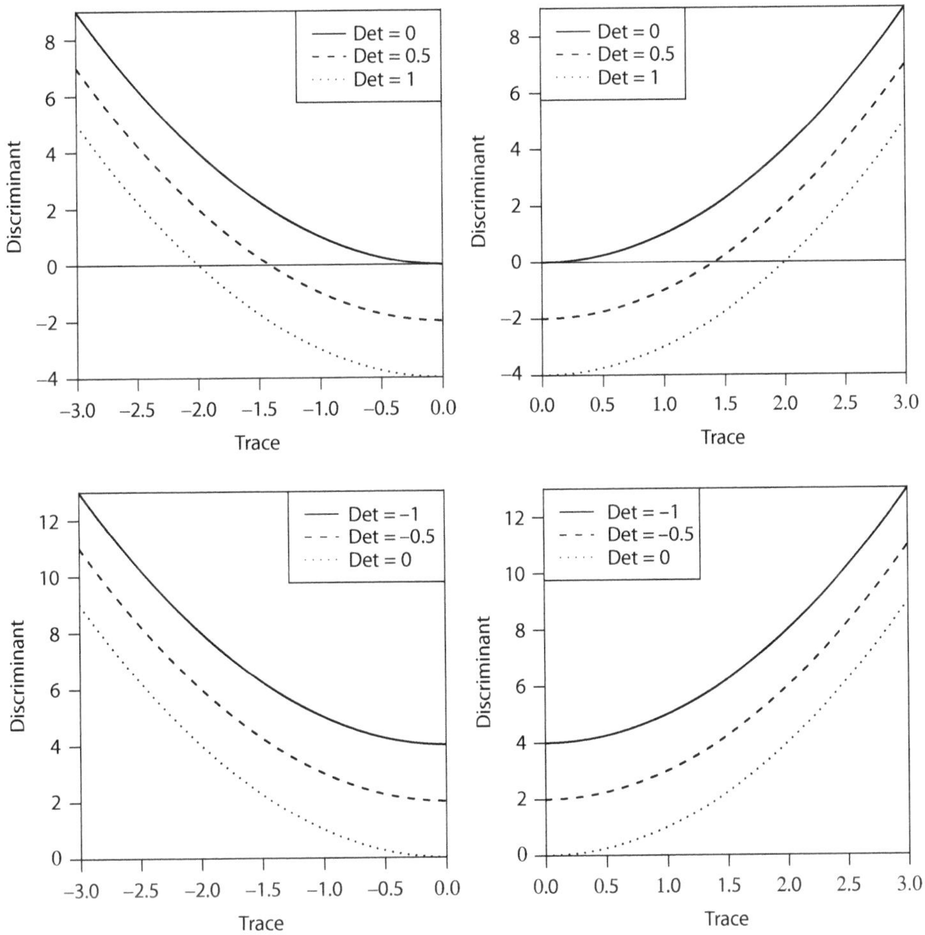

FIGURE 9.7 Discriminant, D, for positive determinant (upper panels), negative determinant (lower panels), negative trace (left-hand side), and positive trace (right-hand side).

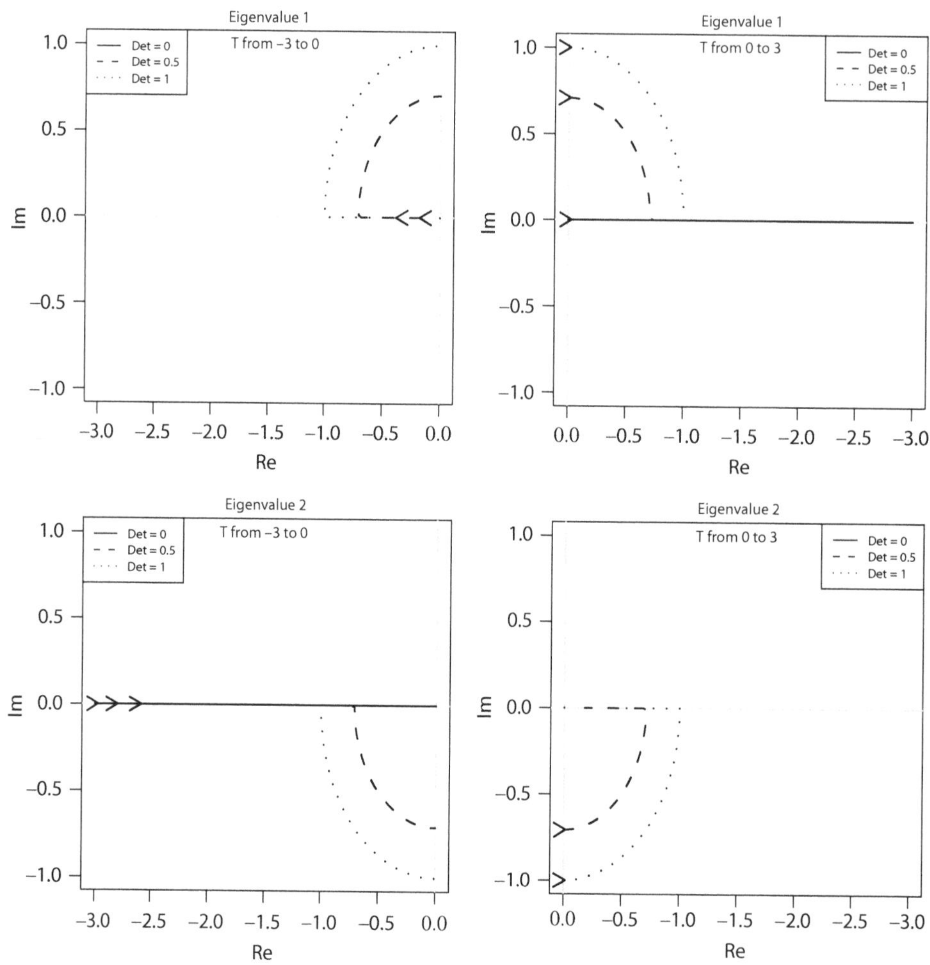

FIGURE 9.8 Eigenvalues for positive and zero determinants, negative trace (left-hand side), and positive trace (right-hand side).

and distinct; if $D = 0$, the eigenvalues will be real negative but repeated, and on the contrary, if $D < 0$, the eigenvalues will be complex conjugate with negative real parts.

- Case 1.2: **Positive nonzero determinant and zero trace**. When $\Delta > 0$ and $T = 0$, we have a positive discriminant, D (Figure 9.7, top panels). This situation will produce complex eigenvalues with zero real parts or pure imaginary parts (root loci in Figure 9.8).
- Case 1.3: **Positive nonzero determinant and positive nonzero trace**. When $\Delta > 0$ and $T > 0$, we can have a positive and negative discriminant, D (Figure 9.7, top right panel). This situation will produce eigenvalues with positive real parts (root loci in Figure 9.8, right-hand side); then if $D > 0$, or $T^2 > 4\Delta$, the eigenvalues will be real (no imaginary part) and distinct; if $D = 0$, the eigenvalues will be real, positive but repeated, and on the contrary, if $D < 0$, the eigenvalues will be complex conjugate with positive real parts.

- Case 2: **Zero determinant**. When $\Delta = 0$ and regardless of the sign of T, the discriminant is positive, except for $T = 0$, for which $D = 0$ (Figure 9.7).
  - Case 2.1: **Zero determinant, negative trace**. Eigenvalue 1 is zero (dot at the origin in Figure 9.8, top left; Figure 9.9, top left), and eigenvalue 2 is real and negative (Figure 9.8, bottom right; Figure 9.9, bottom right).

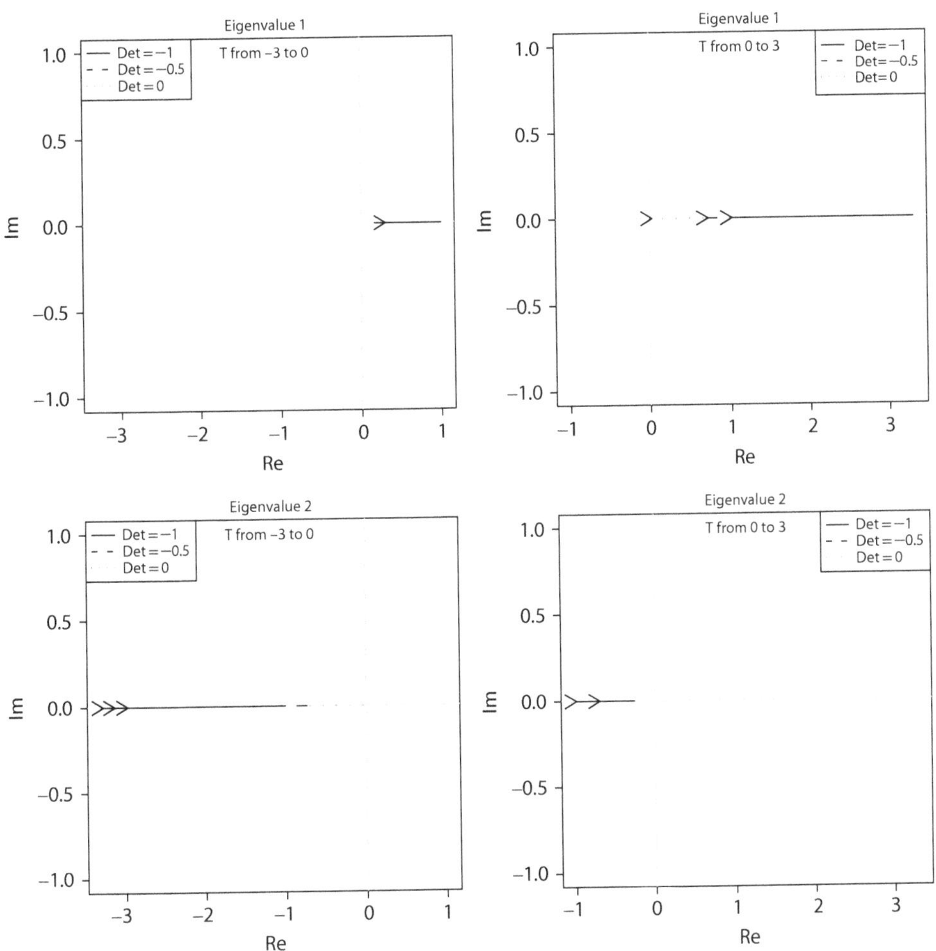

FIGURE 9.9 Eigenvalues for negative and zero determinants, negative trace (left-hand side), and positive trace (right-hand side).

- Case 2.2: **Zero determinant, zero trace**. Eigenvalues are zero.
- Case 2.3: **Zero determinant, positive trace**. The situation is the reverse of case 2.1. Eigenvalue 2 is zero (dot at the origin in Figure 9.8, bottom right; Figure 9.9, bottom right), and eigenvalue 1 is real and positive (Figure 9.8, top right; Figure 9.9, top right).

- Case 3: **Negative nonzero determinant**. When $\Delta < 0$ and regardless of the sign of T, we have positive discriminant D (Figure 9.7, bottom panels), except at $T = 0$.
  - Case 3.1: **Negative nonzero determinant and negative trace**. This situation will produce real and distinct eigenvalues, for each pair one is positive and the other negative (Figure 9.9).
  - Case 3.2: **Negative nonzero determinant and zero trace**. Eigenvalues are real and distinct, one positive and the other negative, but with the same absolute value.
  - Case 3.3: **Negative nonzero determinant and positive trace**. Same as case 3.1; this situation will produce real eigenvalues, for each pair one is positive and the other negative (Figure 9.9).

Table 9.1 summarizes this catalog of eigenvalue types by signs of determinant, trace, and discriminant.

TABLE 9.1
Eigenvalue Types by Values of Δ, *T*, and *D*

	$T < 0$	$T = 0$	$T > 0$
$\Delta > 0$, Case 1	Case 1.1 $D > 0$, distinct real negative $D = 0$, real negative repeated $D < 0$, complex real negative parts	Case 1.2 $D < 0$, always Complex, purely imaginary	Case 1.3 $D > 0$, distinct real positive $D = 0$, real positive repeated $D < 0$, complex real positive parts
$\Delta = 0$, Case 2	Case 2.1 $D > 0$, one real negative, one zero	Case 2.2 Null $D = 0$ zero eigenvalues	Case 2.3 $D > 0$, one real positive, one zero
$\Delta < 0$, Case 3	Case 3.1 $D > 0$, real opposite signs	Case 3.2 $D > 0$, real opposite signs and same magnitude	Case 3.3 $D > 0$, real opposite signs

9.8 LINEAR DYNAMICAL SYSTEMS: CONSTANT COEFFICIENTS

Now that we know matrices and linear algebra, we will concern ourselves with learning the basics of a linear dynamical system. In continuous time, this is a system of ODEs $\frac{d\boldsymbol{X}(t)}{dt} = \mathbf{A}(t)\boldsymbol{X}(t) + \mathbf{B}(t)\boldsymbol{U}(t)$ with initial condition $\boldsymbol{X}(0)$. Here $\boldsymbol{X}$ is the state vector, $\boldsymbol{U}$ is the vector of forcing terms, and **A**, **B** are the matrices of coefficients. In this chapter, we limit ourselves to the simplest case of constant matrix **A** and nonforced systems $\mathbf{B}(t)\boldsymbol{U}(t) = 0$.

9.8.1 State Transition Matrix and Modes

Let us start with a linear ODE system of dimension n:

$$\underbrace{\frac{d}{dt}\begin{bmatrix} X_1 \\ X_2 \\ \cdots \\ X_n \end{bmatrix}}_{dX/dt} = \underbrace{\begin{bmatrix} a_{11} & a_{12} & \cdots & a_{1n} \\ a_{1n} & a_{22} & \cdots & a_{2n} \\ \cdots & \cdots & \cdots & \cdots \\ a_{n1} & a_{n2} & \cdots & a_{nn} \end{bmatrix}}_{A} \underbrace{\begin{bmatrix} X_1 \\ X_2 \\ \cdots \\ X_n \end{bmatrix}}_{X} \quad (9.46)$$

Or in matrix notation $\frac{d\boldsymbol{X}}{dt} = \mathbf{A}\boldsymbol{X}$, where matrix **A** and vector $\boldsymbol{X}$ are indicated in Equation 9.46. The solution $\boldsymbol{X}(t)$ of this equation is given by

$$\boldsymbol{X}(t) = \exp(\mathbf{A}t)\boldsymbol{X}_0 \quad (9.47)$$

where $\boldsymbol{\Phi}(t) = \exp(\mathbf{A}t)$ is defined as the **state transition matrix**. When we have distinct eigenvalues, this matrix is a combination of exponentials. We will not work the details of $\exp(\mathbf{A}t)$ but just find the solution $\boldsymbol{X}(t)$ as a linear combination of eigenvectors with coefficients given by exponentials of the eigenvalues:

$$\boldsymbol{X}(t) = \sum_{i=1}^{n} c_i(0)\boldsymbol{v}_i \exp(\lambda_i t) \quad (9.48)$$

where $\boldsymbol{v}_i$ is the eigenvector associated to eigenvalue λ_i and $c_i(0)$ is the corresponding coefficient to form the initial condition as a linear combination of the eigenvectors:

$$\boldsymbol{X}(0) = \sum_{i=1}^{n} c_i(0)\boldsymbol{v}_i \quad (9.49)$$

When we select $c_i(0)$ such that to pick only one eigenvector, say $c_k(0) = 1$, and all other $c_i(0) = 0$, then the solution from Equation 9.48 is simply

$$\boldsymbol{X}(t) = \boldsymbol{v}_k \exp(\lambda_k t) \tag{9.50}$$

That is, at each time, t, the state is the eigenvector multiplied by the scalar $\exp(\lambda_k t)$. This simple trajectory is a **mode**. Therefore, in general, when we form a linear combination of eigenvectors, the full trajectory is a linear combination of modes.

For example, consider matrix $\mathbf{A} = \begin{bmatrix} -0.4 & 0.25 \\ 0.2 & -0.15 \end{bmatrix}$ of dimensions 2×2. The eigenvalues are $\lambda_1 = -0.53$ and $\lambda_2 = -0.02$, and the eigenvectors are $\boldsymbol{v}_1 = \begin{bmatrix} -0.88 \\ 0.46 \end{bmatrix}$, $\boldsymbol{v}_2 = \begin{bmatrix} -0.55 \\ -0.84 \end{bmatrix}$. Note that, of course, the trace of $\mathbf{A}$ is -0.55 and equal to the sum of eigenvalues $-0.53 - 0.02 = -0.55$.

Let us use only the first eigenvector $\boldsymbol{v}_l$ then $c_1(0) = 1$, $c_2(0) = 0$, or $\boldsymbol{c}(0) = \begin{bmatrix} 1 \\ 0 \end{bmatrix}$. We obtain the first mode shown in the top panel of Figure 9.10, whereas when we use only the second eigenvector $c_1(0) = 0$, $c_2(0) = 1$, or $\boldsymbol{c}(0) = \begin{bmatrix} 0 \\ 1 \end{bmatrix}$, we obtain the bottom panel in Figure 9.10. Note that the second mode is slower than the first, as we would have expected because the first eigenvalue is much larger than the second eigenvalue. Consequently, the exponential should have faster decay in the first mode.

For further insight on how the modes are related to the eigenvectors, see Figure 9.11. The top panel shows that X_2 versus X_1 follows a straight line with negative slope, which is confirmed by the

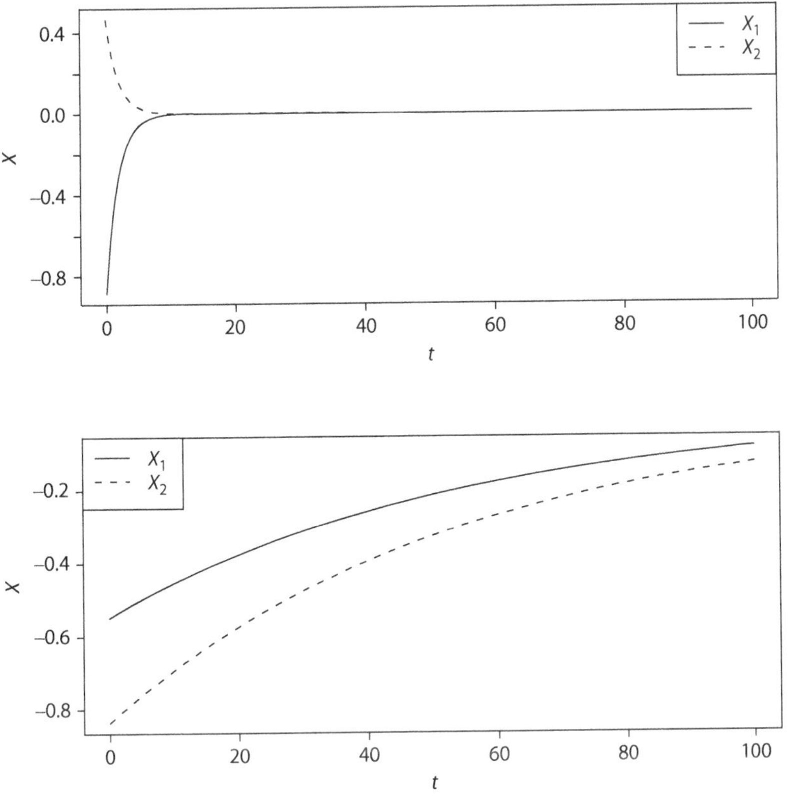

FIGURE 9.10 Dynamics of modes. Top: Mode 1 obtained by only the first eigenvector. Bottom: Mode 2 obtained by only the second eigenvector.

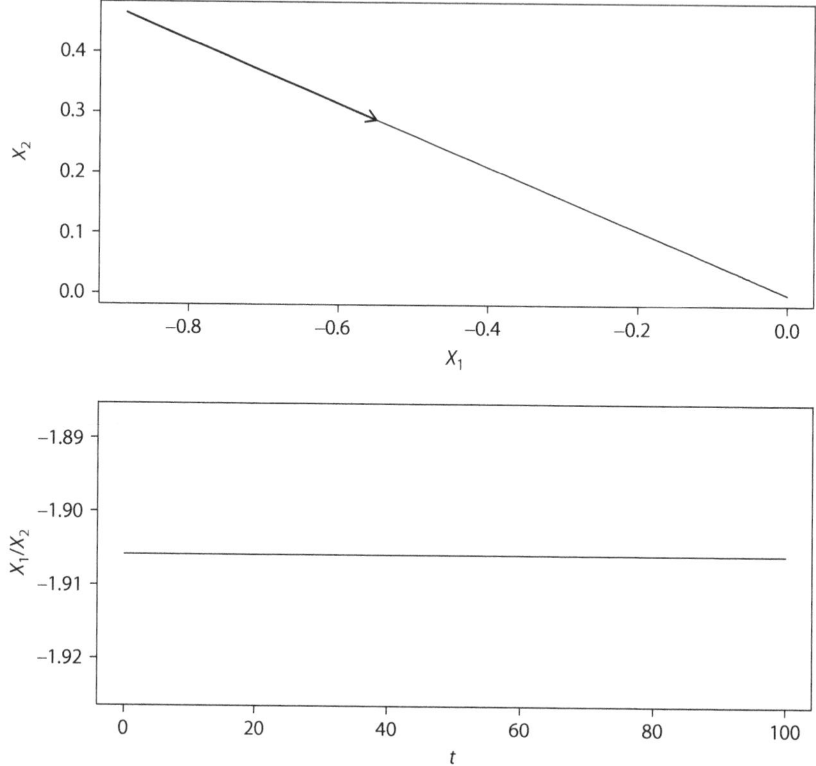

FIGURE 9.11 Phase plane and ratio plot show that mode 1 is just a scaling of the first eigenvector.

constant negative ratio $X_1/X_2 = -1.906$ in the bottom panel; indeed, this ratio is the same as the ratio of the first entry to the second entry of $\boldsymbol{v}_1$, namely $\frac{-0.88}{0.46} = -1.906$. This demonstrates that at all times, mode 1 is just a scaling of the first eigenvector. The scalar is the exponential decay dictated by the first eigenvalue. This is clarified even more in Figure 9.12, where the eigenvector shrinks as time goes on.

Similarly, in Figure 9.13, the top panel shows that X_2 versus X_1 follows a straight line with positive slope, which is confirmed by the constant positive ratio $X_1/X_2 = 0.656$ in the bottom panel and equal to the ratio of the first entry to the second entry of $\boldsymbol{v}_2$, namely $\frac{0.55}{0.84} = 0.656$. This demonstrates that at all times mode 2 is just a scaling of the second eigenvector. The scalar is the exponential decay dictated by the second eigenvalue. This is clarified even more in Figure 9.14, where the eigenvector shrinks as time goes on.

Now, to obtain a particular initial condition we need to find $\boldsymbol{c}(0)$ such that $c_1(0)\boldsymbol{v}_1 + c_2(0)\boldsymbol{v}_2 = \mathbf{V}\boldsymbol{c}(0) = \boldsymbol{X}(0)$. This is an algebraic equation, and we can solve it using the inverse of $\boldsymbol{V}$. For example, take $\boldsymbol{X}(0) = \begin{bmatrix} 100 \\ 50 \end{bmatrix}$; we need to solve

$$\mathbf{V}\boldsymbol{c}(0) = \begin{bmatrix} -0.88 & -0.55 \\ 0.46 & -0.84 \end{bmatrix} \begin{bmatrix} c_1(0) \\ c_2(0) \end{bmatrix} = \begin{bmatrix} 100 \\ 50 \end{bmatrix}$$

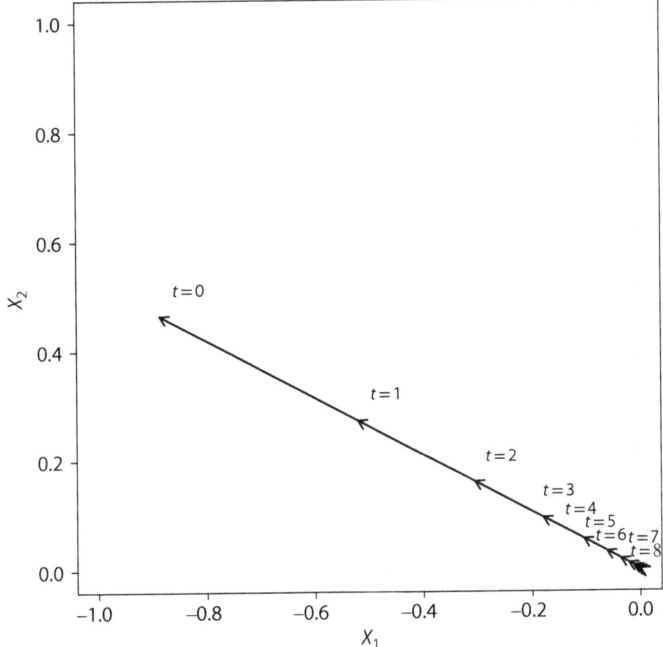

FIGURE 9.12 Geometric interpretation of mode 1. We can see progressive scaling of the first eigenvector as time goes on.

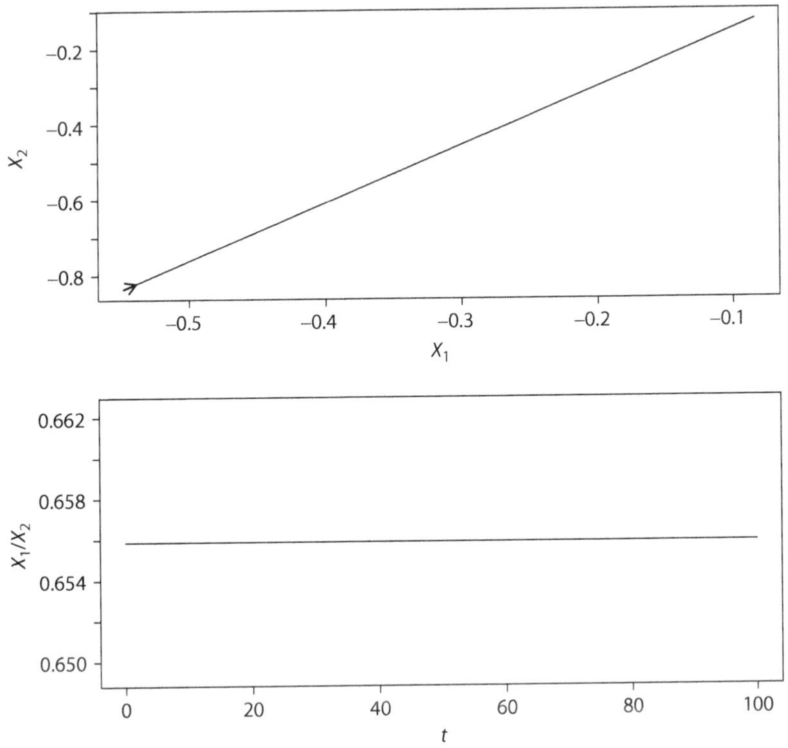

FIGURE 9.13 Phase plane and ratio plot show that mode 2 is just a scaling of the second eigenvector.

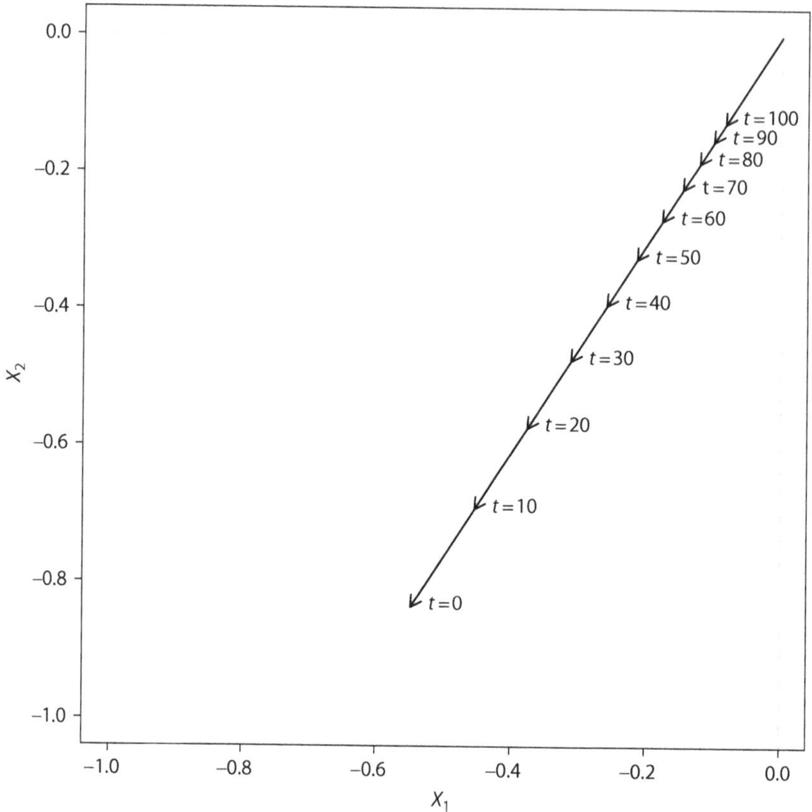

FIGURE 9.14 Geometric interpretation of mode 2. We can see progressive scaling of the second eigenvector as time goes on.

which yields $c_1(0) = -71.66$ and $c_2(0) = -94.55$. In other words, the initial condition is the following linear combination of eigenvectors:

$$\boldsymbol{X}(0) = -0.71\boldsymbol{v}_1 - 94.55\boldsymbol{v}_2 = \begin{bmatrix} 100 \\ 50 \end{bmatrix}$$

When we use this condition in Equation 9.48, we obtain dynamics given by a linear combination of modes as shown by the trajectories in Figure 9.15 and the phase portrait and ratio shown in Figure 9.16. As we can see, the ratio of X_1 to X_2 settles to 0.656, which means that the trajectory becomes that of the slow mode 2 since the fast mode 1 has already decayed.

9.8.2 Stability

Now that we know that eigenvalues determine the solution trajectories, let us go back to the catalog of cases we developed for a 2 × 2 matrix and summarized in Table 9.1. Using this table and what we know about modes, we will now catalog **eight possible dynamic behaviors**. These are organized in Table 9.2 and visualized using example graphs as explained next. The key concept is that exponentials with negative coefficients will decay, producing stability, whereas exponentials with positive coefficients will blow up, producing instability.

- Case 1: **Positive nonzero determinant**. There are five possible types of dynamic behaviors according to the type of eigenvalues obtained, which, in turn, depend on values of the trace and the discriminant.
    - Case 1.1: **Positive nonzero determinant and negative nonzero trace—stable**. When $\Delta > 0$ and $T < 0$, we can lump the behavior for $D = 0$ and $D > 0$ since both yield real

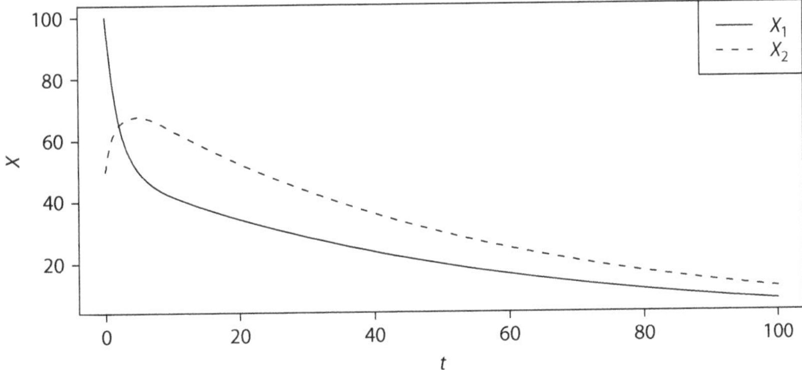

FIGURE 9.15 Mode combination.

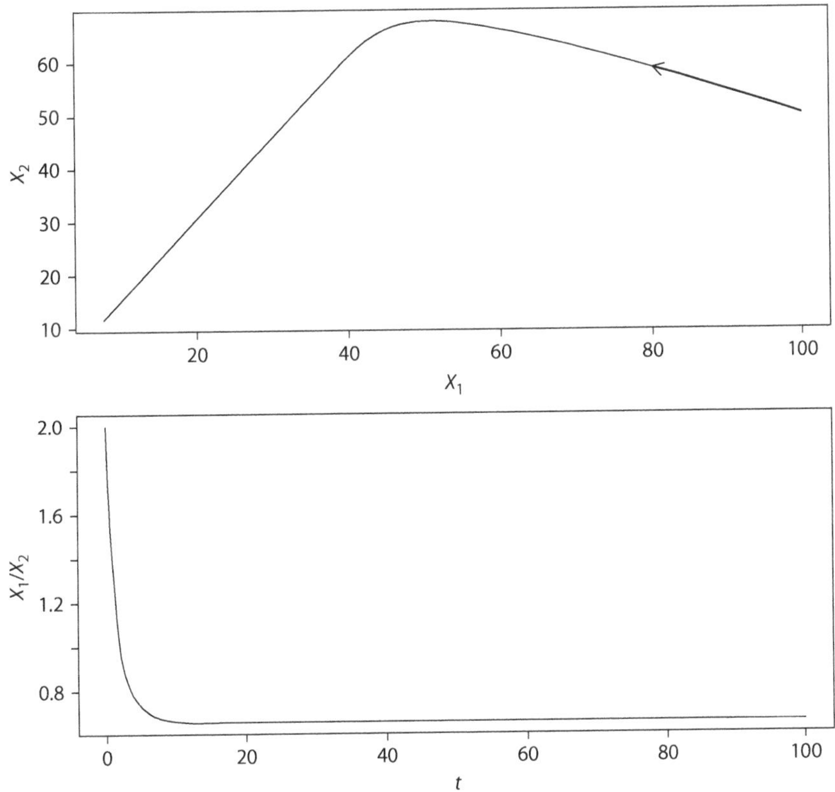

FIGURE 9.16 Phase plane and ratio plot of mode combination shows that the ratio settles to a constant value.

negative eigenvalues leading to stable trajectories (top left of Figures 9.17 and 9.19). On the contrary, if $D < 0$, the eigenvalues are complex conjugate with negative real parts also yielding stable trajectories in the form of damped oscillations (bottom left of Figures 9.17 and 9.19).

- Case 1.2: **Positive nonzero determinant and zero trace—oscillations**. When $\Delta > 0$ and $T = 0$, we have complex eigenvalues with zero real parts or pure imaginary parts yielding oscillatory dynamics (top left of Figures 9.18 and 9.20).
- Case 1.3: **Positive nonzero determinant and positive nonzero trace—unstable**. When $\Delta > 0$ and $T > 0$, we can lump the behavior for $D = 0$ and $D > 0$ since both yield real positive eigenvalues leading to unstable trajectories (top right of Figures 9.17 and 9.19).

TABLE 9.2
Eigenvalue and Stability Types by Values of Δ, *T*, and *D*

	$T < 0$	$T = 0$	$T > 0$
$\Delta > 0$, Case 1	Case 1.1 $D \geq 0$, real negative, stable, nonoscillatory, stable node	Case 1.2 $D < 0$, always complex, purely imaginary, sustained oscillations, center	Case 1.3 $D \geq 0$, real positive, unstable nonoscillatory, unstable node
	Case 1.1 $D < 0$, complex real negative parts, stable damped oscillations, spiral-in		Case 1.3 $D < 0$, complex real positive parts, unstable growing oscillations, spiral-out
$\Delta = 0$, Case 2	Case 2.1 $D > 0$, one real negative, one zero, stable saddle	Case 2.2 Null $D = 0$, zero eigenvalues	Case 2.3 $D > 0$, one real positive, one zero, unstable saddle
$\Delta < 0$, Case 3	For Cases 3.1, 3.2, 3.3, $D > 0$, real opposite signs, unstable		

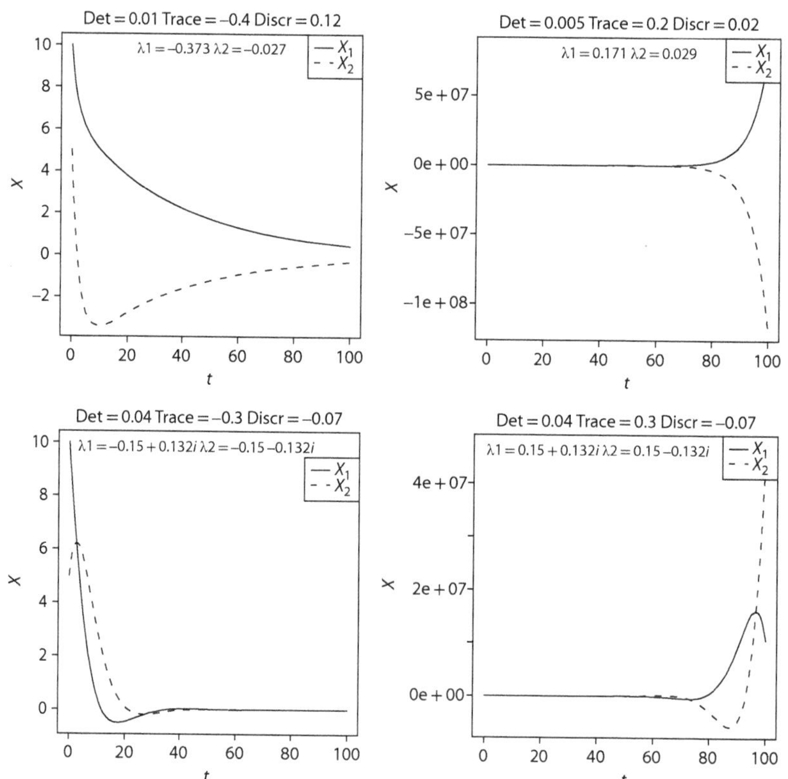

FIGURE 9.17 Dynamic behavior and stability for positive determinant except $T = 0$. Left: $T < 0$; right: $T > 0$; top: $D \geq 0$; bottom: $D < 0$.

On the contrary, if $D < 0$, the eigenvalues are complex conjugate with positive real parts also yielding unstable oscillatory trajectories (bottom right of Figures 9.17 and 9.19).

- Case 2: **Zero determinant**. When $\Delta = 0$, we have one eigenvalue zero and the other real with sign determined by the sign of T.
  - Case 2.1: **Zero determinant and negative trace—stable**. One mode is constant because $\exp(0) = 1$, and the other mode decays because of real negative eigenvalue.

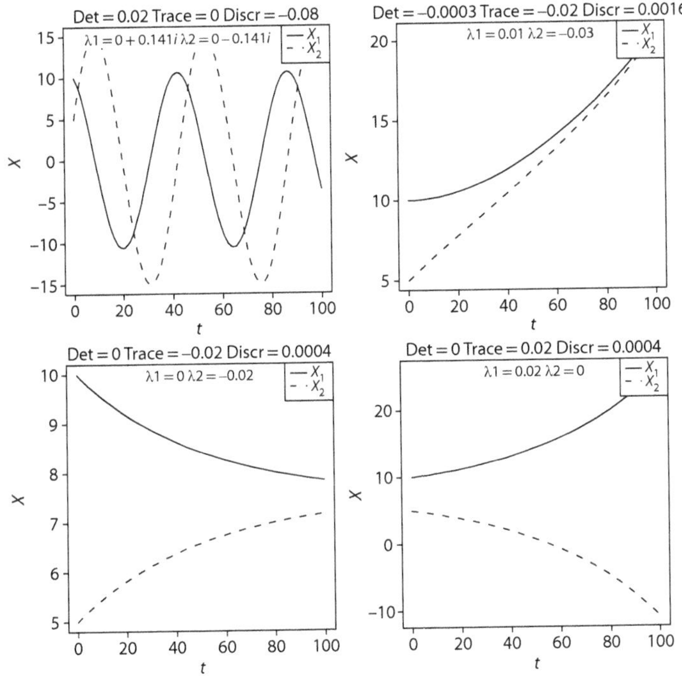

FIGURE 9.18 Dynamic behavior and stability for positive determinant $T = 0$ (top left), negative determinant (top right), zero determinant $T < 0$ (bottom left), and zero determinant $T > 0$ (bottom right).

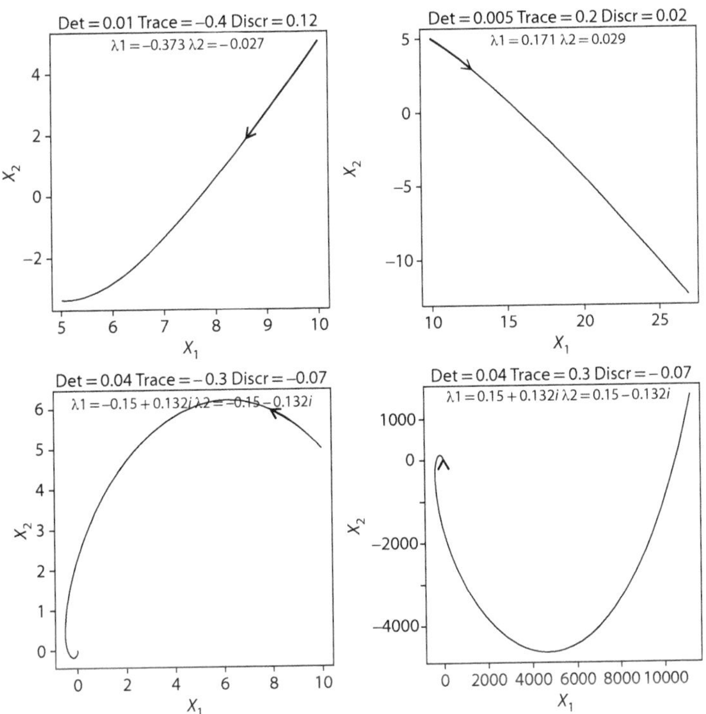

FIGURE 9.19 Phase portraits and stability for positive determinant except $T = 0$. Left: $T < 0$; right: $T > 0$; top: $D \geq 0$; bottom: $D < 0$.

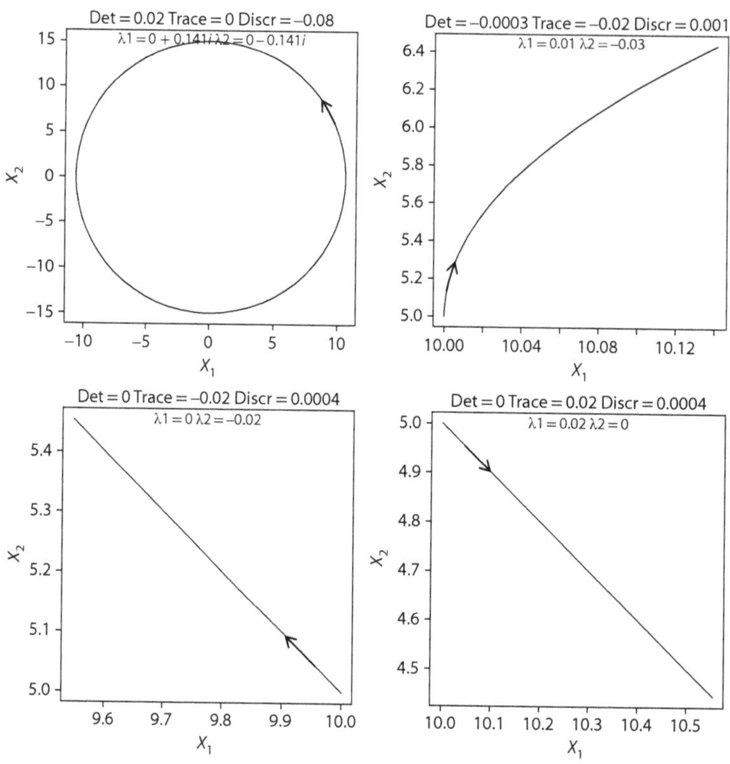

FIGURE 9.20 Phase portraits and stability for positive determinant $T = 0$ (top left), negative determinant (top right), zero determinant $T < 0$ (bottom left), and zero determinant $T > 0$ (bottom right).

So combined, we have a stable behavior, and $X_1 + X_2$ remains constant (bottom left of Figures 9.18 and 9.20).

- Case 2.2: **Zero determinant and positive trace—unstable**. As before, one mode is constant because exp(0) = 1, but the other mode blows up because of real positive eigenvalue. So combined, we have an unstable behavior (bottom right of Figures 9.18 and 9.20).
- Case 3: **Negative nonzero determinant—unstable**. When $\Delta < 0$ and regardless of the value of T, we have real eigenvalues of opposite sign. Thus, we can lump cases 3.1, 3.2, and 3.3 together. The positive eigenvalue makes one mode blow up, and the system is unstable (top right of Figures 9.18 and 9.20).

9.8.3 Discrete-Time Linear Dynamical Systems

When time is discontinuous or discrete, a linear dynamical system is a system of difference equations $\boldsymbol{X}(t+1) = \mathbf{A}\boldsymbol{X}(t)$. It is easy to see, by iterating this equation from $\boldsymbol{X}(0)$, that the solution is just $\boldsymbol{X}(t) = \mathbf{A}^t\boldsymbol{X}(0)$. In a similar manner to ODEs, the solution is a linear combination of eigenvectors, but the coefficients are power terms of the eigenvalues:

$$\boldsymbol{X}(t) = \sum_{i=1}^{n} c_i(0)\boldsymbol{v}_i\lambda_i^t \tag{9.51}$$

where $\boldsymbol{v}_i$ is the eigenvector associated to eigenvalue λ_i and $c_i(0)$ is the corresponding coefficient to form the initial condition as a linear combination of the eigenvectors:

$$\boldsymbol{X}(0) = \sum_{i=1}^{n} c_i(0)\boldsymbol{v}_i \tag{9.52}$$

We have modes as before when we select $c_i(0)$ such that we pick only one eigenvector.

For example, consider matrix $\mathbf{A} = \begin{bmatrix} 0.1 & 0.2 \\ 0.9 & 1.0 \end{bmatrix}$ of dimensions 2×2. The eigenvalues are $\lambda_1 = 1.17$ and $\lambda_2 = -0.07$, and the eigenvectors are $v_1 = \begin{bmatrix} -0.18 \\ -0.98 \end{bmatrix}$, $v_2 = \begin{bmatrix} -0.76 \\ 0.64 \end{bmatrix}$. Note that, of course, the trace of $\mathbf{A}$ is 1.1 and equal to the sum of eigenvalues $1.17 - 0.07 = 1.1$.

We can find modes as before and demonstrate how the solution follows progressive scaling of the eigenvectors; however, for the sake of space, we proceed straight to the solution. For example, taking $\boldsymbol{X}(0) = \begin{bmatrix} 100 \\ 50 \end{bmatrix}$, we find $c_1(0) = -117.96$ and $c_2(0) = -102.37$. When we use $\boldsymbol{c}(0)$, we obtain dynamics given by a linear combination of modes as shown by the trajectories in Figure 9.21 (top) and the phase portrait and ratio shown in the middle and bottom panels, respectively. As we can see, X_1 and X_2 blow up but follow a constant relation to each other, and the ratio of X_1 to X_2 settles to 0.187, which is exactly the ratio of element 1 to element 2 of eigenvector 1, $\frac{-0.184}{-0.983} = 0.187$. This means that the long-term relationship between the state variables is that of eigenvector 1, corresponding to the largest eigenvalue, which dominates the dynamics, that is, it is the **dominant** eigenvalue.

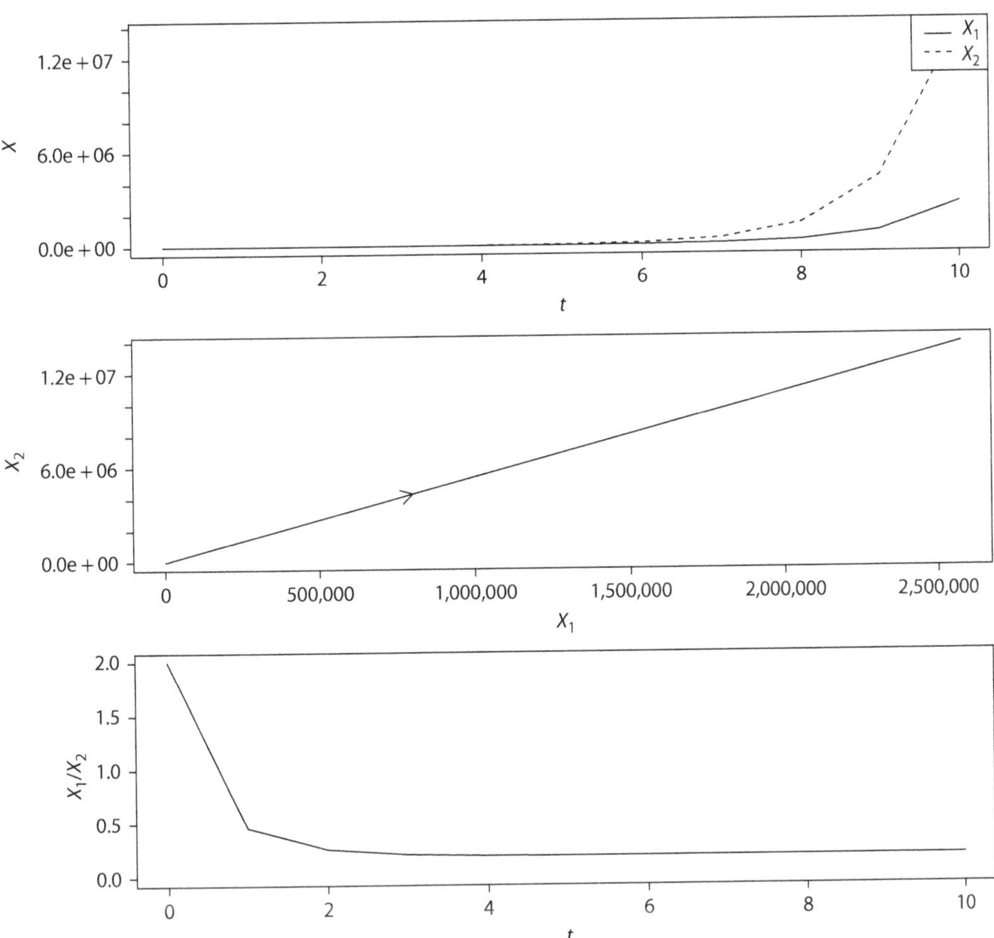

FIGURE 9.21 Discrete time solution. Top: trajectories for both state variables. Middle: phase portraits and stability. Bottom: ratio settles to a constant value.

To analyze all cases of stability, we can proceed as we did for ODEs. However, the key concept now is that powers of eigenvalues smaller than 1 will decay and produce stability, whereas powers of eigenvalues with values larger than 1 will blow up and produce instability. We will not catalog all cases here for the sake of space, but the interested reader can find this in many books (Cadzow, 1973; Lewis, 1977).

9.8.4 Applications

We will encounter linear dynamical systems throughout the remainder of this book. For example, in Chapter 10, we will look at stage-structured populations. In Chapter 11, we will look at compartment models, and in Chapter 12, at interacting populations.

9.9 COMPUTER SESSION: MATRIX ALGEBRA

9.9.1 Creating Matrices

We use the R base matrix library. We already know how to create a matrix from scanning a file. We can also create a matrix using a sequence of numbers. For example:

```
> A <- matrix(1:12, nrow=3, ncol=4)
> A
     [,1] [,2] [,3] [,4]
[1,]    1    4    7   10
[2,]    2    5    8   11
[3,]    3    6    9   12
```

Note that **A** is a rectangular matrix with three rows and four columns. We can obtain the dimensions of matrices with the `dim` function. For example:

```
> dim(A)
[1] 3 4
```

As another example:

```
> B <- matrix(1:9, nrow=3, ncol=3)
> B
     [,1] [,2] [,3]
[1,]    1    4    7
[2,]    2    5    8
[3,]    3    6    9
```

Note that **B** is a square matrix. Confirm with the `dim` function:

```
> dim(B)
[1] 3 3
```

Another way of generating a matrix is by means of the `structure` function and directly declaring the dimensions:

```
> structure(1:9, dim=c(3,3))
     [,1] [,2] [,3]
[1,]    1    4    7
[2,]    2    5    8
[3,]    3    6    9
```

This yields the same result as before.

If we want a matrix with random numbers, we could do

```
> D <- matrix(runif(9),ncol=3)
> round(D,2)
     [,1] [,2] [,3]
[1,] 0.12 0.45 0.60
[2,] 0.36 0.28 0.29
[3,] 0.48 0.85 0.19
>
```

9.9.2 Operations

Operators + and * represent an entry-by-entry sum and multiplication (the Hadamard product) and the dimensions must match. Therefore, we can perform the following:

```
> A + A
     [,1] [,2] [,3] [,4]
[1,]    2    8   14   20
[2,]    4   10   16   22
[3,]    6   12   18   24

> B+B
     [,1] [,2] [,3]
[1,]    2    8   14
[2,]    4   10   16
[3,]    6   12   18
```

But we cannot use + and * on **A** and **B** because the dimensions of **A** and **B** do not match. Try it to see:

```
> A+B
Error in A + B : non-conformable arrays
> A*B
Error in A * B : non-conformable arrays
>
```

Operators *, /, +, and – can also be used with scalars, for example:

```
> C <- 3*B
> C
     [,1] [,2] [,3]
[1,]    3   12   21
[2,]    6   15   24
[3,]    9   18   27

> C/3
     [,1] [,2] [,3]
[1,]    1    4    7
[2,]    2    5    8
[3,]    3    6    9
```

To perform matrix multiplication, we use the operator `%*%` (which is not the same as *). Matrices need to conform for multiplication. Operator `%*%` is not commutative. For example, we can do **BA**, that is, premultiply **A** by **B** because $(3 \times 3) \times (3 \times 4) = (3 \times 4)$:

```
> B%*%A
     [,1] [,2] [,3] [,4]
[1,]   30   66  102  138
[2,]   36   81  126  171
[3,]   42   96  150  204
```

However, we cannot do **AB**, that is, premultiply **B** by **A** because they do not conform, that is, $(3 \times 4) \times (3 \times 3)$ do not match. Thus, if we try, we get an error message:

```
> A%*%B
Error in A %*% B : non-conformable arguments
>
```

Another example is to multiply $3 \times \mathbf{M}$ where $\mathbf{M} = \begin{pmatrix} 1 & 4 \\ 2 & 5 \\ 3 & 6 \end{pmatrix}$:

```
> M <- structure(1:6, dim=c(3,2))
> M
     [,1] [,2]
[1,]    1    4
[2,]    2    5
[3,]    3    6
> 3*M
     [,1] [,2]
[1,]    3   12
[2,]    6   15
[3,]    9   18
```

Calculate $\mathbf{A}^3$ where matrix $\mathbf{A} = \begin{pmatrix} 17 & -6 \\ 45 & -16 \end{pmatrix}$. First, generate the matrix:

```
> A <- matrix(c(17,-6,45,-16), ncol=2, byrow=T)
> A
     [,1] [,2]
[1,]   17   -6
[2,]   45  -16
>
```

Then, multiply three times:

```
> A%*%A%*%A
     [,1] [,2]
[1,]   53  -18
[2,]  135  -46
```

By the way, the power operation ^ does **not** yield the correct result, as you can check by trying the following:

```
> A^3
      [,1]  [,2]
[1,]  4913  -216
[2,] 91125 -4096
```

What this operation does is simply raise each entry to a power, for example, $17^3 = 4913$, and that is not the result of the matrix power operation.

9.9.3 Other Operations

To transpose a matrix, use function `t()`:

```
> t(C)
     [,1] [,2] [,3]
[1,]    3    6    9
[2,]   12   15   18
[3,]   21   24   27
```

To extract the diagonal and the trace, do the following:

```
> diag(C)
[1]  3 15 27
> sum(diag(C))
[1] 45
```

To construct an identity matrix, try the following:

```
> diag(4)
     [,1] [,2] [,3] [,4]
[1,]    1    0    0    0
[2,]    0    1    0    0
[3,]    0    0    1    0
[4,]    0    0    0    1
```

To calculate the determinant of a matrix, use the `det` function. The matrix must be square. For example, try the following:

```
> A
     [,1] [,2] [,3] [,4]
[1,]    1    4    7   10
[2,]    2    5    8   11
[3,]    3    6    9   12
> det(A)
Error in det(A) : x must be a square matrix
```

But, we can apply `det` to matrix **B**:

```
> B
     [,1] [,2] [,3]
[1,]    1    4    7
[2,]    2    5    8
[3,]    3    6    9
> det(B)
[1] 0
```

Note that **B** is singular because det(**B**) is 0. This is because the columns depend on each other.

We can also calculate the det of an identity matrix, and it should be 1:

```
> det(diag(3))
[1] 1
```

Also, the determinant of matrix **D** above is

```
> det(D)
[1] -0.01062473
```

9.9.4 Solving the System of Linear Equations

Recall that to solve a system of linear equations $\mathbf{B}x = c$, we premultiply by the inverse $\mathbf{B}^{-1}\mathbf{B}x = \mathbf{B}^{-1}c$ to obtain the solution $x = \mathbf{B}^{-1}c$. We can also use the function `solve`, for example,

```
> D <- matrix(c(19,2,15,8,18,19,11,17,10), nrow=3, ncol=3)
> D
     [,1] [,2] [,3]
[1,]   19    8   11
[2,]    2   18   17
[3,]   15   19   10
> c <- c(9,5,14)
> x <- solve(D, c)
> round(x,2)
[1]  0.45  0.58 -0.38
>
```

However, try

```
> c <- c(1,2,3)
> x <- solve(B,c)
Error in solve.default(B, c) :
  Lapack routine dgesv: system is exactly singular
>
```

It cannot solve the problem because **B** is singular, and, therefore, det = 0, and it has no inverse.

9.9.5 Inverse

To calculate the inverse of a matrix **B** we also use the function `solve` but make **c** the identity matrix of the same dimension as **B**. This works because

$$\mathbf{B}^{-1}\mathbf{B}x = \mathbf{B}^{-1}\mathbf{I}$$

$$x = \mathbf{B}^{-1}$$

Thus, use **D** above and build an identity matrix and apply `solve`:

```
> I <- diag(3)
> I
     [,1] [,2] [,3]
[1,]    1    0    0
[2,]    0    1    0
[3,]    0    0    1
> D.inv <- solve(D,I)
> round(D.inv,2)
      [,1]  [,2]  [,3]
[1,]  0.04 -0.04  0.02
[2,] -0.07 -0.01  0.09
[3,]  0.07  0.07 -0.10
>
```

9.9.6 Eigenvalues and Eigenvectors

For example, consider matrix $\mathbf{A} = \begin{pmatrix} 1 & 2 \\ -3 & 1.5 \end{pmatrix}$. First, calculate the determinant of **A**:

```
> A <- matrix(c(1,2,-3,1.5), byrow=T, ncol=2)
> A
     [,1] [,2]
[1,]    1  2.0
[2,]   -3  1.5
> det(A)
[1] 7.5
```

`eigen(A)` gives the eigenvalues and eigenvectors:

```
> eigen(A)
$values
[1] 1.25+2.436699i 1.25-2.436699i

$vectors
                     [,1]                  [,2]
[1,] 0.0645497-0.6291529i 0.0645497+0.6291529i
[2,] 0.7745967+0.0000000i 0.7745967+0.0000000i
>
```

Recall that a complex number $a + bi$ is made out of a real part, a, and an imaginary part, b, and that $i = \sqrt{-1}$. Therefore, above the eigenvalues, forms a complex conjugate pair $1.25 \pm 2.44i$. The real part is 1.25 and the imaginary part is $\pm$ 2.44. The eigenvectors are also in complex conjugate pairs.

The following short script plots the eigenvalues in the complex plane (Figure 9.5):

```
A <- matrix(c(1,2,-3,1.5), byrow=T, ncol=2)
L <- eigen(A)$values
plot(Re(L[1]),Im(L[1]),xlab="Re",ylab="Im",xlim=c(-2,2),ylim=c(-3,3))
lines(Re(L[2]),Im(L[2]),type="p")
abline(h=0,v=0, col="grey")
```

Then, we can use the following script to obtain a root locus like the one in Figure 9.6:

```
A <- matrix(c(1,2,-3,1.5), byrow=T, ncol=2)
p <- array(); p[1] <- A[1,1]; np <-100
for(i in 2:np) p[i] <- p[i-1]-1/(i-1)
L <- matrix(nrow=np,ncol=2)
L[1,] <- eigen(A)$values
plot(Re(L[1,1]),Im(L[1,1]),xlab="Re",ylab="Im",xlim=c(-4,2),ylim=c(-3,3))
lines(Re(L[1,2]),Im(L[1,2]),type="p")
abline(h=0,v=0, col="grey")
for(i in 2:np){
 A[1,1] <- p[i]; eigen(A)
 L[i,] <- eigen(A)$values
 lines(Re(L[i,1]),Im(L[i,1]),type="p")
```

```
   lines(Re(L[i,2]),Im(L[i,2]),type="p")
   abline(h=0,v=0, col="grey")
}
arrows(Re(L[1,1]),Im(L[1,1]),Re(L[2,1]),Im(L[2,1]),length=0.1)
arrows(Re(L[1,2]),Im(L[1,2]),Re(L[2,2]),Im(L[2,2]),length=0.1)
```

Exercise 9.13

Calculate the major product matrix $\mathbf{X}^T\mathbf{X}$ where vector $\boldsymbol{x}$ is composed of ten values of $\boldsymbol{X}$ drawn from a standard normal random variable. Hint: Use x <- rnorm(10,0,1).

Exercise 9.14

Calculate the determinant of the major product matrix of the previous exercise. Calculate the inverse.

Exercise 9.15

Use the matrices $\mathbf{A} = \begin{bmatrix} 1 & 3 \\ 2 & 4 \end{bmatrix}$ and $\mathbf{B} = \begin{bmatrix} 5 & 0 \\ 0 & 3 \end{bmatrix}$ and the vectors $\mathbf{c} = \begin{bmatrix} 2 \\ 0 \end{bmatrix}$ and $\boldsymbol{x} = \begin{bmatrix} x_1 \\ x_2 \end{bmatrix}$. Calculate **AI**, **AB**, **BA**, **Bc**, **B**$\boldsymbol{x}$, and **Ic**, where **I** is the identity matrix. Write the equation $\mathbf{B}\boldsymbol{x} = \mathbf{c}$. Solve for $\boldsymbol{x}$.

9.9.7 Linear Dynamical Systems

To find the solution of a linear dynamical system given by matrix **A** and initial condition $\boldsymbol{X}(0)$, simply find the eigenvalues and their eigenvectors:

```
V <- eigen(A)$vectors
L <- eigen(A)$values
```

Then, find a linear combination coefficient $\boldsymbol{c}(0)$ using Equation 9.49:

```
c0 <- solve(V,X0)
```

Define the time sequence, allocate a matrix to store the results, and apply Equation 9.48:

```
t <- seq(0,100,0.1); nt <- length(t)
X <- matrix(nrow=nt,ncol=2)
for(i in 1:length(t))
X[i,] <- c0[1]*V[,1]*exp(L[1]*t[i])+ c0[2]*V[,2]*exp(L[2]*t[i])
```

For example, using $\mathbf{A} = \begin{bmatrix} -0.4 & 0.25 \\ 0.2 & -0.15 \end{bmatrix}$ with $\boldsymbol{X}(0) = \begin{bmatrix} 100 \\ 50 \end{bmatrix}$ will produce Figure 9.15:

```
t <- seq(0,100,0.1); nt <- length(t)
A <- matrix(c(-0.4,0.25,0.2,-0.15), byrow=T, ncol=2)
X0 <-c(100,50); c0 <- solve(V,X0)
for(i in 1:length(t))
X[i,] <- c0[1]*V[,1]*exp(L[1]*t[i])+ c0[2]*V[,2]*exp(L[2]*t[i])
matplot(t,X,type="l", col=1)
legend("topright",leg=c("X1","X2"),lty=1:2,col=1)
```

The following script allows us to explore the dynamics of mode 1 of a linear dynamical system. For this example, $\mathbf{A} = \begin{bmatrix} -0.4 & 0.25 \\ 0.2 & -0.15 \end{bmatrix}$ and the results are in Figure 9.10 (top) and Figure 9.11:

```
t <- seq(0,100,0.1); nt <- length(t)
A <- matrix(c(-0.4,0.25,0.2,-0.15), byrow=T, ncol=2)
V <- eigen(A)$vectors; L <- eigen(A)$values
X <- matrix(nrow=nt,ncol=2)
c0 <-c(1,0); x0 <- solve(V,c0)
for(i in 1:length(t))
X[i,] <- c0[1]*V[,1]*exp(L[1]*t[i])+ c0[2]*V[,2]*exp(L[2]*t[i])
matplot(t,X,type="l", col=1)
legend("topright",leg=c("X1","X2"),lty=1:2,col=1)
plot(X[,1],X[,2],xlab="X1",ylab="X2",type="l")
arrows(X[1,1],X[1,2],X[10,1],X[10,2],length=0.1,lwd=1.7)
plot(t,X[,1]/X[,2],xlab="t",ylab="X1/X2",type="l")
```

We can adjust this easily to study mode 2.

Exercise 9.16

Calculate the dynamics of mode 2 using the script just given. Obtain graphs like the ones shown in Figure 9.10 (bottom) and Figure 9.13. Hint: Use c0 <- c(0,1).

10 Structured Population Models

So far, we have modeled population as nonstructured, that is, we have ignored the differences among individuals according to sex, age, size, and so on. When we discussed the exponential model, we assumed a total population or population density and that the rate coefficient was the same for all individuals. In this chapter, we include differences in individuals to obtain a more detailed and accurate model. In doing so, the model becomes multidimensional since a variable for each class is required. In this chapter, several ways of structuring populations and their corresponding modeling approaches are considered (Nisbet, 1989; Nisbet and Gurney, 1986; Swartzman and Kaluzny, 1987).

10.1 TYPES OF STRUCTURE

There are many ways of modeling populations with structure, the principal ones being by age, size, life cycle stage, and sex. First, for age, use age classes that are in synchrony with the simulation time step. For example, if the simulation is a 1-year time step, then yearly age classes will be required, that is, individuals would be classified as 1 year old, 2 years old, and so forth until n years old, where n is the maximum age. The interval for age classes could also be a multiple of the time step, even though it would have less resolution. For example, if the maximum age is 40 years, then the interval for the age classes would be 10 times larger than the time step to obtain four classes, namely, from 1–10 years old, 11–20 years old, 21–30 years old, and 31–40 years old.

Second, for size, use some measure of size such as the carapace length in zooplankton, diameter at breast height (DBH) in trees, and so on. Here, it is important to have a function that gives growth as a function of time (individual growth or change of the attribute with time, e.g., change of the carapace length or of the DBH with time) so that the passage of time leads to a change in size. Third, for life cycle stage, use a clearly distinguishable stage of a life cycle, for example, eggs, larvae, and adults in insects. Finally, for sex, divide the population into male and female. It is also possible to mix a couple or several of the aforementioned criteria and configure a more complicated classification, for example, eggs, small larvae, large larvae, and female adults.

Exercise 10.1

Assume that the individuals die after the 11th year of age and the age interval is 4 years. How many age classes do you obtain? What would the classes be?

10.2 METHODS OF MODEL STRUCTURE

The most important change with respect to nonstructured models is that the state is not just a scalar. It becomes either a vector for **discrete** classes or a density function with respect to the attributes for **continuous** classes. In discrete classes, each element of the state vector is the total number of individuals in one population class. In continuous classes, the area under the density curve for an attribute interval da (e.g., age interval) is the total number of individuals for the interval da.

Discrete class models include projection matrices by age like the Leslie–Lewis matrix or by size or stage like the Lefkovitch matrix (Caswell, 2001); ODE with states defined by classes (Acevedo, 1980a); delay-differential equations (DDE) by age, size, or stage (Nisbet and Gurney, 1986); and networks by age, size, or stage (Lewis, 1977). Continuous class models include partial differential equations like the McKendrick–von Foerster equation (Kot, 2001). For a variety of methods, see Tuljapurkar and Caswell (1997). We discuss only projection matrices, ODEs, and DDE approaches. We start with the projection matrices, which are the easiest to simulate.

10.3 PROJECTION MATRICES

Projection matrices have been well developed in ecological modeling, and there are numerous applications of this approach (Caswell, 2001).

10.3.1 Age-Structured Leslie–Lewis Matrix

Assume a hypothetical population with the following classes: 0 year (neonates), 1 year (juveniles), and 2 years (mature). Therefore, the state is a vector with three entries. Assume that individuals from ages 2–3 years can reproduce and that the brood at each reproduction event is f_i neonates. In the time period $[t, t + 1]$, an individual in class i can either die or survive. When an individual survives, it is promoted to age class $i + 1$ and produces offspring if $i \geq 2$. We can write the number of individuals in each age class as

$$X_{i+1}(t+1) = s_i \; X_i(t) \tag{10.1}$$

for all age classes, $i = 1, 2, 3$, where s_i is the survival fraction for age class i, and in addition,

$$X_1(t+1) = \sum_{i=2}^{3} f_i \; X_i(t) \tag{10.2}$$

where f_i is the fecundity for age class i (nonzero only when $i = 2, 3$).

Exercise 10.2

What are the units for s_i and f_i?

Equations 10.1 and 10.2 together constitute a system of linear difference equations that can be represented as the matrix equation

$$\begin{bmatrix} X_1(t+1) \\ X_2(t+1) \\ X_3(t+1) \end{bmatrix} = \begin{pmatrix} 0 & f_2 & f_3 \\ s_1 & 0 & 0 \\ 0 & s_2 & 0 \end{pmatrix} \begin{bmatrix} X_1(t) \\ X_2(t) \\ X_3(t) \end{bmatrix} \tag{10.3}$$

or in short,

$$\boldsymbol{X}(t+1) = \mathbf{P} \; \boldsymbol{X}(t) \tag{10.4}$$

where $\boldsymbol{X}$ denotes the vector of the number of individuals in the population and $\mathbf{P}$ is the projection matrix. Vector $\boldsymbol{X}$ is a column vector:

$$\boldsymbol{X}(t) = \begin{bmatrix} X_1(t) \\ X_2(t) \\ X_3(t) \end{bmatrix} \tag{10.5}$$

The projection matrix **P** is named after Leslie and Lewis and contains the parameters s_i and f_i:

$$\mathbf{P} = \begin{pmatrix} 0 & f_2 & f_3 \\ s_1 & 0 & 0 \\ 0 & s_2 & 0 \end{pmatrix} \tag{10.6}$$

More generally, assuming all nonzero values for parameters f_i, we have

$$\mathbf{P} = \begin{pmatrix} f_1 & f_2 & f_3 \\ s_1 & 0 & 0 \\ 0 & s_2 & 0 \end{pmatrix} \tag{10.7}$$

Figure 10.1 illustrates the Leslie projection matrix. It is assumed that only juveniles and adults have nonzero fertility. In the figure, boxes represent the classes and arrows represent the transitions from one class to another.

Now that we know some matrix algebra, we can solve the linear equation (Equation 10.4) and analyze it for stability. For example, the maximal eigenvalue is the growth rate of the population and the stable age distribution is the corresponding eigenvector.

The simulation of this model is very easy: start with the initial condition and premultiply it by the projection matrix to obtain $X(t)$ and then repeat for each time, t. That is, apply Equation 10.4 recursively. Repeating for many time steps, we get the trajectory of the population by age class. To obtain the total population, we add the numbers in all classes.

We will work with a very simple three-age class example of a bird population given in the book by Swartzman and Kaluzny (1987). The birds breed in the second and third years of age with fertilities of 8 and 12 hatched eggs, respectively. When these eggs hatch, there is an equal probability of having a male or a female individual. Survival of the two first age classes is 50%. This model is easy to formulate if the time step is the same as the age class interval. We consider female individuals only and reduce the fecundity coefficients by one-fourth, considering that only half of the eggs will become female individuals and that only half of the population are reproducing females:

$$\mathbf{P} = \begin{bmatrix} 0 & 2 & 3 \\ 0.5 & 0 & 0 \\ 0 & 0.5 & 0 \end{bmatrix} \tag{10.8}$$

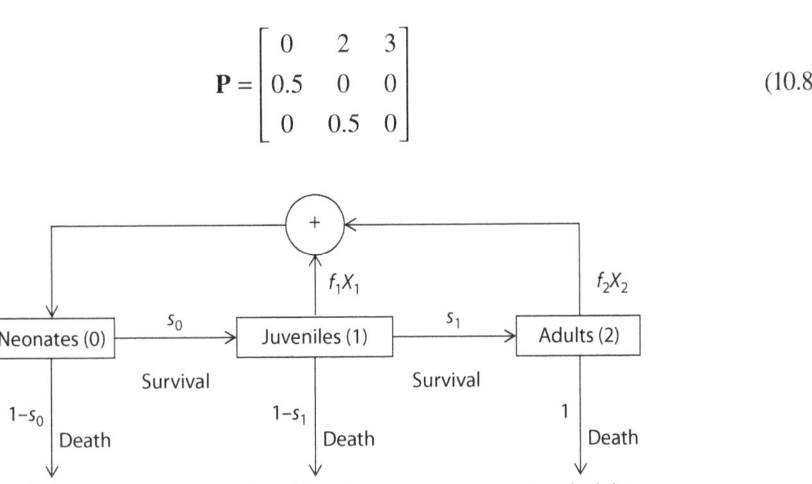

FIGURE 10.1 Schematic representation of Leslie projection matrix for three age classes.

The initial condition for the number of females is

$$\boldsymbol{X}(0) = \begin{bmatrix} 100 \\ 50 \\ 25 \end{bmatrix} \tag{10.9}$$

Exercise 10.3

Calculate $\boldsymbol{X}(1)$ and $\boldsymbol{X}(2)$ using Equation 10.4.

Repeating Equation 10.4 for many time steps, we get the trajectory of the population by the age class and total population (see Figure 10.2). We can see how the total population and each class tend to grow exponentially. The densities are **unstable**; they increase with time in an exponential pattern (top left panel). The total population is analogous to the simple exponential growth as already discussed; in fact, one can estimate r or intrinsic growth rate (bottom right panel) from the **dominant eigenvalue** of the projection matrix. However, the **age class distribution**, that is, the set of proportions in each class with respect to the total, is **stable**; the trajectories approach or settle at a steady-state value as time progresses (top right panel). The ratio of two successive values of the total population also settles to a steady-state value (bottom left panel). We will now see in detail how this happens.

The characteristic equation of matrix $\mathbf{P}$ is found from the determinant of $\lambda\mathbf{I} - \mathbf{P} = 0$,

$$\begin{vmatrix} \lambda & -2 & -3 \\ -0.5 & \lambda & 0 \\ 0 & -0.5 & \lambda \end{vmatrix} = 0 \tag{10.10}$$

that is,

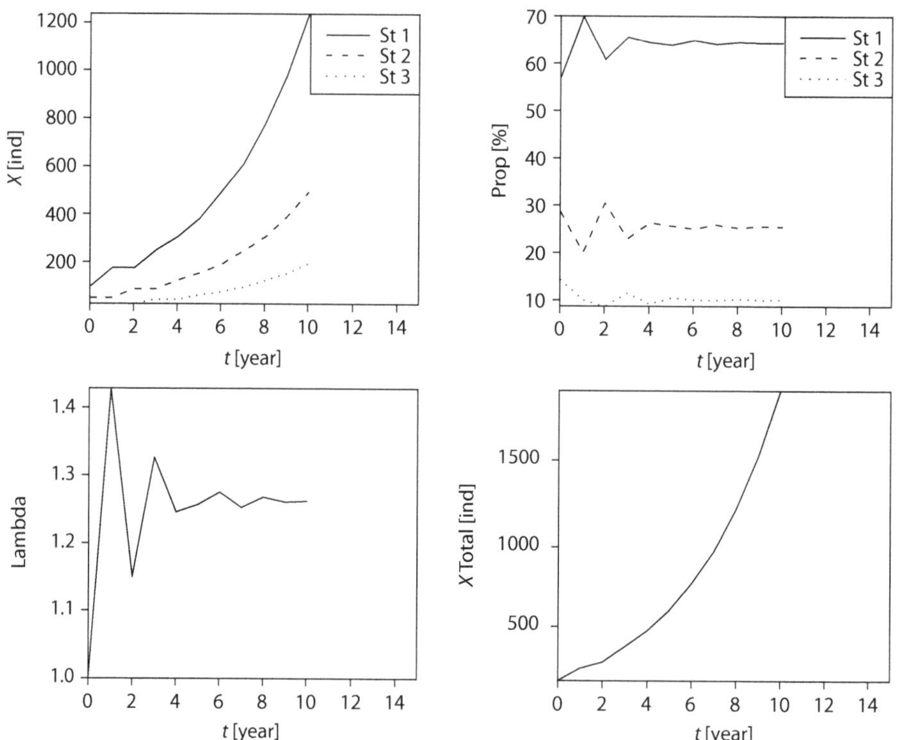

FIGURE 10.2 Population projection by class, proportions (in %), dynamics of λ, and total population.

TABLE 10.1
Population by Class, Total, and Ratio of Total in Successive Steps

t	X1	X2	X3	X_{total}	Ratio
0	100	50	25	175	
1	175	50	25	250	1.429
2	175	88	25	288	1.152
3	250	88	44	381	1.323
4	306	125	44	475	1.247
5	381	153	63	597	1.257
6	494	191	77	761	1.275
7	611	247	95	953	1.252
8	780	305	123	1210	1.270
9	981	390	153	1520	1.256
10	1240	491	195	1920	1.263

$$\lambda^3 - \lambda - 0.75 = 0 \tag{10.11}$$

The real positive solution is $\lambda = 1.263$. It is a nice result that the total population grows with the intrinsic rate determined by the maximal or dominant root or eigenvalue. The relation is $\lambda = e^r$ or $r = \ln(\lambda)$. Therefore, in this example, $r = \ln(1.263) = 0.233$.

As time increases, the ratio of the two successive values of the total population should approach the dominant eigenvalue. Table 10.1 shows the population projection by age class and total. The last column is the ratio of total population for consecutive time steps 2 to 1, 3 to 2, …, 10 to 9. We can appreciate how it converges to the dominant eigenvalue 1.263.

Exercise 10.4

Calculate the ratio of $X_{total}(10)/X_{total}(9)$ and $X_{total}(1)/X_{total}(0)$ and check the values of the ratio mentioned in Table 10.1.

Exercise 10.5

Is the total population growing exponentially with rate coefficient $r = 0.233$ at all times during the simulation?

Now, consider the proportion in each class with respect to the total; the age distribution reaches a steady state or stable age distribution as time increases, as seen in Table 10.2 and Figure 10.2. In Table 10.2, the three columns labeled "Prop" are proportions with respect to the total, and the remaining columns are ratios with respect to age class 3.

An important concept to grasp about Table 10.2 and Figure 10.2 is that there is a stable age distribution. Note that this does not mean that the population in each class, nor the total, stops growing; in fact, the population grows exponentially at a rate indicated by the maximal eigenvalue. Another nice result is that the stable age distribution is given by the eigenvector corresponding to the dominant eigenvalue

$$v_1 = \begin{bmatrix} 6.38 \\ 2.53 \\ 1 \end{bmatrix}$$

TABLE 10.2
Proportions in Each Class with Respect to the Total and Ratio of Each Class with Respect to One Class

t	Prop1	Prop2	Prop3	X1:X3	X2:X3	X3:X3
0	0.571	0.286	0.143	3.999	1.999	1.000
1	0.700	0.200	0.100	7.000	2.000	1.000
2	0.609	0.304	0.087	6.997	3.498	1.000
3	0.656	0.230	0.115	5.712	1.999	1.000
4	0.645	0.263	0.092	7.000	2.858	1.000
5	0.639	0.257	0.105	6.100	2.450	1.000
6	0.649	0.251	0.101	6.450	2.490	1.000
7	0.641	0.259	0.100	6.410	2.590	1.000
8	0.645	0.253	0.102	6.318	2.475	1.000
9	0.644	0.256	0.100	6.426	2.553	1.000
10	0.644	0.255	0.101	6.353	2.518	1.000

Compare the eigenvector with the last row of the last three columns in Table 10.2. An intuitive understanding of this concept is that iteratively multiplying the matrix by the eigenvector yields vectors with the same direction, but the magnitude grows or declines according to the eigenvalue.

Exercise 10.6

Verify that the eigenvector $\boldsymbol{v}_1$ is a solution of the $\mathbf{P}\boldsymbol{v} = \lambda \boldsymbol{v}$ equation corresponding to the maximal eigenvalue.

Exercise 10.7

Consider a Leslie matrix $\mathbf{P} = \begin{bmatrix} 0 & 2 \\ 0.5 & 0 \end{bmatrix}$ and an initial condition $\boldsymbol{X}(0) = \begin{bmatrix} 10 \\ 5 \end{bmatrix}$. (a) Project this population for time 1, 2, and 3, that is, calculate $\boldsymbol{X}(1)$, $\boldsymbol{X}(2)$, and $\boldsymbol{X}(3)$. (b) Calculate the ratio of $X_{total}(3)/X_{total}(2)$, $X_{total}(2)/X_{total}(1)$, and $X_{total}(1)/X_{total}(0)$. Is this population constant with rate coefficient $r = 0$? (c) Calculate the maximal eigenvalue and verify that it is 1, that is, $\lambda = 1.0$. (d) Calculate the eigenvector corresponding to this eigenvalue and verify that it is $\boldsymbol{v}_1 = \begin{bmatrix} 2 \\ 1 \end{bmatrix}$ and that it matches the stable steady-state value.

10.4 EXTENSIONS

One way of modeling the effects of exogenous processes, such as harvesting or exposure to a toxicant, is to make the survival and fertility parameters respond to the forcing factors. For example, harvesting reduces survival parameters s_i, **lethal** exposure levels to a toxicant reduces survival parameters s_i, and **sublethal** exposure levels affect fertility parameters f_i.

We can model density dependence by making the survival and fecundity rates depend on the density or the numbers in the age classes. For example, survival of larvae depends on the number of larvae already present because of resource depletion due to overcrowding. For example, if the survival of neonates from 1 day to 3 days is dependent on their abundance, then model $s_i = s_i(X_1, X_2, X_3)$, where $s_i(., ., .)$ is a decreasing function of the densities X_1, X_2, X_3.

10.4.1 Size or Stage Classes: Lefkovitch Matrix

We now write a matrix similar to the Leslie matrix, but allow for "staying" in a class, that is, besides dying or moving to the next class, an individual can survive and stay in the class by not growing to the required size.

Consider the hypothetical population used as an example in Section 10.3.1. Assume three size classes, for example, 0–1, 1–2, and 2–3 units of length:

$$X_{i+1}(t+1) = s_i\, X_i(t) + r_i\, X_{i+1}(t) \tag{10.12}$$

for $i = 1, 2, 3$, where r_i is the remaining rate; now, the main diagonal entries are not all zeros and contain the remaining rates. Assuming that only classes 1 and 2 reproduce and that all class 0 individuals mature (none remaining),

$$\mathbf{P} = \begin{pmatrix} 0 & f_1 & f_2 \\ s_0 & r_1 & 0 \\ 0 & s_1 & r_2 \end{pmatrix} \tag{10.13}$$

The properties of this matrix are similar to the Leslie matrix. A problem with population matrix models is that it can lead to erroneous results if we do not select classes carefully in regard to the time step. We assume that in a time step, an individual grows in length with respect to the resolution of the size classes; if there is a reasonable growth for the organism, then the model will perform well. However, consider another example: if you model a population and select classes of large size and simulate with a time step of 1 year, you would assume that individuals could grow at an unrealistic rate. Therefore, it is very important to be careful with the selection of size and stage classes and the time step when using projection matrices. Keep the class intervals scaled properly to the time step.

10.5 CONTINUOUS TIME STAGE STRUCTURED MODEL

We can also model stage structure as a system of ODEs where the dependent variables or states correspond to the number of individuals in each class, and there are rates of change to transition among states. For example, consider a two-stage class structure, neonates and mature, where we describe maturation by a simple rate. This is a **lumped** model and is unrealistic because there is no time delay required for the transition from neonate to mature. We have two variables, X_1 for neonates and X_2 for mature. Now we have inflow to neonates by birth calculated as a fertility coefficient, b, multiplied by the number of adults. Outflow from neonates is calculated by the maturation rate parameterized by coefficient m and death given by coefficient d_1. The outflow from neonates from maturation becomes the inflow to mature, and then, outflow from mature is by death parameterized by coefficient d_2:

$$\begin{aligned} \frac{dX_1}{dt} &= bX_2 - mX_1 - d_1X_1 \\ \frac{dX_2}{dt} &= mX_1 - d_2X_2 \end{aligned} \tag{10.14}$$

This can be rewritten as a matrix equation:

$$\frac{d}{dt}\begin{bmatrix} X_1 \\ X_2 \end{bmatrix} = \begin{bmatrix} -m-d_1 & b \\ m & -d_2 \end{bmatrix}\begin{bmatrix} X_1 \\ X_2 \end{bmatrix} \tag{10.15}$$

Selecting parameter values $m = 0.2$, $d_1 = 0.2$, $d_2 = 0.15$, and $b = 0.25$ yields matrix.

$$\begin{bmatrix} -0.4 & 0.25 \\ 0.2 & -0.15 \end{bmatrix}$$

This matrix was used in Chapter 9 to exemplify linear dynamical systems. Recall the eigenvalues, which are $\lambda_1 = -0.53$ and $\lambda_2 = -0.02$, and the eigenvectors are

$$v_1 = \begin{bmatrix} -0.88 \\ 0.46 \end{bmatrix}, \quad v_2 = \begin{bmatrix} -0.55 \\ -0.84 \end{bmatrix}$$

Both eigenvalues are real and negative, and therefore, we know from Chapter 9 that both modes are decaying exponentials, and the system is stable.

Although we can calculate the solution from the eigenvalues and eigenvectors, let us simulate this model using numerical methods so that we will be ready for more complicated situations once we include nonlinear terms and disturbances. The simulation yields the results shown in Figure 10.3 for various values of b. We can see that in particular $b = 0.3$, a value at which birth compensates maturation plus death, yields steady-state values of $X_1 = 56$ and $X_2 = 73$. Values of b lower than 0.3 lead to decline, whereas values greater than 0.3 lead to growth. Recalculating the eigenvalues for $b = 0.3$, we get $\lambda_1 = -0.55$ and $\lambda_2 = 0.0$, which, we know from Chapter 9, produces one steady-state mode and another mode that decays.

This model would be unrealistic in most cases because instantaneous maturation is not possible. To account for maturation time, we can assume that neonates do not become reproductive adults until a fixed delay, t_d, has elapsed.

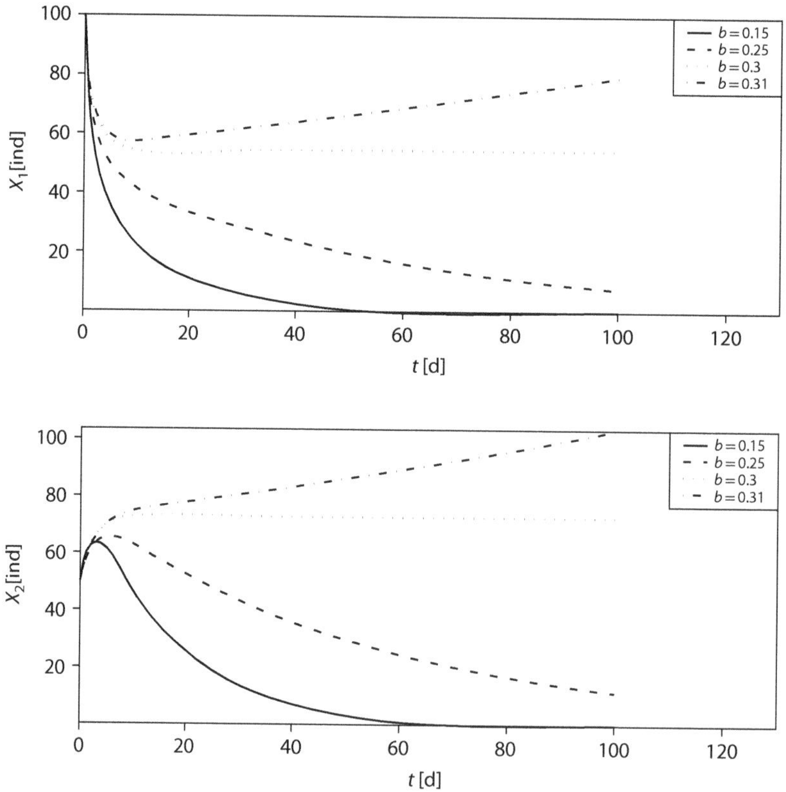

FIGURE 10.3 Two-stage model with no delay.

$$\frac{dX_1}{dt} = bX_2 - mX_1(t - t_d) - d_1X_1$$
$$\frac{dX_2}{dt} = mX_1(t - t_d) - d_2X_2 \tag{10.16}$$

Once we introduce a fixed time delay, the model becomes a DDE, which we discuss in more detail in Section 10.6.

An alternative option to make the model more realistic is to divide the transition from class 1 to 2 into several intermediate stages. Thus, we avoid the unrealistic direct transition from class 1 to 2 since it has to go through the intermediate stages. In this case, the delay is distributed (see Figure 10.4).

For example, assuming one intermediate stage, we now have three states where states denoted by lower case x correspond to the steps within class 1:

$$\frac{dx_1}{dt} = bX_2 - mx_1 - d_1x_1$$
$$\frac{dx_2}{dt} = mx_1 - mx_2 - d_1x_2 \tag{10.17}$$
$$\frac{dX_2}{dt} = mx_2 - d_2X_2$$
and
$$X_1 = x_1 + x_2$$

Now, the total number of class 1 neonates X_1 is the sum of x_1 and x_2. The delay can be generalized to a number, n_d, of intermediate stages (Acevedo, 1980a; Acevedo et al., 1995a).

$$\frac{dx_1}{dt} = bX_2 - mx_1 - d_1x_1$$
$$\frac{dx_2}{dt} = m(x_1 - x_2) - d_1x_2$$
$$\frac{dx_i}{dt} = m(x_{i-1} - x_i) - d_1x_i \quad \text{for } i = 3,\ldots,n_d + 1 \tag{10.18}$$
$$\frac{dX_2}{dt} = mx_{n_d+1} - d_2X_2$$
and
$$X_1 = \sum_{i=1}^{n_d+1} x_i$$

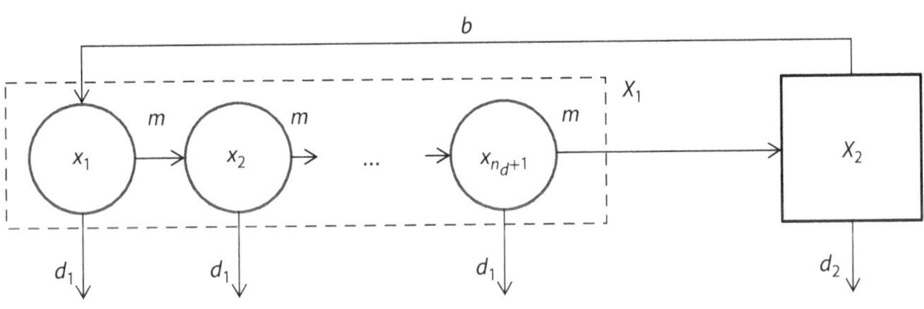

FIGURE 10.4 Diagram of intermediate stages to simulate distributed delay.

The set of first-order ODE for x_i emulates the distributed delay and is analogous to the Erlang pdf (Acevedo et al., 1996). This equivalence has been referred to as the linear chain trick (McDonald, 1978), catenary system (van Hulst, 1979), or pseudocompartments (Matis et al., 1992). We will discuss this in detail in Chapter 16 when we study semi-Markov models.

We can rewrite Equation 10.18 as a matrix equation, for example, for $n_d = 2$,

$$\frac{d}{dt}\begin{bmatrix} x_1 \\ x_2 \\ x_3 \\ X_2 \end{bmatrix} = \begin{pmatrix} -m-d_1 & 0 & 0 & b \\ m & -m-d_1 & 0 & 0 \\ 0 & m & -m-d_1 & 0 \\ 0 & 0 & m & -d_2 \end{pmatrix} \begin{bmatrix} x_1 \\ x_2 \\ x_3 \\ X_2 \end{bmatrix} \tag{10.19}$$

The longer the delay, the more mortality will occur at the juvenile stage, and therefore, it is reasonable to expect that for sustaining the population with the same birth rate, b, we need to scale down the death rate. For example, using a third of the previous rate since we have three times $(n_d + 1)$ more mortality exposure, we get $d_1 = 0.2/3 = 0.067$. The eigenvalues are $\lambda_1 = -0.45$, $\lambda_2 = -0.24 + 0.21i$, $\lambda_3 = -0.24 - 0.21i$, and $\lambda_4 = -0.019$. All have negative real parts and the system is stable, but because two eigenvalues are a complex pair, there should be damped oscillations. The smallest eigenvalue corresponds to the slower mode $\exp(-0.019t)$, and it should be the one persisting after the transients die out.

Increasing the value of b to 0.3 produces all decaying modes, except $\lambda_4 \sim 0.00$, and it is negative and therefore a steady-state mode, and as we increase b, for example, $b = 0.4$, $\lambda_4 > 0.00$, indicating an unstable system.

Figure 10.5 shows illustrative simulation results for four values of n_d. Here, we have used $b = 0.4$, $m = 0.2$, $d_1 = 0.2/(n_d + 1)$, and $d_2 = 0.15$. We can note transitory oscillations introduced by this

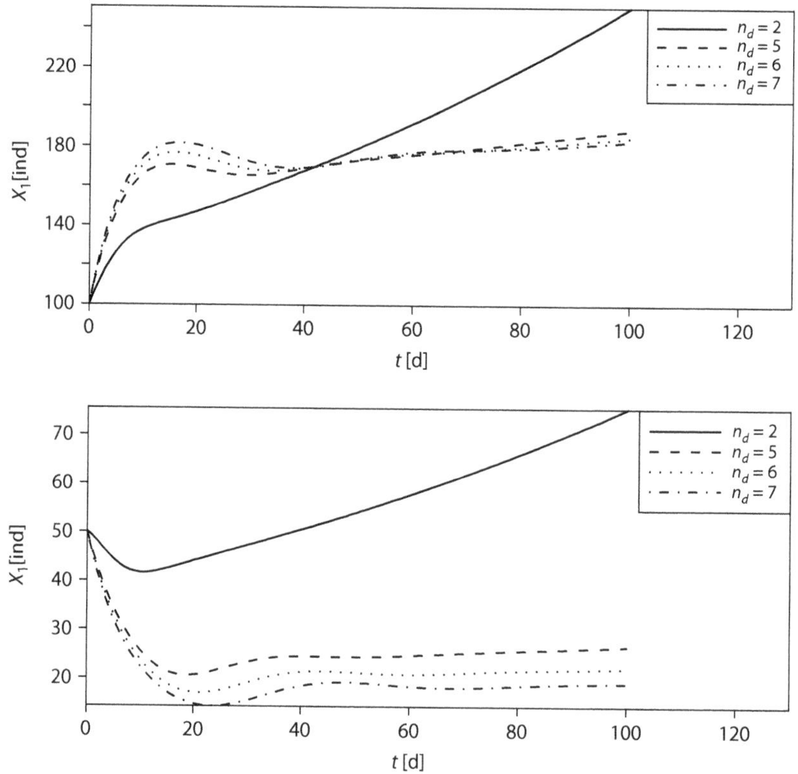

FIGURE 10.5 Two-stage model with distributed delay for various values of delay order n_d.

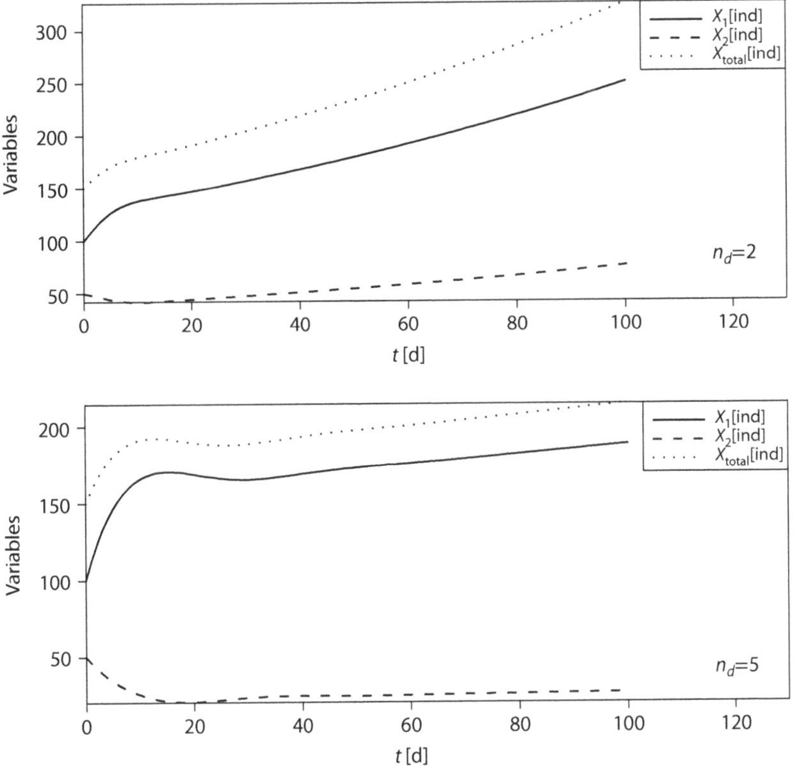

FIGURE 10.6 Two-stage model with distributed delay for two values of delay order n_d.

distributed time delay and that we have more waves for higher values of n_d. In addition, increasing the delay makes the population grow at a slower pace and the mature classes remain nearly constant after the oscillations die down. The total growth is dominated by the growth of juveniles, while the mature class remains nearly constant (Figure 10.6).

10.6 DELAY-DIFFERENTIAL EQUATIONS

The basis of this model is to assume that growing from size or stage class i to class $i + 1$ takes more time than the time step. This time is the **delay** to change class. In addition, we will assume that time runs in a continuous manner, not discretely as in the projection matrix approach (Nisbet and Gurney, 1986).

For each class i, write a balance of net maturation, death, and recruitment rates (Figure 10.7) as defined below, where all units are [ind] T^{-1}

$$\frac{dX_i}{dt} = R_i - D_i - M_i \tag{10.20}$$

The **recruitment rate, R_i**, is the number of individuals recruited to this class per unit time. There are several ways of modeling this process. For eggs, recruitment is due to the process of egg laying by adults; for larvae, recruitment is due to the hatching of eggs; and for adults, recruitment is by growth to a given size. The **death rate, D_i**, is the number of individuals that die per unit of time according to the mortality factors specific to this class, for example, the predation of larvae. The **maturation rate, M_i**, is the number of individuals moving to the next class per unit time. For eggs, maturation is by hatching, and for neonates or adults, it is due to the growth to the next size.

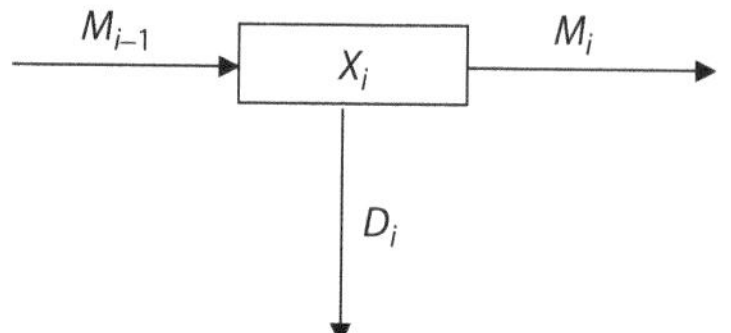

FIGURE 10.7 Balance of rates in class i.

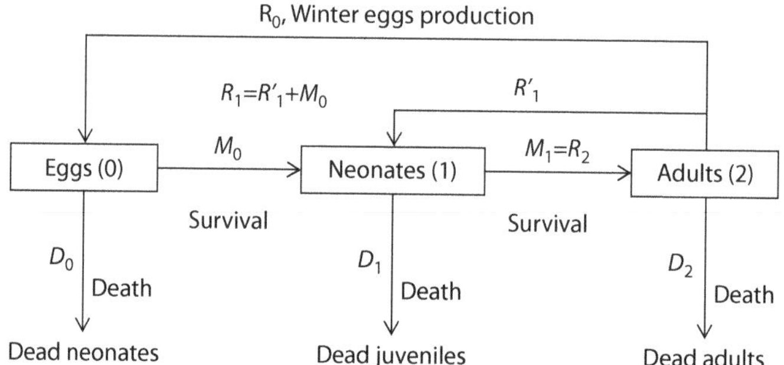

FIGURE 10.8 Three zooplankton classes based on the life cycle.

You can see that growth plays an important role in determining the delay to change classes and therefore controls the dynamics; therefore, the population model needs a growth submodel describing how the individuals of each class grow in time. This growth model should include available resources or food. There is a time delay associated with all maturation processes, and therefore, we no longer have the unrealistic possibility of maturing instantaneously as in the projection matrix. You also need a reproduction submodel, and this should take into account the fertility of various classes; it can potentially be a function of the resources.

The net R, D, M rates should also be modeled with functions involving a per capita rate, for example (an easy one), net death rate $D_i = d_i X_i$, where d_i is a per capita death rate coefficient in [1/T] units.

Consider a zooplankton example, more specifically a cladoceran: we have eggs, neonates, and adults (Figure 10.8). Take class 0, eggs: R_0 is the recruitment by sexual reproduction and emergence of winter eggs; M_0 is the maturation to neonates by hatching; and D_0 is death with a per capita rate specific to eggs. Now consider class 1, neonates: R_1 is recruitment by the hatching of eggs (M_0) or parthenogenesis (R'_1); M_1 is the maturation to adults by growth rate specific of neonates; and D_1 is their death rate. Finally, consider class 2, adults: R_2 is recruitment by growth of neonates (note that $R_2 = M_1$), and D_2 is their death rate.

We will cover an example of this type of model using a case study. The analytical and numerical solutions to Equation 10.20 are complicated because of the delay. In this book, we will not cover the mathematical details but concentrate on the simulations. We developed this model to study population-level effects of multiple stresses on a cladoceran, specifically *Ceriodaphnia dubia*, a species used in toxicity tests (Acevedo, 1998; Acevedo et al., 1995b; Acevedo and Waller, 2000). The model is based on population classes or stages selected based on asexual–sexual reproduction: resting eggs, males, female neonates, and female adults (see Figure 10.9).

There are two different versions of this model: one for natural conditions where the algae (the food) growth rate varies with environmental conditions and another for artificially controlled environmental conditions, including food supply. The model follows a delay-differential approach (Gurney et al., 1986, 1990). The rate of change of the population density $X_i(t)$ ind (which in this case is individuals per unit volume in milliliter) of each class i, at time t (in d), is given by the balance of recruitment $R_i(t)$, maturation $M_i(t)$, and death $D_i(t)$ rates, all in ind (which in this case is individuals per unit volume in milliliter) d^{-1}:

$$\frac{dX_i(t)}{dt} = R_i(t) - M_i(t) - D_i(t) \tag{10.21}$$

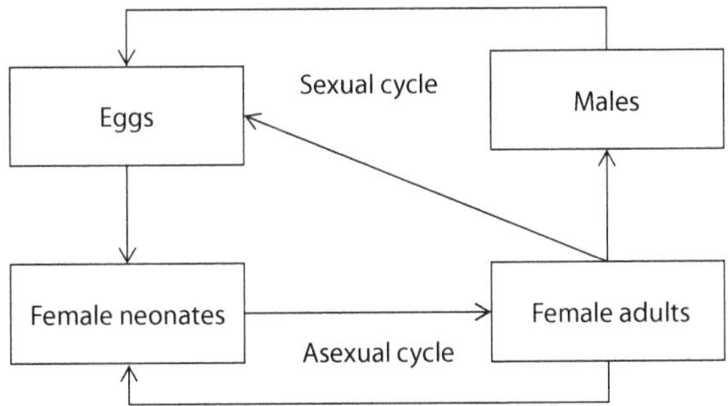

FIGURE 10.9 Life cycle for modeling *Ceriodaphnia dubia* population.

Each one of the three rates for each class is built from models for various processes. Recruitment rates of eggs (R_0), female neonates (R_1), males (R_m), and female adults (R_2) are given, respectively, by

$$R_0(t) = F_0\left(C_2(t)\right) Y \frac{X_m(t)}{X_2(t)} \left(1 - \mathrm{Sw}\left(C_2(t)\right)\right) X_2(t) + I(t) \tag{10.22}$$

$$R_1(t) = F_1\left(C_2(t)\right) \mathrm{Sw}\left(C_2(t)\right) X_2(t) + M_0(t) \tag{10.23}$$

$$R_m(t) = F_1\left(C_2(t)\right) F_m \left(1 - \mathrm{Sw}\left(C_2(t)\right)\right) X_2(t) \tag{10.24}$$

$$R_2(t) = M_1(t) \tag{10.25}$$

where the fertilities $F_i(C_2(t))$ are in (d^{-1}), for $i = 0, 1$, and the male productivity fraction $F_m(C_2(t))$, as well as the switching fraction $\mathrm{Sw}(C_2(t))$ are nonlinear functions of the ingestion rate $C_2(t)$ (in 10^6 cells d^{-1} ind^{-1} or Mcell d^{-1} ind^{-1}) of the female adults.

The mating factor $Y[X_m(t)/X_2(t)]$ (unitless) is a hyperbolic function of the ratio of the male density $N_m(t)$ to the adult female density $X_2(t)$:

$$Y(X_m/X_2) = \frac{X_m/X_2}{y + X_m/X_2} \tag{10.26}$$

where the coefficient y is the sex ratio for one-half of the maximum mating factor. Eggs can additionally be inoculated at rate $I(t)$, whereas neonates are also recruited at the maturation rate $M_0(t)$ or hatching rate of eggs, which includes a time delay. Finally, note that the recruitment of female adults $R_2(t)$ is simply the maturation rate of neonates $M_1(t)$.

Fertility rates are affected by environmental conditions, for example, temperature and food density. This last effect is modeled as a switch to sexual reproduction induced by a decrease of the ingestion rate below a given threshold. Thus, fertilities and switching are modeled, respectively, as

$$F_j(C_2) = W_j \left[1 + \exp\left(\frac{-C_2 + c_j aU_2}{gU_2}\right)\right]^{-1} \tag{10.27}$$

$$\mathrm{Sw}(C_2) = \left[1 + \exp\left(\frac{-C_2 + aU_2}{bU_2}\right)\right]^{-1} \tag{10.28}$$

parameterized by W_j = maximum fertility ($j = 0$ sexual, $j = 1$ asexual; in the same units as F_j) and U_2 = maximum adult female ingestion rate (in the same units as C_2) and several coefficients (unitless). These are c_j = half fertility decrease, g = fertility sensitivity to ingestion, a = switch threshold (as a fraction of ingestion rate), and b = switch sensitivity.

A **growth-dependent delay,** τ_i, for each class i controls the maturation rate and is given by

$$\frac{d\tau_i(t)}{dt} = 1 - \frac{G_i(t)}{G_i(t-\tau_i)} \tag{10.29}$$

where G_i = growth rate of class i (mm d^{-1}). For example, the maturation rate from neonates (i = 1) to adults is

$$M_1(t) = \frac{G_1(t)}{G_1(t-\tau_1)} R_1(t-\tau_1) S_1 \tag{10.30}$$

where S_1 = survival fraction (described later). Maturation is controlled by growth because neonates become reproductive adults at a given body length. The maximum length, L_i, and maximum growth, G_{maxi}, are given by

$$L_i = \int_{t-\tau_i(t)}^{t} G_i(\sigma)\, d\sigma, \quad G_{\max i} = L_i/\tau_i(0) \tag{10.31}$$

The growth rate of each class i is made proportional to its ingestion rate

$$G_i(t) = \varepsilon_i C_i(t) \tag{10.32}$$

by an efficiency coefficient ε_i (mm $cell^{-1}$) of food ingested for class i. The ingestion rate is dependent on body length and food density as well as physical and chemical stresses.

The maximum ingestion rate is multiplied by several limiting factors (values between 0 and 1). For example, for food density, f, temperature, T, and a chemical stress, X,

$$C_i(t) = U_i Q_f(t) Q_T(t) Q_X(t) \tag{10.33}$$

The multipliers Q_f, Q_T, and Q_X are given by functions, such as an M3 hyperbolic function of food quantity, a parabolic function of temperature, and a sigmoid function of chemical stress. For example, the food factor Q_f is modeled by an M3 hyperbolic function of food density to yield the following ingestion rate:

$$C_i(t) = U_i Q_f\left(f(t)\right) = U_i \frac{f(t)}{f(t) + K_h} \tag{10.34}$$

where $f(t)$ is the food density (cells ml^{-1}) and K_h is the half-rate coefficient in the same units as f (Kooijman, 1993). The product of multipliers as in Equation 10.33 constrain ingestion by all factors. Other formulations are possible, such as taking the minimum or most limiting of the factors.

The dependence of the hatching of eggs on temperature and photoperiod is modeled by

$$M_0 = \mathrm{GF}_0(T,P)\, R_0\left(t - \tau_0(t)\right) S_0 \tag{10.35}$$

using a hatching fraction GF_0 that depends on the accumulated temperature, T, and photoperiod, P. We also include a developmental time delay, τ_0, as we did for the other classes.

The death rate, D_i, for class i is proportional to the density, X_i, with coefficient δ_i (d^{-1}) and allows calculation of the survival fraction by solving

$$\frac{dS_i(t)}{dt} = S_i(t)\left(\frac{G_i \delta_i[t-\tau_i(t)]}{G_i[t-\tau_i(t)]} - \delta_i\right) \tag{10.36}$$

Mortality increases directly by starvation, thermal, and other stress. These equations are not shown here for the sake of brevity. A harvest rate coefficient $h(t)$ can be added to the death rate coefficients δ_1 and δ_2 for neonates and adults to calculate the net death rate of these classes.

Food density, f, is calculated as the balance between the supply rate, $f_s(t)$, and consumption rate, $f_c(t)$, both in Mcell d^{-1} ml^{-1}:

$$\frac{df(t)}{dt} = f_s(t) - f_c(t) \tag{10.37}$$

Consumption rate is calculated as the individual ingestion rate, $C_i(t)$, for class i multiplied by the population density in that class and summed over all classes i:

$$f_c(t) = \sum_i C_i(t) X_i(t) = \sum_i U_i Q_f\left[f(t)\right] X_i(t) \tag{10.38}$$

Recall that $X_i(t)$ is given by the solution of the overall balance, Equation 10.21, and $C_i(t)$ is calculated from food density using Equation 10.34.

When applied to natural systems, the food supply rate is the algae population growth rate. We will discuss algae growth later in this book. In artificially maintained systems, the food supply rate is a controlled variable typically limited by a maximum supply rate $f_s(t) \le f_{smax}$. The supply is set to a maximum, $f_s(t) = f_{smax}$, when food density is below a reference value f_d, and to 0, $f_s = 0$, when the food density equals or exceeds the reference, $f(t) \ge f_d$.

A typical simulation starts with zero animal density $X_i(0) = 0$ for all i, and a certain number of resting eggs ready to hatch, that is, an inoculation $I(t)$ in Equation 10.22 modeled as an impulse function, $I(t) = I_0$ ind at $t = 0$ and $I(t) = 0$ for $t > 0$. After the hatching delay, neonate and adult densities increase, leading to increasing food consumption and diminishing food density because of the restricted food supply rate. Scarce food triggers sexual reproduction to produce resting eggs (see Figure 10.10). This situation corresponds to a laboratory run with constant temperature and

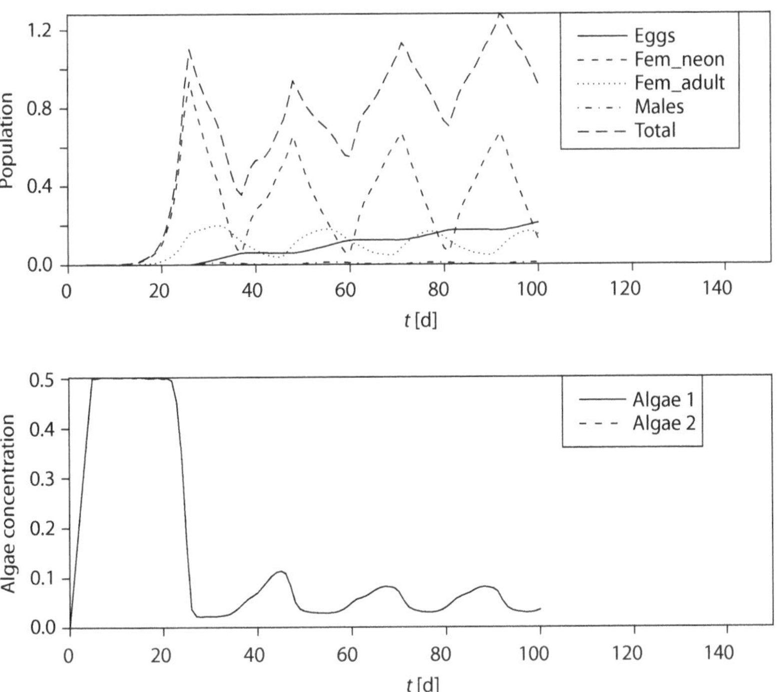

FIGURE 10.10 Simulation runs illustrating limited food supply conditions when the population is started by inoculation.

photoperiod. Food density reaches a maximum and stays limited up until about day 20 because the model limits the supply when the food density reaches the maximum allowed.

As the population explodes toward day 20, food density crashes because there are too many animals trying to eat the limited amount of food. This drives a burst of males and winter eggs and therefore, a reduction in female neonates. The unstable population growth is therefore controlled, but preserves winter eggs for the future. Food density builds up because the number of animals decreases.

10.7 COMPUTER SESSION

10.7.1 Projection Matrices: Three Age Classes

Let us use the example of a Leslie–Lewis matrix for three age classes of a bird population with matrix **P** given in Equation 10.8 (Swartzman and Kaluzny, 1987). Define matrix `P` and calculate the determinant of `P`:

```
> P <- matrix(c(0,2,3, 0.5,0,0, 0,0.5,0), byrow=T, ncol=3)
> P
     [,1] [,2] [,3]
[1,]  0.0  2.0    3
[2,]  0.5  0.0    0
[3,]  0.0  0.5    0
> det(P)
[1] 0.75
```

`eigen(P)` gives the eigenvalues and the eigenvectors:

```
> eigen(P)
$values
[1] 1.2625511+0.0000000i -0.6312756+0.4421838i -0.6312756-0.4421838i

$vectors
                [,1]                   [,2]                   [,3]
[1,] -0.9200166+0i  0.7910725+0.0000000i  0.7910725+0.0000000i
[2,] -0.3643483+0i -0.4203325-0.2944265i -0.4203325+0.2944265i
[3,] -0.1442905+0i  0.1137601+0.3128841i  0.1137601-0.3128841i
```

From the preceding result, the dominant eigenvalue is real with value 1.263 rounded to three decimal places. The population grows because this maximal eigenvalue is larger than 1. The eigenvector for this eigenvalue is the first column of component `$vectors`. We can extract the real part and rounded to three decimal places.

```
> ev <- round(Re(eigen(P)$vectors[,1]),3)
> ev
[1] -0.920 -0.364 -0.144
```

The eigenvector for the dominant eigenvalue is the stable age distribution. We can scale all entries with respect to last entry. After rounding to two decimal places,

```
> s.age <- round(ev/ev[length(ev)],2)
[1] 6.39 2.53 1.00
```

or in mathematical form, $\boldsymbol{x} = \begin{bmatrix} 6.39 \\ 2.53 \\ 1.00 \end{bmatrix}$. Another way to look at this distribution is as a proportion of the total:

```
> round(s.age/sum(s.age),2)
[1] 0.64 0.26 0.10
```

This is 64% of the population in class 1, 26% in class 2, and 10% in class 3. The other two eigenvalues constitute a complex conjugate pair, $-0.63 \pm 0.44i$, and contain information about oscillations and damping of the transients as they settle into the steady state.

Exercise 10.8

Premultiply $\boldsymbol{x} = \begin{bmatrix} 1 \\ 0.5 \\ 0.25 \end{bmatrix}$ by the **P** matrix. Repeat 10 times with a loop. Demonstrate that the final value of x approaches the stable age distribution. Hint: `x < -c(1, 0.5, 0.25); for(i in 1:10) x < -P%*%x`.

We will use **seem**'s function `sim.proj` (see explanation at the end of the chapter) to simulate three age classes with the initial condition and parameters given above. The arguments are time sequence, initial condition, projection matrix, and units.

```
x0 <- c(100,50,25); xunit="(Indiv)"
t=seq(0,10,1); tunit="(Year)"
x <- sim.proj(t, x0, P, vars.plot=c(1:3),tunit, xunit)
```

The output can be requested on the console by typing x and obtaining a list with several components: `$x` has the results for all classes, `$xp` has the classes, a proportion of total, `$xs` has the ratio of classes over the last one, `$lambda` has the ratio of two consecutive values of `xtot`, and finally `xtot` is listed.

```
> x
$x
      t
 [1,]  0  100.000  50.000  25.000
 [2,]  1  175.000  50.000  25.000
 [3,]  2  175.000  87.500  25.000
 [4,]  3  250.000  87.500  43.750
 [5,]  4  306.250 125.000  43.750
 [6,]  5  381.250 153.125  62.500
 [7,]  6  493.750 190.625  76.562
 [8,]  7  610.938 246.875  95.312
 [9,]  8  779.688 305.469 123.438
[10,]  9  981.250 389.844 152.734
[11,] 10 1237.891 490.625 194.922
$xp
      t
 [1,] 0 57.143 28.571 14.286
 [2,] 1 70.000 20.000 10.000
 [3,] 2 60.870 30.435  8.696
 [4,] 3 65.574 22.951 11.475
 [5,] 4 64.474 26.316  9.211
 [6,] 5 63.874 25.654 10.471
 [7,] 6 64.887 25.051 10.062
```

```
 [8,]  7 64.098 25.902 10.000
 [9,]  8 64.512 25.275 10.213
[10,]  9 64.394 25.583 10.023
[11,] 10 64.358 25.508 10.134

$xs
      t
 [1,]  0 400.000 200.000 100
 [2,]  1 700.000 200.000 100
 [3,]  2 700.000 350.000 100
 [4,]  3 571.429 200.000 100
 [5,]  4 700.000 285.714 100
 [6,]  5 610.000 245.000 100
 [7,]  6 644.898 248.980 100
 [8,]  7 640.984 259.016 100
 [9,]  8 631.646 247.468 100
[10,]  9 642.455 255.243 100
[11,] 10 635.070 251.703 100

$lambda
      t lambda
 [1,]  0  1.000
 [2,]  1  1.429
 [3,]  2  1.150
 [4,]  3  1.326
 [5,]  4  1.246
 [6,]  5  1.257
 [7,]  6  1.275
 [8,]  7  1.253
 [9,]  8  1.268
[10,]  9  1.261
[11,] 10  1.262

$xtot
      t     xtot
 [1,]  0  175.000
 [2,]  1  250.000
 [3,]  2  287.500
 [4,]  3  381.250
 [5,]  4  475.000
 [6,]  5  596.875
 [7,]  6  760.938
 [8,]  7  953.125
 [9,]  8 1208.594
[10,]  9 1523.828
[11,] 10 1923.438
>
```

Exercise 10.9

If you had a 2 × 2 Leslie matrix for two age classes, how many output variables do you have? Explain how many for the population in each class, for the proportions (<1.00) in each class, and for the total population (all classes added).

The graphics are the same as shown in Figure 10.2 clarifying the difference in the growth pattern of the total population and the stability of the age or size distribution.

Exercise 10.10

Determine the **stable age distribution** in this model. Write it down as a vector. Compare to the eigenvector results.

Exercise 10.11

Produce multiple runs with systematic changes in the values of one parameter. For example, explore the effects of a lethal exposure decreasing the survival of age class 1. Explore scenarios of decreasing survival of age class 1 to a 20%, 40%, and 60% decrease with respect to nominal 0.5. Use sensitivity. Does the population grow at any of these lethal exposure levels? Discuss.

10.7.2 Projection Matrices: Seven Age Classes

Let us analyze a larger matrix: a 7 × 7 projection for blue whales (Swartzman and Kaluzny, 1987).

$$\mathbf{P} = \begin{bmatrix} f_1 & f_2 & f_3 & f_4 & f_5 & f_6 & f_7 \\ s_1 & 0 & 0 & 0 & 0 & 0 & 0 \\ 0 & s_2 & 0 & 0 & 0 & 0 & 0 \\ 0 & 0 & s_3 & 0 & 0 & 0 & 0 \\ 0 & 0 & 0 & s_4 & 0 & 0 & 0 \\ 0 & 0 & 0 & 0 & s_5 & 0 & 0 \\ 0 & 0 & 0 & 0 & 0 & s_6 & s_7 \end{bmatrix}$$

The time step is 2 years. Note that this is a modified Leslie matrix, because entry 7,7 is nonzero. Class 7 represents individuals with ages of 12 years or more.

The matrix entries are available in the file **chp10/les7.txt**. Read it and convert it to a matrix. Double-check the result.

```
>les7 <- matrix(scan("chp10/les7.txt"), ncol=7, byrow=T)
>les7
     [,1] [,2] [,3] [,4] [,5] [,6] [,7]
[1,] 0.00 0.00 0.19 0.44 0.50 0.50 0.45
[2,] 0.87 0.00 0.00 0.00 0.00 0.00 0.00
[3,] 0.00 0.87 0.00 0.00 0.00 0.00 0.00
[4,] 0.00 0.00 0.87 0.00 0.00 0.00 0.00
[5,] 0.00 0.00 0.00 0.87 0.00 0.00 0.00
[6,] 0.00 0.00 0.00 0.00 0.87 0.00 0.00
[7,] 0.00 0.00 0.00 0.00 0.00 0.87 0.80
```

Perform eigendecomposition. The first eigenvalue is dominant, and the first eigenvector is the stable age distribution.

```
> eval <- round(Re(eigen(les7)$values),3)
> ev <- round(Re(eigen(les7)$vectors[,1]),3)
> e.val
[1]  1.099   0.200   0.200 -0.464 -0.185 -0.185  0.136
> ev
[1] 0.547 0.433 0.343 0.272 0.215 0.170 0.496
```

The first eigenvalue 1.10 is the dominant one. It is greater than 1; therefore, the population grows. The first eigenvector is the stable age distribution.

```
> s.age <- round(ev/ev[length(ev)],2)
> s.age
[1] 1.10 0.87 0.69 0.55 0.43 0.34 1.00
```

Here, we have scaled with respect to the last entry to obtain the stable proportions in each class.

```
> round(s.age/sum(s.age),2)
[1] 0.22 0.17 0.14 0.11 0.09 0.07 0.20
```

Exercise 10.12

Assume a simple initial condition: for example, 1 in the first age class and 0 in the others. Premultiply it by the les7 matrix. Repeat five times with a loop as in Exercise 10.11. Change to 10, 15, and then 20 times. What is the shortest time when you can claim that the final value of x approximates the stable age distribution? Recall that the time step is 2 years.

We will simulate the whale Leslie matrix model given above using matrix `les7` in the call to `sim.proj` and a stepping time of 2 years instead of 1:

```
x0 <- rep(800,7); xunit="(Ind)"
t=seq(0,20,2); tunit="(Year)"
P <- les7
x <- sim.proj(t, x0, P, vars.plot=c(1,3,5,7),tunit, xunit)
```

Note that each time step in the output corresponds to 2 years. That is, for example, position 10 is $t = 18$ years. We can see that λ settles around 1.09 as the steady value. This corresponds to the maximal eigenvalue of P as calculated previously. The plots are shown in Figure 10.11. All classes increase with oscillations first and then settle in an exponential pattern toward the end of the 20-year run (top left). The proportions settle toward the end of the run (top right), lambda tends to 1.09 (bottom left), and the total population increases and acquires an exponential pattern toward the end of the run (bottom right).

Let us explore the effect of harvesting this population. Change all adult survival fractions by –20% of the nominal (0.87 and 0.80) to simulate harvest.

```
P <- les7; harv <- 20
for (i in 1:6) P[i+1,i] <- (1-harv/100)*les7[i+1,i]
P[7,7] <- (1-harv/100)*les7[7,7]
x.harv.all <- sim.proj(t, x0, P, vars.plot=c(1,3,5,7), tunit, xunit)
```

The results are shown in Figure 10.12. Now we have a declining population. The value of `lambda` is ~0.91, which is less than 1. In fact, we could easily confirm this value by calculating the eigenvalue with `eigen(P)`. Note that the proportions settle in steady state while the population declines.

Now harvest only old whales, and decrease survival by –20% of only the last age class (age 12 or more) from the nominal 0.8 in the same manner as before.

```
P <- les7; harv <- 20
P[7,7] <- (1-harv/100)*les7[7,7]
x.harv.old <- sim.proj(t, x0, P, vars.plot=c(1,3,5,7), tunit, xunit)
```

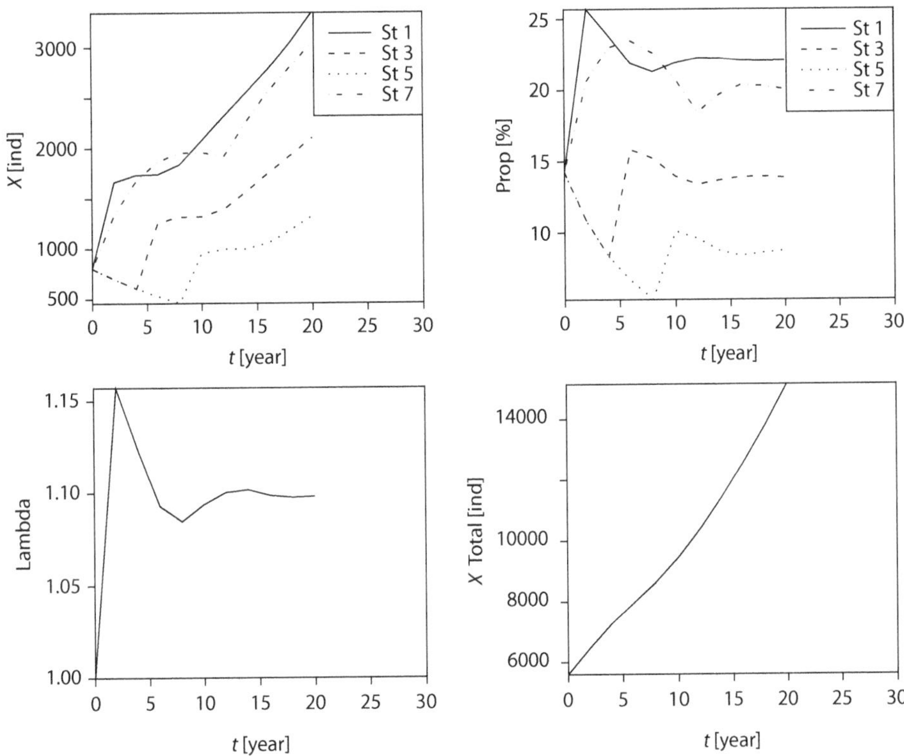

FIGURE 10.11 Dynamics of Leslie model for whale population.

The graphs are shown in Figure 10.13. Now we preserve population growth with `lambda` around 1.07.

To understand the effect of harvesting on the stable age distribution, we can plot it as a bar graph with different bar types for each run (one bar type for normal and one for harvesting). We do the same for the proportions and the density.

```
mat <- matrix(1:2,2,1,byrow=T)
nf <- layout(mat, widths=rep(5,2), heights=rep(7/2,2), TRUE)
par(mar=c(4,4,3,.5),xaxs="i",yaxs="i")

x.comp <- cbind(x$xp[10,-1], x.harv.old$xp[10,-1])
barplot(t(x.comp), beside=T, names.arg=c(1:7), space=c(0,1),
        density=c(0,50), xlab="Age Classes", ylab="Proportions %",
col=1)

x.comp <- cbind(x$x[10,-1], x.harv.old$x[10,-1])
barplot(t(x.comp), beside=T, names.arg=c(1:7), space=c(0,1),
          density=c(0,50), xlab="Age Classes", ylab="Ind", col=1)
```

The results are shown in Figure 10.14. We can see how harvesting old individuals reduces the number of individuals in all age classes (reduced recruitment due to decrease of reproducing individuals) but proportionally more in the older age class (reduced survival).

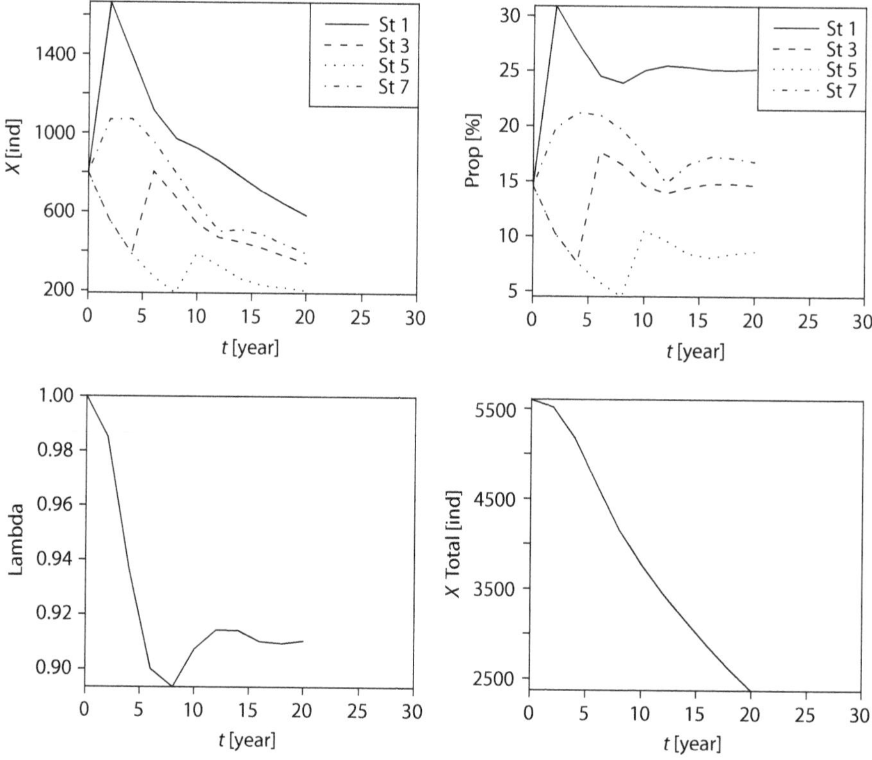

FIGURE 10.12 Dynamics of Leslie model for harvested whale population.

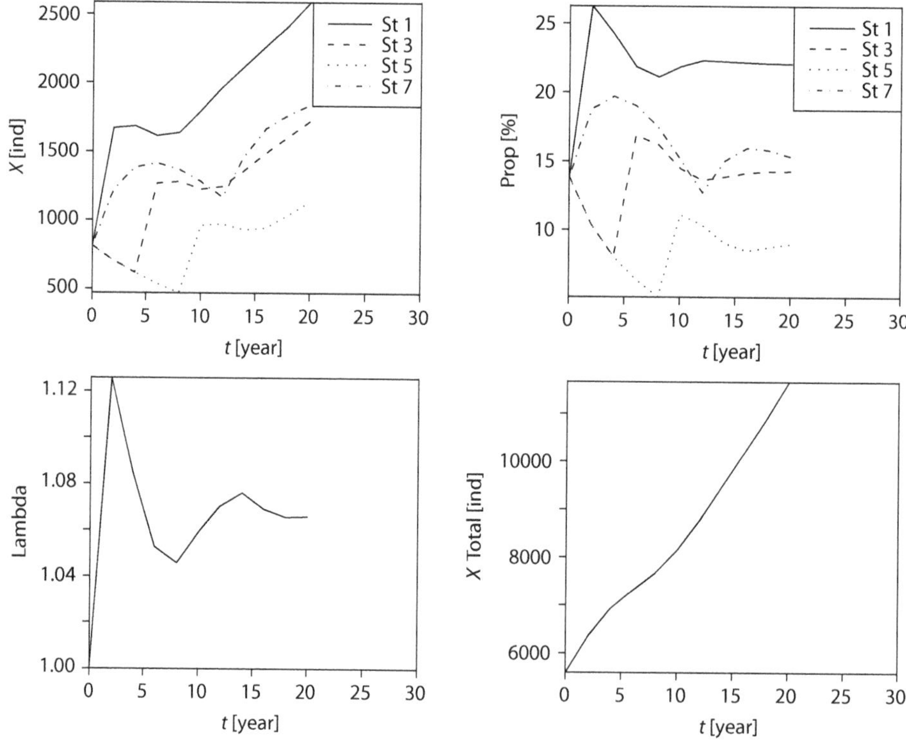

FIGURE 10.13 Dynamics of whale population with harvesting of older individuals.

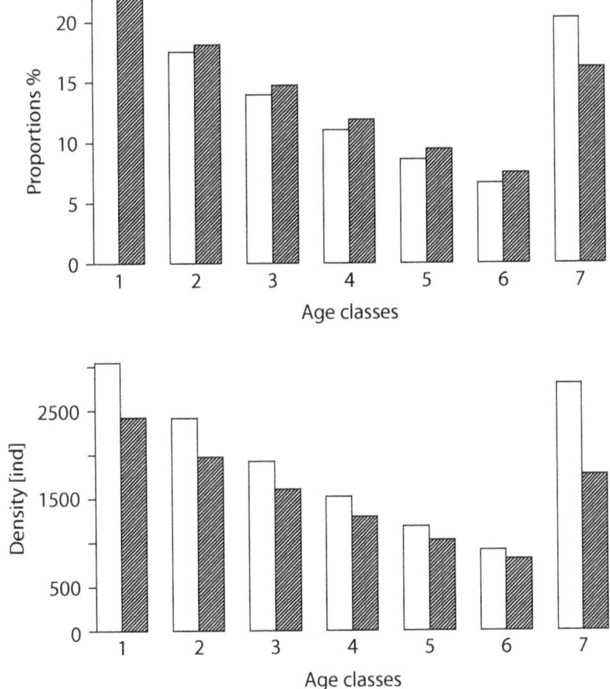

FIGURE 10.14 Comparing effect of harvesting (dark bars) on steady-state age structure.

Exercise 10.13

Make two more runs to simulate harvesting at 10% and 15% of survival of older individuals. Compare to the example just presented that was simulated with 20%. Discuss changes in age class distribution as you decrease survival by harvesting to these lower values of 10 and 15%. Generate bar graphs like Figure 10.14 but include these new runs. Estimate the steady state growth rate λ (confirm with value of dominant eigenvalue) for each run and plot this value as a function of harvest rate. In addition, plot sensitivity of the growth rate (as a metric) to the perturbed survival parameter.

This section should have helped you understand the concept of stable structure (age or size) and exponential growth, and how to impose an exogenous force (like harvesting) through the model parameters.

10.7.3 Structured Population Models in Continuous Time

We will start with a two-stage class structure, juveniles and mature, with no fixed or discrete time delay. We have two variables, X_1 for juveniles and X_2 for mature. We have modified the function `sim` to include a calculation of the total population and a possible time delay for population class 1. The new function is `simt` and it is explained at the end of the chapter. Now we define a model `two.stage.cont` similar to functions we have written before and explained at the end of the chapter.

The input file **chp10/two-stage-inp.csv** contains

```
Label,Value,Units,Description
t0,0.00,d,time zero
tf,100.00,d,time final
dt,0.1,d,time step
```

```
tw,1.0,d,time to write
b,0.3,1/d,birth rate coeff
m,0.2,1/d,maturation juveniles rate coeff
d1,0.2,1/d,death juveniles rate coeff
d2,0.15,1/d,death adults rate coeff
X1.0,100.00,ind,juveniles initial condition
X2.0,50.00,ind,adults initial condition
digX,4,None,significant digits for output file
```

where $b = 0.3$, the maturation rate coefficient has a value of $m = 0.2$, and the death rate coefficients are $d_1 = 0.2$, $d_2 = 0.15$. Variables are initialized to 100 and 50 mature. We are ready to specify the model and call the simulation function with b varied in 0.15, 0.25, 0.3, and 0.31.

```
two.stage <-list(f=two.stage.cont)
param <- list(plab="b", pval = c(0.15,0.25,0.3,0.31))
t.X <- simt(two.stage,"chp10/two-stage-inp.csv", param,pdfout=T)
```

The simulation yields the results shown in Figure 10.3 for both X_1 and X_2 for these various values of b. Other pages of the output PDF file, not displayed here, include the total population (the sum of two model variables) and phase plane graphs.

Exercise 10.14

Vary the maturation rate, m, around the nominal $m = 0.2$. Perform sensitivity analysis.

Next, we practice the concept of distributed delay by using the model functions `two.stage.cont.delay` and `two.stage.x0`. The input file **chp10/two-stage-delay-inp.csv** is the same as before except that we have increased the birth rate to $b = 0.4$ and introduced a new parameter, n_d (the order of the delay). Internally, in the function, we scale down the death rate of class 1 according to the delay length; in the simulation, $d_1 = 0.2/(n_d + 1)$:

```
b,0.4,1/d,birth rate coeff
m,0.2,1/d,maturation juveniles rate coeff
d1,0.2,1/d,death juveniles rate coeff
d2,0.15,1/d,death adults rate coeff
nd,5,None,order of delay
```

Now, we call the function varying the order of the delay to obtain graphs as shown in Figures 10.5 and 10.6.

```
two.stage.delay <-list(f=two.stage.cont.delay,x0=two.stage.x0)
param <- list(plab="nd", pval = c(2,5,6,7))
t.X <- simt(two.stage.delay,"chp10/two-stage-delay-inp.csv",
param,pdfout=T)
```

Exercise 10.15

Using a distributed delay of order 5, vary b to have values 0.1, 0.3, and 0.5. Leave the other parameter values as in the input file.

10.7.4 Stage Structure: Delay-Differential Equations

We will work with a cladocera population growing in laboratory conditions. This model accounts for the following stages: eggs, neonates, and adults. This model can run under different modalities for food supply: constant, variable, and controlled. We will work with constant food, temperature. and photoperiod. The default values for the parameters are as given in the work by Acevedo et al. (1995b) and are in the input file **chp10/cerio-inp.txt**. The entire file is not reproduced here for the sake of space. Only the first four lines direct to perform sensitivity analysis for one parameter is mentioned as follows:

```
par_sens_name     none
par_sens_max      0.00
par_sens_min      0.00
par_sens_step     0.00
```

We declare laboratory conditions and constant food supply modality here:

```
lab_mode          1
food_mode         0
```

Just to exemplify, here we have the maximum and half-rates in the M3 formulation of food ingestion rate for all classes, that is, `cmax` and `pkhalf` for all classes (lines 26–33):

```
cmax(1)           0.00
cmax(2)           0.50
cmax(3)           5.00
cmax_unit         [Mc/day/ind]
pkhalf(1)         0.00
pkhalf(2)         0.20
pkhalf(3)         0.20
pkhalf_unit       [Mcell/ml]
```

In lines 82–84, we have

```
fsmax1            0.1000
fsmax2            0.1000
fs_unit           [Mc/day/ml]
```

We will use seem's function `cerio.F` which is an interface to a Fortran program `cerio` that executes the simulation. To be able to use a Fortran program we first dynamically load the **seem.dll** which resides in the seem library directory. Assuming a 32 bit system we use the following statement:

```
> dyn.load(paste(.libPaths(),"/seem/libs/i386/seem.dll",sep="")).
```

Recall from chapter 1 that you can automatically load this dll when you start `R` by editing the file **Rprofile.site** in the **etc** directory of your `R` installation. To make a run, we call function `cerio.F` with two arguments—the folder name and the prefix for input and output files:

```
> fileout <- cerio.F("chp10","cerio")
```

which goes to the desired directory; renames the input file **chp10/cerio-inp.txt** as needed; runs the executable, which is a Fortran program will read the renamed input file and produces output, which is copied to files **cerio-out.txt** and **cerio-dex.txt** in folder **chp10**; and does some cleanup.

Function cerio.F is given and explained at the end of the chapter. It is interesting to study this short script because it shows the type of commands we can put together to execute a precompiled simulation program that accesses files with preassigned names. In this way, we can make use of various programs without rewriting the code.

The output file **cerio-out.txt** contains 11 columns corresponding to time stamp (first column); values of the population in each stage/sex (next four columns, because there are four classes: eggs, female neonates, female adults, and males); two food items (algae 1 and algae 2); total population; and laboratory conditions, water temperature, photoperiod, and light. For this execution, we ignore light because under this condition it does not control algae growth. The other output file **cerio-dex.txt** contains values of the sensitivity metrics for each run. In this case we have performed only one run and the file is not very useful.

Exercise 10.16

What is the water temperature in degrees Celsius? Is it constant throughout the run? What is the photoperiod in hours? Is it constant throughout the run? Are the food items (algae 1 and algae 2) equal? Why? What is the final value for eggs? And males?

We now apply the function `read.plot.cerio.out` included in seem designed to scan the output file fileout produced by the call to cerio.F and extract the population numbers in each class, the food items, temperature, photoperiod, and light:

```
>x <- read.plot.cerio.out(fileout)
```

This allows us to obtain graphical results from the run, which we already showed in Figure 10.10. For the sake of space, we have not shown graphs for temperature, photoperiod, and light. From the figure, one can identify the increase in the eggs every time food becomes scarce.

Exercise 10.17

Now make a run with fsmax2 = 0.05 by using the corresponding input file **chp10/cerio-fsmax2-inp.txt**. Use the cerio.F to run an execution and the `read.plot.cerio.out` function to produce graphs. Discuss the results. Hint: in this case the prefix for input and output filenames is cerio-fsmax2.

The function `read.plot.cerio.out` includes a sensitivity optional argument `sens=T` to produce multiple runs with systematic changes in the values of one parameter. For example, we subject the population to a change in food regime by establishing this design in the input file where parameter `fsmax2` will vary from 0.05 to 2.00 in steps of 0.05, that is, 0.05, 0.10, 0.15, and 0.20.

```
par_sens_name    fsmax2
par_sens_max     0.20
par_sens_min     0.05
par_sens_step    0.05
sub_title        lab_fsmax2_sens
```

This is already done in **chp10/cerio-fsmax2-sens-inp.txt**. We can execute and produce graphs using

```
fileout <- cerio.F("chp10","cerio-fsmax2-sens")
x <- read.plot.cerio.out(fileout,sens=T, pdfout=T)
```

We obtain a PDF file with several pages of graphics; we show only the first two pages here. Figure 10.15 shows eggs and female neonates, and Figure 10.16 illustrates female adults and males.

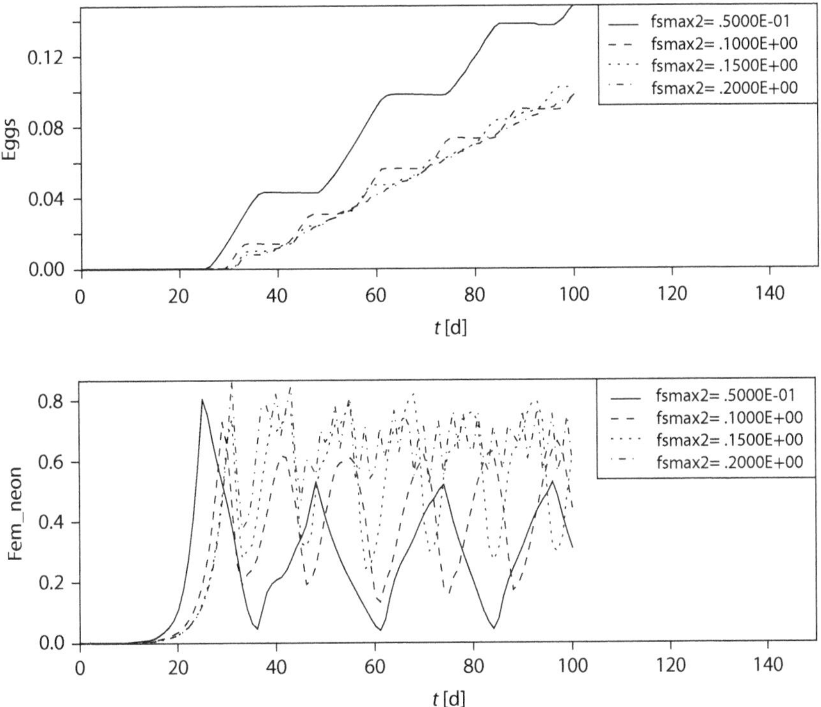

FIGURE 10.15 Ceriodaphnia dynamics at various levels of food supply. Eggs and female neonates.

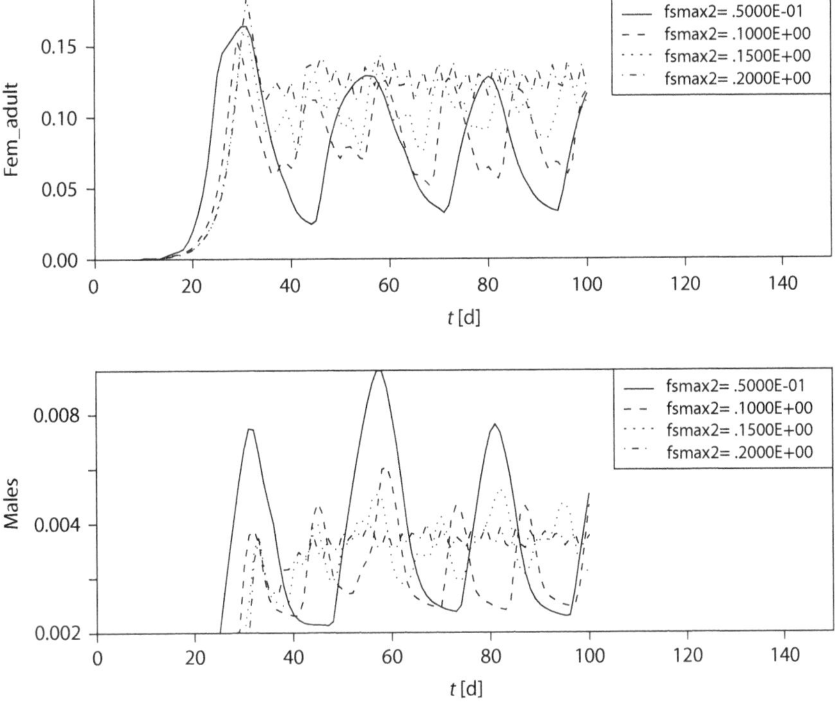

FIGURE 10.16 Ceriodaphnia dynamics at various levels of food supply. Adult females and males.

Exercise 10.18

Discuss the effect of increasing the food supply `fsmax2` as shown in the example. How does the number of winter eggs change at these supply rates? How sensitive is the adult female maximum value to the food supply?

10.7.5 Build Your Own

Exercise 10.19

Following the article by Crouse et al. (1987), build a Lefkovitch matrix model for sea turtles composed of seven classes: one (eggs, hatchlings), two (small juveniles), three (large juveniles), four (subadults), five (novice breeders), six (1-year remigrants), and seven (mature breeders). Use parameter values given in that article as follows: f_5, f_6, f_7 = 127, 4, 80; s_1, …, s_6 = 0.675, 0.049, 0.015, 0.052, 0.809, 0.809; r_2, r_3 = 0.737, 0.661, r_7 = 0.809. All other elements of the matrix are zero. Also see Kot (2001). Analyze by eigendecomposition and use the function `sim.proj` to simulate.

10.8 SEEM FUNCTIONS EXPLAINED

10.8.1 Function sim.proj

To perform the simulation, we write a loop for time to iterate the multiplication of the matrix `proj.mat` by the state vector `x`. As time progresses in the loop, it calculates the total population `xtot`, the proportions `xp` in each class with respect to the total, and an approximation `lambda` to the dominant eigenvalue as the ratio of the total population at adjacent times, and the population is scaled with respect to the last class. Note that the calculation of `lambda` is the ratio of current year to previous, for example, λ = xtot(20)/xtot(18).

```
sim.proj <- function(t, x0, proj.mat, vars.plot, tunit, xunit,
pdfout=F){
 nx= length(x0); nt =length(t)
 varlab=paste("St",as.character(c(1:nx)))

x <- matrix(nx*nt, ncol=nx, nrow=nt)
x.p <- x; xs <- x
lambda <- t;xtot <- t

x[1,] <- x0
xtot[1] <- sum(x[1,])
lambda[1] <- 1
# proportions as percent
x.p[1,] <- 100*(x[1,]/xtot[1])
# scaled with  respect to last as percent
xs[1,] <- 100*(x[1,]/x[1,nx])

for(i in 2:nt) {
  x[i,]<- proj.mat%*%x[i-1,]
  xtot[i] <- sum(x[i,])
  lambda[i] <- xtot[i]/xtot[i-1]
  x.p[i,] <- 100*(x[i,]/xtot[i])
  xs[i,] <- 100*(x[i,]/x[i,nx])
}
```

Then we plot, for example, to PDF:

```
mat <- matrix(1:4,2,2,byrow=T)
nf <- layout(mat, widths=rep(7/2,2), heights=rep(7/2,2), TRUE)
par(mar=c(4,4,3,.5),xaxs="i",yaxs="i")

if(pdfout==T){
pdf("chp10/proj-mat-out.pdf")

matplot(t,x[,vars.plot], type="l", xlim=c(0,1.5*max(t)),lty=1:nx,
col=1,
         xlab=paste("t",tunit), ylab=paste("X",xunit))
legend (1.01*max(t), max(x), legend=varlab[vars.plot], lty=1:nx,col=1)

matplot(t,x.p[,vars.plot], type="l", xlim=c(0,1.5*max(t)),lty=1:nx,
col=1,
         xlab=paste("t",tunit), ylab="Prop (%)")
legend (1.01*max(t), max(x.p), legend=varlab[vars.plot],
lty=1:nx,col=1)

plot(t,lambda, type="l", xlim=c(0,1.5*max(t)),lty=1:nx, col=1,
         xlab=paste("t",tunit), ylab="Lambda")

plot(t,xtot, type="l", xlim=c(0,1.5*max(t)),lty=1:nx, col=1,
         xlab=paste("t",tunit), ylab=paste("X Total",xunit))

dev.off()
```

Then we finalize, returning all calculations:

```
x <- round(cbind(t,x),3)
xp <- round(cbind(t,x.p),3)
lambda <- round(cbind(t, lambda),3)
xs <- round(cbind(t,xs),3)
xtot <- round(cbind(t,xtot),3)

return(list(x=x, xp=xp, xs=xs,lambda=lambda, xtot=xtot))
}
```

10.8.2 Models two.stage.cont and two.stage.cont.delay

These functions are similar to many already explained, like the logistic. First, with no delay, the function to integrate is

```
two.stage.cont <- function(t,p,x){
# x[1]  juveniles, x[2]  mature adults
# b, m, d1, d2 are p[1:4]
dx1 <- p[1]*x[2] - p[2]*x[1] - p[3]*x[1]
dx2 <- p[2]*x[1] - p[4]*x[2]
dx <- c(dx1,dx2)
return(dx)
}
```

Including a delay introduces several changes because we have to add the delay ODE after array component 2. We use a loop to make it general according to the value of n, which is given by `p[5]`.

```
two.stage.cont.delay <- function(t,p,x){
#x[1] juveniles, x[2] mature adults
#x[3],…,x[n+2] delay vars
# b, m, d1, d2 are p[1:4]
n <- p[5]; p[3] <- p[3]/(n+1); y <- array()
y[1] <- p[1]*x[2] - p[2]*x[1] - p[3]*x[1]
y[3] <- p[2]*(x[1] - x[3]) - p[3]*x[3]
for(i in 4:(n+2))
  y[i] <- p[2]*(x[i-1] - x[i]) - p[3]*x[i]
  y[2] <- p[2]*x[n+2] - p[4]*x[2]
dx <- y
return(dx)
}
```

Then, the companion function `two.stage.x0` establishes the initial conditions of the delay states by proportional allocation:

```
two.stage.x0 <- function(p,X0){
x0 <- c( X0[1]/(p[5]+1), X0[2], rep(X0[1]/(p[5]+1),p[5]) )
return(x0)
}
```

When this function is not included in the model list, it takes the value `NULL`. Consequently, we use the `NULL` condition in `simt` to decide whether there is a delay or not.

10.8.3 Function simt

What we have done is to modify the function `sim` to include intermediate stages of a delay and the total for all classes. These changes only affect the lines within the loop for parameter `p` in the following manner. When there is a delay, (1) call `model$x0` component to compose extended initial conditions, (2) declare a 3-D array, `Y`, to store the intermediate stages, and (3) sum over the intermediate stages to get the total for class 1. These three modifications are executed only when `model$x0` is not `NULL`.

```
    # sub-states delay
   if(is.null(model$x0)==F){
    nd <- p[length(mp)]
    X0 <- model$x0(p,X0)
    Y  <- structure(1:(nt*nd*np), dim=c(nt,nd,np));Y[,,] <- 0
   }
   # integration
   out <- RK4(X0, t, model$f, p, dt)
   # rename x part of out for simpler use
   if(NS>1) {
   for(k in 1:NS) X[,i,k] <- out[,k]
```

```
      # sub-states delay
      if(is.null(model$x0)==F){
       Y[,,i] <- out[,(NS+1):(nd+NS)]
       for(j in 1:nt) X[j,i,1] <- out[j,1] + sum(Y[j,1:nd,i])
      }
      # total
      for(j in 1:nt) X[j,i,(NS+1)] <- sum(X[j,i,(1:NS)])
      }else{
      # calculate rate for one-state systems
       X[,i,1] <- out
       for(j in 1:nt) dX.dt[j,i] <- model$f(t[j],p,X[j,i,1])
      }
  } # end of parameter p loop
```

Then, regardless of whether there is a delay or not, we calculate the total population after integration by summing over all stages.

10.8.4 Function read.plot.cerio.out

This is a lengthier function and it is not discussed here because it is very specific to the particular kind of file we had to read. If you are interested to see how it works, please study it from the console.

10.8.5 Function cerio.F

The first two lines are to capture the working directory so we can restore it later, change the directory to the folder specified by argument 1. Then the third line is to copy a file named according to the prefix given by argument number 2 to a temporary file **cerio_inp.txt**, which is the filename expected by the Fortran program **cerio.exe**. Then, we execute the Fortran program with R function .Fortran cerio.exe and rename the output files produced by cerio.exe to our filename specified by argument number 2. Finally, we delete the temporary file and change back to the work directory. As explained in section 10.7.5, to be able to use a Fortran program we first dynamically load the **seem.dll** which resides in the seem library directory. Assuming a 32 bit system we use the following statement

```
> dyn.load(paste(.libPaths(),"/seem/libs/i386/seem.dll",sep="")).
```

As explained in chapter 1, you can automatically execute this statement every time you start R by adding this line to the **Rprofile.site file** in the **etc** directory of your R installation. Further instructions are on the book's website.

```
cerio.F <- function(x,fileprefix){
y <- getwd()
  setwd(x)
  file.copy(paste(fileprefix,"-inp.txt",sep=""), "cerio_inp.txt")
  .Fortran("cerio", package="seem")
  fileout <- paste(fileprefix,"-out.txt",sep="")
  filedex <- paste(fileprefix,"-dex.txt",sep="")
  file.rename("cerio_out.txt", fileout)
  file.rename("cerio_dex.txt", filedex)
  file.remove("cerio_inp.txt")
  setwd(y)
  return(paste(x,"/",fileout,sep=""))
}
```

11 Ecotoxicological Modeling

We have already encountered examples of models quantifying how the presence of a chemical in the environment may affect organisms. In all cases, we assumed that we knew the concentration of the chemical in the environment (water, air, or soil). Such knowledge may come from the measurements or modeling of the transport and fate of chemicals in the environment and how the environmental concentration converts into **exposure** to the organism. In this book, we do not cover transport and fate models (water and air quality) to estimate the pollutant concentration.

In this chapter, we proceed by building models of **exposure** using the **bioavailability** of the chemical as a toxicant, of **bioaccumulation** using uptake and depuration kinetics, and of the **effects** of the pollutant accumulated in the body of organisms (Figure 11.1). Often these types of models use the compartment approach by defining compartments within the body. Many times, the models are used to understand the equilibrium concentration in the body (Jorgensen, 1988; O'Connor et al., 1989; Thomann, 1989).

11.1 BIOAVAILABILITY

Not all chemical forms of a compound in the environment affect plants and animals. For example, free ionic forms dissolved in water are easier to transfer to organisms than those bound to suspended sediments, salts, and other particles. Therefore, we need to calculate the part of the toxicant concentration that is **bioavailable**; that is, the fraction of the concentration that represents a true exposure to the organism.

Bioavailable concentrations depend on the conditions of the medium (e.g., conditions of the stream water) and the organism (e.g., is it a filter feeder or a grazer). In water, for example, bioavailability depends on the concentration of the suspended solids, organic and inorganic ligands, and mineral concentration (i.e., water hardness).

A useful method is to multiply the concentration by **partition coefficients**, equilibrium constants, or equilibrium isotherms (e.g., Langmuir). In many cases, these components are included in a water quality model and thus bioavailable fractions are already calculated. For bioavailability estimations, a mechanistic approach would be to derive the kinetics of the compound in water, whereas a more empirical approach would be to derive regressions between the bioavailable fraction and water conditions, for example, salt concentration.

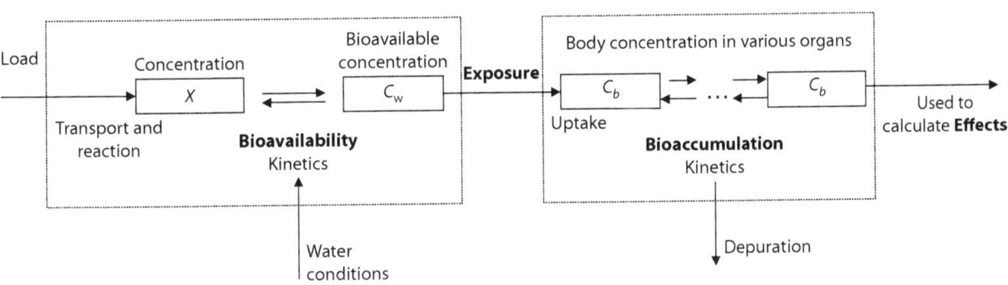

FIGURE 11.1 Major components of ecotoxicological models: bioavailability, bioaccumulation, and effects. Water is shown as an example.

Because it is difficult to have partition coefficients for all types of tissues, a surrogate for a partition coefficient is derived from a generic partition coefficient to organic carbon:

$$K_p = f_{oc} K_{oc} \tag{11.1}$$

where f_{oc} is the organic carbon fraction of the particulate matter, and K_{oc} is the organic carbon partition coefficient. K_{oc} can be derived empirically, by regression, from the octanol–water partition coefficient K_{ow}:

$$\log(K_{oc}) = a \log(K_{ow}) + b \tag{11.2}$$

We can combine partition coefficients of various chemical species according to concentration. For example, consider exposure to silver (Ag) in stream water. The octanol–water partition coefficient K_{ow} for Ag^+ (~0.02) and $AgCl_{aq}$ (~0.09) can be combined at their stream chloride concentrations to determine uptake rates. Suppose that at some chloride concentration we have [Ag] = 70% of total and $AgCl_{aq}$ = 30%, then $0.7 \times 0.02 + 0.3 \times 0.09 = 0.041$.

The K_{ow} is an approximate calculation of the equilibrium portioning of the chemical in the body of an organism (Hemond and Fechner, 1994). For example, $\log(K_{ow})$ of atrazine (an herbicide) is 2.68 and thus the K_{ow} is $10^{2.68} = 479$. Assume that a fish is 5% fatty tissue, 85% water, and the rest is air and tissue that do not absorb the chemical. Denote C_{tot} as the total atrazine concentration. The partitioning to fatty tissue is

$$C_{fat} \simeq \frac{0.05 \times K_{ow} \times C_{tot}}{0.05 \times K_{ow} \times C_{tot} + 0.85 \times C_{tot}} = \frac{0.05 \times 479}{0.05 \times 479 + 0.85} = 0.97$$

Therefore, there will be approximately 97% of the atrazine in the fatty tissue and 3% of the atrazine in the body water. This estimate is only an approximation because we ignored air and other tissue.

Exercise 11.1

Calculate the percent of atrazine in the fat if the organism is 80% water and 10% fat.

The concentration of total silver can be calculated with a water quality model (mixing, transport, and fate models). The silver concentration in the environment ranges from 10 to 200 pM.

To exemplify further, let us examine with more detail how silver bioavailability varies with chloride concentration in water. Silver accumulation in aquatic organisms is due to the bioavailability of the free Ag ion (Ag^+) and also of the $AgCl_{aq}$. Which one of these two may dominate depends on the ambient chloride concentrations. Let us work with Ag^+.

The reaction is given by

$$AgCl \rightleftarrows Ag^+ + Cl^- \tag{11.3}$$

with equilibrium constant $Ksp = 1.56 \times 10^{-10}$. First, calculate the solubility. We have at equilibrium

$$Ksp = X^2 \tag{11.4}$$

where X is the concentration of Cl^- and Ag^+. Solving for it, we get

$$[Ag^+] = \sqrt{Ksp}\ M = 1.29 \times 10^{-5}\ M = 12.9\ \mu M \tag{11.5}$$

Now assume that you have Y extra Cl from salinity. In Y M of NaCl, the Cl^- from NaCl solution shifts the equilibrium:

$$Ksp = X(Y + X) \tag{11.6}$$

Because X is much smaller than Y, we have

$$Ksp = YX \tag{11.7}$$

and solving for X we get

$$X = Ksp/Y \tag{11.8}$$

Let us assume relatively low chloride concentrations (0 to 50 mM) typical of fresh to brackish waters. For example, in 10 mM, or $[Cl^-] = 0.01$ M

$$[Ag^+] = Ksp/0.01 = 1.56 \times 10^{-8}\ M = 15.60\ nM = 15600\ pM \tag{11.9}$$

The concentration of Ag^+ for a range of Cl from 10 to 100 mM is shown in Figure 11.2.

Exercise 11.2

Calculate the molar solubility of AgCl at 55 mM chloride.

In the absence of a detailed chemical fate model, the percent bioavailability of silver as a function of chloride concentration could be calculated from an empirical relation derived by regression. Just for the sake of an example, assume that

$$Ag_{per} = -0.674[Cl] + 81.76 \tag{11.10}$$

and then calculate the concentration of the bioavailable fraction C_w:

$$C_w = (Ag_{per}/100) \times Ag_{tot} = Ag_{tot} \times (-0.674[Cl] + 81.76)$$

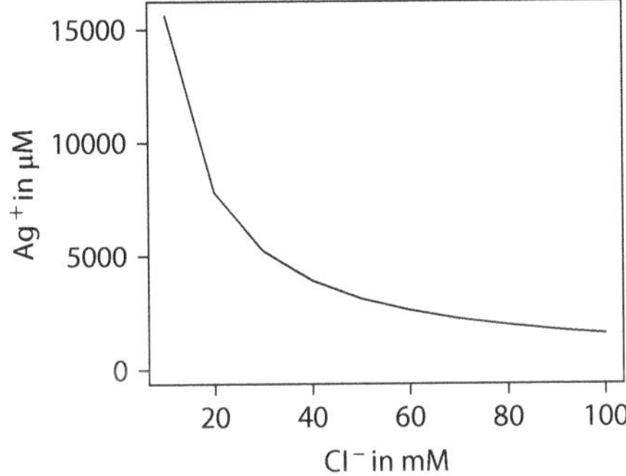

FIGURE 11.2 Free Ag^+ ion as a function of chloride concentration.

For $Ag_{tot} = 0.1$ mg/liter, as chloride varies, say 70, 80, and 90, the exposure C_w changes to 0.0341, 0.0274, 0.0207 mg/liter. Higher chloride concentration decreases the bioavailable silver. The stream chloride concentration is 12.5%, around the nominal of 80. The bioavailability of silver decreases by ~25%.

11.2 BIOACCUMULATION

To estimate the effects of a pollutant or toxicant on an organism, we have to model how fast the pollutant enters the organism (i.e., uptake) and how fast it leaves the organism, that is, depuration or clearance (Calabrese and Baldwin, 1993; Jorgensen, 1988; O'Connor et al., 1989). The balance between these two processes determines how much of it accumulates in the organism, for example, the net rate of accumulation in the body of an animal. To determine the uptake, we examine the entry pathways of the toxic substance into the body, for example, whether it enters with the food or directly from the environment (air or water). To determine the depuration, we examine the ways of getting rid of the compound, for example, metabolic breakdown or excretion rate.

For example, for fish, assuming no body weight change, the rate of change of toxicant concentration in the body is an ordinary differential equation (ODE):

$$\frac{dX(t)}{dt} = U - E \tag{11.11}$$

where $X(t)$ is the concentration of the toxicant in the body as a function of time, t, and given in units of toxicant mass per unit of body weight, say mg/kg.

A simple assumption is to consider an uptake rate linear with respect to the exposure, that is, concentration in water C_w, given as toxicant mass per unit volume (say mg/liter) and assumed constant (Calabrese and Baldwin, 1993). Then $U = uC_w$, where u is the uptake rate coefficient. Note that u must have units of (liter/kg)/h when time is given in hours. This rate coefficient, u, can be modeled with an uptake efficiency parameter, e_w, (unitless) and the ratio of V = water volume flow rate through the gills (volume/time, liter/h) to W = body weight (in kg) (Jorgensen, 1988):

$$u = e_w \frac{V}{W} \tag{11.12}$$

Also assume that depuration is linear with respect to X with the excretion rate coefficient, e_x, and therefore, for this simple model

$$\frac{dX(t)}{dt} = e_w \frac{V}{W} C_w - e_x X(t) \tag{11.13}$$

This is a simple forced ODE, which we examined in Chapter 7. The forcing term is $U = uC_w = e_w \frac{V}{W} C_w$:

$$\frac{dX(t)}{dt} = U - e_x X(t) \tag{11.14}$$

At steady state, the concentration in the body is constant and therefore $dX/dt = 0$; substituting and solving, we can find the equilibrium value:

$$X^* = \frac{U}{e_x} = \left(\frac{u}{e_x}\right) C_w = \left(\frac{e_w V}{e_x W}\right) C_w \tag{11.15}$$

The **bioconcentration factor** ($B(t)$) is the ratio of the body toxicant concentration ($X(t)$) to the ambient (water) toxicant concentration (C_w) and has units of volume to mass (liter/kg):

$$B(t) = \frac{X(t)}{C_w} \tag{11.16}$$

From Equation 11.15 at steady state, $B(t)$ is a constant and equal to B^* in the following:

$$B^* = \frac{X^*}{C_w} = \frac{u}{e_x} = \frac{e_w V}{e_x W} \tag{11.17}$$

that is to say, B^* is given by the ratio of uptake and depuration rate coefficients or also by the ratio of uptake efficiency to excretion rate coefficient times the ratio of water flow to body weight.

The transient is an exponential,

$$X(t) = X^*\left(1 - \exp(e_x t)\right) \tag{11.18}$$

For example, assuming $e_x = 0.05\ \text{h}^{-1}$, $e_w = 0.1$, $V = 0.5$ liter/h, and $W = 0.5$ kg, the steady state $X^* = 2 \times C_w$. For concentrations of C_w of 0, 0.2, and 0.5, 1 mg/liter, we obtain $X^* = 0$, 0.4, and 1, 2 mg/kg. The runs to simulate transients $X(t)$ for 96 h = 4 d are shown in Figure 11.3, where we can see that the transient and steady state X^* correspond to the values used. The 4-day time frame chosen is the standard toxicity test duration.

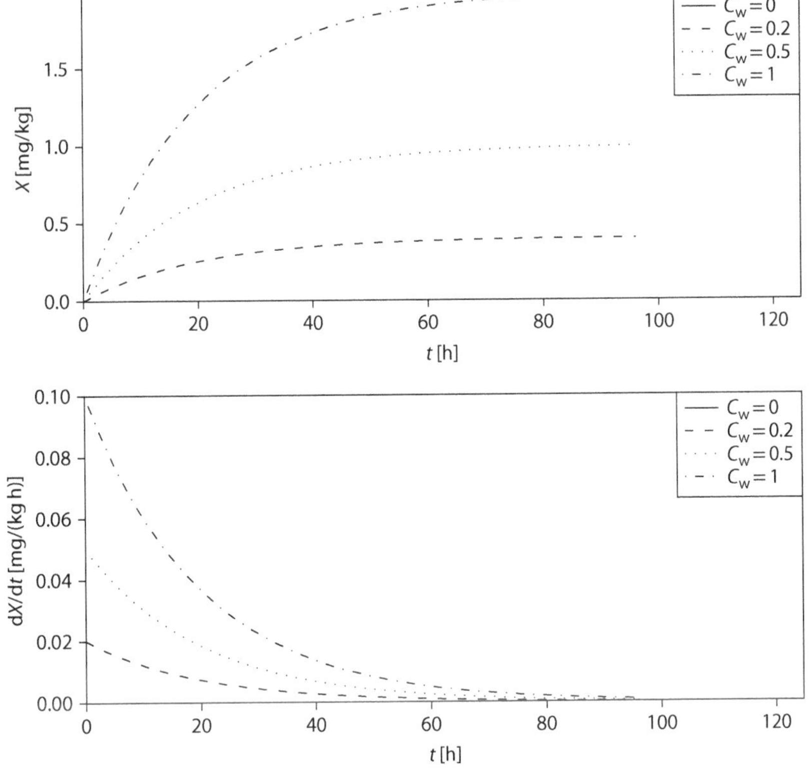

FIGURE 11.3 Bioaccumulation during 96 hours for several exposure levels.

11.3 EFFECTS: LETHAL AND SUBLETHAL

By counting the deaths in a toxicity test at various ambient toxicant concentrations, C_w, we can determine the lethal dose LC_{50} defined as the concentration C_w at which only 50% of the individuals survive. Similarly, by measuring an effect on reproduction, we can determine the sublethal EC_{50} dose as the concentration at which 50% of the individuals are affected.

It is more involved to measure body burden or toxicant concentration at the end of a test, but when done multiple times, it can help establish a relationship of body concentration to lethal and sublethal effects (Calabrese and Baldwin, 1993). With these data, we can derive the empirical relationships to link body concentration at the end of the 96 h period ($X(t = 96)$) and the effects of ambient toxicity. Lower lethal doses should correspond to higher bioaccumulation; therefore, for example, the lethal ambient concentration should relate to the inverse of $X_L(96)$ at which lethality occurs:

$$LC_{50} = \frac{k}{X_L(96)} \tag{11.19}$$

where k is a constant to be obtained from regression. This $X_L(96)$ is referred to as the CBR (critical or median lethal body residue) and is expressed in mass concentration (toxicant mass per unit of body weight).

More generally, we can assume a power function in the X term to obtain a better fit:

$$LC_{50} = \frac{k}{X_L(96)^a} \tag{11.20}$$

If the body burden reaches a steady state by the end of the 96 h period, as it does in the example of the previous section, then $X_L(96) = X^*$. For the example above, if $LC_{50} = 1$ mg/liter and $a = 1$, then $X_L(96) = X^* = 2$ mg/kg is the CBR. But if $LC_{50} = 0.5$ mg/liter, then CBR = 1 mg/kg.

11.4 INCLUDING FOOD PATHWAYS

Organisms can also take the toxicant from their food when contaminated (Jorgensen, 1988). The uptake rate, U, can be modeled with two uptake efficiency parameters, one for water, e_w, and one for food, e_f, and by separating the two concentrations of the chemical in food (C_f) in (mass toxic)/(mass of food) and in water (C_w) in (mass toxic)/(volume of water):

$$U = \frac{e_f C_f F + e_w C_w V}{W} \tag{11.21}$$

F is the food ingestion rate in MT^{-1}, V is water volume flow rate through the gills L^3T^{-1}, and W is the body weight. This can also be used for air, by using C_w for the concentration in air, V for volume flow rate through the lungs, and e_w for the efficiency of uptake from air.

We model the excretion rate as linear with respect to body concentration as in the simpler model above, that is, $e_x X$, where e_x = excretion rate coefficient:

$$\frac{dX(t)}{dt} = \frac{e_f C_f F + e_w C_w V}{W} - e_x X(t) \tag{11.22}$$

Bioconcentration $B(t)$ is now defined by the ratio to lumped ambient concentration: water concentration plus food concentration, that is,

$$B(t) = \frac{X(t)}{C_w + C_f}$$

In particular, at a steady state

$$B^* = \frac{X^*}{C_w + C_f} \tag{11.23}$$

The model (Equation 11.22) is a forced ODE as before. The forcing term is more complicated requiring more parameters. However, the dynamics are still the same and determined by $\exp(-e_x t)$.

11.5 COMPARTMENT MODELS

It is also possible to model concentrations in various organs using a **compartment** approach. Examples are two- and three-compartment models (Calabrese and Baldwin, 1993; Jorgensen, 1988). Compartment models allow for the calculation of how the contaminant distributes itself in several organs. State variables can be toxicant mass in the compartment or normalized to concentration dividing by the weight or volume of each compartment.

Compartment models are very important in a variety of environmental and ecological problems and work in the following manner: The material (e.g., nutrient, toxicant) can be present in a number n of "stocks," "pools," or "compartments." The net rate of change of the amount or the concentration X_i in the pool or compartment i is given by a **balance** of inflow minus outflow (Figure 11.4):

$$\mathrm{d}X_i/\mathrm{d}t = \mathrm{IF}_i - \mathrm{OF}_i \tag{11.24}$$

where IF_i is the inflow or the rate of transfer **into** compartment i from all other compartments j, $\mathrm{IF}_i = \sum_{j=1}^{n} f_{ij}$, and OF_i is the outflow or the rate of transfer **out of** compartment i and into all other compartments j, $\mathrm{OF}_i = \sum_{j=1}^{n} f_{ji}$. Please note that the order of subscripts is important f_{ij} is from j to i. In addition, note the f_{ij} are the net rates and therefore its units are $[\text{unit of } X]/\mathrm{T}$ for example, (mg/liter)/day or $\mathrm{mg} \cdot \mathrm{liter}^{-1} \cdot \mathrm{d}^{-1}$.

The linear compartment model assumes that each net rate of transfer is proportional to the concentration in the source compartment:

$$f_{ij} = k_{ij} X_j \tag{11.25}$$

where k_{ij} is a transfer rate coefficient from compartment j to compartment i and it should have units of T^{-1} (Figure 11.5). You should recall that this is similar to the linear assumption of an exponential model and thus conclude that the material in each compartment should change as a combination of exponentials because of the various flows in and out of the compartment.

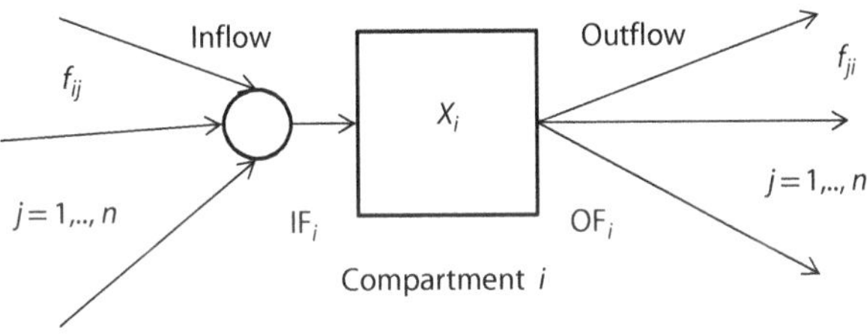

FIGURE 11.4 Element of a compartment model. Balance at compartment i.

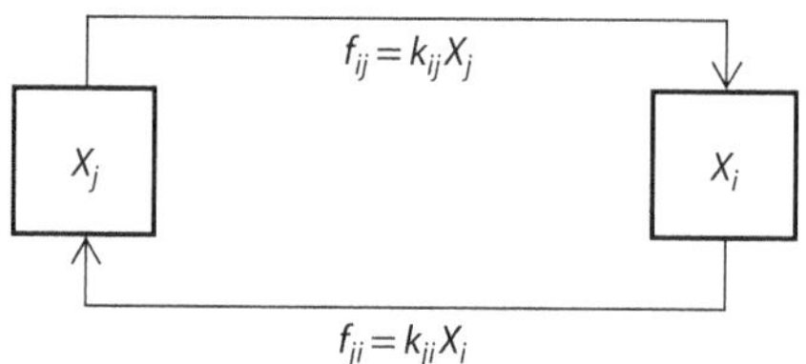

FIGURE 11.5 Flows proportional to stock yield a linear model.

Substitute equations like Equation 11.25 in each one of the f_{ij} to obtain inflow and outflow as linear combinations of the concentrations in the source compartments: $\mathrm{IF}_i = \sum_{j=1}^{n} k_{ij}X_j$ and $\mathrm{OF}_i = \sum_{j=1}^{n} k_{ji}X_i$, and therefore, we can rewrite Equation 11.24 in matrix form as a **system** of ODEs or a linear dynamical system:

$$\mathrm{d}\boldsymbol{X}(t)/\mathrm{d}t = \mathbf{K}\boldsymbol{X}(t) \tag{11.26}$$

where $\boldsymbol{X}(t)$ is a column vector of dimension $n \times 1$ and $\mathbf{K}$ is a $n \times n$ square matrix with entries k_{ij}.

11.5.1 Closed System

In the absence of compartment gains or leaks other than these transfers, the sum of all the outflows of one compartment (say compartment i) is equal to the sum of all the inflows of all compartments receiving material from compartment i. This condition leads to a matrix **K** with the property that the sum of all the entries in a column should add up to zero, and mathematically this means that the matrix **K** is singular, that is, the determinant is zero.

In other words, the material is **conservative**; there are no exchanges from the compartments to "outside" the system. Therefore, the equations are subject to the additional condition that $\sum_{i=1}^{n} X_i = X_t$, where X_t is the total in all compartments and remains constant. The system will be at equilibriumwhen there is no change; recall that this requires $\mathrm{d}\boldsymbol{X}/\mathrm{d}t = \mathbf{0}$.

For example, consider two compartments, X_1 and X_2, with linear kinetics and a closed system as shown in Figure 11.6. The system of ODEs is

$$\begin{aligned} \frac{\mathrm{d}X_1}{\mathrm{d}t} &= -k_{21}X_1 + k_{12}X_2 \\ \frac{\mathrm{d}X_2}{\mathrm{d}t} &= k_{21}X_1 - k_{12}X_2 \end{aligned} \tag{11.27}$$

with the condition $X_1 + X_2 = C$. In this example, the matrix of coefficients is

$$\mathbf{K} = \begin{bmatrix} -k_{21} & k_{12} \\ k_{21} & -k_{12} \end{bmatrix} \tag{11.28}$$

Note that each one of the two columns indeed add up to 0 because this is a closed system with no inflow from or outflow to the outside. Therefore, the matrix is singular and we need to use the extra equation $X_1 + X_2 = C$, where C is a constant, meaning that the material is conserved. It is practical to treat X_1 and X_2 as fractions of the total, $X_1 + X_2 = 1$. Using this relation to get $X_2 = 1 - X_1$ and substituting for X_2 in the first equation of system (Equation 11.27), we obtain

$$\frac{\mathrm{d}X_1}{\mathrm{d}t} = -k_{21}X_1 + k_{12}(1 - X_1) = k_{12} - (k_{12} + k_{21})X_1 \tag{11.29}$$

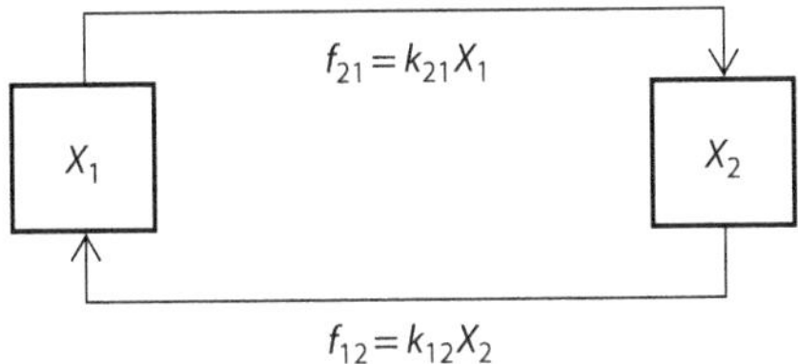

FIGURE 11.6 A closed system of two compartments.

Then find the equilibrium of Equation 11.29 by using $dX_1/dt = 0$, yielding

$$0 = k_{12} - (k_{12} + k_{21})X_1^* \tag{11.30}$$

and solve for the equilibrium to obtain

$$X_1{}^* = \frac{k_{12}}{k_{12} + k_{21}} \tag{11.31}$$

This means that the equilibrium value in compartment 1 is the ratio of inflow into this compartment to the total transfer (inflow + outflow) for the compartment.

Now $X_2{}^* = 1 - X_1{}^*$, that is, the remainder of the material or the fraction in compartment 2, and thus

$$X_2{}^* = \frac{k_{21}}{k_{12} + k_{21}} \tag{11.32}$$

This means that the value in compartment 2 is the ratio of inflow into this compartment to the total transfer (inflow plus outflow) for the compartment.

The equilibrium for a linear compartment model is stable. The variables X_1 and X_2 follow exponential time functions with the same rate coefficient, which is the nonzero eigenvalue of **K**. In this two-compartment example, the eigenvalues are $\lambda_1 = 0$ and $\lambda_2 = -(k_{12} + k_{21})$.

Exercise 11.3

Demonstrate that the eigenvalues of **K** in Equation 11.28 are $\lambda_1 = 0$ and $\lambda_2 = -(k_{12} + k_{21})$.

Thus, X_1 follows the exponential kinetics with a rate coefficient given by $\lambda_2 = -(k_{12} + k_{21})$ and the solution is

$$X_1(t) = \left(X_1(0) - X_1{}^*\right)\exp(-(k_{12} + k_{21})t) + X_1{}^* \tag{11.33}$$

$X_1(t)$ varies from the initial condition $X_1(0)$ to the equilibrium point in exponential fashion and $X_2(t)$ is given as $X_2(t) = 1 - X_1(t)$.

For example, assume $k_{12} = 0.01\ h^{-1}$ and $k_{21} = 0.02\ h^{-1}$ and initial conditions such that all the material are in compartment 1 at time zero, $X_1(0) = 1$, $X_2(0) = 0$. We obtain the dynamics shown in Figure 11.7.

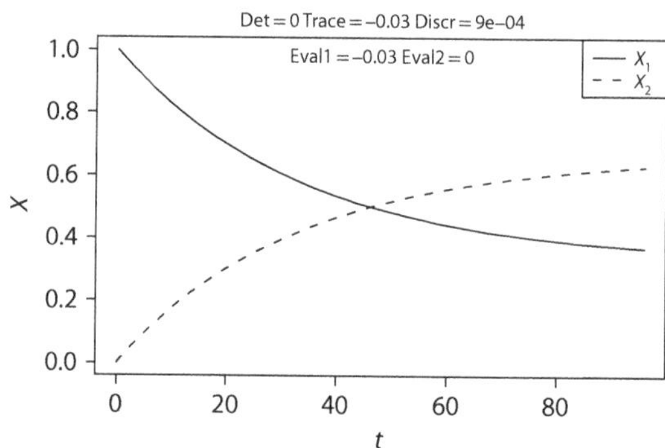

FIGURE 11.7 Dynamics of a closed two-compartment system.

Exercise 11.4

Given that $k_{12} = 0.01\ \text{h}^{-1}$ and $k_{21} = 0.02\ \text{h}^{-1}$, assume that the total material is 1 mg/liter. Calculate the equilibrium values X_2^*, X_1^* in mg/liter. Compare the results with Figure 11.7.

Exercise 11.5

Use the values of Exercise 11.4. Put all the material in compartment 2 at time = 0. Sketch a graph of $X_1(t)$ and $X_2(t)$ versus time.

11.5.2 Open Systems

This system is open when there is transfer **into (or from)** at least one pool or compartments **from (or to)** "outside" the system, as happens for the uptake and depuration of toxicants. At least one column of the matrix does not add up to zero because of a donor-controlled output. In addition, if the input or output is independent of $\boldsymbol{X}$, then we have a forced ODE:

$$\frac{\mathrm{d}\boldsymbol{X}}{\mathrm{d}t} = \mathbf{K}\boldsymbol{X} + \boldsymbol{U} \tag{11.34}$$

$\boldsymbol{U}$ is the vector of inflows or outflows if these do not depend on the concentrations in the compartments.

11.6 BIOACCUMULATION: TWO-COMPARTMENT MODEL

We will use a two-compartment open system as shown in Figure 11.8. We divide the body into two compartments, X_1 and X_2, with linear kinetics. It receives fixed inflow (beginning at time 0, i.e., a **step function**) in the form of uptake from ambient concentration into pool 1 and has linear donor-controlled depuration from pool 2:

$$\frac{\mathrm{d}X_1}{\mathrm{d}t} = -k_{21}X_1 + k_{12}X_2 + U \tag{11.35}$$

$$\frac{\mathrm{d}X_2}{\mathrm{d}t} = k_{21}X_1 - k_{12}X_2 - e_x X_2 \tag{11.36}$$

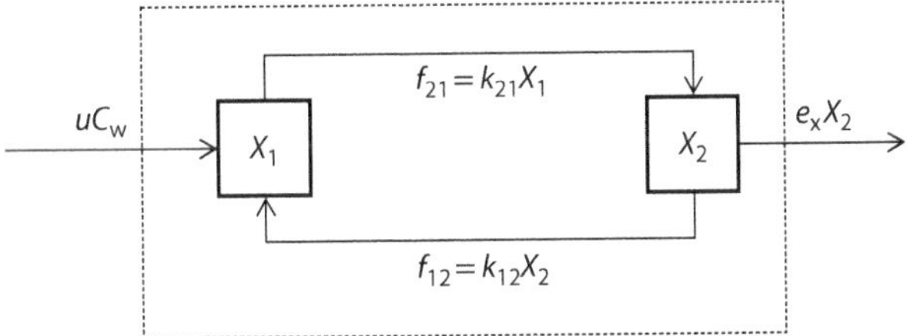

FIGURE 11.8 An open two-compartment system.

We can rewrite as

$$\frac{\mathrm{d}\boldsymbol{X}}{\mathrm{d}t} = \mathbf{A}\boldsymbol{X} + \boldsymbol{U} \tag{11.37}$$

where matrix **A** is

$$\mathbf{A} = \begin{bmatrix} -k_{21} & k_{12} \\ k_{21} & -k_{12} - e_x \end{bmatrix} \tag{11.38}$$

and vector $\boldsymbol{U}$ is

$$\boldsymbol{U} = \begin{bmatrix} U \\ 0 \end{bmatrix} \tag{11.39}$$

It is important to note that e_x is a rate coefficient, whereas U is the net rate.

In matrix **A**, one column does not add up to 0; therefore, the condition $\boldsymbol{X}_1 + \boldsymbol{X}_2 = C$ does not hold, in other words, the material is not conserved at all times. The determinant det (**A**) can be calculated to get

$$\begin{vmatrix} -k_{21} & k_{12} \\ k_{21} & -k_{12} - e_x \end{vmatrix} = k_{21}e_x \tag{11.40}$$

We can calculate the equilibrium values $\boldsymbol{X}_1^*$, $\boldsymbol{X}_2^*$ by making Equations 11.35 and 11.36 equal to zero. Alternatively, making the left-hand side of Equation 11.37 equal to zero, we have

$$\mathbf{A}\boldsymbol{X}^* = -\boldsymbol{U} \tag{11.41}$$

and

$$\boldsymbol{X}^* = -\mathbf{A}^{-1}\boldsymbol{U} \tag{11.42}$$

After solving, we have

$$\boldsymbol{X}^* = \begin{bmatrix} \dfrac{(k_{12} + e_x)U}{k_{21}e_x} \\ \dfrac{U}{e_x} \end{bmatrix} \tag{11.43}$$

We can find this solution without inverting the matrix by first noting that at steady state the net flow from/to the outside is zero; therefore, $U = e_x X_2^*$ or $X_2^* = \frac{U}{e_x}$. This means that the steady state value in compartment 2 is the ratio of the input and output rates. Substituting this value in the dX_2/dt Equation 11.36 and equating to 0 gives

$$\frac{dX_2}{dt} = 0 = k_{21}X_1^* - (k_{12} + e_x)\left(\frac{U}{e_x}\right) \tag{11.44}$$

yielding

$$X_1^* = \left(\frac{k_{12} + e_x}{k_{21}}\right)\left(\frac{U}{e_x}\right) \tag{11.45}$$

This means that the steady value in the first compartment is the ratio of the transfer coefficients times the ratio of the input and output rates. Variables X_1 and X_2 follow exponential time functions with coefficients given by the eigenvalues of **A**.

For example, assume we have $k_{12} = 0.01\ h^{-1}$, $k_{21} = 0.03\ h^{-1}$, $e_x = 0.04\ h^{-1}$, and $u = 0.01\ h^{-1}$. Varying C_w to have values 1, 5, and 10 (mg/liter), we obtain the dynamics shown in Figure 11.9.

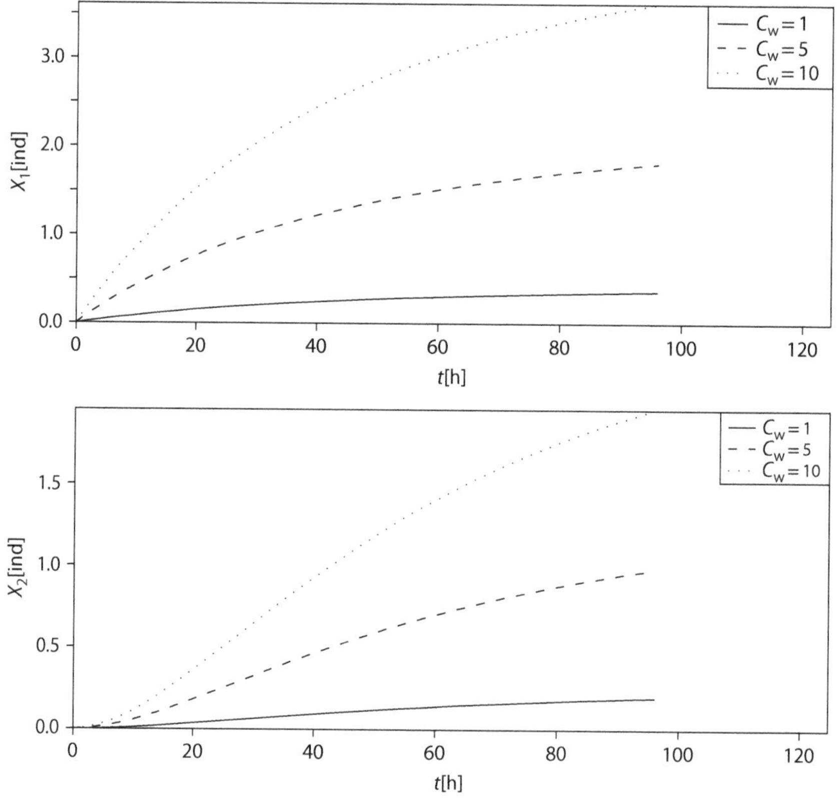

FIGURE 11.9 Dynamics of an open two-pool model.

Exercise 11.6

Given $k_{12} = 0.01\ h^{-1}$, $u = 0.01\ h^{-1}$, $k_{21} = 0.03\ h^{-1}$, $e_x = 0.04\ h^{-1}$ as in Figure 11.9, calculate the equilibrium values for each compartment for $C_w = 1$ mg/liter and compare with the ones shown in the figure.

Exercise 11.7

Using the parameter values of the previous exercise, calculate the eigenvalues of **A**. Predict the dynamics of $X_1(t)$ and $X_2(t)$ from the eigenvalues.

Exercise 11.8

Assume that compartments 1 and 2 start at their steady-state values for $C_w = 1$ mg/liter at time = 0 h and that there is a step increase in C_w input to 5 mg/liter, but e_x stays the same. What are the new steady states? Is there a change in the eigenvalues?

Exercise 11.9

Using Exercise 11.8, describe the dynamics of $X_1(t)$ and $X_2(t)$ and draw a graph for each one.

Experimentally, in order to estimate rate coefficients, the material is marked with radioisotopes (tracers). This assumes that the rate of transfer of the tracer is the same as the rate of transfer for the material.

Exercise 11.10

Suppose a two-compartment system has $k_{21} = 0.3$ and $k_{12} = 0.5$ with no inputs or outputs. (a) Is the system open or closed? (b) Assume X_1 and X_2 are the fractions of the total. What are the values of X_1 and X_2 at equilibrium? (c) Write the associated matrix of coefficients. (d) What are the eigenvalues? (e) If all the material is in compartment 2 at time 0, what is the exponential kinetics for X_1 and X_2? Sketch a graph of X_1 and X_2 versus time.

The net rates are not necessarily linear combinations of the compartment state. For example, we could have M3 or hyperbolic kinetics in the flow from i to j, where

$$f_{ji} = \left[\frac{X_i}{X_i + K_h}\right] X_j \tag{11.46}$$

In other words, the flow from i to j depends on the concentration of the donor compartment X_i and is rate limited, achieving a maximum or saturation value for large X_i. In these cases, the system of equations is not linear. We will study this case in the next model.

11.7 BIOACCUMULATION: MULTICOMPARTMENT

We will consider an example consisting of three compartments (Jorgensen, 1988): compartment 1 (e.g., the bloodstream) corresponds to uptake and depuration kinetics; compartment 2 is the organ where metabolic destruction of the toxicant occurs, for example, the liver; and finally, compartment 3 corresponds to long-term accumulation (e.g., bones); see Figure 11.10. Use X_i to represent toxicant mass per unit body weight, say mg/kg, for each compartment. Later, we can normalize them by dividing by the ratio of the weight of each compartment with respect to the total weight.

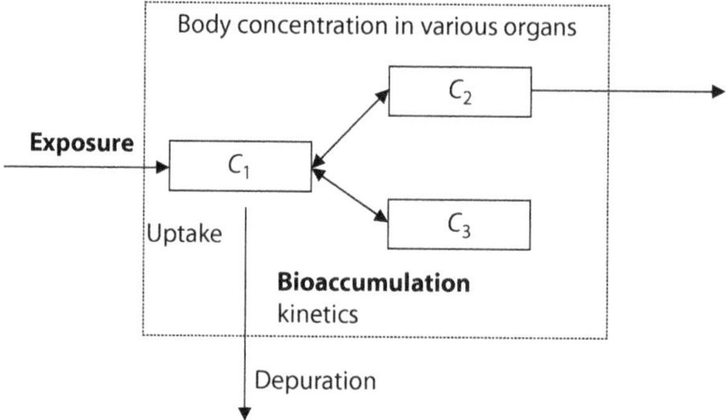

FIGURE 11.10 Three-compartment model.

The first compartment includes uptake and depuration (linear rates) as before and also transfers to the other two compartments and returns

$$dX_1/dt = uC_w - e_x X_1 - k_{31}X_1 - k_{21}X_1 + k_{12}X_2 + k_{13}X_3 \tag{11.47}$$

Note that the depuration rate is proportional to the amount in this compartment via the rate coefficient (e_x) determined by excretion via the kidney. The product of an uptake rate coefficient (u) and ambient concentration (C_w) gives the uptake net rate. The second compartment is where the metabolic loss occurs and this is modeled using M3 kinetics:

$$dX_2/dt = k_{21}X_1 - k_{12}X_2 - K_{max}\frac{X_2}{K_h + X_2} \tag{11.48}$$

Finally, the third compartment includes only transfer rates to and from compartment 1 and tends to accumulate without an explicit mechanism for clearance:

$$dX_3/dt = k_{31}X_1 - k_{13}X_3 \tag{11.49}$$

A compact matrix equation formed by all three previous expressions is

$$\frac{d}{dt}\begin{bmatrix} X_1 \\ X_2 \\ X_3 \end{bmatrix} = \underbrace{\begin{bmatrix} -k_{31} - k_{21} - e_x & k_{12} & k_{13} \\ k_{21} & -k_{12} - \dfrac{K_{max}}{K_h + X_2} & 0 \\ k_{31} & 0 & -k_{13} \end{bmatrix}}_{\mathbf{A}} \begin{bmatrix} X_1 \\ X_2 \\ X_3 \end{bmatrix} + \underbrace{\begin{bmatrix} u \\ 0 \\ 0 \end{bmatrix}}_{\mathbf{u}} C_w \tag{11.50}$$

Assigning matrices names as shown, we can write it more succinctly as

$$\frac{d\boldsymbol{X}}{dt} = \mathbf{A}\boldsymbol{X} + \boldsymbol{u}C_w \tag{11.51}$$

Note that Equation 11.51 includes a nonlinear term and thus we cannot solve for the equilibrium directly as a system of linear equations. A starting point is to recognize that at steady state the uptake should be equal to the sum of the depuration and breakdown; therefore,

$$uC_w = e_x X_1^* + \frac{K_{max} X_2^*}{K_h + X_2^*}$$

or (11.52)

$$X_1^* = \frac{uC_w}{e_x} - \frac{K_{max} X_2^*}{e_x \left(K_h + X_2^*\right)}$$

When X_2 reaches saturation for breakdown, the kinetics would be proceeding at maximum rate K_{max}; therefore, this last equation reduces to

$$X_1^* = \frac{uC_w}{e_x} - \frac{K_{max}}{e_x} = \frac{uC_w - K_{max}}{e_x} \tag{11.53}$$

Then, under the same conditions from the second Equation 11.48 of the system, we must have

$$X_2^* = \frac{k_{21} X_1^* - K_{max}}{k_{12}} = \frac{k_{21} \frac{uC_w - K_{max}}{e_x} - K_{max}}{k_{12}} = \frac{k_{21}(uC_w - K_{max}) - e_x K_{max}}{e_x k_{12}} \tag{11.54}$$

and substituting in the third equation of the system (Equation 11.49) to find X_3,

$$X_3^* = \frac{k_{31}}{k_{13}} \frac{uC_w - K_{max}}{e_x} \tag{11.55}$$

Defining a weight ratio $w_i = W_i/W$, we can scale each compartment to obtain the bioconcentration factor, B_i in each compartment X_i using the ratio $B_i = \frac{X_i}{w_i C_w}$. Note that the lighter the compartment is with respect to the total, the larger the bioconcentration. For example, using Equation 11.55, we get for compartment 3

$$B_3^* = \frac{k_{31}}{k_{13}} \frac{uC_w - K_{max}}{e_x C_w w_3} \tag{11.56}$$

For a specific numeric example, suppose we have a final time of 4 days (or 96 hours) with zero concentration for the initial condition in all compartments, exposure $C_w = 10$ mg/liter. The parameter values are $u = 0.1\ h^{-1}$, $k_{21} = k_{31} = 0.005\ h^{-1}$, $k_{12} = k_{13} = 0.01\ h^{-1}$, $e_x = 0.05$ liter/h, $K_{max} = 0.10\ mg \cdot h^{-1}$, and $K_h = 3$ mg.

Exercise 11.11

Assume that breakdown kinetics is proceeding at a maximum rate at steady state. Determine the steady-state values of this three-compartment model for the parameter values given above. Determine the bioconcentration factor at steady state if the weight of compartments 2 and 3 are 25% of the total body weight and compartment 1 is 50%.

We will now study dynamics using simulations. The final time will be 4 days (or 96 hours) with zero concentration for the initial condition in all compartments. The parameter values are $k_{21} = k_{31} = 0.005\ h^{-1}$, $k_{12} = k_{13} = 0.01\ h^{-1}$, $e_x = 0.05$ liter/h, $K_{max} = 0.10\ h^{-1}$, and $K_h = 3$ mg. Then the matrix for transfer rates between compartments is

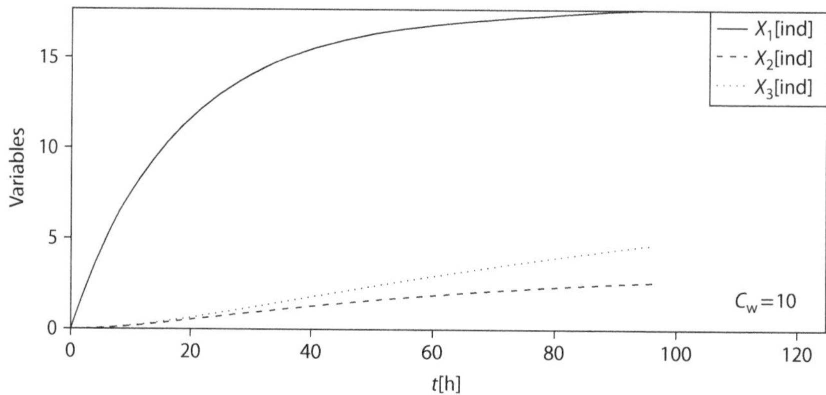

FIGURE 11.11 Toxi-kinetics in three organs.

$$\begin{bmatrix} -0.005-0.005 & 0.01 & 0.01 \\ 0.005 & -0.01 & 0 \\ 0.005 & 0 & -0.01 \end{bmatrix} = \begin{bmatrix} -0.01 & 0.01 & 0.01 \\ 0.005 & -0.01 & 0 \\ 0.005 & 0 & -0.01 \end{bmatrix} \tag{11.57}$$

With an exposure of $C_w = 10$ mg/liter, we obtain the dynamics shown in Figure 11.11. The steady state of X_1 is around 18. Compartments 2 and 3 show accumulation because the loss rates from those compartments are low.

Once a bioconcentration in various compartments is calculated, we can relate the lethal (LC_{50}) and sublethal (EC_{50}) values of ambient concentrations to the X_3 ($t = 96$) as we did for the simple bioaccumulation models (Calabrese and Baldwin, 1993).

Exercise 11.12

We know that the LC_{50} (concentration in water) from a 96 h constant exposure test is 5 mg/liter, and assume that lethal effects occur by reaching a given threshold value in the toxicant mass in compartment 3. Assuming that steady state is reached in 96 h, estimate the CBR, that is, in toxicant mass, the threshold in compartment 3 inducing death in 50% of the individuals. Hint: Use the results from the previous exercise.

11.8 BIOACCUMULATION: VARYING EXPOSURE

Many toxicity tests are conducted at C_w = constant; the above models work well in this situation. However, often the ambient concentration is time varying, for example, when contaminant transport is subject to tidal fluctuations.

The above models can be easily modified by making C_w a function of time, say a train of pulses of high (C_H) exposure for a fraction t_{on} of the period (T) and of low exposure (C_L) during the remainder fraction t_{off} of the period (T):

$$C_w(t) = \begin{cases} C_H & \text{if} \quad iT < t < iT + t_{on}T \\ C_L & \text{if} \quad (iT + t_{on}T) < t < (i+1)T \end{cases} \tag{11.58}$$

Several papers discuss how to analyze this time varying exposure using toxicity test data derived for constant exposure (Breck, 1988; Mancini, 1983). In other words, ambient concentration is a rectangular wave or pulse train, characterized by four parameters: the period T and the **duty cycle** or ratio t_{on}/T is the fraction of T with high concentration. This time varying exposure is in the top panel of Figure 11.12.

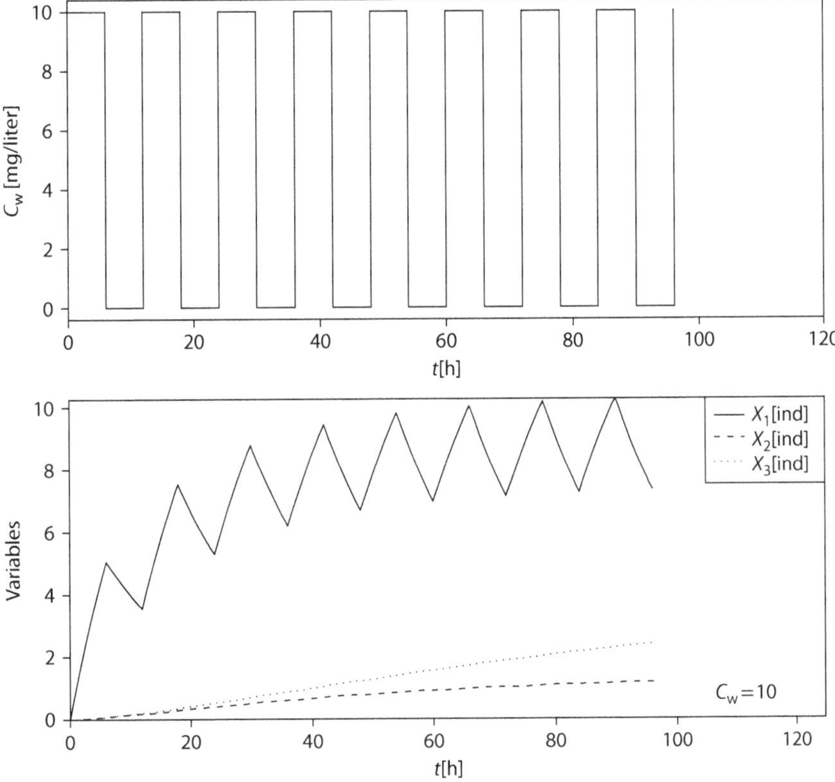

FIGURE 11.12 Pulse exposure and effects of pulse exposure in three organs.

The simulation results are in the lower panel of Figure 11.12. You can see how only the first compartment tracks the varying water concentration, because it has the fastest exchange with the ambient and the fastest losses; the second and third compartment show a much lower response to the pulse variation. Also, note that the first compartment settles to a varying "sawtooth" pattern ranging between 7 and 10. The time average is lower (about a half) than the steady state reached for constant exposure.

11.9 POPULATION EFFECTS

A complicated question is that of extrapolating individual effects to the population level (Kooijman, 1993). For this, we need to link the biological effects to population model parameters. An example of this approach is the cladoceran population model in the work by Acevedo et al. (1995b).

The response to the toxicant is modeled with a sigmoid:

$$Q_{\text{tox}}(t) = \frac{1}{1 + \exp\left(\dfrac{C_w - c}{\zeta}\right)} \tag{11.59}$$

where $C_w(t)$ = concentration of the toxicant (μg/liter) at time t. c is a threshold or stress level for half of the maximum limiting factor due to stress (μg/liter), and ζ is the nonlethal sensitivity to stress (μg/liter). These are not based on a specific chemical; for purposes of illustration, arbitrary values of $c = 10$ μg/liter and $\zeta = 0.1$ μg/liter are selected (Figure 11.13, top panel). If all other factors are nonlimiting, the threshold of the switching factor would be attained at the critical value slightly above the stress threshold. Experimental evidence for decreasing the ingestion rate as a function of toxicant concentration has been documented for cladocerans (see references in Acevedo et al., 1995b).

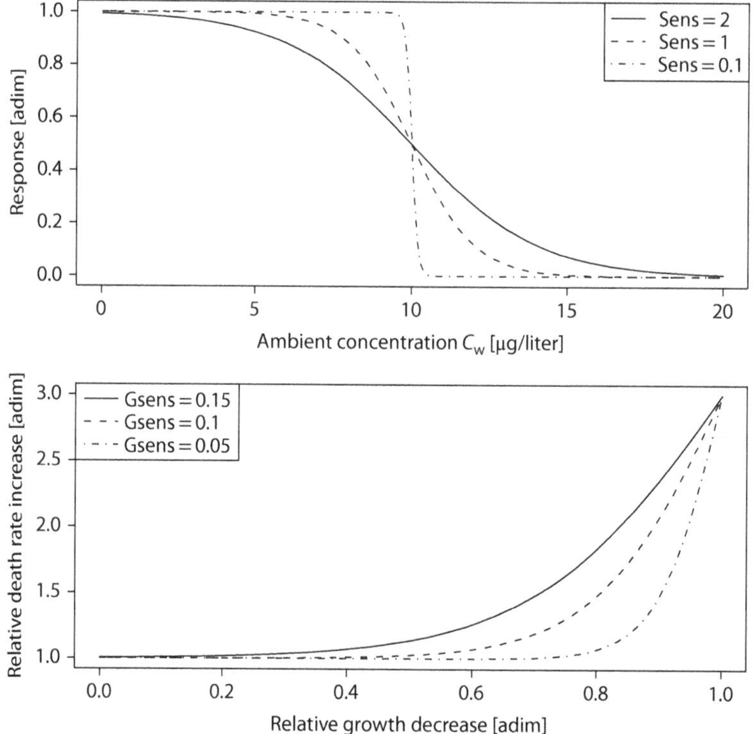

FIGURE 11.13 Sigmoid form of the toxicant response as a multiplier. A 50% value is obtained at the stress threshold. The steepness of the curve is controlled with the parameter ζ.

The model has a relative increase of mortality due to stress (adimensional), which is calculated from the accumulated growth deficit $u_i(t)$ below the a fraction of the maximum growth rate as a sigmoid function:

$$\gamma_i(t) = \frac{\gamma_{\max}}{1 + \exp\left(\dfrac{vG_{\max i} - u_i(t)}{\xi G_{\max i}}\right)} \tag{11.60}$$

where $\gamma_{\max}$ = maximum relative increase of mortality with stress (adimensional), v is the period in days for which a deficit of growth rate will induce half of the maximum increased mortality, and ξ = lethal sensitivity to stress (also in days of growth deficit); see the bottom panel of Figure 11.13. The accumulated starvation is calculated as

$$u_i = \int_{t-\tau_s}^{t} \left(aG_{\max i} - G_i(\sigma)\right)d\sigma \quad \text{for} \quad aG_{\max i} > G_i(\sigma) \tag{11.61}$$

where $\tau_s(t)$ is the starvation period.

These expressions make increasing mortality a function of decreasing growth rates, which in turn depend on toxicant concentration, temperature, and food as described in Chapter 10. The parameter $\gamma_{\max}$ is set to 4 to produce a fivefold increase in the death rates after long periods of starvation. The value of v = 1.00 is chosen to yield 50% of the maximum relative increase after a 3-day starvation period. The sensitivity level is set at ξ = 0.1. The mortality increase term is general and allows the study of other lethal effects and thresholds of temperature, starvation, and toxicants.

11.10 RISK ASSESSMENT

Risk assessment methodology hinges on the definition of exposure and effects and emphasizes quantitative measures based on probabilities. There is an increased recognition of the importance of using models within the risk assessment framework for estimating the exposure and the effects (Pastorok, 2002). The previous two sections of this chapter are illustrations of the use of models within the risk assessment framework (Calabrese and Baldwin, 1993).

11.11 COMPUTER SESSION

11.11.1 Bioavailability

The following few lines of code in script **chp11/bioavailable.R** calculate and plot Ag+ for a range of Cl from 10 to 100 mM (Figure 11.2):

```
# equilibrium constant
K <- 1.56*10^(-10)
# seq of chloride in mM
Cl.mM <- seq(10,100,10)
# convert to M
Cl <- Cl.mM*10^(-3)
# calculate Ag ion
Ag <- K/Cl
# convert to pM
Ag.pM <- Ag*10^12
plot(Cl.mM,Ag.pM, type="l", ylim=c(0,max(Ag.pM)),
     xlab="Cl- in mM", ylab="Ag+ in uM")
cbind(Cl.mM,Ag.pM)
```

After running, we get

```
> cbind(Cl.mM,Ag.pM)
      Cl.mM      Ag.pM
 [1,]    10 15600.0000
 [2,]    20  7800.0000
 [3,]    30  5200.0000
 [4,]    40  3900.0000
 [5,]    50  3120.0000
 [6,]    60  2600.0000
 [7,]    70  2228.5714
 [8,]    80  1950.0000
 [9,]    90  1733.3333
```

The following few lines of code calculate the exposure for varying chloride concentration assuming a constant total silver $Ag_{total} = 0.1$:

```
# calculate Cw concentration in water for varying Cl
Agtotal <- 0.1
# chloride
Cl <- c(70,80,90)
perc.avail <- 81 - 0.67 * Cl
Cw <- (perc.avail/100) * Agtotal
```

Here, we used three values of chloride, 70, 80, and 90, and obtained

```
> Cw
[1] 0.0341 0.0274 0.0207
```

Higher chloride decreases bioavailable silver and this in turn produces a decrease in uptake.

11.11.2 Bioaccumulation: One Compartment

The model function tox1.bioacc simulates one-compartment kinetics as given in the bioaccumulation section earlier in this chapter. The first-order depuration rate is given by the excretion rate coefficient (e_x), and the uptake rate coefficient (u) takes into account one pathway (water) using e_w, V, and W. The input file is **chp11/simple-bioaccu-inp.csv** containing

```
Label,Value,Units,Description
t0,0.00,h,time zero
tf,96.00,h,time final
dt,0.10,h,time step
tw,1.00,h,time to write
ex,0.05,1/h,depuration rate coeff
ew,0.1,None,uptake efficiency
V,0.5,liter/h,flow water
W,0.5,kg,body weight
Cw,0.5,mg/liter,water concentration
X0,0.0,mg/kg,initial concentration
digX,4,None,significant digits for output file
```

Now make a set of runs varying the exposure:

```
tox1 <- list(f=toxi1.bioacc)
param <- list(plab="Cw",pval=c(0,0.2,0.5,1))
t.X <- sim(tox1,"chp11/simple-bioaccu-inp.csv",param,pdfout=T)
```

Examine the resulting PDF file. The first page of this file contains Figure 11.3, which has already been presented and discussed.

Exercise 11.13

What is the percent change of X^* for a 10% increase and decrease in exposure around 0.5 mg/liter?

11.11.3 Bioaccumulation: Two Compartments

The model function tox2.bioacc simulates two-compartment kinetics as given in the corresponding section earlier in this chapter. This model has a first-order depuration rate from compartment 2 given by the excretion rate coefficient (e_x), a linear uptake into compartment 1 given by an uptake rate coefficient (u), and two transfer coefficients k_{12} and k_{21}. The input file is **chp11/two-compart-open-inp.csv** containing

```
Label,Value,Units,Description
t0,0.00,h,time zero
tf,96.00,h,time final
dt,0.01,h,time step
tw,0.1,h,time to write
```

```
ex,0.04,1/h,depuration rate coeff
u,0.01,1/h,uptake rate coeff
Cw,1.0,mg/liter,fixed uptake
k12,0.01,1/h,transfer coeff
k21,0.03,1/h,transfer coeff
X1.0,0.0,ind,initial condition
X2.0,0.0,ind,initial condition
digX,4,None,significant digits for output file
```

Now make a set of runs by varying the exposure:

```
tox2 <- list(f=toxi2.bioacc)
param <- list(plab="Cw",pval=c(1,5,10))
t.X <- sim(tox2,"chp11/two-compart-open-inp.csv",param,pdfout=T)
```

Examine the resulting PDF file. The first page of this file contains Figure 11.9, which has already been presented and discussed.

Exercise 11.14

What would happen if the excretion rate decreases? What would be the percent change of X_2^* for a 10% increase and decrease in e_x around the nominal 0.04 h^{-1}?

11.11.4 Bioaccumulation: Multiple Compartments

We will use the model function `tox.multi.bioacc` that calculates constant and pulsed exposure. This three-compartment model corresponds to the one explained in the corresponding section earlier in this chapter. First, we will use constant exposure.

The input file is **chp11/multi-compart-inp.csv** containing

```
Label,Value,Units,Description
t0,0.00,h,time zero
tf,96.00,h,time final
dt,0.01,h,time step
tw,0.1,h,time to write
ex,0.05,1/h,depuration rate coeff
u,0.10,1/h,uptake rate coeff
Cw,10.0,mg/liter,fixed uptake
k12,0.010,1/h,transfer coeff
k13,0.010,1/h,transfer coeff
k21,0.005,1/h,transfer coeff
k23,0.000,1/h,transfer coeff
k31,0.005,1/h,transfer coeff
k32,0.000,1/h,transfer coeff
Kmax,0.1,1/h,breakdown max-rate
Kh,3.0,mg/g,breakdown half-rate
Te,0,h,period exposure
Du,0.0,Fraction,Duty cycle exposure
X1.0,0.0,ind,initial condition
X2.0,0.0,ind,initial condition
X3.0,0.0,ind,initial condition
digX,4,None,significant digits for output file
```

The exposure is constant in time because the period (T_e) and duty cycle (D_u) were made 0. Now make a set of runs for several values of the constant exposure:

```
toxm <- list(f=toxi.multi.bioacc)
param <- list(plab="Cw",pval=c(1,5,10))
t.X <- sim(toxm,"chp11/multi-compart-inp.csv",param,pdfout=T)
```

Examine the resulting PDF file. The last page of this file contains Figure 11.11, which has already been presented and discussed.

Exercise 11.15

Use sensitivity analysis to calculate steady state X_3 as a function of the uptake rate coefficient. What is the percent change of steady state X_3 for a 10% increase in the uptake rate coefficient? Discuss the result.

11.11.5 Pulse Exposure

Next, we will use the same function `tox.multi.bioacc` but with time-varying ambient concentration. The input file is **chp11/multi-compart-pulse-inp.csv**, which is the same as before except that the period (T_e) and duty cycle (D_u) are not zero:

```
Te,12,h,period exposure
Du,0.5,Fraction,Duty cycle exposure
```

The period is 12 h and the duty cycle is 0.5 for the fraction of the period with high exposure, which is set at 10 mg/liter. In other words, every half a day, there is exposure for 6 h and no exposure for the other 6 h. The low exposure is no exposure or 0 mg/liter.

Execute simulations changing the exposure:

```
param <- list(plab="Cw",pval=c(1,5,10))
t.X <- sim(toxm,"chp11/multi-compart-pulse-inp.csv",param,pdfout=T)
```

The last page of the PDF file has the results shown in Figure 11.12. The oscillations are noticeable in compartment 1, which is subject to the variations in C_w. From the first page of the PDF file, we can see that the oscillations decrease in magnitude as C_w decreases (Figure 11.14).

Exercise 11.16

Assume that lethality occurs by reaching a given threshold value in the toxicant mass in compartment 3 as calculated in the previous exercise. Derive an empirical linear formula for the "time to death" versus "duty" fraction for the median level of toxicity and at a high water concentration of 10 mg/liter (default). Hint: Use the threshold value calculated in Exercise 11.12; run a sensitivity analysis on the "duty" parameter; for each run, find the time to death by reading when the mass reaches the threshold; perform regression of time-to-death versus duty-fraction.

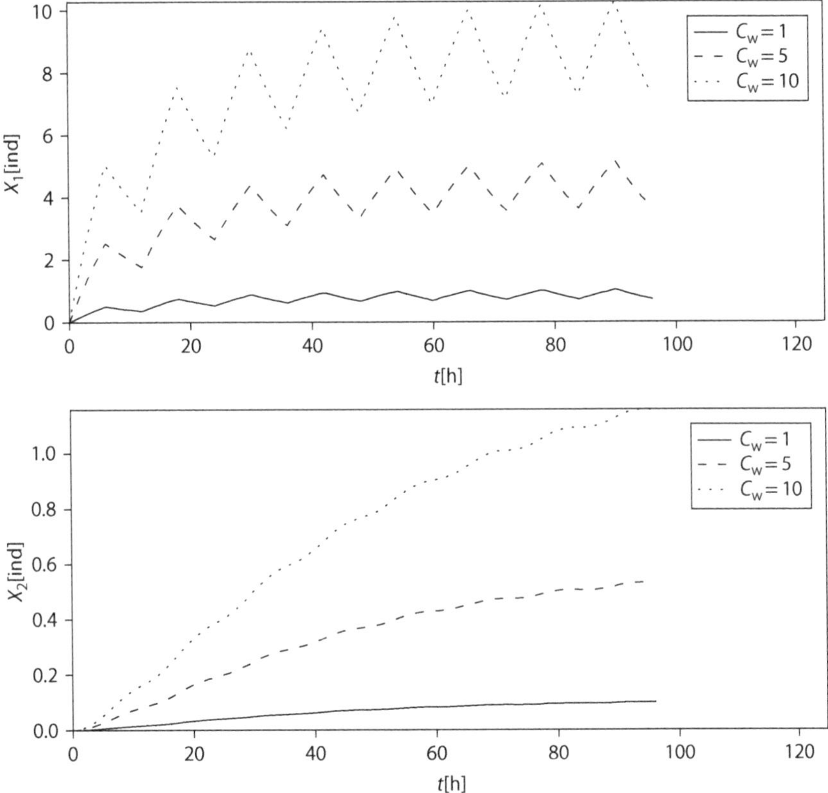

FIGURE 11.14 Pulse exposure and effects of pulse exposure in compartment 1 for several values of exposure.

11.11.6 Population Effects

Using various levels of exposure controlled by parameter `tox` in the *Ceriodaphnia* model introduced in the previous chapter, we can simulate lethal and sublethal effects. The model is stage-structured by eggs, neonates, and adults. The default parameter values were discussed in Chapter 10. In this section, we subject the population to the effect of a toxic chemical compound. We already discussed part of the input file in the previous chapter; thus, we now focus on the parts related to toxicant response; see **chp11/cerio-inp.txt**. Parameter `tox` corresponds to the bioavailable fraction C_w and is given towards the end of the file:

```
tox                 0.0000
tox_unit            [ug/l]
```

Toxicant stress causes a mortality rate increase at lethal dose levels or the death rate increases with toxicant concentration; this is controllable with three parameters: `ampmax`, `thresh`, and `restre` (lines 64–74 of the input file) corresponding to γ_{max}, ν, and ξ, respectively, and defined in the above section on population effects:

```
ampmax(1)           4.0000
ampmax(2)           4.0000
ampmax(3)           4.0000
amp_unit            [adim]
thresh(1)           1.0000
thresh(2)           1.0000
```

```
thresh(3)        1.0000
restre(1)        0.1000
restre(2)        0.1000
restre(3)        0.1000
units            [days]
```

The value for the lethal threshold parameter v is derived from the LC_{50}.

Sublethal toxicant levels affect growth via an ingestion rate decrease with toxicant concentration, controllable with two parameters: $c = 10$ µg/liter and $\zeta = 0.1$ µg/liter, corresponding to `stcri` and `stsen` (lines 40–46 of input file):

```
stcri(1)   0.0000
stcri(2)   10.0000
stcri(3)   10.0000
stsen(1)   0.0000
stsen(2)   0.1000
stsen(3)   0.1000
st_unit   [ug/L]
```

From Chapter 10, the changes in ingestion rate at sublethal dose levels affect fertility and reproductive mode via two functions with several coefficients (unitless). These are c_j = half fertility decrease, g = fertility sensitivity to ingestion, a = switch threshold (as a fraction of ingestion rate), and b = switch sensitivity. These are given in the input file as `bdcri, bdsen` (lines 54–55), `fercri, fersen,` (lines 52–53), and `swsex, swsen` (lines 56–57).

Example: To vary the `tox` parameter in the sublethal range (e.g., tox = 0.0, 0.5, and 1.0), edit the first four lines of the input file in the following manner and save as **cerio-tox-sens-inp.txt**:

```
par_sens_name     tox
par_sens_max      1.00
par_sens_min      0.00
par_sens_step     0.50
```

First recall to dynamically load the `seem.dll` unless you have automated it in the `Rprofile.site` file by using `dyn.load(paste(.libPaths(),"/seem/libs/i386/seem.dll",sep=""))`. We run the Fortran program `cerio` using `cerio.F` and read and plot the results as explained in Chapter 10. Namely,

```
fileout  <- cerio.F("chp11","cerio-tox-sens")
tox.sens <- read.plot.cerio.out(fileout,sens=T,pdfout=T)
```

Here we have used the read and plot fileout function in sensitivity mode (`sens = T`) to obtain a PDF file **cerio-tox-sens-out.txt.pdf** with several pages. See Figure 11.15 for the dynamics of eggs and female neonates and Figure 11.16 for female adults and males.

Exercise 11.17

Review Figure 11.15. How do the oscillations change as you increase the value of parameter `tox`? How does the number of winter eggs change at these doses?

Exercise 11.18

Review Figure 11.16. Discuss the effect of increasing the toxicant. How sensitive is adult female density to the exposure?

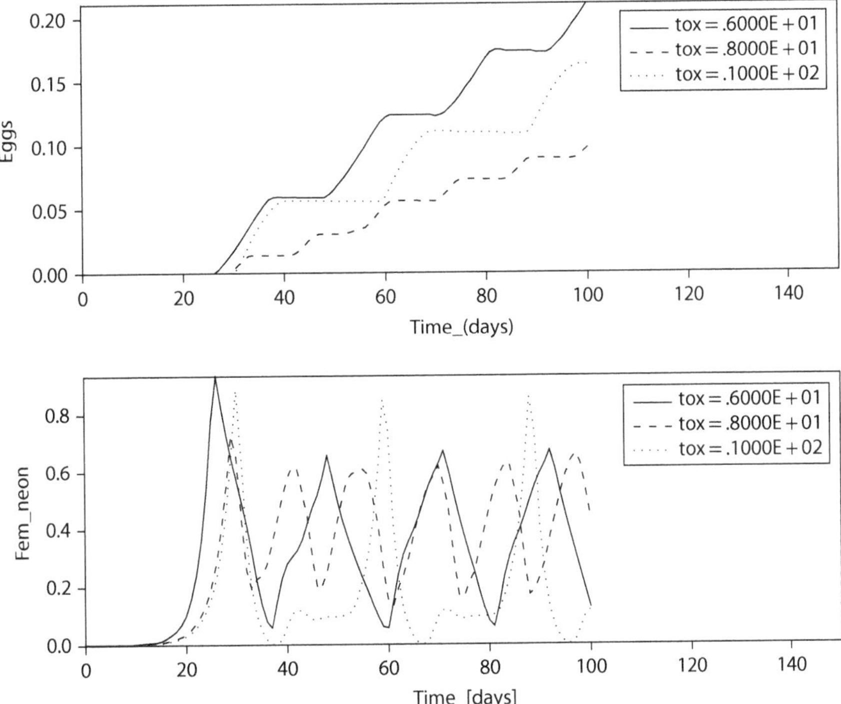

FIGURE 11.15 Sublethal effects, eggs and female neonates.

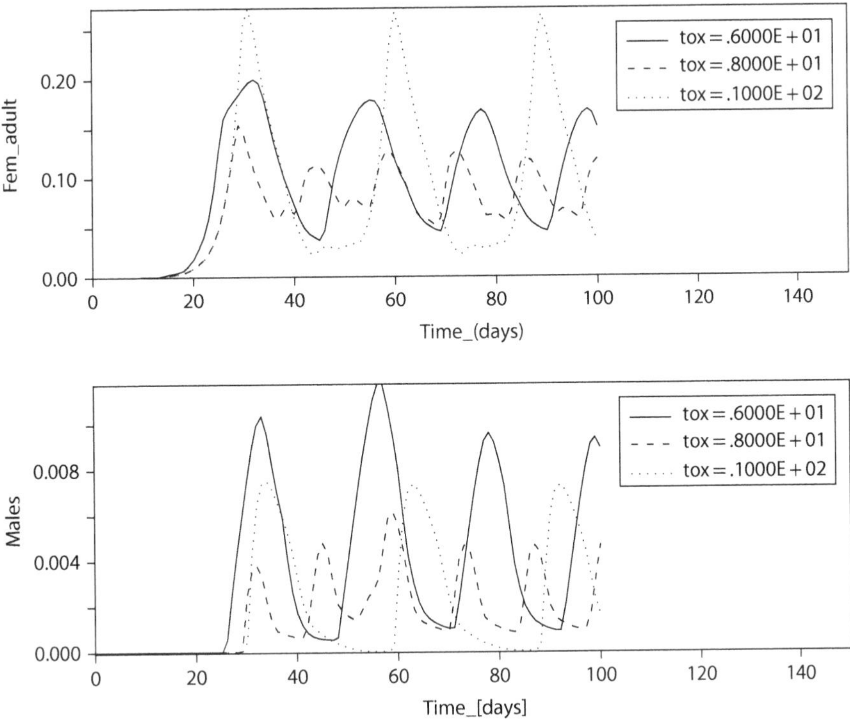

FIGURE 11.16 Sublethal effects, female adults and males.

Exercise 11.19

Change parameter `tox` to higher values. Generate the results and discuss.

Example: To vary **tox** parameter in lethal range (e.g., tox = 8.0, 9.0, and 10.0), edit the first four lines of the input file in the following manner and save as **cerio-lethal-sens-inp.txt**.

```
par_sens_name     tox
par_sens_max      12.00
par_sens_min      8.00
par_sens_step     2.00
```

We run **cerio.F** and read and plot the results as explained before. Namely,

```
fileout <- cerio.F("chp11","cerio-lethal-sens")
lethal.sens <- read.plot.cerio.out(fileout,sens=T,pdfout=T)
```

Here we have used the read and plot fileout function in sensitivity mode `(sens = T)` to obtain a file **cerio-lethal-sens-out.txt.pdf** with several pages. See Figure 11.17 for the dynamics of eggs and female neonates and Figure 11.18 for female adults and males. There are only two traces in each figure because all individuals die at the highest exposure.

Exercise 11.20

Review Figure 11.17. How do the dynamics change as you increase the value of parameter `tox`? How does the number of winter eggs change at these doses?

Exercise 11.21

Review Figure 11.18. Discuss the effect of the increasing toxicant. How sensitive is the adult female density to the exposure?

11.11.7 Build Your Own

Exercise 11.22

Follow the four-step method to develop a model of one-compartment bioaccumulation for which the depuration rate is not continuous but occurs episodically at random times distributed as an exponential. Hint: Step 1—Define model functions `toxi.bioacc.z` and `toxi.bioacc.g`. Use the logistic example in Chapter 6 as the example. Step 2—Source these functions. Step 3—Edit the input file simple-bioaccu-inp.csv to include one extra line for the period and decrease the time step and `tw` to show the sudden change. Step 4—Define the model list, specify the `param` set for multiple runs to be repeated runs with $C_w = 0.5$, and call the simulation function `simd`. The result should be plots of several realizations as in Figure 11.19.

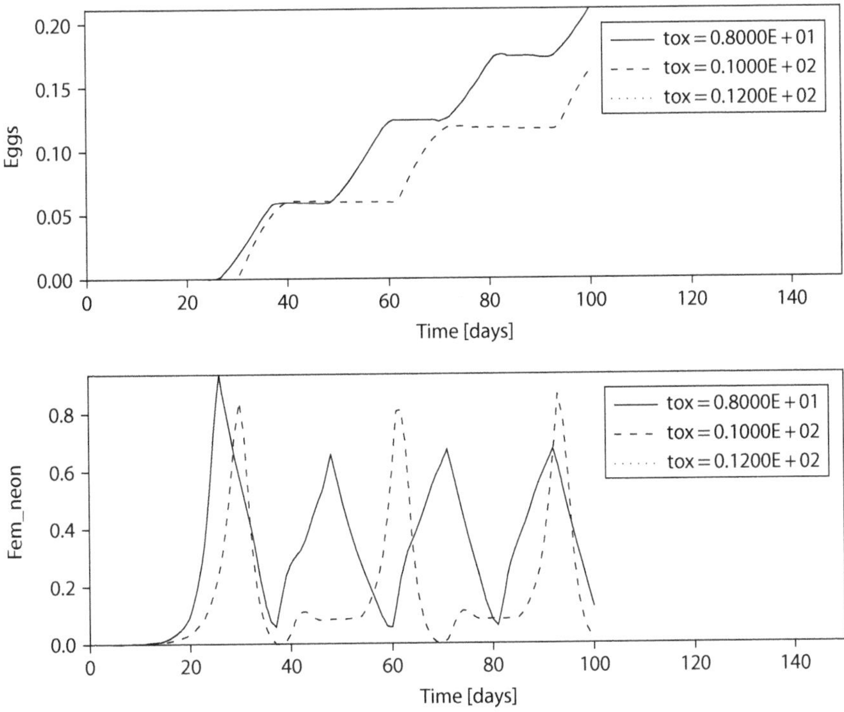

FIGURE 11.17 Lethal effects, eggs and female neonates.

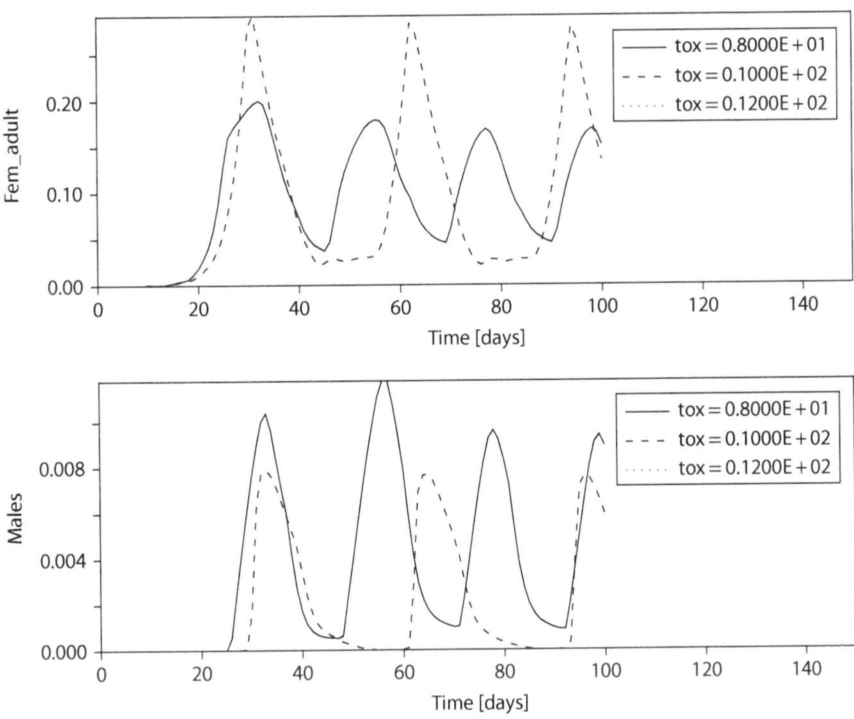

FIGURE 11.18 Lethal effects, female adults and males.

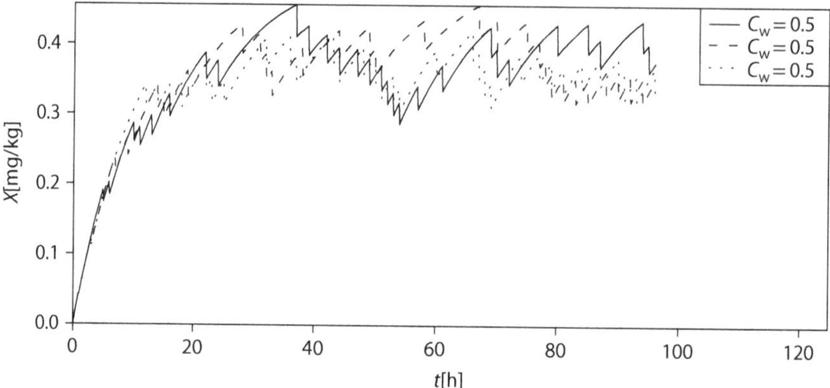

FIGURE 11.19 Bioaccumulation when depuration occurs as a Poisson process.

11.12 SEEM FUNCTIONS EXPLAINED

11.12.1 Function tox1.bioacc

This function is essentially the same as the `expon.forced` function in Chapter 7, but adapted to more parameters to calculate the uptake rate.

```
toxi1.bioacc <- function(t,p,x){
#x  body concentration
#"ex","ew","V","W","Cw" p[1:5]
dx <- (p[2]*p[3]/p[4])*p[5] - p[1]*x
return(dx)
}
```

11.12.2 Function tox2.bioacc

This function is similar to the `two.stage.cont` population model of Chapter 10. Here the model to integrate is

```
toxi2.bioacc<- function(t,p,x){
# x[1], x[2], compartmnents
# 1"ex",2"u",3"Cw",4"k12",5"k21"
k <- matrix(ncol=2,nrow=2) # transfers
k[1,1] <- 0; k[1,2] <- p[4]
k[2,1] <- p[5]; k[2,2] <- 0
k[1,1] <- -sum(k[,1]); k[2,2] <- -sum(k[,2])
u <- c(p[2], 0.0)*p[3]
e <- c(0.0,p[1])
dX <- c(k%*%x + u - e*x)
return(dX)
}
```

The diagonal elements of the intercompartmental transfer matrix **K** denoted by **K** in the code, are initialized to zero and then completed assuming that they should be equal to the negative of the sum of the column elements. Uptake and depuration are calculated separately and then included in the line dX <- c(k%*%x + u - e*x), which has the matrix multiplication of **K** and ***x***, the ***u*** vector, and the entry-wise or Hadamard multiplication of ***e*** and ***x***.

11.12.3 Function tox.multi.bioacc

This function is a variation of the tox2.bioacc function but includes the pulse exposure, one more compartment, and M3 breakdown kinetics.

```
toxi.multi.bioacc <- function(t,p,x){
# x[1], x[2], x[3] compartmnents
# parameters 1"ex",2"u",3"Cw",4"k12",5"k13",6"k21",
# 7"k23",8"k31",9"k33",10"Kmax",11"Kh",12"Te",13"Du"

# pulse
if(p[12]>0){
 j <- floor(t/p[12])+1
 if(t< p[12]*(j+p[13]-1)) u1 <- p[3]
 if(t>=p[12]*(j+p[13]-1)) u1 <- 0
} else u1 <- p[3]
u <- c(p[2], 0.0, 0.0)*u1

k <- matrix(ncol=3,nrow=3) # transfers
k[1,1] <- 0;     k[1,2] <- p[4]; k[1,3] <- p[5]
k[2,1] <- p[6]; k[2,2] <- 0;     k[2,3] <- p[7]
k[3,1] <- p[8]; k[3,2] <- p[9]; k[3,3] <- 0
k[1,1] <- -sum(k[,1]); k[2,2] <- -sum(k[,2]); k[3,3] <- -sum(k[,3])

# excretion and breakdown
e <- c(p[1], 0.0, 0.0)
b <- c(0.0, p[10]/(p[11]+x[2]), 0.0)
dX <- c(k%*%x + u -(e+b)*x)
 for(i in 1:3) if(x[i] < 0) x[i] <- 0
 return(dX)
 }
```

Here two logical if statements are written in sequence to determine whether the time corresponds to exposure or not, thereby assigning nonzero or zero exposure. Next, we complete the k matrix as in the previous function tox2 and then we expand depuration to include breakdown defined by nonlinear kinetics.

12 Community Dynamics

So far, we have treated populations by themselves, excluding the interaction with other populations. Now, we consider these interactions explicitly. In this chapter, we examine the dynamics of several interacting species. To get started, we will consider only two species and lumped (nonstructured) models for each species and then add more species and interactions. As the number of species increases, so does the complication of the analysis, forcing one to make simplifying assumptions. Many books and articles have examined the mathematical complications and the biological significances of these models (Hallam, 1986a; Maurer, 1999).

12.1 TWO-STATE SYSTEM CONCEPTS

Now, let us take a moment to explain some system concepts focusing on the simple two-species case. In general, the model will be nonlinear functions f_1 and f_2 of X_1 and X_2, that is, vector $\boldsymbol{X}$, with parameters $\boldsymbol{p}$:

$$\begin{aligned} \frac{\mathrm{d}X_1}{\mathrm{d}t} &= f_1(t, \boldsymbol{p}, \boldsymbol{X}) \\ \frac{\mathrm{d}X_2}{\mathrm{d}t} &= f_2(t, \boldsymbol{p}, \boldsymbol{X}) \end{aligned} \tag{12.1}$$

Here we have used $\boldsymbol{p}$ also as vector to indicate that we can have several parameters. Long-term coexistence requires that a stable nontrivial equilibrium point exists. The nontrivial equilibrium points are those when both functions f_i are 0 at points different from those that include a vanishing population $X_1 = 0$ and $X_2 = 0$.

Recall that the **phase plane** is formed by two state variables plotted together in the same graph, X_2 versus X_1. In this plane, we see the trajectory of the state vector, but time is implicit. The state moves along the trajectory as time unfolds.

Linearization consists of taking partial derivatives of the net rates around an equilibrium point, say $\boldsymbol{X}^*$, with respect to the states to form a new system consisting of excursions or deviations around the neighborhood of the equilibrium point. This is **local** analysis, in contrast to the **global** analysis (all-phase plane). Linearization then yields a system of linear equations based on the **Jacobian** matrix **J** evaluated at the equilibrium point:

$$\mathbf{J}^* = \begin{bmatrix} \left.\frac{\partial f_1}{\partial X_1}\right|_{X^*} & \left.\frac{\partial f_1}{\partial X_2}\right|_{X^*} \\ \left.\frac{\partial f_2}{\partial X_1}\right|_{X^*} & \left.\frac{\partial f_2}{\partial X_2}\right|_{X^*} \end{bmatrix} \tag{12.2}$$

Entries i, j of the Jacobian matrix consist of the partial derivatives of each function f_i with respect to the dependent variable X_j. For brevity, we will denote this matrix by **C**.

Then, the linear system of perturbations x_1 and x_2 around the equilibrium point is given by

$$\frac{\mathrm{d}}{\mathrm{d}t}\begin{bmatrix} x_1 \\ x_2 \end{bmatrix} = \begin{bmatrix} c_{11} & c_{12} \\ c_{21} & c_{22} \end{bmatrix}\begin{bmatrix} x_1 \\ x_2 \end{bmatrix} \tag{12.3}$$

or $\frac{d\boldsymbol{x}}{dt} = \mathbf{C}\boldsymbol{x}$ for short. Matrix **C** is known as the **community matrix** (May 1973). We are using lowercase x to denote perturbations. Now, we calculate the eigenvalues λ_1, λ_2 to determine the local stability of the equilibrium point as described in Chapter 9. In the phase plane, the trajectories would converge on a stable equilibrium point, diverge from an unstable equilibrium point, or cycle around the equilibrium point (Lewis and Othmer, 1997).

One additional important behavior of the nonlinear system is a stable orbit or cycle, that is, trajectories that would converge to the orbit from inside or from outside the orbit. This type of orbit is called a **limit cycle**.

12.2 TWO SPECIES: SIMPLE INTERACTIONS

Interactions can be represented by a **digraph** (**di**rectional **graph**), in which circles represent the states and arcs connecting the circles indicate the interaction (Figure 12.1). Each arc has two labels: the direction and the sign of the interaction. Self-interactions in this digraph will look like loops coming out and going back to each circle. The signs are very important because they determine the nature of the interaction and represent the signs of the community matrix, **C**, or Jacobian matrix evaluated at equilibrium. Qualitative stability analysis is based solely on the signs of entries of **C** (May 1973, Levins 1974).

We should note that the entries of **C** are only coefficients that represent the effect of one species on another, found by the derivatives of the functions f_i. Therefore, they should not be confused with the parameters representing the mechanism of the interaction.

In general, the model will be nonlinear functions f_1 and f_2 of X_1 and X_2, that is, vector $\boldsymbol{X}$, with parameters $\boldsymbol{p}$ as in Equation 12.1, but we are interested in the special case

$$\frac{dX_1}{dt} = f_1(t, \boldsymbol{p}, \boldsymbol{X}) = g_1(t, \boldsymbol{p}, \boldsymbol{X})X_1$$
$$\frac{dX_2}{dt} = f_2(t, \boldsymbol{p}, \boldsymbol{X}) = g_2(t, \boldsymbol{p}, \boldsymbol{X})X_2 \tag{12.4}$$

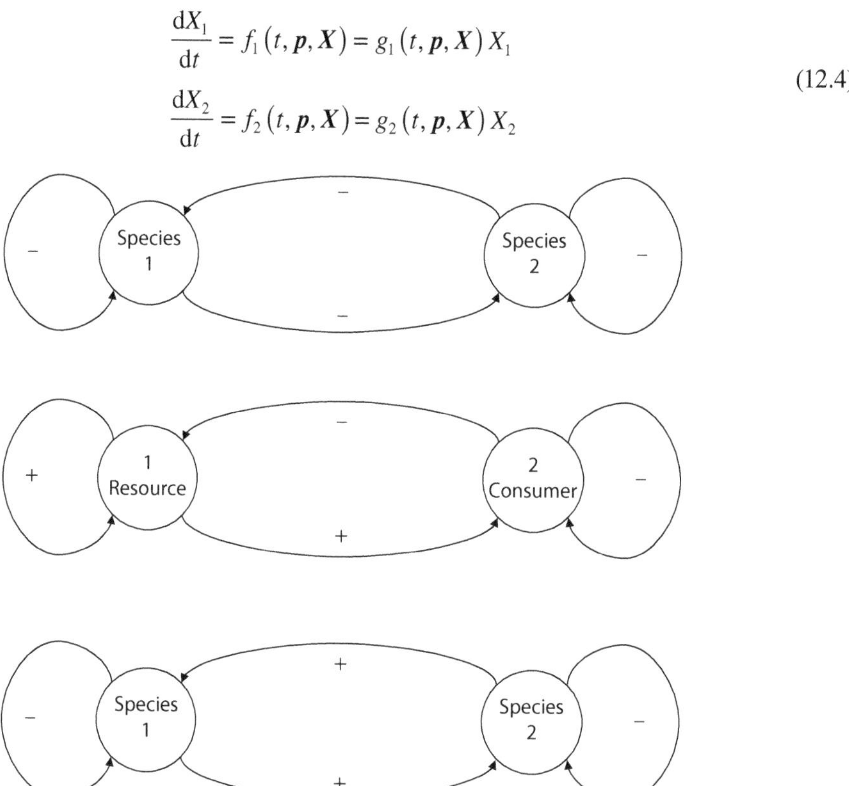

FIGURE 12.1 Digraphs of two-species communities based on interactions: competition, consumer–resource, and mutualism.

where f_i can be decomposed as a product of a per capita rate function g_i and the population X_i, and therefore the nontrivial equilibrium points are the roots of the system formed by g_1, g_2. In the simplest models, the functions g_i are linear functions of $\boldsymbol{X}$, and therefore there is one nontrivial equilibrium point given by the unique solution of a linear system conformed by g_1, g_2. These are called **Lotka–Volterra** type of models. In other words, we specify the system as

$$\begin{aligned}\frac{dX_1}{dt} &= (r_1 + a_{11}X_1 + a_{12}X_2)X_1 \\ \frac{dX_2}{dt} &= (r_2 + a_{22}X_2 + a_{21}X_1)X_2\end{aligned} \tag{12.5}$$

The per capita rates, g_i, are

$$\begin{aligned}g_1(t,\boldsymbol{p},\boldsymbol{X}) &= r_1 + a_{11}X_1 + a_{12}X_2 \\ g_2(t,\boldsymbol{p},\boldsymbol{X}) &= r_2 + a_{22}X_2 + a_{21}X_1\end{aligned} \tag{12.6}$$

$$\begin{bmatrix} g_1 \\ g_2 \end{bmatrix} = \begin{bmatrix} a_{11} & a_{12} \\ a_{21} & a_{22} \end{bmatrix}\begin{bmatrix} X_1 \\ X_2 \end{bmatrix} + \begin{bmatrix} r_1 \\ r_2 \end{bmatrix} \tag{12.7}$$

or $\boldsymbol{g} = \mathbf{A}\boldsymbol{X} + \boldsymbol{r}$, where $\mathbf{A}$ is the matrix of interaction coefficients a_{ij}, and vector $\boldsymbol{r}$ has intrinsic rates. Therefore,

$$\frac{d\boldsymbol{X}}{dt} = \left(\mathbf{A}\boldsymbol{X} + \boldsymbol{r}\right)\circ\boldsymbol{X} \tag{12.8}$$

Nontrivial equilibrium occurs when both per capita rate functions g_i are 0 or the solution of the linear equation is

$$\begin{bmatrix} a_{11} & a_{12} \\ a_{21} & a_{22} \end{bmatrix}\begin{bmatrix} X_1^* \\ X_2^* \end{bmatrix} = -\begin{bmatrix} r_1 \\ r_2 \end{bmatrix} \tag{12.9}$$

or $\mathbf{A}\boldsymbol{X}^* = -\boldsymbol{r}$ for short. This linear equation has a solution $\boldsymbol{X}^* = -\mathbf{A}^{-1}\boldsymbol{r}$ or $\boldsymbol{X}^* = -\frac{\text{adj}\mathbf{A}}{|\mathbf{A}|}\boldsymbol{r}$:

$$\begin{bmatrix} X_1^* \\ X_2^* \end{bmatrix} = \frac{-1}{\Delta_A}\begin{bmatrix} a_{22} & -a_{12} \\ -a_{21} & a_{11} \end{bmatrix}\begin{bmatrix} r_1 \\ r_2 \end{bmatrix} = \frac{1}{\Delta_A}\begin{bmatrix} -a_{22}r_1 + a_{12}r_2 \\ a_{21}r_1 - a_{11}r_2 \end{bmatrix} \tag{12.10}$$

where Δ_A is the determinant of the matrix $\mathbf{A}$ or $\Delta_A = \det(\mathbf{A}) = a_{11}a_{22} - a_{12}a_{21}$, which is positive when $a_{11}a_{22} > a_{12}a_{21}$. Coexistence of the two species requires that both the elements of this nontrivial equilibrium point be positive. From Equation 12.10, to obtain positive equilibrium values for both species, we must have

$$\begin{aligned}r_2a_{12} - r_1a_{22} &> 0 \\ r_1a_{21} - r_2a_{11} &> 0\end{aligned} \tag{12.11}$$

Then, for long-term coexistence, this equilibrium should be stable. The Jacobian would be

$$\mathbf{J} = \begin{bmatrix} \frac{\partial f_1}{\partial X_1} & \frac{\partial f_1}{\partial X_2} \\ \frac{\partial f_2}{\partial X_1} & \frac{\partial f_2}{\partial X_2} \end{bmatrix} = \begin{bmatrix} g_1 + \frac{\partial g_1}{\partial X_1}X_1 & \frac{\partial g_1}{\partial X_2}X_1 \\ \frac{\partial g_2}{\partial X_1}X_2 & g_2 + \frac{\partial g_2}{\partial X_2}X_2 \end{bmatrix} = \begin{bmatrix} g_1 + a_{11}X_1 & a_{12}X_1 \\ a_{21}X_2 & g_2 + a_{22}X_2 \end{bmatrix} \tag{12.12}$$

Evaluating the Jacobian at the equilibrium point, we get **C**, and at this point, g_1 and g_2 would become 0:

$$\mathbf{C} = \mathbf{J}^* = \begin{bmatrix} a_{11}X_1^* & a_{12}X_1^* \\ a_{21}X_2^* & a_{22}X_2^* \end{bmatrix} = \begin{bmatrix} X_1^* & 0 \\ 0 & X_2^* \end{bmatrix} \begin{bmatrix} a_{11} & a_{12} \\ a_{21} & a_{22} \end{bmatrix} \tag{12.13}$$

Using the notation $\text{diag}(X_i^*)$ to represent forming a square diagonal matrix with elements X_i, we can rewrite Equation 12.13 as

$$\mathbf{C} = \text{diag}(X_i^*)\,\mathbf{A} \tag{12.14}$$

The eigenvalues of matrix **C** determine the local stability properties. These eigenvalues depend on **A** and the equilibrium point. We will now be more specific by looking at several types of interactions.

12.2.1 Competition

When two species compete for the same resource, the interspecific interaction signs are negative for both species. For example, two plant species compete for light, nutrients, or other resources.

The classical Lotka–Volterra competition model includes density dependence in both populations, and therefore the intraspecific interaction signs are also negative for both species. Matrix **A** has all negative entries: $\mathbf{A} = \begin{bmatrix} -a_{11} & -a_{12} \\ -a_{21} & -a_{22} \end{bmatrix}$; its determinant is positive when $a_{11}a_{22} > a_{12}a_{21}$. A coexistence equilibrium point occurs when conditions in Equation 12.11 are met because all signs are reversed. The Jacobian at equilibrium is given by

$$\mathbf{J}^* = \mathbf{C} = \text{diag}(\boldsymbol{X}^*) \begin{bmatrix} -a_{11} & -a_{12} \\ -a_{21} & -a_{22} \end{bmatrix} \tag{12.15}$$

which has a negative trace and a positive discriminant; therefore, the nontrivial point is a stable node. Thus, the outcomes are that either one species can outcompete the other or that the two species can coexist at the equilibrium point.

For specific examples, consider $\mathbf{A} = \begin{pmatrix} -0.1 & -0.01 \\ -0.01 & -0.1 \end{pmatrix}$ and $\boldsymbol{r} = \begin{pmatrix} 0.15 \\ 0.1 \end{pmatrix}$, which satisfy Equation 12.11. The equilibrium is 1.41, 0.86. The community matrix **C** has a positive determinant, negative trace, and positive discriminant, which leads to real negative eigenvalues −0.14, −0.08. For these conditions, the coexistence equilibrium point should be a stable node (upper panel of Figure 12.2).

On the other hand, consider $\mathbf{A} = \begin{pmatrix} -0.1 & -0.01 \\ -0.17 & -0.1 \end{pmatrix}$ and $\boldsymbol{r} = \begin{pmatrix} 0.15 \\ 0.1 \end{pmatrix}$, which do not satisfy Equation 12.11. There is no coexistence equilibrium point; one of the species (species 1 in this case) outcompetes the other (species 2) as shown in the lower panel of Figure 12.2.

12.2.2 Cooperation: Mutualism

In this case, both species benefit from the interaction, and therefore this is modeled using positive signs for the interaction terms and thus matrix $\mathbf{A} = \begin{bmatrix} -a_{11} & a_{12} \\ a_{21} & -a_{22} \end{bmatrix}$; its determinant is positive when $a_{11}a_{22} > a_{12}a_{21}$, which requires that the intraspecific competition term should be larger than the interspecific cooperation. When it occurs, there is always a coexistence equilibrium

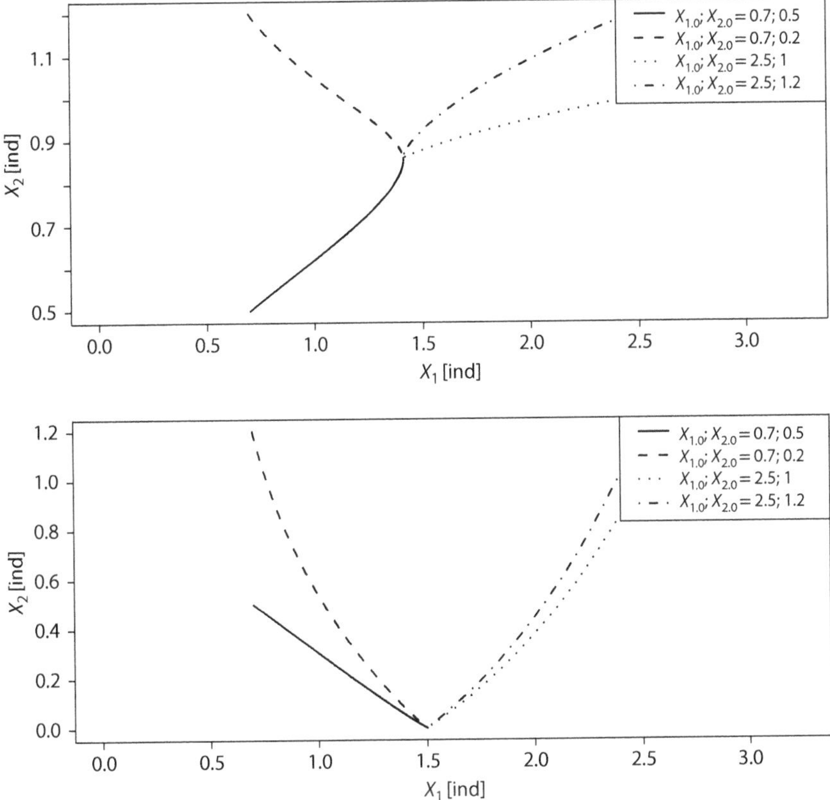

FIGURE 12.2 Competition dynamics: stable or coexistence (top panel) and unstable or extinction of one species (bottom panel).

point because conditions in Equation 12.11 are met. The equilibrium point is such that each species now has a larger density than what it would have by itself. The community matrix

$$\mathbf{J}^* = \mathbf{C} = \text{diag}(\boldsymbol{X}^*)\begin{bmatrix} -a_{11} & a_{12} \\ a_{21} & -a_{22} \end{bmatrix} \tag{12.16}$$

has a negative trace and a positive discriminant, therefore, the equilibrium is a stable node. For a specific example, consider $\mathbf{A} = \begin{pmatrix} -0.1 & 0.01 \\ 0.01 & -0.1 \end{pmatrix}$ and $\boldsymbol{r} = \begin{pmatrix} 0.15 \\ 0.1 \end{pmatrix}$. The equilibrium point is 1.616, 1.162, and the eigenvalues are real negative −0.165, −0.112 (see Figure 12.3).

12.2.3 Commensalism

One species benefits, and the other does not benefit but is not negatively affected either. In this case, $\mathbf{A} = \begin{bmatrix} -a_{11} & a_{12} \\ 0 & -a_{22} \end{bmatrix}$, each species grows logistically, but species 1 is helped by the cooperation of species 2.

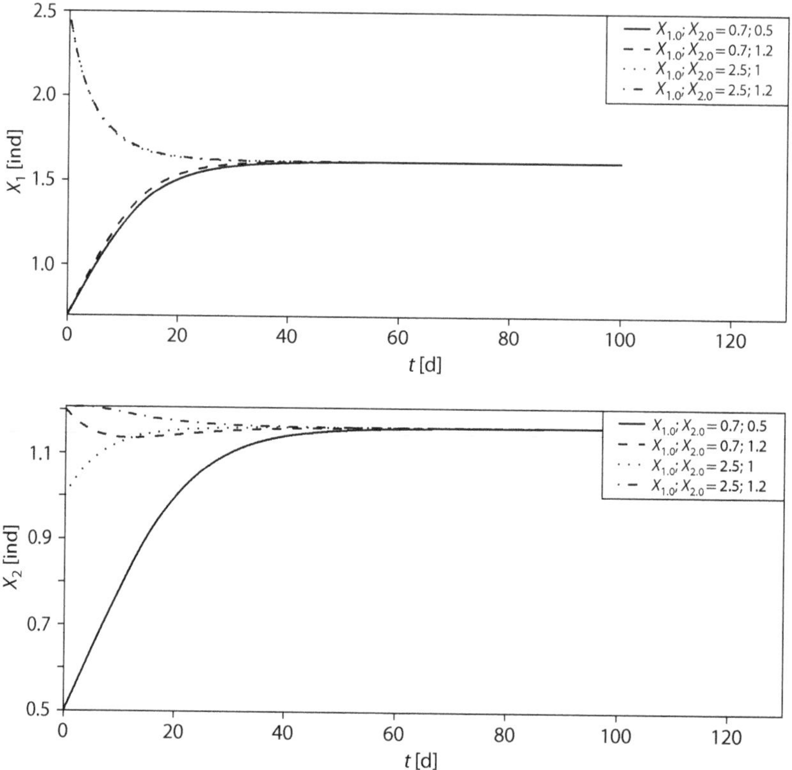

FIGURE 12.3 Mutualistic cooperation dynamics.

Exercise 12.1

Draw a digraph of commensalism between two species. Determine matrix **A**. Determine the nontrivial equilibrium point, community matrix, and conditions for its stability.

12.2.4 Consumer Resource

One species consumes the other or "benefits" from the other, as occurs in **predator–prey** and **parasite–host** interactions (Hallam, 1986a). The consumer X_2 affects the net rate of growth for the resource X_1 negatively. In addition, the consumer population declines if resource population is depleted and the resource population grows if no predation pressure is applied. In this case, $\mathbf{A} = \begin{bmatrix} -a_{11} & -a_{12} \\ a_{21} & -a_{22} \end{bmatrix}$; its determinant $\Delta_A = \det(\mathbf{A}) = a_{11}a_{22} + a_{12}a_{21}$ is always positive. Conditions for the positive nontrivial point are reduced to $r_2 a_{12} < r_1 a_{22}$. The community matrix is expressed as

$$\mathbf{J}^* = \mathbf{C} = \text{diag}(\boldsymbol{X}^*) \begin{bmatrix} -a_{11} & -a_{12} \\ a_{21} & -a_{22} \end{bmatrix} \tag{12.17}$$

With a negative trace and negative discriminant, the equilibrium is a stable spiral. For example, consider $\mathbf{A} = \begin{pmatrix} -0.2 & -0.1 \\ 0.1 & -0.1 \end{pmatrix}$ and $\boldsymbol{r} = \begin{pmatrix} 0.6 \\ 0.1 \end{pmatrix}$. The equilibrium point is 1.66, 2.66. The trace of **C** is −0.60 and the determinant is −0.17; thus, the eigenvalues are complex, with negative real part $-0.300 \pm 0.208i$. We obtain the trajectories for each species in Figure 12.4 and the phase portrait in Figure 12.5.

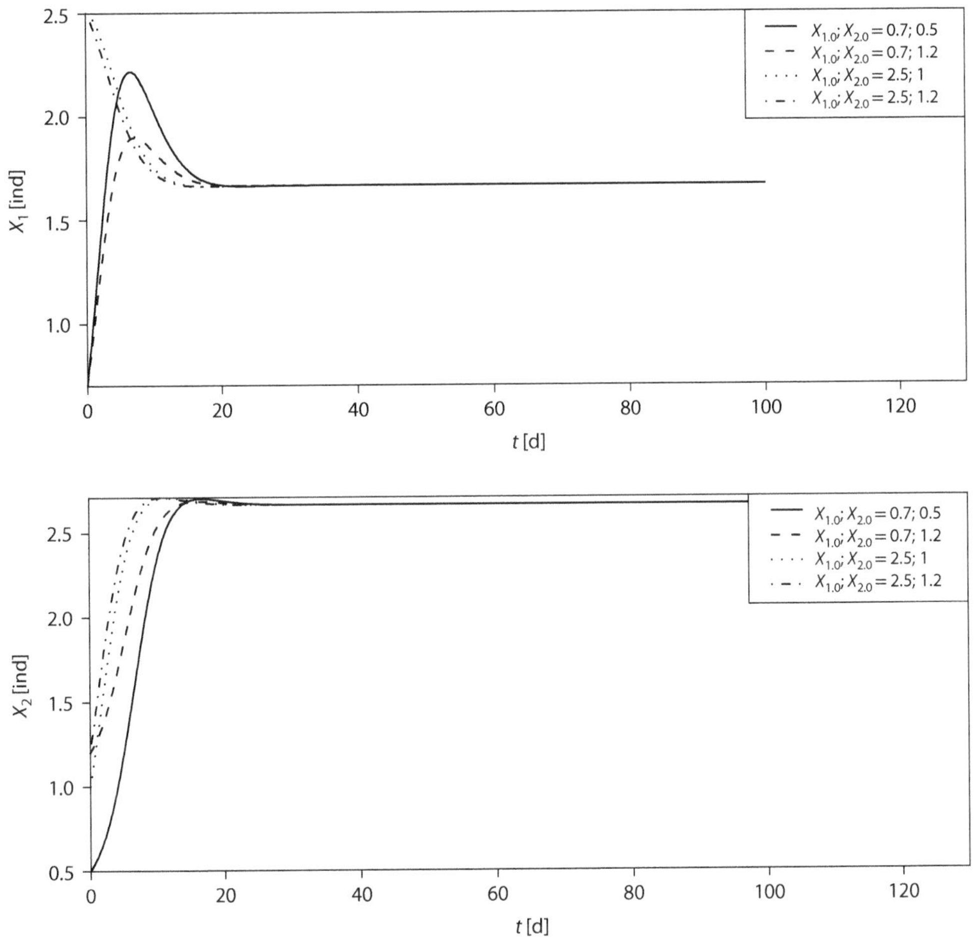

FIGURE 12.4 Consumer–resource dynamics: two species versus time.

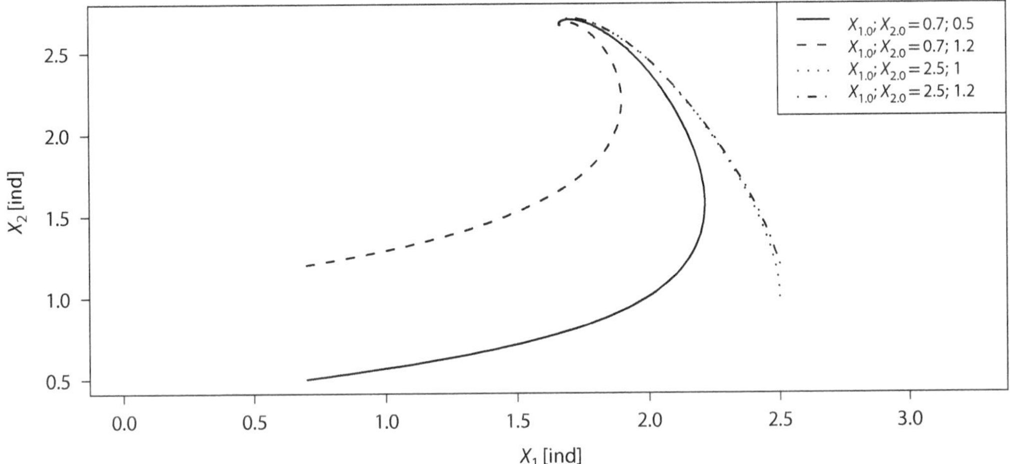

FIGURE 12.5 Consumer–resource dynamics: phase plane portrait.

12.3 CYCLES IN TWO-SPECIES CONSUMER–RESOURCE INTERACTIONS

Sustained consumer–resource oscillations are observed in nature, and therefore one important capability of consumer–resource models should be to exhibit sustained oscillatory dynamics under certain conditions. An increase in the resource is followed by the growth of the consumer; therefore, the resource declines due to consumption, and consequently, the consumer declines. Now, the resource can recover because of the release of predation pressure; therefore, the consumer recovers, and so on.

However, as shown in Section 12.2, for a simple Lotka–Volterra type of model, the oscillations are damped, and hence we need an extension of this type of model. One important model extension is the Holling–Tanner model, which includes density-dependent effects in the resource and consumer and rate-limited consumption by the consumer. If there is no predation pressure $X_2 = 0$, then the resource grows according to the logistic model where $1/a_{11}$ is equivalent to the carrying capacity, K_1.

The dynamics are modeled by

$$\begin{aligned} \frac{dX_1}{dt} &= \left(r_1 - a_{11} X_1 - a_{12}(X_1) X_2\right) X_1 \\ \frac{dX_2}{dt} &= \left(r_2 - a_{22}\frac{X_2}{X_1}\right) X_2 \end{aligned} \tag{12.18}$$

Note that the per capita rates are no longer linear. The equivalent carrying capacity of the consumer is X_1/a_{22}, that is, the abundance of the resource affects the carrying capacity. The consumer consumption rate reaches a saturation value for large resource density, for example, ingestion limited by a filtering rate. Instead of constant consumption rate, use a **functional response** $a_{12}(X_1)$ as shown in Figure 12.6 to make the rate change with resource abundance (Nisbet et al., 1997; Swartzman and Kaluzny, 1987):

$$a_{12}(X_1) = \frac{k_{max}}{k_h + X_1} \tag{12.19}$$

where k_{max} is the maximum consumption rate and k_h is the saturation **efficiency**. Equation 12.18 can be rewritten as

$$\begin{aligned} \frac{dX_1}{dt} &= \left(r_1 - \frac{X_1}{K_1}\right) X_1 - \frac{k_{max} X_1}{k_h + X_1} X_2 \\ \frac{dX_2}{dt} &= \left(r_2 - a_{22}\frac{X_2}{X_1}\right) X_2 \end{aligned} \tag{12.20}$$

This functional response termed "Holling type II" in this context is similar to the Michaelis–Menten–Monod or M3 or hyperbolic response (Chapter 6). When the resource is scarce, or X_1 is low, $a_{12}(X_1) \sim k_{max}/k_h$, but when the resource is very abundant, $a_{12}(X_1) \sim k_{max}$. A small value of k_h means that the consumer satiates at low resource density, and a large value of k_h means that the consumer requires high resource density before satiating.

Analysis is beyond the scope of this introductory book. Under many conditions, this model results in trajectories that oscillate transiently and settle at an equilibrium point. However, this model also produces a **limit cycle** around the equilibrium point under certain combinations of parameter values. Therefore, it predicts a cyclic fluctuation of consumer–resource densities. The cycle depends

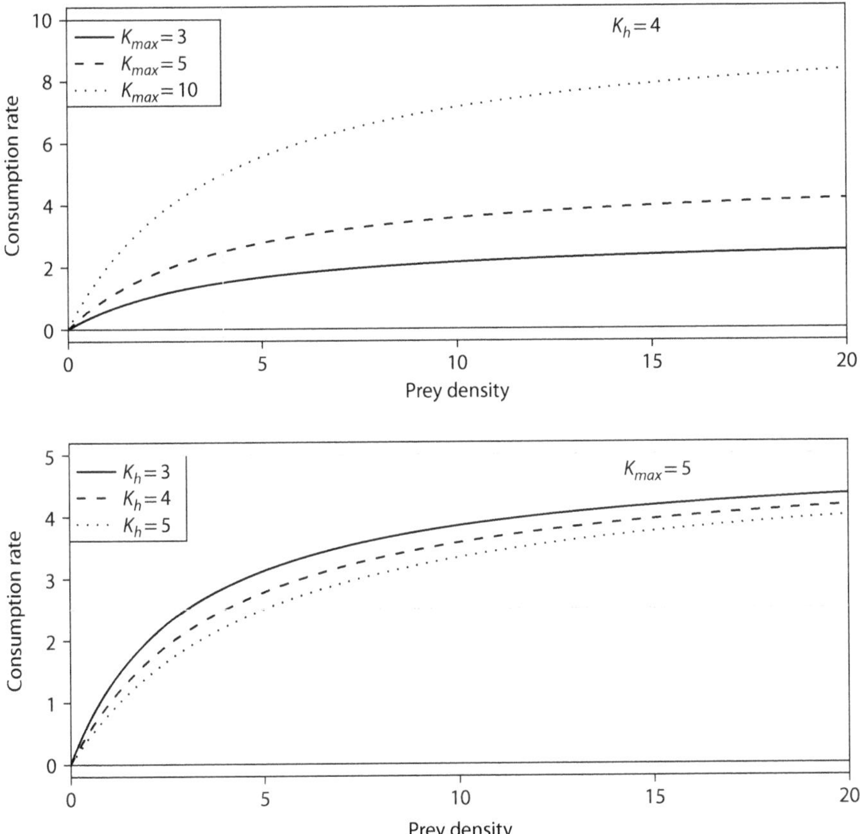

FIGURE 12.6 Consumption rate: hyperbolic form. Varying k_{max} (top panel) and varying k_h (bottom panel).

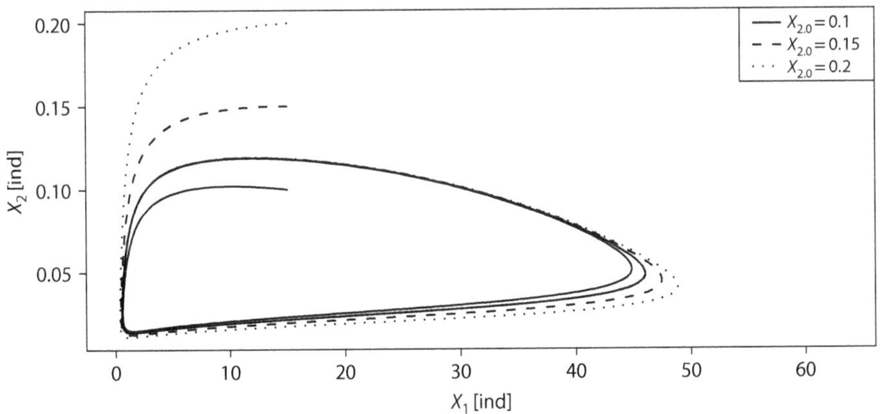

FIGURE 12.7 Approaching the limit cycle from the outside and from the inside.

on the parameters and not on the initial condition; therefore, the oscillation is recovered after a perturbation. When the initial condition is outside the limit cycle, the trajectories spiral inward and go back to the cycle. When the initial condition is inside the limit cycle, the trajectory spirals outwards back to the cycle (Figure 12.7). As a function of time, the oscillations look like those in Figure 12.8 for initial conditions inside and outside the limit cycle.

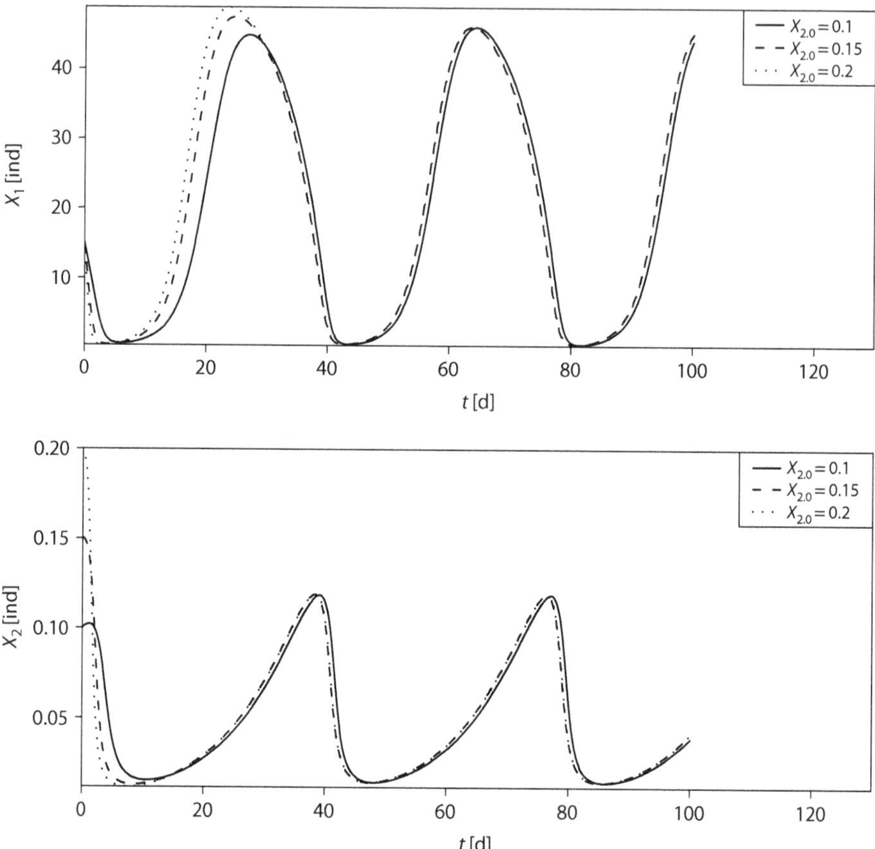

FIGURE 12.8 Temporal view of the limit cycle.

12.4 TWO SPECIES: DISTURBANCES

What would happen when one or two of the populations in a two-species community is disturbed (e.g., harvested)? For example, suppose the predator and prey species are both fish: What would happen to these species if you harvest them with the same effort? In fact, this question was the original motivation for Volterra's work.

In general,

$$\begin{aligned}\frac{dX_1}{dt} &= f_1(X_1, X_2) - h_1 \\ \frac{dX_2}{dt} &= f_2(X_1, X_2) - h_2\end{aligned} \tag{12.21}$$

For example, take the Holling–Tanner model and make the "harvest" term the same for both species, $h_1 = h_2 = h$:

$$\begin{aligned}\frac{dX_1}{dt} &= \left(r_1 - \frac{X_1}{K_1}\right)X_1 - \frac{k_{max}X_1}{k_h + X_1}X_2 - h \\ \frac{dX_2}{dt} &= \left(r_2 - a_{22}\frac{X_2}{X_1}\right)X_2 - h\end{aligned} \tag{12.22}$$

Here, h is the harvest term; it is a net type of disturbance (not per capita). Harvesting makes the predator population decrease and the prey population increase. This is important because if the prey is a valuable fish, harvesting removes the predation pressure exerted on the prey. In Figure 12.9, we show one possible outcome. Here, the limit cycle changes as we increase harvesting intensity, h, and eventually breaks down, throwing the resource toward a stable value (its carrying capacity) and extinguishing the consumer.

Therefore, one could increase the resource by harvesting both species. However, if the prey was a pest (e.g., an insect attacking some important crop) that has a natural enemy (a predator), the application of a pesticide affecting both populations indiscriminately could cause the reverse effect. The pest could actually increase by removal of the pressure exerted by the natural enemy. Using sensitivity tools, we can explore the effect of the harvest on metrics of the system (Figure 12.10).

Now consider more generally Lotka–Volterra systems with harvest:

$$\frac{d}{dt}\begin{bmatrix} X_1 \\ X_2 \end{bmatrix} = \left(\begin{bmatrix} a_{11} & a_{12} \\ a_{21} & a_{22} \end{bmatrix} \begin{bmatrix} X_1 \\ X_2 \end{bmatrix} + \begin{bmatrix} r_1 \\ r_2 \end{bmatrix} \right) \circ \begin{bmatrix} X_1 \\ X_2 \end{bmatrix} - \begin{bmatrix} h_1 \\ h_2 \end{bmatrix} \tag{12.23}$$

or succinctly

$$\frac{d\boldsymbol{X}}{dt} = \left(\mathbf{A}\boldsymbol{X} + \boldsymbol{r}\right) \circ \boldsymbol{X} - \boldsymbol{h} \tag{12.24}$$

It is now more complicated to find the equilibrium points because the algebraic system

$$\left(\mathbf{A}\boldsymbol{X}^* + \boldsymbol{r}\right) \circ \boldsymbol{X}^* = \boldsymbol{h} \tag{12.25}$$

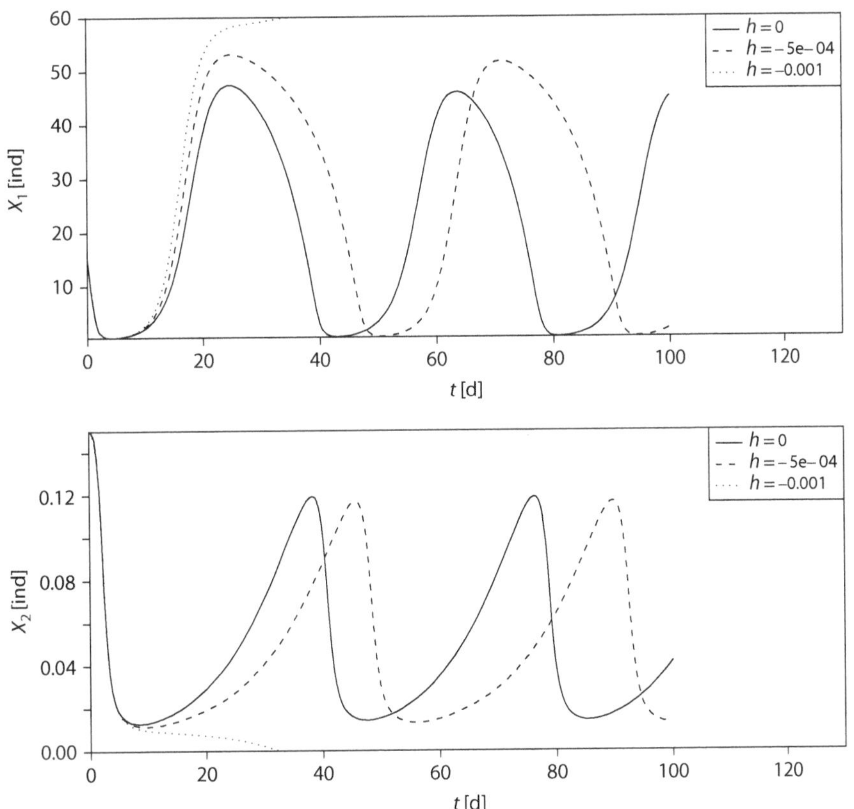

FIGURE 12.9 Disturbance of a predator–prey model.

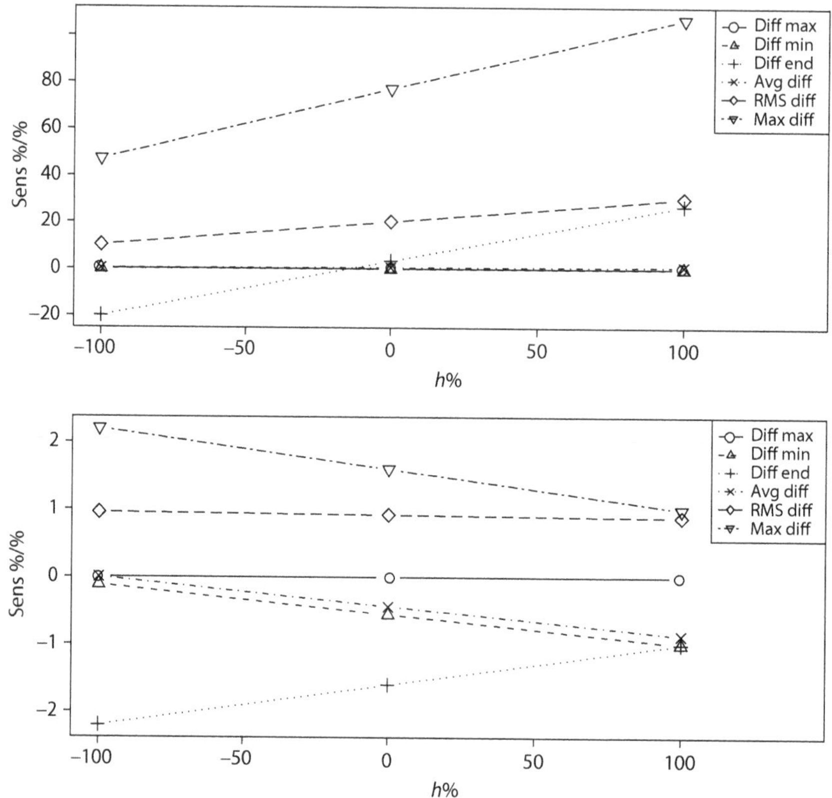

FIGURE 12.10 Sensitivity analysis of a disturbance consumer–resource model.

is nonlinear. It is also more complicated to generalize the behavior of the community matrix. Numerical analysis and simulations are now more helpful.

For example, consider competition with a stable coexistence. When the disturbance intensity is different for each species, the disturbance can modify the levels of coexistence. On applying harvest intensity to the strongest competitor (species 1) only, its density decreases, while the density of species 2 is increased. Further increases of the harvest intensity cause the demise of species 1 and dominance of species 2 (Figure 12.11). Moreover, the disturbance can force coexistence, even if the nonforced system fails to have a stable coexistence.

Variations of this theme include disturbances of mutualistic species, periodic disturbances in predator–prey, impulsive disturbance in competition, and delays in interacting populations (Cushing, 1977). Another example is a cladoceran population feeding on two competing algae populations (Acevedo et al., 1995b); the cladocera can change the outcome of competition between these two phytoplankton species.

12.5 TWO SPECIES: STRUCTURED POPULATIONS

Now assume that one or both of the species have age, size, or stage structure. This situation leads to more interesting and detailed models. We can make the survival or fecundity parameters depend on the other species. For example, the predator may attack the young of the prey species but not the adults. An animal can prey on the seeds of a tree but not on the seedlings or the mature trees.

We will not cover this type of model in this introductory book. There are various ways of modeling population structure for interacting species. One way is to use Leslie matrices. For example, consider one matrix for the consumer and one for the prey. The survival and fecundities of one species will

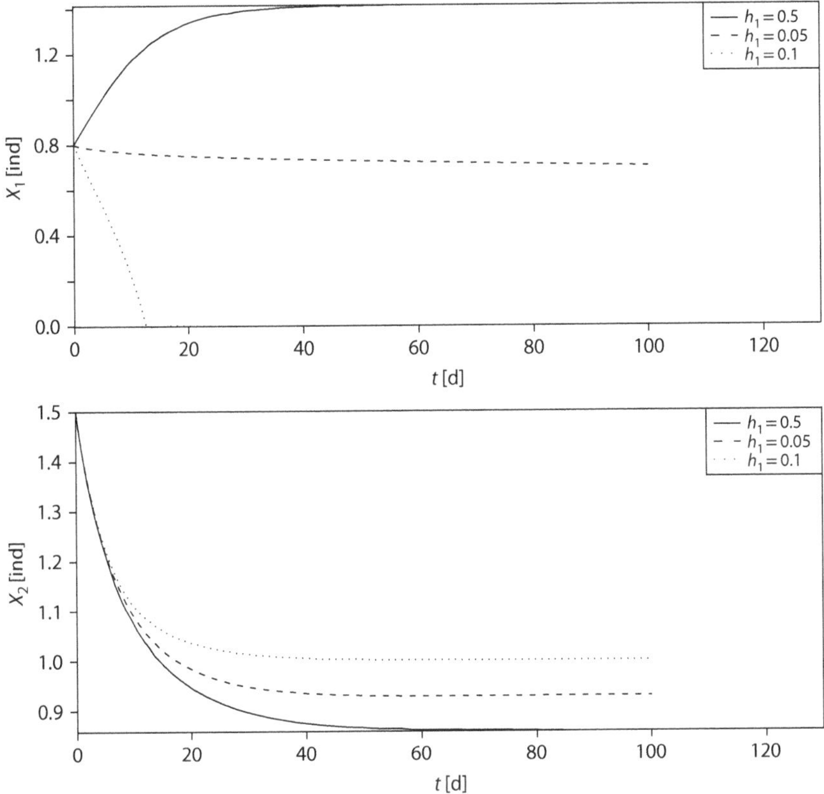

FIGURE 12.11 Effect of disturbance on competition outcome. It promotes the dominance of the weak competitor.

depend on the state vector of the other species. More specifically, model the resource population X as structured in two age classes—neonates X_1 and adults X_2—and the consumer Y as structured in two age classes—young Y_1 and adults Y_2. Assume that the consumer likes to feed on young individuals but not the adults of the resource. Low food supply will also mean lower survival. Make the survival of neonates of the resource depend on the total number (adults and young ones) of the predator $sx_1 = \dfrac{a}{(Y_1 + Y_2)}$, $sy_1 = \dfrac{b}{(X_1)}$, and $sy_2 = \dfrac{c}{(X_1)}$. It is more difficult to have an intuitive feeling for the results, except for some general predictions. The mathematical analysis of these problems is more difficult. You can appreciate the value of the simulations for these cases. Another example is a cladoceran population structured by stage and feeding on the algae according to the stage (Acevedo et al., 1995b).

12.6 MORE THAN TWO SPECIES, BUT NOT TOO MANY

Complicated dynamics occur for more than two species. Let us think of just a few species first, that is, three or four. We could easily think of many combinations: three competitors; two competing prey and one predator; three mutualist species; two mutualist prey and one predator; two predators and one prey, and one resource, one prey, and one predator (i.e., three-level trophic chain). We will explore three cases: three competitors, two competing prey and one predator, and a three-level trophic chain (Abrams and Roth, 1994).

Here, trajectories provide a three-dimensional (3-D) phase portrait of the three species X_1, X_2, and X_3. A trajectory or orbit is a curve in this space. There are also three important planes (phase planes): (X_1, X_2) for $X_3 = 0$, (X_2, X_3) for $X_1 = 0$, and (X_1, X_3) for $X_2 = 0$. Assume that the equilibria are X_1^*, X_2^*, and X_3^* obtained as points where the per capita rates are 0. The community matrix **C** is now 3×3.

12.6.1 Three Competitors

We use Equation 12.24 with a dimension of 3 and a matrix **A** with all negative entries. A digraph is shown in Figure 12.12 for the nonforced interactions. Under certain conditions, this system can achieve a stable coexistence of the three species. For example, a simulation run yields the results shown in Figure 12.13. Here, we see the result for no harvest on the upper panel and harvest of species 2 in the bottom panel. The harvest has the effect of increasing oscillations. An important issue is determining the effects of disturbance, harvest in this case, on the relative distribution of the species or diversity.

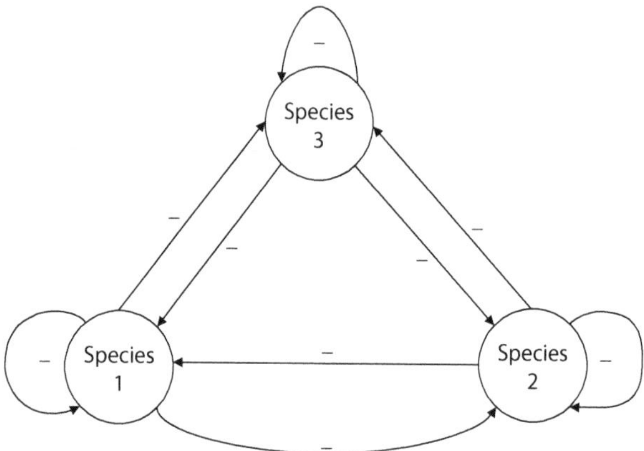

FIGURE 12.12 Digraph for three competitors.

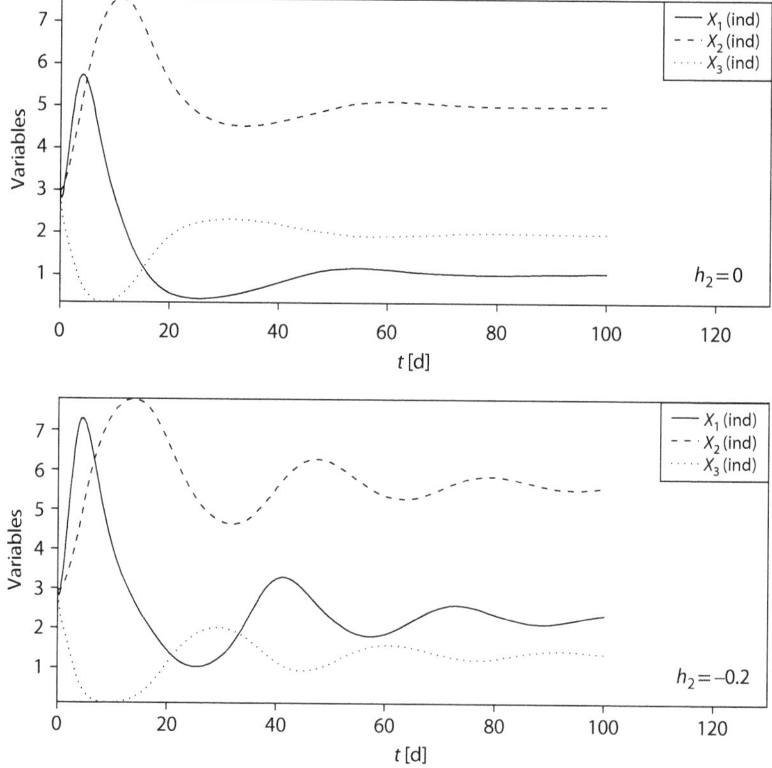

FIGURE 12.13 Dynamics for three competitors: stable coexistence.

12.6.2 Species Diversity

Species diversity is measured by species richness, n, and by some index of the relative abundances of the species. Richness is simply the number of coexisting species. In the case of three-species communities, richness at the equilibrium point can be 1, 2, or 3. For an index of relative abundances, we need the species composition given by the distribution of densities or the relative abundances of the species. That is, the proportions of each species with respect to the total, calculated by

$$p_i = \frac{X_i}{\sum_{i=1}^{n} X_i} \tag{12.26}$$

The diversity of species composition is calculated by indices based on functions of the species distribution, p_i. Several indices are common; among them are the Simpson diversity index and the Shannon diversity index. The Shannon diversity index is popular and used frequently and termed as "evenness." It is derived from the concept of information:

$$E = -\sum_{i=1}^{n} p_i \ln(p_i) \tag{12.27}$$

Here, we have used natural logarithms, but actually, the base of the logarithms determines the units. For example, if we select base 2, we obtain binary units.

For example, let us calculate evenness for three species, evenly distributed, with a total density $X\text{tot} = 10$. In this case, $\boldsymbol{p} = \begin{bmatrix} 1/3 \\ 1/3 \\ 1/3 \end{bmatrix}$. Equation 12.27 yields 1.098 ~ 1.1.

Exercise 12.2

Calculate evenness for three different cases of distribution of total density $X\text{tot} = 10$. (a) Coexistence of three species with slight dominance of species 3, $\boldsymbol{X} = \begin{bmatrix} 3 \\ 2 \\ 5 \end{bmatrix}$, (b) species 2 excluded, dominance of species 3, $\boldsymbol{X} = \begin{bmatrix} 2 \\ 0 \\ 8 \end{bmatrix}$, and (c) species 2 and 3 excluded, total dominance of species 3, $\boldsymbol{X} = \begin{bmatrix} 0 \\ 0 \\ 10 \end{bmatrix}$.

As we saw in Section 12.6.1, the pattern of community interaction coefficients determines the distribution at steady state and therefore determines diversity. In addition, external factors such as harvesting affect diversity. Figure 12.14 illustrates the dynamics of the Shannon diversity index as we increase the harvesting intensity on species 2. We can see how diversity oscillates as species composition oscillates and settles to a steady-state value as species composition achieves steady state. We approach steady-state diversity more quickly for extreme cases, that is, those cases of no harvesting and maximum harvesting. Using these models, we can study questions such as what level of disturbance intensity leads to larger diversity. One way of exploring the question is performing sensitivity analysis of diversity on disturbance intensity. For example, sensitivity analysis of the simulations just discussed results in the plot of Figure 12.15, which illustrates that the end value of diversity increases slightly for low intensity and then decreases as we increase the intensity.

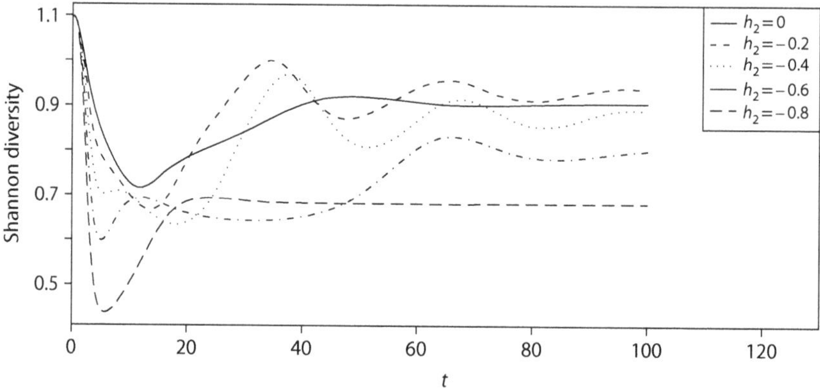

FIGURE 12.14 Dynamics of Shannon diversity for three competitors: stable coexistence.

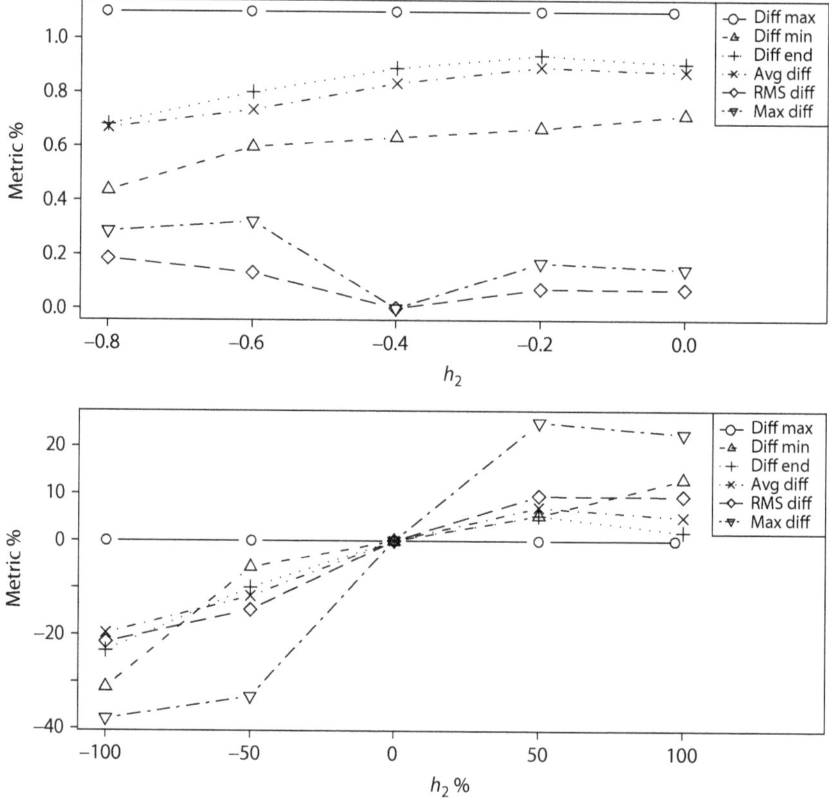

FIGURE 12.15 Sensitivity analysis of diversity versus disturbance intensity for three competitors: stable coexistence.

12.6.3 Two Competing Prey and One Predator

Assign two prey species to X_1 and X_2 and the predator to species 3. Matrix **A** is expressed as

$$\mathbf{A} = \begin{pmatrix} -a_{11} & -a_{12} & -a_{13} \\ -a_{21} & -a_{22} & -a_{23} \\ a_{31} & a_{32} & -a_{33} \end{pmatrix} \tag{12.28}$$

There is a negative effect of the predator on both prey species and in turn positive terms for both prey species on the predator (Figure 12.16). Harvesting the prey can diminish its abundance and allow the two prey species to reach the carrying capacity (Figure 12.17). Another example is a cladoceran population (predator) feeding on two competing algae populations (two prey) (Acevedo et al., 1995b). Therefore, the cladocera can change the outcome of competition between these two phytoplankton species.

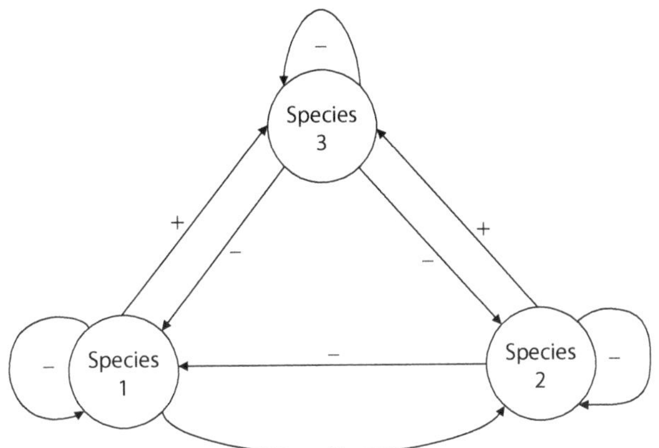

FIGURE 12.16 Digraph for two prey and one predator.

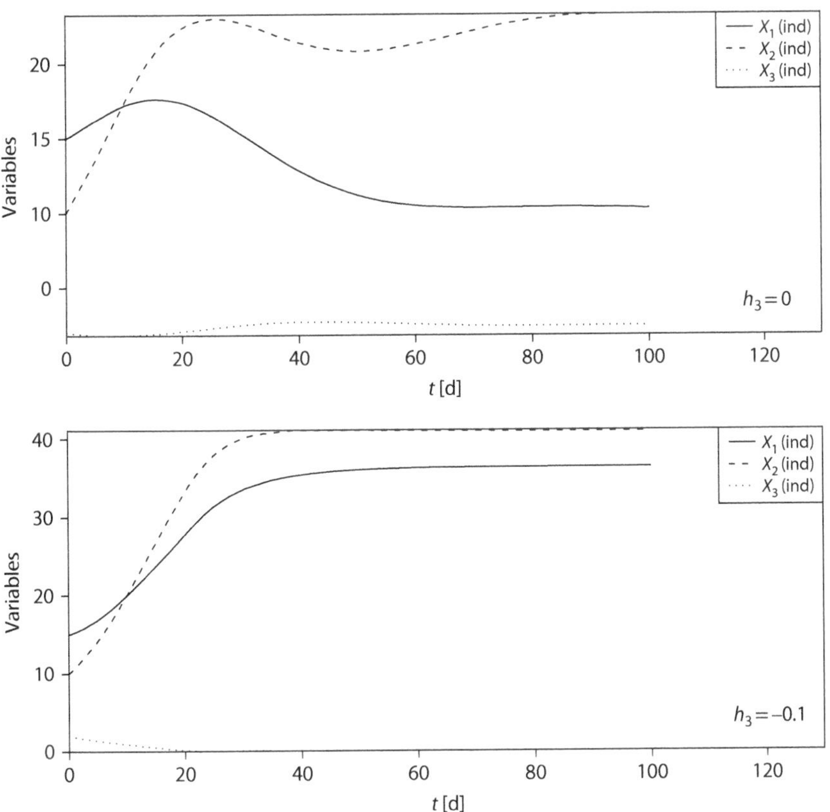

FIGURE 12.17 Dynamics for two prey and one predator.

12.6.4 Three-Level Trophic Chain

For this case, we have a resource X_1, a consumer X_2 of the resource, and a predator X_3 of the consumer:

$$\mathbf{A} = \begin{pmatrix} -a_{11} & -a_{12} & 0 \\ a_{21} & -a_{22} & -a_{23} \\ 0 & a_{32} & -a_{33} \end{pmatrix} \tag{12.29}$$

These interactions can be represented schematically with the digraph of Figure 12.18. As an example, the results of one simulation run are given in Figure 12.19. A type of disturbance that can

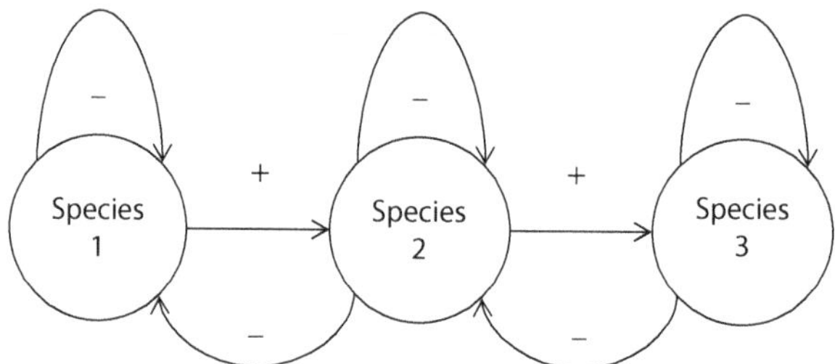

FIGURE 12.18 Digraph for three-level trophic chain.

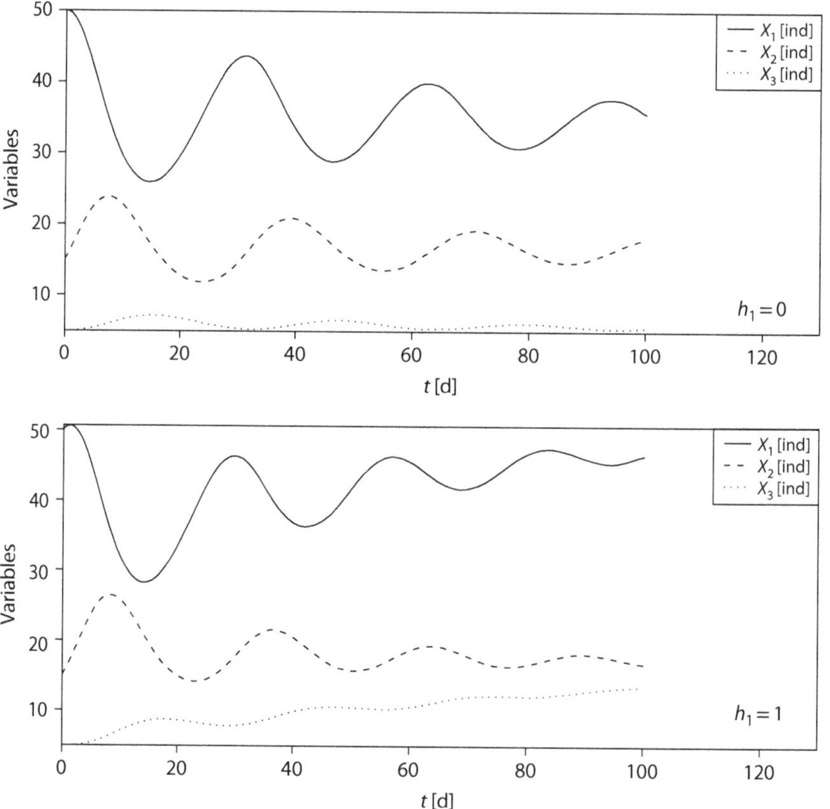

FIGURE 12.19 Dynamics for three-level trophic chain.

be examined in this model is enrichment of the resource, say by nutrient pollution, leading to an increase of productivity at the base of the chain (bottom panel of Figure 12.19).

More detailed models have been developed, for example, by using M3 functional responses on all four interactions of the food chain (Ford, 1999, pp. 58–68). There are results indicating more complex impacts of enrichment: chaotic behavior for intermediate enrichment and nonmonotonic change with carrying capacity.

12.6.5 Four Species

As we increase the number of species, we increase the number of combinations of the interactions. Possibilities include four competitors, three competing prey–one predator, four mutualist species; three mutualist prey–one predator; two predators–two prey (with special cases of competing prey or mutualistic prey); two resources–one prey and one predator (three-level trophic chain with two resources); one resource–two prey and one predator (three-level trophic chain).

Exercise 12.3

Write the matrix A for the case of the three-level trophic chain with two resources $X1$ and $X2$ at the base, a prey $X3$ at level 2, and a predator $X4$ at level 3.

12.6.6 Succession

Now consider the process of succession or transition from the dominance of one species to another. Succession results from underlying processes such as competition among the species. Each state variable is a fraction or proportion of land cover of a species or combination of species. Because the total fraction remains constant, this is a compartment model (Figure 12.20). It includes disturbance by fire or some other event of a given intensity and period.

One case is prairie succession (Acevedo et al., 1997) with cover type definition following a succession model for Oklahoma old fields after left fallow (Bledsoe and Van Dyne, 1971). These state variables include cover of (1) *Helianthus annuus* (sunflower) and *Digitaria sanguinalis*; (2) *Aristida oligantha*, three-awn grass; (3) *A. oligantha* and *Aristida basiramea* (*Aristida dichotoma* in north Texas); (4) *Eragrostis secundiflora* (lovegrass); and (5) *Andropogon scoparius* (little bluestem) and *Bouteloua curtipendula* (sideoats grama).

Simulations are done as a compartment model with rate coefficients as given in Bledsoe and Van Dyne (1971):

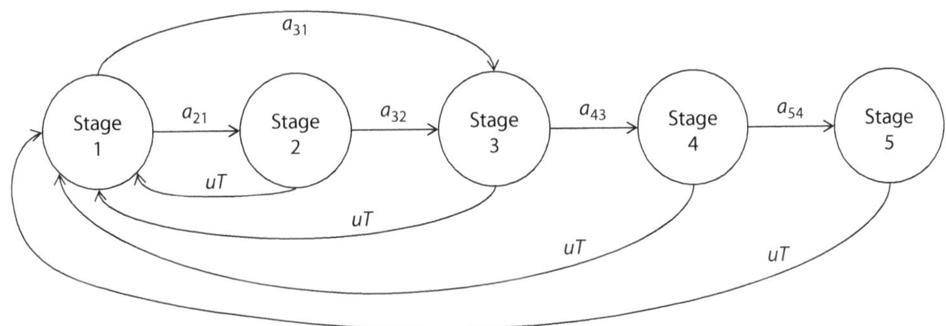

FIGURE 12.20 Succession system with disturbance reset.

$$\frac{dX}{dt} = \begin{bmatrix} -1.0 & 0.0 & 0.0 & 0.0 & 0.0 \\ 0.8 & -1.0 & 0.0 & 0.0 & 0.0 \\ 0.2 & 1.0 & -0.7 & 0.0 & 0.0 \\ 0.0 & 0.0 & 0.7 & -0.8 & 0.0 \\ 0.0 & 0.0 & 0.0 & 0.8 & 0.0 \end{bmatrix} X \tag{12.30}$$

A nondisturbed condition is illustrated in the top panel of Figure 12.21, displaying the typical succession pattern and dominance of cover type 5, *A. scoparius* and *B. curtipendula*. All other stages decline in the end. In particular, stage 1, sunflower, goes to 0 by year 5.

We can also include interrupting succession every few years by disturbance, that is, a disturbance parameterized by frequency and intensity. This is a periodic and sudden disturbance such as harvesting, grazing, or seasonal fires. Disturbed conditions with intensity of 0.4 with a 2-year period are illustrated in the bottom panel of Figure 12.21, showing a sawtooth pattern and recovery of early succession stages due to the disturbance. To clarify further, Figure 12.22 shows the effect of disturbance for the first stage (top panel) and the last stage (bottom panel) of succession. We see how the disturbance preserves stage 1 and diminishes stage 5.

On calculating the diversity, we see that in the absence of disturbance, the Shannon index reaches a maximum (~1.1) in about 1 year as several species are established and then declines to a minimum (~0.2) as stage 5 dominates and excludes all the other stages (Figure 12.23). However, when

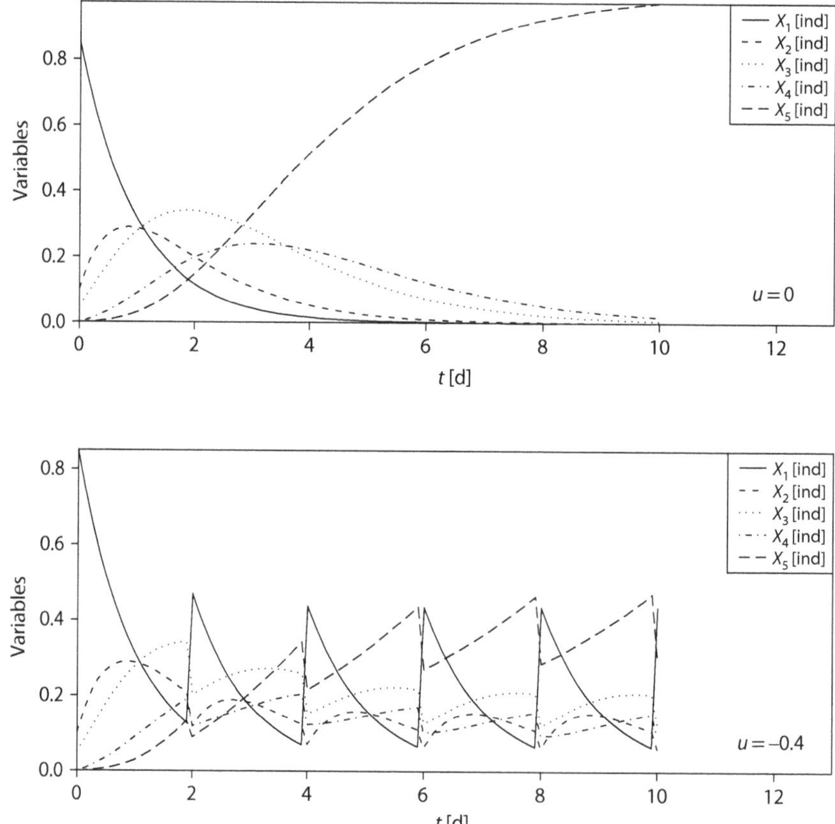

FIGURE 12.21 Dynamics of prairie succession. Top panel: no disturbance. Bottom panel: succession reset by disturbance.

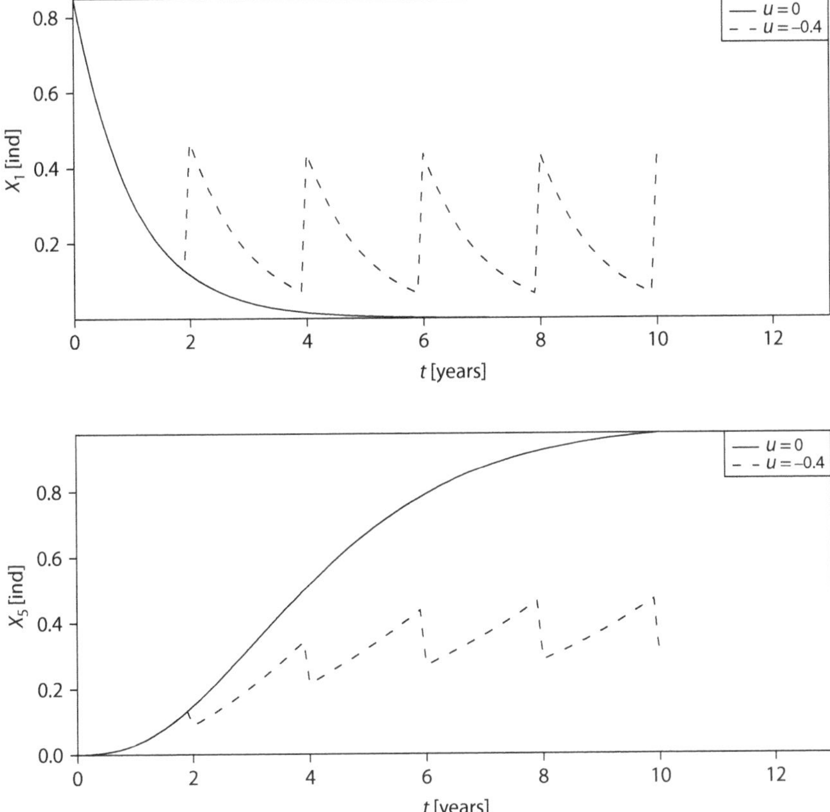

FIGURE 12.22 First and last stage of succession, for intensity $u = 0.4$ and period $t = 2$ years.

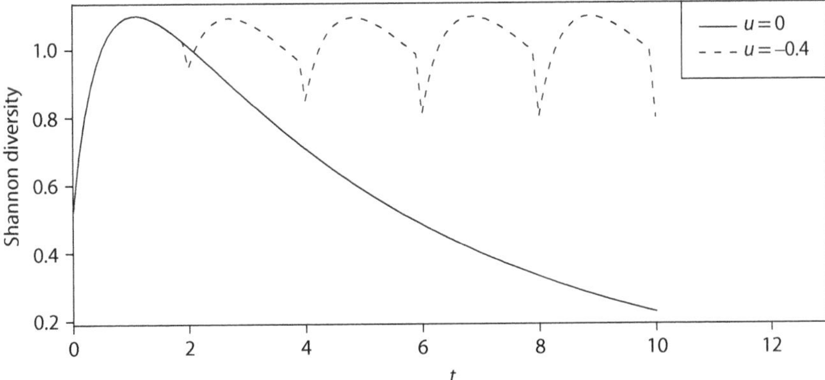

FIGURE 12.23 Diversity dynamics for succession, with no disturbance and for intensity $u = -0.4$ and period $t = 2$ years.

disturbing the system, we are able to preserve diversity at higher levels (~1.0) due to the reset of succession to earlier stages (Figure 12.23). By sensitivity analysis of diversity versus intensity, we see that further increases of intensity lower the diversity only slightly for the time average but substantially for the end value (Figure 12.24). Average cover values for late stages decrease as the disturbance period decreases (Acevedo et al., 1997). Sustaining early succession cover types requires the interruption of transition to late stages and eventual woody plant encroachment.

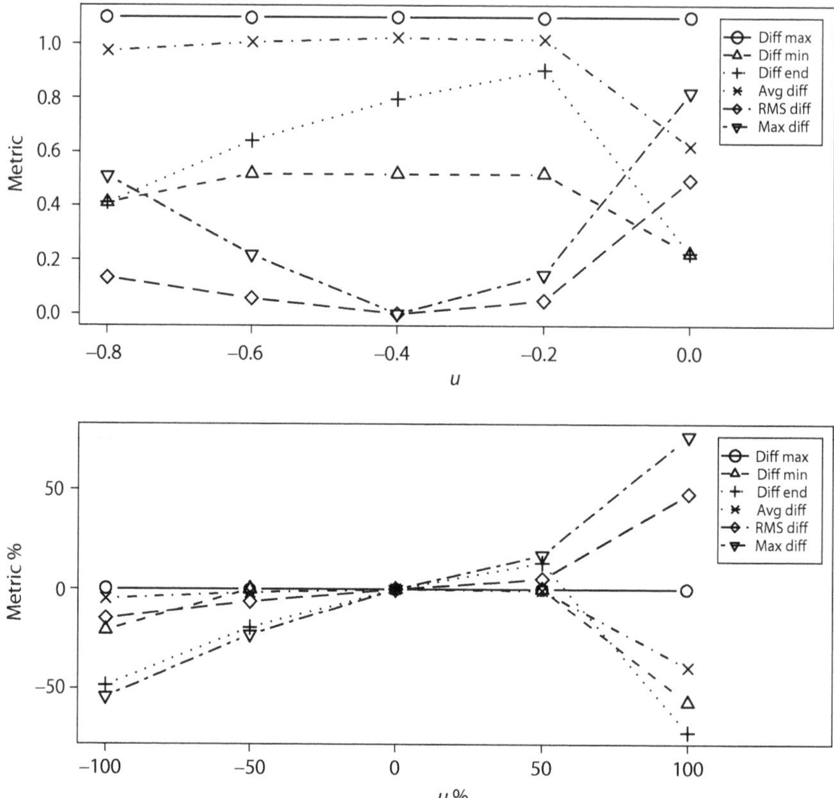

FIGURE 12.24 Sensitivity analysis of diversity versus disturbance intensity for a period $t = 2$ years.

12.6.7 Multiple Species

As we increase the number of species and go beyond just a few, mathematical complications increase. There have been developments for many years (Anosov and Arnold, 1988; Arnold, 1978; Hallam, 1986a; Hirsch and Smale, 1974; Maurer, 1999). The structure of the interaction is important. Local behavior around one equilibrium point is relatively easier to analyze by using linearization. However, global analysis of the nonlinear system is complicated. For local analysis, a useful tool is qualitative stability analysis based on the signs of the community matrix, that is, the "sign of **C**," as denoted by sign(**C**). This is a matrix with signs of the entries of **C** and can be derived straight from the digraph.

The following "Quirk-Ruppert" conditions determine qualitative stability from the community matrix (May, 1973) and were first derived in economic models:

1. No species with positive self-interaction coefficient, $c_{ii} \leq 0$ for all i.
2. At least one species is self-interacting, $c_{ii} \neq 0$ for at least one i.
3. The members of any pair of interacting species must have opposite signs (opposite effects on one another); $c_{ij} \times c_{ji} \leq 0$ for all i different from j.
4. No closed loops of interactions among three or more species, that is $c_{ij} \times c_{jk} \times \ldots \times c_{qr} \times c_{ri} = 0$ for any sequence of $i, j, \ldots q, r, i$.
5. No species unaffected by interactions with itself or other species; det(**C**) is nonzero.

When all conditions are met the system is stable, but if not all conditions are met there is no conclusive result; the system can be stable or unstable depending on the magnitude of the entries.

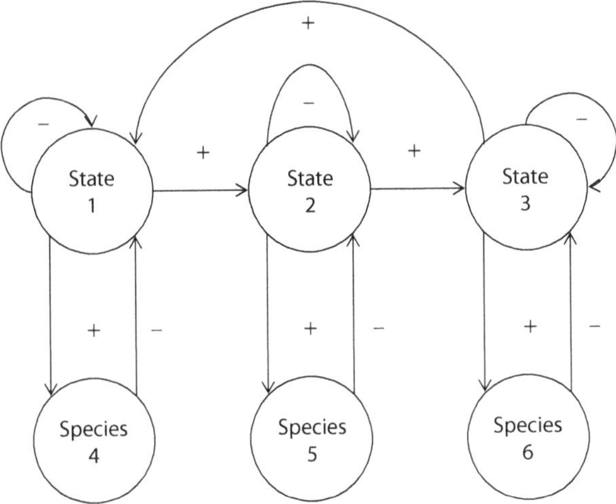

FIGURE 12.25 Host–parasite or predator–prey system with three stages.

For example, consider six states: three parasites or predator species X_4, X_5, and X_6 attack three stages of host or prey species X_1, X_2, and X_3. A digraph is shown in Figure 12.25. The matrix of signs is

$$\text{sign}\,\mathbf{C} = \begin{bmatrix} - & 0 & + & - & 0 & 0 \\ + & - & 0 & 0 & - & 0 \\ 0 & + & - & 0 & 0 & - \\ + & 0 & 0 & 0 & 0 & 0 \\ 0 & + & 0 & 0 & 0 & 0 \\ 0 & 0 & + & 0 & 0 & 0 \end{bmatrix} \tag{12.31}$$

Checking the conditions given above, condition 1: satisfied, condition 2: satisfied, condition 3: satisfied, condition 4: fails. In conclusion, the system is not qualitatively stable. This does not mean that the system is unstable; we would have to work with the magnitudes for a conclusive result. Qualitative stability analysis has been extended to include other criteria and metrics (Dambacher et al., 2003).

The concept of species diversity becomes particularly important for multiple species. The following questions are relevant. What community structural factors (biotic) lead to greater diversity? What factors make diversity decrease?

When we add a disturbance to multiple species, the models become more complicated to analyze. As before, disturbances can be added as proportional and nonproportional. An important issue is to analyze the effects of a disturbance on diversity. For example, determining what level of disturbance intensity leads to larger diversity.

12.7 COMPUTER SESSION

12.7.1 Two Species: Competition

First, let us analyze a simple competition system with the following values:

$$\mathbf{A} = \begin{bmatrix} -0.1 & -0.01 \\ -0.01 & -0.1 \end{bmatrix} \quad \boldsymbol{r} = \begin{bmatrix} 0.15 \\ 0.1 \end{bmatrix} \tag{12.32}$$

We use the function eq.2sp to check the matrix and find the eigenvalues of the community matrix

```
A = matrix(c(-0.1,-0.01,-0.01,-0.1),ncol=2,byrow=T)
r = c(0.15,0.1)
eq.2sp(A,r)
```

to obtain

```
$detpos
[1] TRUE
$coex
[1] TRUE
$xeq
[1] 1.414 0.859
$Det.Tra.Disc
[1] 0.012 -0.227 0.004
$eval
[1] -0.144 -0.084
```

The determinant is positive and the conditions for coexistence are met (coex=T). The equilibrium point is 1.41, 0.86, and it is a stable node because the eigenvalues are real and negative. Indeed, from the values in Det.Tra.Disc, $\Delta > 0$, $\mathrm{Tr} < 0$, and $D > 0$, this corresponds to case 1.1 in Table 9.2 of Chapter 9.

These parameter values are in the input file for the simulation of a simple competition model; file **comp-inp.csv**:

```
dt,0.01,d,time step
tw,0.1,d,time to write
r1,0.15,1/yr,intrinsic rate for sp1
r2,0.1,ind,intrinsic rate for sp2
a11,-0.1,1/yr,intra-sp coeff sp1
a12,-0.01,1/yr,inter-sp coeff sp2 on sp1
a21,-0.01,1/yr,inter-sp coeff sp1 on sp2
a22,-0.1,1/yr,intrasp coeff sp2 h,0,ind/yr,harvest
X1.0,0.8,ind,sp1 initial condition
X2.0,1.5,ind,sp2 initial condition
digX,4,None,significant digits for output file
```

Now, we use model function **int.2sp** and vary the initial conditions to visualize the trajectories converging to the node.

```
LV.2sp <- list(f=LVint.2sp)
param <- list(plab=c("X1.0","X2.0"), pval = cbind(c(0.7,0.7,2.5,2.5),
c(0.5,1.2,1.0,1.2)))
t.X <- sim(LV.2sp,"chp12/comp-inp.csv", param,pdfout=T)
```

The output PDF contains several pages including the trajectories for each variable versus time, phase plane portraits, and the trajectories of all variables together versus time. Figure 12.26 shows graphs of species density versus time. We already showed the phase plane portrait in the upper panel of Figure 12.2.

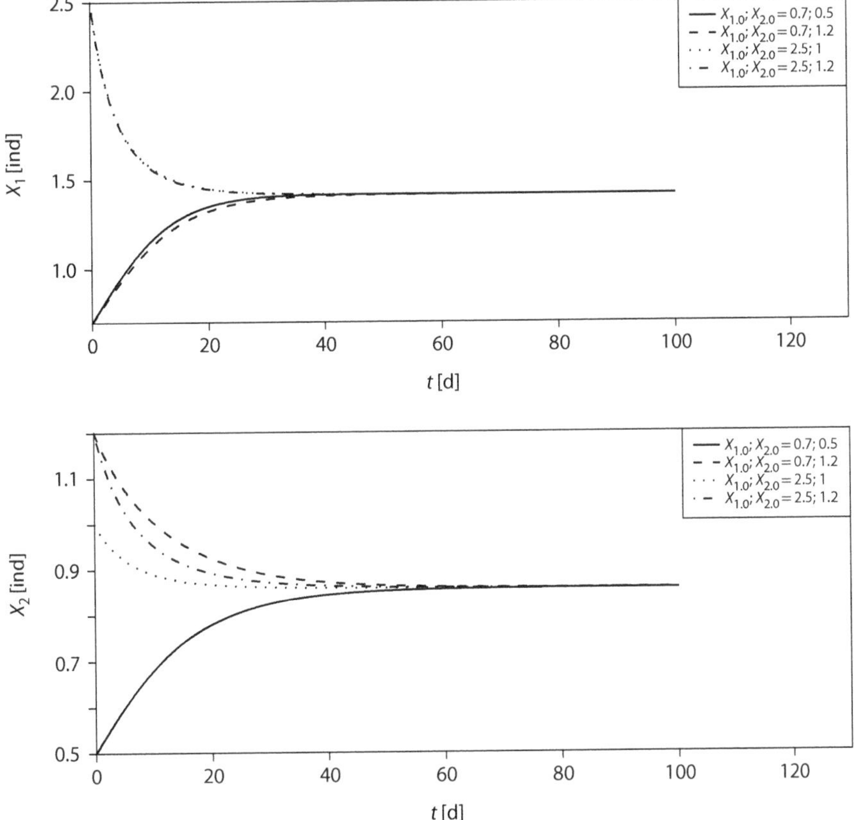

FIGURE 12.26 Dynamics of simple competition: stable coexistence.

Now, let us analyze a simple competition system exhibiting the dominance of one species; for example:

$$\mathbf{A} = \begin{bmatrix} -0.1 & -0.01 \\ -0.17 & -0.1 \end{bmatrix} \quad \boldsymbol{r} = \begin{bmatrix} 0.15 \\ 0.1 \end{bmatrix} \tag{12.33}$$

We use function `eq.2sp` to check the matrix:

```
A = matrix(c(-0.1,-0.01,-0.17,-0.1),ncol=2,byrow=T)
r = c(0.15,0.1)
eq.2sp(A,r)
```

We obtain

```
$detpos
[1] TRUE
$coex
[1] FALSE
```

The determinant is positive, but conditions for coexistence are not met (`coex=F`). These parameter values are in the input file for the simulation of a simple competition model, file **comp-dom-inp.csv**. The only difference with respect to the previous file is the line for `a21` that now reads

```
a21,-0.17,1/yr,inter-sp coeff sp1 on sp2
```

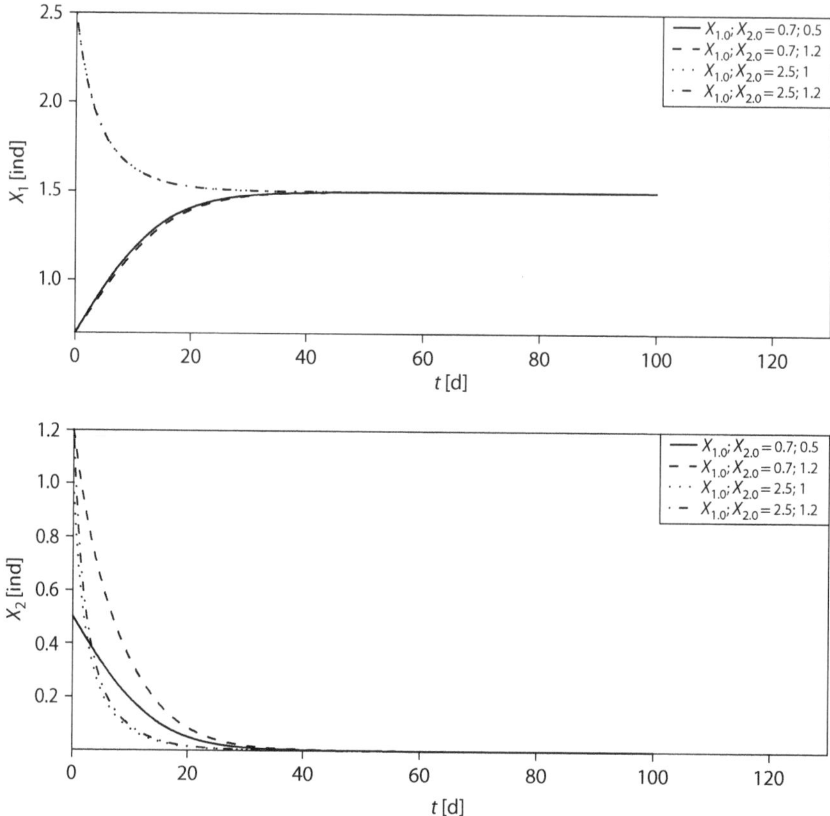

FIGURE 12.27 Dynamics of simple competition: dominance of species 1.

Again, we use model `LV.2sp` and vary the initial conditions to visualize trajectories:

```
param <- list(plab=c("X1.0","X2.0"), pval = cbind(c(0.7,0.7,2.5,2.5),
c(0.5,1.2,1.0,1.2)))
t.X <- sim(LV.2sp,"chp12/comp-dom-inp.csv", param, pdfout=T)
```

Figure 12.27 shows graphs of species density versus time, and Figure 12.2, bottom panel, shows the phase plane portrait.

12.7.2 Two Species: Mutualism

Consider $\mathbf{A} = \begin{pmatrix} -0.1 & 0.01 \\ 0.01 & -0.1 \end{pmatrix}$ and $\boldsymbol{r} = \begin{pmatrix} 0.15 \\ 0.1 \end{pmatrix}$.

Exercise 12.4

Use the above matrix A. Use the function `eq.2sp` to determine whether the conditions for coexistence equilibrium are met, find properties of the community matrix, and determine stability. Run simulations for various initial conditions using `param` and `LV.2sp` as in Section 12.7.1. Demonstrate that you get the results shown in Figure 12.3.

12.7.3 Two Species: Consumer–Resource

Consider $\mathbf{A} = \begin{pmatrix} -0.2 & -0.1 \\ 0.1 & -0.1 \end{pmatrix}$ and $r = \begin{pmatrix} 0.6 \\ 0.1 \end{pmatrix}$.

Exercise 12.5

Use matrix **A** from Section 12.7.3. Use the function `eq.2sp` to determine whether conditions for coexistence equilibrium are met, find the properties of the community matrix, and determine stability. Run simulations for various initial conditions using `param` and `LV.2sp` as in Section 12.7.1. Demonstrate that you get the results shown in Figures 12.4 and 12.5.

12.7.4 Holling–Tanner: Limit Cycle

In this section, we will use the function `holling.tanner` that simulates a Holling–Tanner model; the input file is **holl-tann-inp.csv** with the following contents:

```
Label,Value,Units,Description
t0,0.00,d,time zero
tf,100.00,d,time final
dt,0.01,d,time step
tw,0.1,d,time to write
r1,0.6,1/yr,intrinsic rate for sp resource
r2,0.1,1/yr,intrinsic rate for sp consumer
a11,0.01,1/(indxyr),intra-sp coeff for resource - inverse of carrying cap
a22,10,1/yr,intra-sp coeff for resource - scale resource as carrying cap
Kmax,150,1/yr,max-rate of consumption
Kh,6,ind,half-rate consumption
h,0,ind/yr,harvest
X1.0,15.00,ind,substrate initial condition
X2.0,0.15,ind,population initial condition
digX,4,None,significant digits for output file
```

First, let us make runs varying the initial condition of the consumer to demonstrate how the orbits converge on the limit cycle regardless of the initial state:

```
holl.tann <- list(f=holling.tanner)
param <- list(plab="X2.0", pval = seq(0.1,0.2,0.05))
t.X <- sim(holl.tann,"chp12/holl-tann-inp.csv", param,pdfout=T)
```

The PDF file should contain graphics like the phase portrait of Figure 12.7 and the trajectories of Figure 12.8.

Second, harvesting both the resource and consumer should lead to an increase of the resource and the eventual breakdown of the limit cycle. We can set up a sensitivity analysis for each state variable to obtain the results shown in Figures 12.9 and 12.10:

```
pnom=0.0005; plab = c("h"); runs=3
pval=c(0,0.0005,0.001)
param <- list(plab=plab, pval=pval, pnom=pnom, fact=T)
t.X <- sim(holl.tann,"chp12/holl-tann-inp.csv", param,pdfout=T)
# use variable 1
```

```
v2=1;i =1; v1<- v2+1;v2 <- runs+(v1-1); v <-c(1,v1:v2)
s.y <- sens(t.X$output[,v], param, fileout="chp12/
holl-tann-sens-1Hout",pdfout=T)
# use variable 2
i =2; v1<- v2+1;v2 <- runs+(v1-1); v <-c(1,v1:v2)
s.y <- sens(t.X$output[,v], param, fileout="chp12/
holl-tann-sens-2Hout",pdfout=T)
```

12.7.5 Two-Species Competition with Harvest

Again, let us analyze a simple competition system with the following values:

$$\mathbf{A} = \begin{bmatrix} -0.1 & -0.01 \\ -0.01 & -0.1 \end{bmatrix} \quad r = \begin{bmatrix} 0.15 \\ 0.1 \end{bmatrix} \tag{12.34}$$

but subject to a harvest of species 1, $h_1 < 0$.

Exercise 12.6

Use matrix **A** from Section 12.7.5. Use values of $h_1 = 0, -0.05$, and -0.1 and perform runs to obtain the graphs in Figure 12.11.

12.7.6 Three Species

We use the function `LVint.3sp` to simulate a three-species system. We consider three cases: all competitors (input file **comp-3sp-inp.csv**), two competing prey and one predator (input file **prey2-pred1-inp.csv**), and a trophic chain (input file **trophic-chain-inp.csv**). We explore the effect of harvest for all these three systems.

First, take the three-competitor system and harvest the second species:

```
LV.3sp <- list(f=LVint.3sp)
param <- list(plab="h2", pval=c(0,-0.2,-0.5))
t.X <- sim(LV.3sp,"chp12/comp-3sp-inp.csv", param,pdfout=T)
```

The PDF file should contain the results shown in Figure 12.13.

Exercise 12.7

Use the input files for two prey, one predator and harvest the predator with values of $h_3 = 0, -0.1$, and -0.2 to perform runs to obtain the graphs in Figure 12.17.

Exercise 12.8

Use the input files for the trophic chain and values of $h_1 = 0$, 1, and 1.5 to perform runs to obtain the graphs in Figure 12.19. Please note that the disturbance h_1 is positive because we are promoting growth of species 1, not harvesting.

To calculate the Shannon index, use the function `diversity` and apply it to the output of a simulation. For example, we set up the nominal and perturbation, run the simulation, and then calculate diversity to obtain the graphical result shown in Figure 12.14:

```
pnom=-0.4; plab = c("h2"); runs=5
pval=c(0,-0.2,-0.4,-0.6,-0.8)
param <- list(plab=plab, pval=pval, pnom=pnom, fact=T)
t.X <- int.3sp("chp12/comp-3sp-inp.csv", param,pdfout=T)
t.H <- diversity(t.X$output,param,fileout="chp12/
comp3sp-div",pdfout=T)
```

to obtain the graphical result shown in Figure 12.14, and then perform sensitivity using function `sens`

```
s.y <- sens(t.H, param, fileout="chp12/comp3sp-div-sens",pdfout=T)
```

Some of the results are shown in Figure 12.15.

12.7.7 Successional Dynamics

The functions `succession`, `succession.z`, and `succession.g` form a compartment model with discontinuous disturbances. The input file is **succession-inp.csv**. Set the disturbance period to 2 years and vary the disturbance intensity, u, to demonstrate the effect of disturbance. Use the simulation function `simd`.

```
succ <- list(f=succession,z=succession.z,g=succession.g)
param <- list(plab="u", pval=c(0,-0.4))
t.X <- simd(succ,"chp12/succession-inp.csv", param,pdfout=T)
```

The output PDF contains the graphical results shown in Figures 12.21 and 12.22.

Exercise 12.9

Produce a graph of diversity like that in Figure 12.23 for the values of $T = 2$ years, $u = 0$, and $u = -0.4$. Perform sensitivity analysis on disturbance intensity $u = 0$ to $u = -0.8$ with nominal -0.4. Demonstrate that you get the results shown in Figure 12.24.

Exercise 12.10

Perform sensitivity analysis on the disturbance period (2–10 years in steps of 2 years). Look at the average (avg) metric of stage 5 ("climax" *Andropogon* or bluestem grass) cover versus disturbance periodicity. Discuss the results; do you reduce the bluestem cover by disturbing more often? How often is it possible to disturb the site and still maintain at least a 50% average of bluestem grass cover?

12.7.8 Build Your Own

Exercise 12.11

Develop a simulation for a community of four species using the results of Exercise 12.3 and extend function `LVint.3sp` to another function `LVint.4sp` that will simulate one more species for a total of 4. Develop an appropriate input file based on the trophic chain (input file **trophic-chain-inp.csv**). Include the harvest of one species at the base of the chain.

12.8 SEEM FUNCTIONS EXPLAINED

12.8.1 Function eq.2sp

This function is a tool to check the condition for a nontrivial coexistence equilibrium point and gives the community matrix properties if the equilibrium exists.

```
eq.2sp <- function(A,r){
# is determinant positive
if(A[1,1]*A[2,2]>A[1,2]*A[2,1]) detpos <- TRUE else detpos <- FALSE
# is there a coexistence equilibrium
if(((-r[1]*A[2,2]+r[2]*A[1,2]) >0)
&& ((-r[2]*A[1,1]+r[1]*A[2,1])>0)) coex <- TRUE else coex <- FALSE
if(coex==T){
  # find equilibrium
  xeq <- -solve(A,r)
  # community matrix
  C <- diag(xeq)%*%A
  # det, Tr, and Disc
  delta <- det(C); Tr<- sum(diag(C));D <- Tr^2-4*det(C)
  Det.Tra.Disc <- c(delta, Tr, D)
  # eigenvalues
  eval <- eigen(C)$values
  out <- list(detpos=detpos,coex=coex,xeq=round(xeq,3),
                Det.Tra.Disc=round(Det.Tra.Disc,3),eval=round(eval,3))
} else out <- list(detpos=detpos,coex=coex)
return(out)
}
```

12.8.2 Function LVint.2sp

This model function is for systems of two species in interaction. It is similar to the function we used before for two compartments.

```
LVint.2sp <- function(t,p,x){
  # we write model with all positive coefficients
  # and let signs of parameters values determine type of interaction
  r <- p[1:2]; A <- matrix(p[3:6],ncol=2,byrow=T); u <- p[7:8]
  dX <- c(r +A%*%x)*x +u
  return(dX)
}
```

12.8.3 Function holling.tanner

This function is similar to `LVint.2sp` but includes the nonlinear terms.

```
holling.tanner <- function(t,p,x){
  # x[1] resource, x[2] consumer
  # p[1] r1, p[2] r2, p[3] a11, p[4] a22, p[5] Kmax, p[6] Kh, p[7] h
  dX1 <- (p[1] - p[3]*x[1]-x[2]*p[5]/(x[1]+p[6]))*x[1] + p[7]
```

```
  dX2 <- (p[2] - p[4]*x[2]/x[1])*x[2] + p[7]
  dX<-c(dX1, dX2)
  return(dX)
}
```

12.8.4 Function LVint.3sp

This function is very similar to `LVint.2sp` but is extended to three variables.

```
LVint.3sp <- function(t,p,x){
  r <- p[1:3]; A <- matrix(p[4:12],ncol=3,byrow=T); u <- p[13:15]
  dX <- (r + A%*%x)*x + u
  return(dX)
}
```

12.8.5 Function diversity

This function includes preparations to convert simulation output to a list.

```
if(nv>1){
  nvars <- matrix(nrow=nruns, ncol=nv)
  outX <- output[,-1]
  for(i in 1:nruns){for(j in 1:nv) nvars[i,j] <- i+nruns*(j-1)}
  Y <- list(); for(i in 1:nruns) Y[[i]] <- outX[,nvars[i,]]
}
```

The function then uses this list to calculate the proportions and the Shannon index.

```
# allocate list to store proportions and matrix to store Shannon's
prop <- Y
H <- matrix(nrow=nt,ncol=nruns)
# loop thru runs, time, and variables
for(i in 1:nruns){
 for(j in 1:nt){
   for(k in 1:nv){
     # calculate proportions
     prop[[i]][j,k] <- Y[[i]][j,k]/sum(Y[[i]][j,])
   }
   # calculate Shannon's index
   H[j,i] <- sum(-prop[[i]][j,]*log(prop[[i]][j,]))
 }
}
```

The remainder of the function consists of opening a PDF file and plotting the results.

12.8.6 Function succession

This model function is similar to the compartment models discussed in Chapter 11; the value in the diagonal is computed, taking into account that the sum of elements in a column should be 0.

```
succession <- function(t,p,x){
 k <- matrix(ncol=5,nrow=5) # transfers
 k[1,1:5] <- 0
 k[2,1] <- p[1]; k[2,2:5] <- 0
 k[3,1] <- p[2]; k[3,2] <- p[3]; k[3,3:5] <- 0
 k[4,1:2] <- 0.0; k[4,3] <- p[4]; k[4,4:5] <- 0
 k[5,1:3] <- 0.0; k[5,4] <- p[5]; k[5,5] <- 0
 for(i in 1:5) k[i,i] <- -sum(k[,i])
 dX <- c(k%*%x)
 return(dX)
}
```

Because the disturbance regime is discontinuous, we define a set of times and a function g to implement the discontinuities as we did in Chapter 7. In function g, we impact stages 2–5 negatively by the proportion given in u and then add all the cover taken to stage 1 to reset succession.

```
succession.z <- function(t,p,x) {
 # periodic impulse train
 tz <- seq(t[1],t[length(t)],p[7])
 return(tz)
}

succession.g <- function(t,p,x,tz){
 u <- rep(0,5)
 for(j in 1:length(tz))
  if(abs(t - tz[j])<0.00001) {
    u[2:5] <- rep(p[6],4)*x[2:5]
    u[1] <- -sum(u[2:5])
  }
 return(u)
}
```

13 Ecosystems
Nutrients and Energy

Our study on communities in the last chapter emphasized **biotic** interactions among species comprising the community without explicitly considering the **abiotic** factors involved in such a relationship, such as air, light, soil, and water. We now turn our attention to functional relationships among species by focusing on the transfer of material and energy among them and their interactions with the abiotic factors (Figure 13.1). Thus, we turn our attention to **ecosystems**, defined as the interaction of the biotic, organisms and species living in an area, and abiotic factors. Furthermore, we will be less concerned about the species themselves and more concerned about their functions and their role in ecosystem processes. Many times, we will lump species into compartments or pools according to their functions.

As a generalization, we will be concerned with modeling how materials (particularly nutrients) **cycle** and how energy **flows** among compartments (Figure 13.2). Modeling energy flow and nutrient cycling is key at various trophic levels: (1) the primary production level: uptake of available nutrients and generation of detritus; (2) the secondary production level: consumption of nutrients and generation of detritus; and (3) the decomposition level: bacterial, chemical, and physical breakdown of detritus and converting nutrients into the available form. Many of the introductory books we already cited are good references for ecosystem models (Ford, 1999; Huggett, 1993; Jorgensen, 1988; Keen and Spain, 1992; Swartzman and Kaluzny, 1987).

13.1 NUTRIENT CYCLES

Nutrient cycles are commonly modeled using the **compartment** approach introduced in Chapter 11. For each compartment, we write a balanced equation for inflows and outflows of a material, and once we have accounted for all the compartments, we have a dynamic system model:

$$\mathrm{d}\boldsymbol{X}(t)/\mathrm{d}t = \mathbf{K}\boldsymbol{X}(t) \tag{13.1}$$

where $\boldsymbol{X}(t)$ is a column vector of dimension $n \times 1$ giving the amount of material in each compartment and $\mathbf{K}$ is a $n \times n$ square matrix with entries k_{ij} giving the rate coefficient of transfer or flow of the material among the compartments.

Defining the compartments is crucial to nutrient dynamics modeling. It involves scales in terms of time, space, and model resolution. Therefore, we can model cycles in a patch of a forest, in a pond, in a watershed, or in the entire globe. In addition, we can model at coarse resolution, such as nutrients in the compartments of biomass, organic matter, and minerals, or be more specific and model different forms of organic matter, different molecular weight fractions, and other finer resolution forms in which the material can be present in the system. Concomitant to the definition of the compartment is the definition of the outside or the exterior to the system. For example, in a pond, input transfer would be as nutrients carried in the water that runoffs from the surrounding land and into the pond.

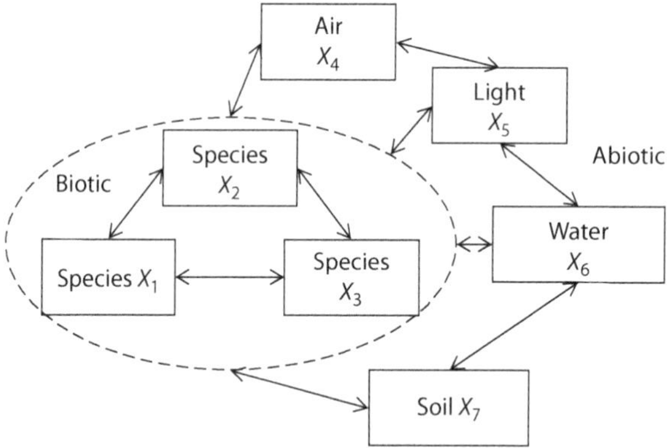

FIGURE 13.1 Communities (biotic) and its environmental variables (abiotic) form an ecosystem.

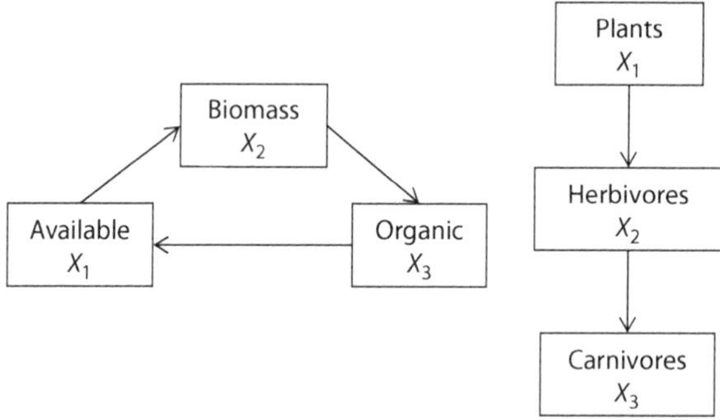

FIGURE 13.2 Ecosystem nutrient cycle (left) and energy flow (right).

13.2 CLOSED CYCLES

In a closed nutrient cycle, the material is **conservative**, that is, no loss or gain from the compartments to the "outside" of the system. Recall that matrix **K** has the property that the sum of all the entries in a column should add up to zero, and mathematically this means that the matrix **K** is singular. Therefore, we use the additional condition that $\sum_{i=1}^{n} X_i = X_t$, where X_t is the total material in all compartments and it remains constant. The system will be at an equilibrium point $\boldsymbol{X}^*$ when there is no change at that point, that is, $\mathrm{d}\boldsymbol{X}/\mathrm{d}t = 0$ at $\boldsymbol{X} = \boldsymbol{X}^*$.

13.3 OPEN CYCLES

There is transfer **into** (or **from**) at least one pool or compartment from (or to) "outside" the system, for example, input of nutrients due to water runoff loaded with fertilizer or output from soil erosion. The output is donor-controlled or dependent on at least one compartment, and when linear, its coefficient becomes part of the matrix **K**. However, the input is independent of $\boldsymbol{X}$ and then we have a forced system:

$$\frac{\mathrm{d}\boldsymbol{X}}{\mathrm{d}t} = \mathbf{K}\boldsymbol{X} + \boldsymbol{U} \tag{13.2}$$

$\boldsymbol{U}$ is the vector of inflows that do not depend on the concentrations in the compartments.

13.4 NUTRIENT CYCLE EXAMPLES

We will consider two important nutrient cycles: nitrogen and phosphorous. At the coarse scale of compartment definition, consider Figure 13.3 as a model for the nitrogen cycle. Here, we are only concerned with the movement of material from biomass (in plants and animals) X_2 to organic matter X_3 by death and decay (rate f_{32}), to the mineral form X_1 available for plant uptake (rate f_{13}), and to a slowly decomposing form in humus X_4 (rate f_{14}). Then importantly, it is the return transfer from X_1 to X_2 by uptake f_{21} and from X_1 to air by denitrification that closes the loop ($-u_a$). Input from the exterior is by fixation from air (almost 80% of the atmosphere is nitrogen as N_2 gas) requiring microbial mediation U_a and from sedimentary rocks by leaching (U_r). Indeed, there is recent research indicating that contribution from rocks is in some cases more important than once thought in some forest ecosystems (Morford et al., 2011). We have also included an input from fertilization (U_f) to account for nitrogen amendments in agricultural ecosystems or addition of nutrients carried in the runoff and losses from X_2 by harvest (u_h), from X_3 by erosion (u_e), and from X_4 by leaching (u_L).

For the purposes of illustration, we will model the simple cycle of Figure 13.3 and also assume linear transfer rates for simplicity of analysis such that $f_{ij} = k_{ij}X_j$:

$$\frac{d}{dt}\begin{pmatrix} X_1 \\ X_2 \\ X_3 \\ X_4 \end{pmatrix} = \begin{pmatrix} -k_{21}-u_a & 0 & k_{13} & k_{14} \\ k_{21} & -k_{32}-u_h & 0 & 0 \\ 0 & k_{32} & -k_{13}-k_{43}-u_e & 0 \\ 0 & 0 & k_{43} & -k_{14}-u_L \end{pmatrix}\begin{pmatrix} X_1 \\ X_2 \\ X_3 \\ X_4 \end{pmatrix} + \begin{pmatrix} U_a+U_r+U_f \\ 0 \\ 0 \\ 0 \end{pmatrix} \tag{13.3}$$

For practical purposes, we will model the harvest and fertilization rates with time-discontinuous functions and more specifically with a $\delta(t-T_h)$ Dirac impulse function, that is, it is zero for all t except at the discontinuity times dictated by the harvest period T_h. To simplify, we assume that the fertilization has the same period as that for the harvest. We will use some arbitrary parameter values for illustration: $U_a = 10$, $U_f = 5$, $U_r = 1$ all in g/(m^2 · year), $u_h = 0.5$ year^{-1}, and $u_L = 0.01$ year^{-1}. Transfers are given as $k_{21} = 0.9$, $k_{32} = 0.3$, $k_{43} = 0.2$, $k_{13} = 0.2$, and $k_{14} = 0.1$, all in year^{-1}.

Simulating for the two values of harvest intensity and using $T_h = 0.5$ year, we obtain the results shown in Figure 13.4. There we can see how N in biomass adopts a sawtooth pattern, indicating a swing from event to event, and addition of N by fertilization generates abrupt increases in the available N that decrease during the growing season due to uptake.

To this very coarse resolution cycle, we can add details by dividing the X_3 compartment into four compartments corresponding to ammonia, ammonium, nitrites, and nitrates (Figure 13.5). These four forms are part of the nitrification process starting from decomposing of matter and ending in

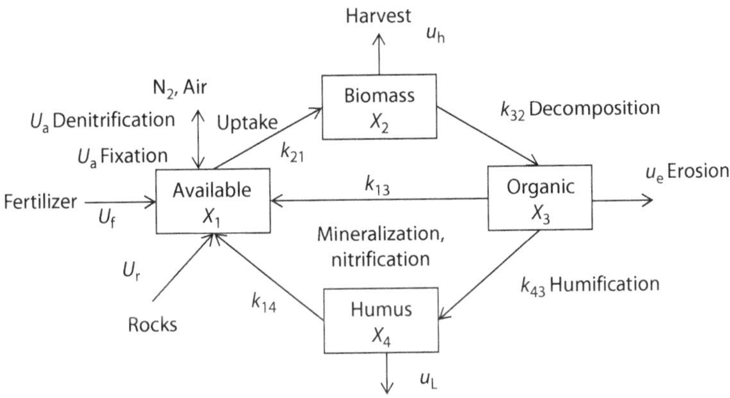

FIGURE 13.3 Nitrogen cycle terrestrial ecosystems model.

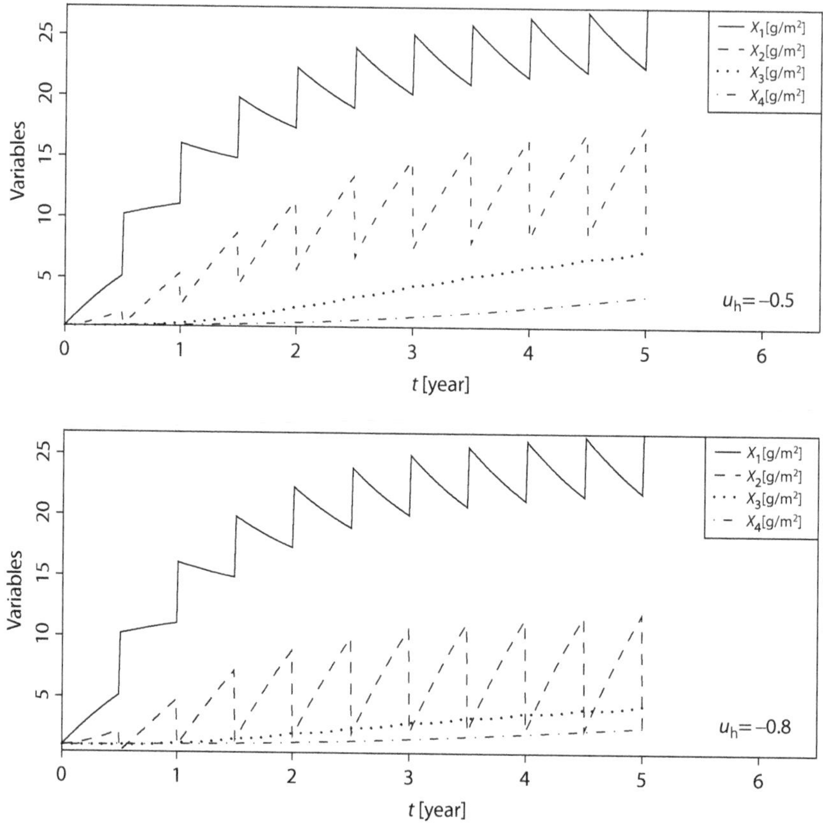

FIGURE 13.4 Simulation of the nitrogen cycle terrestrial ecosystems model.

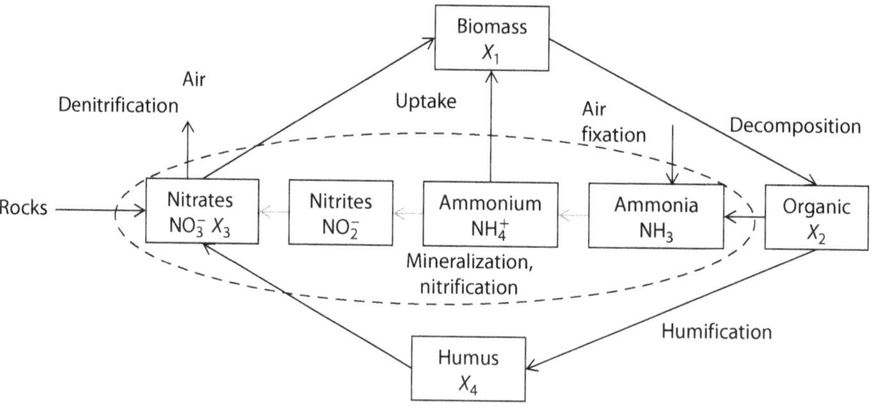

FIGURE 13.5 Nitrogen cycle terrestrial ecosystems model: more details on the nitrification process.

nitrate formation. Plants can uptake ammonium; this exchange with air is now broken down into two parts: fixation is to ammonia and denitrification is from nitrates. Models of nitrogen and carbon are useful for soil management (Shaffer et al., 2001).

Exercise 13.1

Assign X_5, X_6, and X_7, to ammonia, ammonium, and nitrites, respectively. Assume simple linear transfers from X_2 to X_5, X_5 to X_6, and X_6 to X_7. Rewrite Equation 13.3 to include these new compartments.

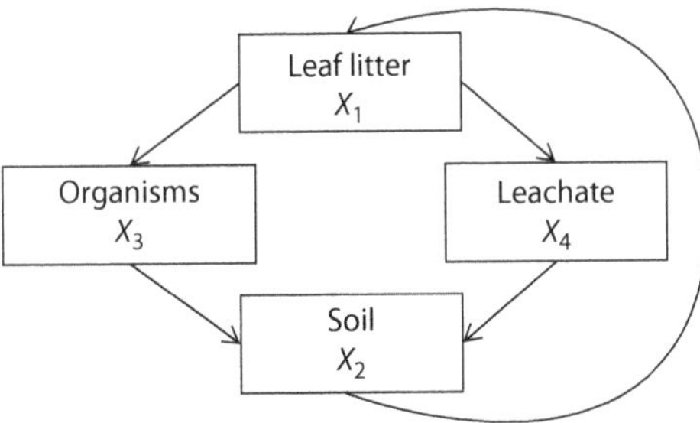

FIGURE 13.6 Five-compartment terrestrial cycle of leaf litter to soil.

This example is adapted from the book by Swartzmann and Kaluzny (1987). It represents the flow of nutrients from the leaf litter in the forest floor by leaching and decomposition. Figure 13.6 displays only the most important transfer coefficients. Here we assume X_1 = leaf litter, X_2 = soil, X_3 = microflora and millipedes, and X_4 = leachate. The matrix of coefficients is

$$\mathbf{K} = \begin{pmatrix} -0.054 & 0.01 & 0 & 0 \\ 0 & -0.001 & 0.04 & 0.02 \\ 0.037 & 0 & -0.04 & 0 \\ 0.017 & 0 & 0 & -0.02 \end{pmatrix} \tag{13.4}$$

Exercise 13.2

Write a balance equation of all the transfers into and out of the soil compartment. Is the system closed? Which compartment has most of the material at steady state?

13.5 NUTRIENT CYCLE EXAMPLE: TERRESTRIAL ECOSYSTEMS

In this example, we will consider a generic phosphorous cycle in aquatic ecosystems (Figure 13.7). This issue is of great interest because phosphorous contributes to eutrophication in water bodies. At a coarse-resolution compartment definition, we consider X_1 = inorganic soluble form, as phosphate receives input from runoff; X_2 = particulate form in algae and bacteria; X_3 = low molecular weight compounds; and X_4 = colloidal form (Lean, 1973).

Most of the input as phosphate goes into compartment 1: particulate, leading to algae bloom (and possibly eutrophication). Return flows from all the compartments to soluble form constitute cycling of the nutrient inside the system. Assume linear kinetics for all the compartments (Lean, 1973).

The following rate coefficients are for illustration only. They were calibrated to obtain steady state values close to the values reported in work by Lean (1973). Assume output $u_L = -0.01$ and input by runoff modeled as episodic with intensity U_f and periodicity T_f:

$$\mathbf{K} = \begin{bmatrix} -0.9 & 0 & 0.1 & 0.04 \\ 0.9 & -0.001 & 0 & 0 \\ 0 & 0.001 & -0.4 & 0 \\ 0 & 0 & 0.3 & -0.04 \end{bmatrix} \tag{13.5}$$

Figure 13.8 shows simulations for two input values. The top panel is for zero input and the bottom panel is for an input of $U_f = 0.1$ at 0.1 year.

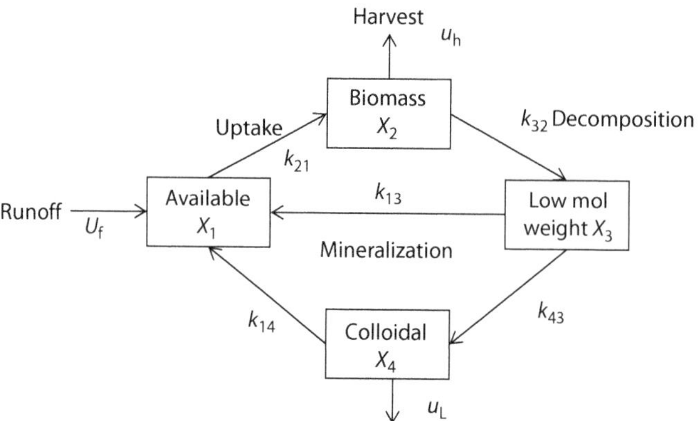

FIGURE 13.7 Phosphorous cycling in a lake.

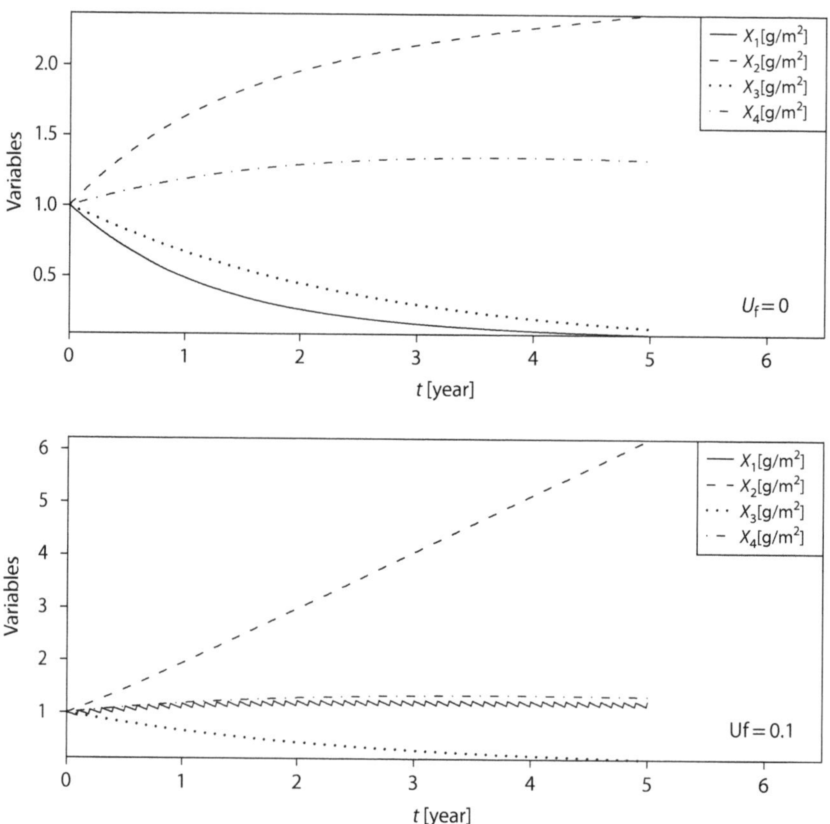

FIGURE 13.8 Dynamics of phosphorous.

Exercise 13.3

For the zero input case of Figure 13.8, why is most of the material in X_2 at steady state? Why is X_2 still growing at the end of the simulation in the nonzero input case (bottom panel of Figure 13.8)?

13.6 GLOBAL BIOGEOCHEMICAL CYCLES

Compartment models can be used to represent global planetary cycles, like carbon (Harte, 1988; Keen and Spain, 1992; Walker, 1991). Carbon is an important part of the biosphere and plays a key role in global climate control because the concentration of carbon dioxide (CO_2) in the atmosphere contributes in determining the greenhouse effect; thus, we include a compartment X_1 for the atmosphere (Figure 13.9) (Gates, 1993). Carbon dioxide (CO_2) is used by primary producers (e.g., terrestrial plants and algae) to make carbohydrates by photosynthesis using sunlight. Therefore, we have two compartments, one for terrestrial biota X_2 and another for the upper layer of the ocean X_3, and two fluxes f_{21} and f_{31}. We will study photosynthesis with more detail in the next section.

Some of the carbon goes back to CO_2 by respiration and by emission from these compartments as fluxes f_{12} and f_{13}; the rest is stored, consumed, and decomposed (fluxes f_{42} and f_{73}) and a part is recycled (fluxes f_{24} and f_{37}). At slower rates, the carbon transfers to fossil organic matter, sediments, and sedimentary rocks (fluxes f_{54}, f_{64}, and f_{67}). Sedimentary deposits contain most of the carbon. Human action accelerates the release of CO_2 by burning fossil fuels (coal, oil, and gas) at rate u_f and from terrestrial biota by deforestation at rate u_d. The time scale is mixed; some processes occur rapidly such as the exchanges between atmosphere and biota and others occur slowly such as sedimentation. We will use a time scale of years in the simulation and gigatons or Gt for carbon amount in the compartments. The fluxes are then given in Gt/year. One gigaton is one peta g = 10^{15} g.

Estimated values for the compartments $X_1, \ldots, X_7$ are 740, 1760, 1000, 1500, 10,000, 20 million, and 39,000 Gt, respectively. These values are just approximations.

Exercise 13.4

Assume $f_{12} = 100$, $f_{21} = 102$, $f_{13} = 90$, $f_{31} = 92$, $u_d = 2$, and $u_f = 5.3$ Gt/year. Also assume that the flow from the rocks to the atmosphere is negligible compared to the fluxes from the biota and fossil fuel combustion. What is the total flow of carbon into the atmosphere and what is the total flow of carbon out of the atmosphere? Assuming that these fluxes remain constant, what would be the carbon content in the atmosphere in 50 years? Assume that we reduce carbon emission from fossil use by 1% with respect to 5.3 Gt/year. What would be the carbon content in the atmosphere in 50 years? What is the percent difference between this value and the one calculated previously?

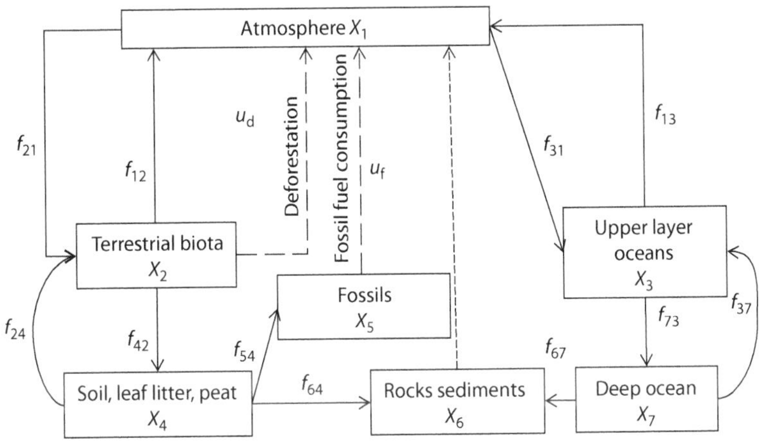

FIGURE 13.9 Global carbon cycle.

13.7 PRIMARY PRODUCTIVITY

This occurs at the base of the trophic level chain and is based on the process of converting solar energy to chemical energy.

13.7.1 REVIEW OF PHOTOSYNTHESIS

Primary producers convert solar energy to chemical energy stored in the form of carbohydrates. This process requires carbon dioxide (CO_2) and water (H_2O). The chemical reaction is given as

$$6CO_2 + 6H_2O \xrightarrow{\text{light}} C_6H_{12}O_6 + 6O_2 \tag{13.6}$$

To convert one mole of CO_2 to chemical energy requires 8 einsteins (E) of solar radiation. In terms of joules, it takes 0.472 MJ to convert one mole of CO_2 or 0.00004 MJ to fix 1 mg of carbon.

Photosynthesis occurs in two steps inside the leaf cells at the chloroplasts using chlorophyll. In the first step, light-dependent reactions, the solar energy is transferred to energy carriers, then in the second step, light-independent reactions or Calvin cycle, these energy carriers are used to reduce the carbon dioxide.

C3 and C4 types of photosynthesis processes in plants result in different responses to sunlight levels, hot and dry conditions, and CO_2 concentration. C4 plants capture CO_2 differently, increasing the carbon fixation rate at lower CO_2 concentrations. An additional type, crassulacean acid metabolism (CAM), allows for capturing CO_2 at night and storing it for use during the daytime.

Photosynthesis is wavelength-dependent. Photosynthetically active radiation (PAR) is the radiation in the 400–700 nm range. The efficiency of photosynthesis as a percent of light energy converted to chemical energy is about 18%. The rate of photosynthesis can be expressed as a function of photosynthetic mass, for example, the CO_2 fixed per unit time per chlorophyll concentration ($\text{mmole}\cdot CO_2\cdot s^{-1}/\text{mg}\cdot \text{Chl}\cdot m^{-3}$).

13.7.2 PHOTOSYNTHESIS MODEL: LIGHT

Photosynthesis rate varies with light intensity; at lower light levels, the light reaction rates increase linearly with light. At higher levels, there is saturation of the Calvin cycle enzymes and the photosynthesis rate drops at higher light levels by photoinhibition.

There are several popular models describing this rate versus light relationship. We will describe two models that do not include photoinhibition and two that do include it.

The Smith model does not include photoinhibition:

$$P(L) = \frac{\alpha L P_{max}}{\sqrt{P_{max}^2 + (\alpha L)^2}} = \frac{L P_{max}}{\sqrt{\left(P_{max}/\alpha\right)^2 + L^2}} \tag{13.7}$$

where P_{max} = maximum rate (assuming it is not CO_2-limited), L = light intensity, PAR = 400–700 nm in $E/m^2/s$ or W/m^2, and α = maximum efficiency to light in moles/E or moles/W/sec = slope of the curve zero light intensity or $L = 0$. The effect of varying P_{max} and α is illustrated in the top panels of Figure 13.10.

The Thornley model is a nonrectangular hyperbola or a variation of the M3 response (Thornley, 2002):

$$P(L) = \frac{1}{2\xi}\left(\alpha L + P_{max} - \sqrt{\left(P_{max} + \alpha L\right)^2 - 4\xi\alpha L P_{max}}\right) \tag{13.8}$$

where an additional parameter ξ with values between 0 and 1 controls the shape of the curve. Low values of ξ approximate a M3 response, while high values produce a sharp increase to the maximum. Its behavior as a function of parameters α and ξ is shown in the bottom panels of Figure 13.10.

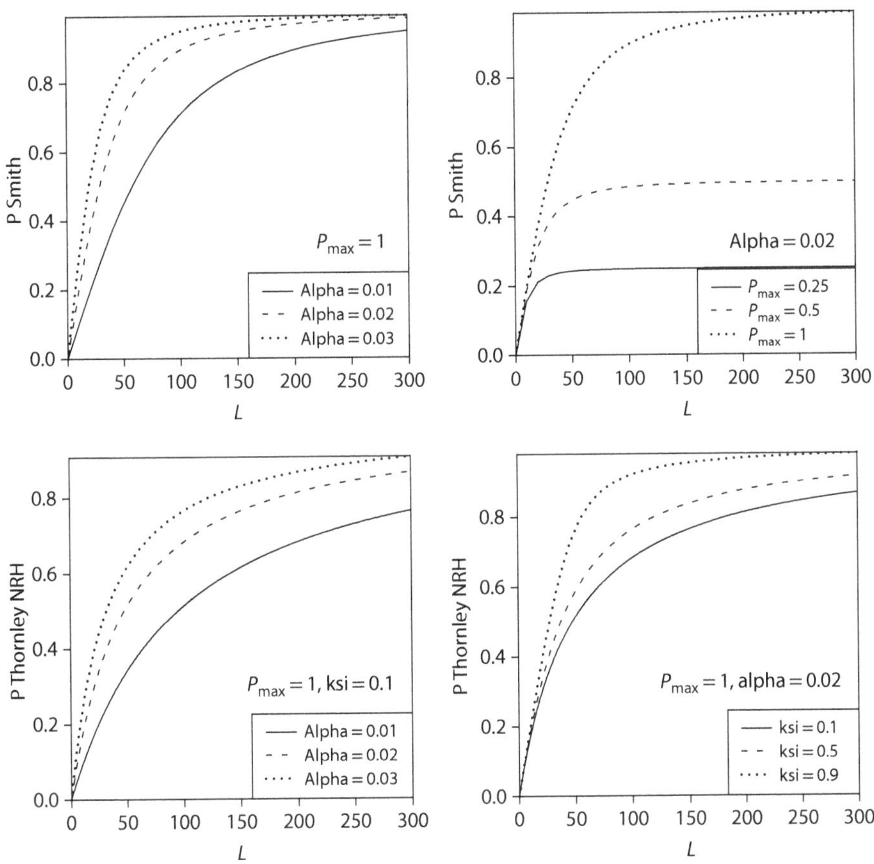

FIGURE 13.10 Smith and Thornley PP models.

To see the relationship of this function to the M3 response, start with the M3 $P = \frac{P_{max} L}{K_h + L}$, its derivative is $\frac{dP}{dL} = \frac{K_h P_{max}}{(K_h + L)^2}$, which at $L = 0$ evaluates to P_{max}/K_h which should be equal to α, defined as the slope at zero light or $L = 0$. Therefore, we can write M3 as $P = \frac{P_{max} L}{\frac{P_{max}}{\alpha} + L}$ and rearrange it as $-P(P_{max} + \alpha L) + P_{max}\alpha L = 0$ Now add a second-order term ξP^2 to get $\xi P^2 - P(P_{max} + \alpha L) + P_{max}\alpha L = 0$, which when solved for P yields Equation 13.8.

Higher efficiency of C4 plants is represented by a higher value of parameter α; possible values of α are 0.02 for C3 and 0.04 for C4.

Exercise 13.5

Which curve in the left-side panels of Figure 13.10 is closer to a C4 plant? Which is closer to a C3 plant? Which curve in the bottom right-side panel of Figure 13.10 is closer to a M3 response?

The simplest model accounting for photoinhibition is the Steele model:

$$P(L) = P_{max} \frac{L}{L_{opt}} \exp\left(1 - \frac{L}{L_{opt}}\right) \tag{13.9}$$

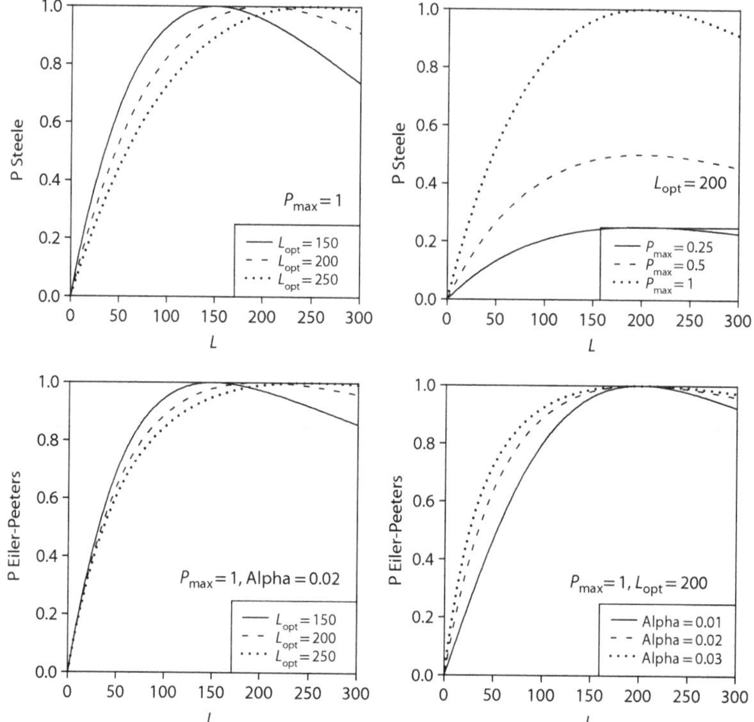

FIGURE 13.11 Steele and Eilers–Peeters PP models.

where P_{max} = maximum rate as before, and L_{opt} = optimal light level, such that when $L = L_{opt}$, we get the maximum rate $P = P_{max}$. The rate P increases for $L < L_{opt}$, but decreases when the light is above the optimum, $L > L_{opt}$. The effect of varying P_{max} and L_{opt} is illustrated in the top panels of Figure 13.11.

A more complicated model is the Eilers–Peeters model (Eilers and Peeters, 1993; Macedo et al., 1998):

$$P(L) = P_{max} \frac{L}{aL^2 + bL + c} \tag{13.10}$$

where parameters a, b, and c are calculated from the basic parameters L_{opt} and α as $a = \frac{1}{\alpha\left(L_{opt}\right)^2}$, $b = \frac{1}{P_{max}} - \frac{2}{\alpha L_{opt}}$, and $c = \frac{1}{\alpha}$. The response behavior is given in the bottom panels of Figure 13.11.

Exercise 13.6

Which model shows a faster decrease of the rate past the optimum: Steele or Eilers–Peeters?

13.7.3 Simple Diffusion: Fick's First Law

A very important component in modeling is **Fick's first law of diffusion**, in which the flow is determined by the gradient of the concentration and goes from high to low concentration (Hemond and Fechner, 1994). For one dimension,

$$F_d = -D\frac{dX}{dx} \tag{13.11}$$

F_d is the **flux density** or rate of mass transported per unit area (cross sectional) ($M \cdot L^{-2} \cdot T^{-1}$), D is the transport coefficient ($L^2 \cdot T^{-1}$), X is the concentration ($M \cdot L^{-3}$), and x is distance (L). Do not confuse concentration X with distance x. The negative sign indicates that flow is directed down the gradient (from high to low).

Exercise 13.7

Concentration of a compound varies linearly from $X = 2$ mg/liter at $x = 0$ cm to $X = 1$ mg/liter at $x = 1$ cm. The coefficient $D = 1$ cm²/s. Calculate the flux density magnitude and direction.

13.7.4 Photosynthesis Model: Ambient CO_2 Concentration

We will use a simple one-compartment model to derive a function of the photosynthesis rate versus CO_2 ambient concentration. The compartment will be inside the leaf cells at the chloroplast where CO_2 is processed. Input to the compartment is a flow rate (F_d) controlled by diffusion and the output is controlled by the photosynthesis rate (P):

$$\frac{dX}{dt} = F_d - P \tag{13.12}$$

For terrestrial plants, ambient CO_2 is controlled by its concentration in air. Fick's first law of diffusion for CO_2 influx can be written as a function of the difference in concentration:

$$F_d = \frac{C_a - X}{r} \tag{13.13}$$

F_d is the CO_2 flux density, C_a is the concentration in air [CO_2] in moles/m³, and X is the concentration in chloroplast (CO_2) in moles/m³. Units for concentration can be parts per million (ppm), where 300 ppm = 12.5 mmole/m³. We do not specify the sign in Equation 13.13 because we assume that the ambient concentration exceeds the internal concentration. In this model, $D = 1/r$ is a lumped resistance to diffusion in s/m; this lumps resistance from air and through leaf boundary layer, stomata, cell wall, intercellular space, cell, wall, and cytoplasm.

Exercise 13.8

What is the flux density when the atmospheric concentration is suddenly doubled to 600 ppm and X is at 300 ppm? Assume $r = 125$ s/m.

Assume M3 kinetics of CO_2 fixation rate at the chloroplast:

$$P = P_{max} \frac{X}{K_h + X} \tag{13.14}$$

where P is the photosynthetic rate [mmoles/m²/sec], P_{max} is the saturation in the same units as P, and K_h is the half-rate concentration in the same units as X.

Substituting Equations 13.14 and 13.13 in Equation 13.12, we get

$$\frac{dX}{dt} = \frac{C_a}{r} - X\left(\frac{P_{max}}{K_h + X} + \frac{1}{r}\right) \tag{13.15}$$

which is an ODE forced by the ambient concentration C_a. At steady state, all incoming CO_2 is fixated, this means that CO_2 influx = F_d must be equal to P, so using X^* at equilibrium, equate influx from Equation 13.13 and kinetics from Equation 13.14 to solve for X^*:

$$\frac{C_a - X^*}{r} = P_{max} \frac{X^*}{K_h + X^*} \tag{13.16}$$

After rearranging, we get a quadratic equation which has two solutions,

$$(X^*)^2 + X^*\left(rP_{\max} - C_a + K_h\right) - C_a K_h = 0 \tag{13.17}$$

and we select the positive one:

$$X^* = \frac{-(rP_{\max} - C_a + K_h) + \sqrt{(rP_{\max} - C_a + K_h)^2 + 4C_a K_h}}{2} \tag{13.18}$$

Once we have X^*, then we use it in Equation 13.14 to express P as a function of C_a and the parameters K_h, $P_{\max}$, and r:

$$P = \frac{C_a + K_h + rP_{\max}}{2r} - \frac{\sqrt{(C_a + K_h + rP_{\max})^2 - 4rC_a P_{\max}}}{2r} \tag{13.19}$$

Given the biophysical parameter r and the biochemical parameters $P_{\max}$ and K_h, we can evaluate P as a function of C_a. For simplicity use $P_{\max} = 1$ and let us see the effect of varying C_a for various values of K_h and r. In Figure 13.12, we plot the photosynthesis rate as a fraction $P/P_{\max}$.

For C4 plants, the photosynthesis rate increases faster as a function of the internal CO_2 concentration. We will use a simplified Collatz model (Cox et al., 1998) as an extension of M3 in the form of a nonrectangular hyperbola as we did with the light response. We start with Equation 13.14 and modify $P\left(\frac{P_{\max}}{k} + X\right) = P_{\max}X$, where k is the rate constant determined by the slope at $X = 0$ and is equal to $P_{\max}/K_h$. Rewrite $-P(P_{\max} + kX) + P_{\max}kX = 0$ and then add a second-order term P^2 with coefficient β to get $\beta P^2 - P(P_{\max} + kX) + P_{\max}kX = 0$, which can be solved for P:

$$P(X) = \frac{1}{2\beta}\left(kX + P_{\max} - \sqrt{\left(P_{\max} + kX\right)^2 - 4\beta kXP_{\max}}\right) \tag{13.20}$$

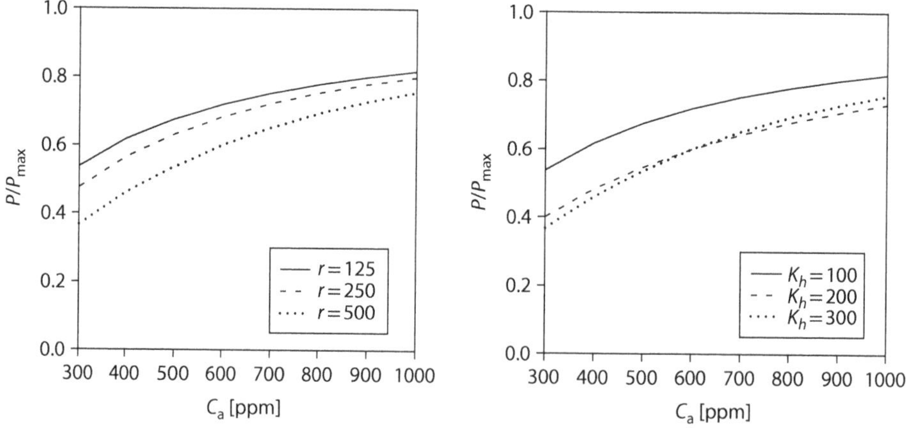

FIGURE 13.12 Effect of ambient CO_2 for various values of resistance (left panel) at fixed $K_h = 100$ ppm and for various values of half-rate coefficient (right panel) at fixed $r = 250$ s/m.

Exercise 13.9

Interpret Figure 13.12 and discuss. What is the effect of C_a on X^* and P? How is that effect modified as we increase the resistance? How is the effect modified as we increase K_h?

13.7.5 Photosynthesis Model: Temperature

Photosynthesis increases linearly with temperature for a moderate change in the values of temperature T:

$$P_{max} = pT \tag{13.21}$$

where P_{max} is the rate when CO_2 and light are not limited and p is a coefficient. However, for larger changes of temperature, we need to extend the response function; for example, the rate can decrease with temperature for higher temperature values. The effect of temperature is commonly modeled with a parabola:

$$P = \begin{cases} 4P_{max}\dfrac{(T_u - T)(T - T_l)}{(T_u - T_l)^2} & \text{when} \quad T_l < T < T_u \\ 0 & \text{when} \quad T_l \geq T \text{ or } T_u \leq T \end{cases} \tag{13.22}$$

where T_u = upper threshold and T_l = lower threshold. Note that when T is midway between T_u and T_l, P reaches P_{max}.

This model of temperature dependence is symmetric around the optimal temperature (Figure 13.13, left panel). In the figure, curves for three values of T_u are shown, all with $T_l = 0$. It gives the same rate of increase to the maximum as the rate of decrease from the optimum. For practical purposes, below or above the respective thresholds, the effect is made equal to zero. Otherwise, it would have negative values.

Exercise 13.10

Is the optimum temperature at half the difference of the sum of the two thresholds?

A simple method to describe the effect of temperature on chemical reactions is the Q_{10} approximation. Define parameter q_{10} = change in the reaction rate due to a 10°C change in T. The rate $k(T)$ at a temperature T can be calculated from the rate at a known temperature k_0 using

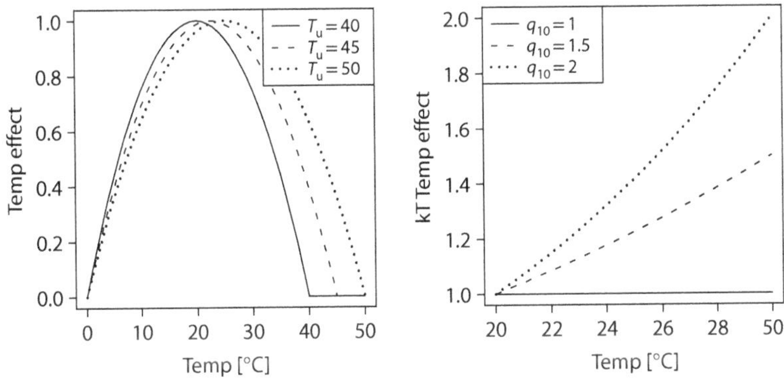

FIGURE 13.13 Left: Symmetrical effect of temperature modeled by a parabolic function. Right: Increasing rate with temperature using the q_{10} parameter.

$$k(T) = k_0 q_{10}^{T/10} \tag{13.23}$$

See the right-hand side panel of Figure 13.13, which shows an increase of 0% for $q_{10} = 1$, of 50% for $q_{10} = 1.5$, and of 100% for $q_{10} = 2$. The Q_{10} formulation is valid for relatively small changes in temperature (Keen and Spain, 1992; Swartzman and Kaluzny, 1987).

13.7.6 Productivity: Soil Water

The amount of water in the soil is a very important factor for productivity rates of terrestrial plants. Soil water depends on a balance between infiltration, evaporation, and transpiration. We will cover more details in Chapter 15. For now, just consider a simple relationship of productivity to soil water deficit. To calculate this deficit, start with the water demand or potential evapotranspiration (PET) and subtract the amount that can be supplied or the actual evapotranspiration (AET):

$$W_D = \frac{\text{PET} - \text{AET}}{\text{PET}} \tag{13.24}$$

Here, we have divided by PET to normalize as a fraction between 0 and 1. When AET << PET, then W_D is approximately 1; however, if AET is close to PET, then W_D is approximately 0.

Productivity is then modeled to decrease linearly with the relative water deficit (W_D) as the rate of change of biomass [g/m²/day] is

$$P = P_{\max}(1 - W_D) \tag{13.25}$$

when all other factors are nonlimiting. Prolonged water stress can be included by the cumulative deficit, that is, the integral of W_D over time.

Exercise 13.11

Calculate the percent relative productivity P/P_{max} for a day when PET = 5 mm/d and AET = 2 mm/d.

13.7.7 Photosynthesis Model: Nutrients

Plants make carbohydrates with CO_2 and water. However, plants also need minerals to make protein, lipids, and other biochemical compounds. A simple way of modeling this factor is to use a M3 hyperbolic response assuming no limitations due to light, temperature, or CO_2:

$$P = P_{\max} \frac{X}{K_h + X} \tag{13.26}$$

where P = productivity rate, P_{max} = maximum P for saturating X, K_h = half-rate parameter, and X = concentration of the available nutrient (mg/liter).

In this expression, we use the available form of the nutrient. For example, for nitrogen, we would use the concentration of ammonium. In addition, we would use the nutrient to which the organism studied is most responsive. For example, for algae we would use response to silica or to inorganic phosphorous if limiting.

Exercise 13.12

Calculate the percent relative productivity P/P_{max} when the limiting nutrient concentration is 50% more than the half-rate value.

13.7.8 Net Primary Productivity

Net primary productivity (NPP) rate is the gross productivity minus respiration:

$$\text{NPP} = P - R \tag{13.27}$$

where P = gross primary productivity and R = respiration.

We already have models for P as developed in the previous sections; now we need a respiration model. To include respiration, most models assume that dark and photorespiration increase linearly with temperature. The simplest assumption is

$$R = cT \tag{13.28}$$

where c is the coefficient and T is the temperature.

To obtain data for these rates in aquatic systems, we can use dissolved oxygen (DO) as an indicator and the light–dark bottle method or DO sensors read from automatic data-loggers to infer DO dynamics. The light–dark bottle method consists of: incubating a sample of water about four hours (this is IT = incubation time) during peak sunlight and then measure DO_i (DO initial, before incubation), DO_D (DO in dark bottle after incubation), DO_L (DO in light bottle after incubation). Then, calculate productivity and respiration rates [mg·C/m^3/hr] in the following manner:

$$\begin{aligned} R &= \frac{DO_i - DO_D}{IT} \\ P &= \frac{DO_L - DO_D}{IT} \\ \text{NPP} &= \frac{DO_L - DO_i}{IT} \end{aligned} \tag{13.29}$$

13.8 SECONDARY AND TERTIARY PRODUCTIVITY

In this section, we will cover some basic models of productivity for herbivores or secondary producers and tertiary producers or consumers of herbivores (Jorgensen, 1988; Kooijman, 1993; Swartzman and Kaluzny, 1987).

13.8.1 Feeding and Growth

Net productivity of a consumer is translated to growth rate or rate of change of weight with time dw/dt (g/day), which is a balance of assimilation minus respiration:

$$dw/dt = A - R \tag{13.30}$$

where A = rate of food assimilation and R = respiration rate.

Assimilation is proportional to consumption and a function of weight:

$$A = \varepsilon C w \tag{13.31}$$

where ε = assimilation efficiency in units of (1/[food units]) is a coefficient, C = per unit weight of feeding rate in units of [food units]$\cdot g \cdot d^{-1}$, and w is weight in [g]. For example, for zooplankton, the unit of food is Mcell/liter, and so the unit of C is (Mcell/liter)$\cdot g \cdot d^{-1}$.

In the same manner as PP, we can study the effect of several factors on the ingestion rate C.

Consumption rate is proportional to a power of weight:

$$C = cw^b \tag{13.32}$$

where c is the feeding rate coefficient or the feeding rate for $w = 1$ g, and b is an exponent between 0 and −1 to represent the concept that larger animals will eat a proportionally smaller fraction of their body weight than smaller animals. The power coefficient b is approximately −1/3.

Substituting Equation 13.32 into the assimilation equation for the net rate (Equation 13.31), we obtain

$$A = \varepsilon c w^{1-1/3} = \varepsilon c w^{2/3} \tag{13.33}$$

The 2/3 exponent can be interpreted as a result from the surface/volume ratio.

The q_{10} parameter is used in more complex models of temperature dependence, such as

$$C(T) = a(T)^d \exp\left(d - a(T)d\right) \tag{13.34}$$

where the function $a(T)$ and parameter d are in turn modeled as a function of temperature using three parameters T_{max}, T_{opt}, and q_{10}:

$$a(T) = \frac{T_{max} - T}{T_{max} - T_{opt}} \tag{13.35}$$

$$d = (T_{max} - T_{opt})\left[\frac{\log q_{10}}{10}\right] \tag{13.36}$$

where T_{max} = the maximum temperature tolerance (i.e., feeding ceases for temperatures above this value), T_{opt} = the optimal temperature for feeding, T = temperature, and q_{10} = the change in feeding rate due to a 10°C change in T.

The effect is asymmetrical around the optimum temperature with fast declines past the optimum, as it gets closer to the maximum value tolerated, especially when the optimum is close to the maximum; see Figure 13.14.

Exercise 13.13

What would be the form of the curve if the optimum temperature is much lower than the maximum tolerance, say Temp = 5°C in Figure 13.14?

We already discussed this as the consumer functional response. This can be modeled using the M3 function of resource density Y:

$$C = C_{max}\frac{Y}{k + Y} \tag{13.37}$$

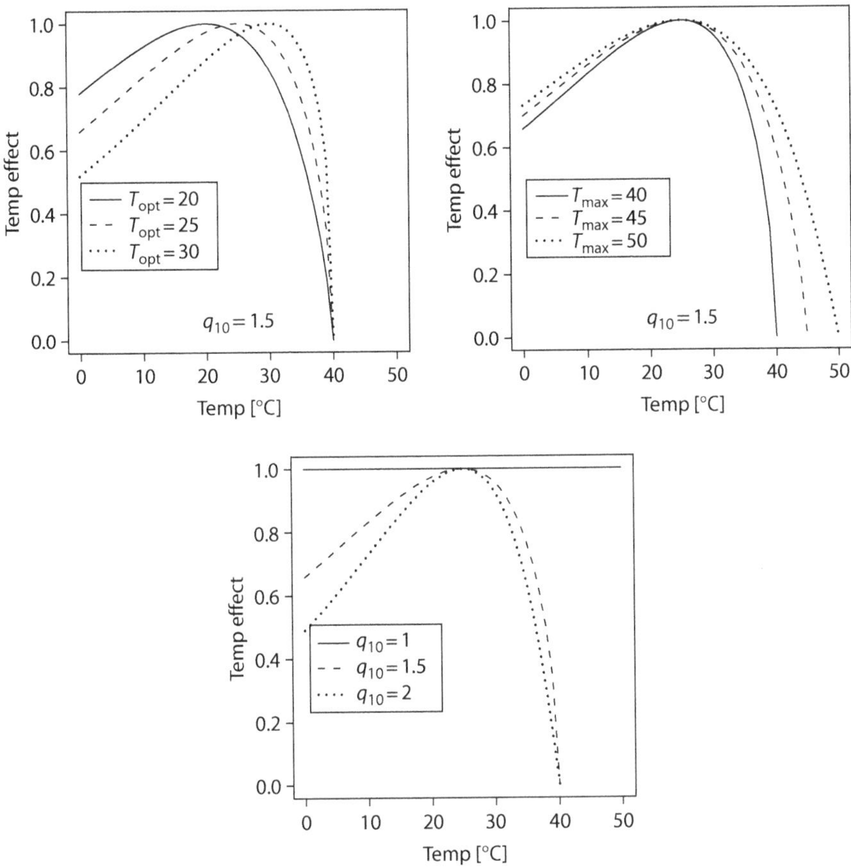

FIGURE 13.14 Asymmetrical temperature effect based on gamma-like function and Q_{10} parameter.

If there is no preference for a resource, simply add all resource densities when substituting Y in the M3 expression above. In other words, **food quantity** is the only variable of interest. If there is preference, that is, when **food quality** makes a difference, then use a selectivity s_j coefficient for resource item j, multiply s_j by the item abundance Y_j, and then divide by the sum of all abundances weighted by the selectivity coefficients. In other words,

$$C_j = \frac{s_j Y_j}{\sum_{i=1}^{n} s_i Y_i} \tag{13.38}$$

All these models for feeding and growth are simple and do not include allocation dynamics of assimilated food to either growth or reproduction.

13.9 COMPUTER SESSION

13.9.1 Nutrient Cycles: Nitrogen

We will use a model function `nut.cycle` (explained at the end of this chapter), which is compartment-based, including a matrix of linear coefficients as well as inputs and outputs. First, we will exercise it for the nitrogen cycle model. The input file is **Ncycle-inp.csv**, which contains

```
Label,Value,Units,Description
t0,0.00,yr,time zero
tf,5.00,yr,time final
dt,0.01,yr,time step
tw,0.01,yr,time to write
k13,0.2,1/yr,Nitrification coeff comp3 to comp1
k14,0.1,1/yr,Slow Mineralization comp4 to comp1
k21,0.9,1/yr,Uptake coeff comp1 to comp2
k32,0.3,1/yr,Decomposition coeff comp2 to comp3
k43,0.2,1/yr,Humification coeff comp3 to comp4
uh,-0.7,1/yr,harvest intensity coeff
uL,-0.01,1/yr,loss rate coeff
Ua,10,g/(m2yr),fixation from from air
Ur,1,g/(m2yr),leaching from rocks
Uf,5,g/(m2yr),harvest
Th,0.5,yr,period of harvest and fertilization
X1.0,1.0,g/m2,comp1 initial condition
X2.0,1.0,g/m2,comp2 initial condition
X3.0,1.0,g/m2,comp3 initial condition
X4.0,1.0,g/m2,comp4 initial condition
digX,4,None,significant digits for output file
```

Define the model list using components z and g and execute simulations with simd by varying the harvest rate uh:

```
cycle <- list(f=nut.cycle, z=nut.cycle.z, g=nut.cycle.g)
param <- list(plab="uh",pval=c(-0.5,-0.8))
t.X <- simd(cycle,"chp13/Ncycle-inp.csv",param, pdfout=T)
```

The resulting PDF file has an output including the one shown in Figure 13.4.

Exercise 13.14

Simulate the N cycle by exploring the effect of input from rocks. Use Ur = 0.0, 0.1, and 0.2. Produce graphs like the one shown in Figure 13.4.

We will now calculate the eigenvalues of the matrix of coefficients to infer dynamics behavior. First, set up the matrix from Equation 13.3 assuming $u_e = 0$.

$$\mathbf{A} = \begin{pmatrix} -k_{21} - u_a & 0 & k_{13} & k_{14} \\ k_{21} & -k_{32} - u_h & 0 & 0 \\ 0 & k_{32} & -k_{13} - k_{43} - u_e & 0 \\ 0 & 0 & k_{43} & -k_{14} - u_L \end{pmatrix}$$

$$= \begin{pmatrix} -0.9 - 10 & 0 & 0.2 & 0.1 \\ 0.9 & -0.3 + 0.7 & 0 & 0 \\ 0 & 0.3 & -0.2 - 0.2 & 0 \\ 0 & 0 & 0.2 & -0.1 + 0.01 \end{pmatrix}$$

```
> A <- matrix(c(-10.9,0,0.2,0.1, 0.9,-0.4,0,0, 0,0.3,-0.4,0,
  0,0,0.2,-0.09), ncol=4, byrow=T)
> A

> round(eigen(A)$values,3
-10.900 -0.461 -0.344 -0.085
```

We see that all eigenvalues are negative and real; the largest dominates the dynamics of the biomass change and the smallest determines the slow dynamics of accumulation in the humus compartment.

Exercise 13.15

Calculate the eigenvalues of matrix **K** in Equation 13.4 and infer the dynamics.

13.9.2 Nutrient Cycles: Phosphorous

Next, let us study the dynamics for phosphorous. The input file is **Pcycle-inp.csv**. We use the same model list `cycle` and explore the effects of input by runoff U_f.

```
param <- list(plab="Uf",pval=c(0.0,0.1,0.2))
t.X <- simd(cycle,"chp13/Pcycle-inp.csv",param, pdfout=T)
```

Simulation results include the graph shown in Figure 13.8.

Exercise 13.16

Follow the same procedure for the terrestrial example. How would you use the eigenvalues to infer the dynamic behavior? Do all the eigenvalues have negative real parts?

Exercise 13.17

What would be the consequences of dosing a pond with phosphate? Hint: To explore this question, you would increase the initial condition of X_1. What is the effect of changing the transfer rate coefficient from X_3 to X_1? Perform sensitivity analysis on this rate coefficient using values from 0 to 0.9. Why is Diff end(X_4) versus k_{13} a decreasing curve? Why is Diff end(X_1) versus k_{13} an increasing curve? Thus, closing the cycle decreases the washout of colloidal P and increases the eutrophication.

13.9.3 Ecosystems: Primary Productivity

We write the functions **PP.Smith**, **PP.Thornley**, **PP.Steele**, and **PP.Eilers** to calculate the response to sunlight by the Smith, Thornley, Steele, and Eilers–Peeters models. For example, for the Thornley model, we write the following function:

```
PP.Thornley <- function(L, Pmax, alpha, ksi){
 a <- ksi; b <- alpha*L + Pmax; c <- alpha*L*Pmax
 PP <- (1/(2*a))* (b-sqrt(b^2-4*a*c))
 return(PP)
}
```

The other functions are explained at the end of this chapter. Now use these functions to generate graphs like the ones shown in Figure 13.10. For example, for the Thornley model, the following script will produce a set of graphs varying alpha and a set varying ksi:

```
# Thornley NRH model
L <- seq(0,300,1);nL <- length(L)
Pmax=1; alpha <-c(0.01,0.02,0.03); ksi <- 0.1
np <- length(alpha)
PP <- matrix(nrow=nL,ncol=np)
for(i in 1:np){
 PP[,i] <- PP.Thornley(L,Pmax,alpha[i],ksi)
}
matplot(L, PP, type="l", ylab="P Thornley NRH",col=1)
legend("bottomright", leg=paste("alpha=",alpha), lty=c(1:3))
mtext(side=4,line=-2,text="Pmax=1, ksi=0.1")

L <- seq(0,300,1);nL <- length(L)
Pmax=1; alpha <-0.02; ksi <- c(0.1,0.5,0.9)
np <- length(sig)
PP <- matrix(nrow=nL,ncol=np)
for(i in 1:np){
 PP[,i] <- PP.Thornley(L,Pmax,alpha,ksi[i])
}
matplot(L, PP, type="l", ylab="P Thornley NRH",col=1)
legend("bottomright", leg=paste("Ksi=",ksi), lty=c(1:3))
mtext(side=4,line=-2,text="Pmax=1, alpha=0.02")
```

Exercise 13.18

Write a script to produce the graphs as in Figure 13.11 for the Eilers–Peeters model.

13.9.4 Ecosystems: Response to Temperature

The `parab` function calculates the parabolic temperature response:

```
# parabolic function
parab <- function(Temp,a,b){
 y <- 4*(b-Temp)*(Temp-a)/(b-a)^2
 if(y <0) y <- 0
 return(y)
}
```

We have truncated the negative values. The following script applies this function `parab` to explore the effect of varying the upper limit, yielding the results shown in the left-hand side of Figure 13.13.

```
Temp <- seq(0,50); nT <- length(Temp)
Tu <- c(40,45,50);np <- length(Tu)
Tl <- 0
# dimension array
y <- matrix(nrow=nT,ncol=np)
# loop for the various optimal temperatures and then for temp range
```

```
for(j in 1:np){
for(i in 1:nT) y[i,j] <- parab(Temp[i],Tl,Tu[j])
}
matplot(Temp, y, type="l", lty=1:np, col=1, ylab="Temp effect",
xlab="Temp(deg C)")
legend("topright", lty=1:np, col=1, leg=paste("Tu=",Tu),cex=0.8)
```

Exercise 13.19

Write a script that would use the function `kT.rate` to produce graphs as in the right-hand side of Figure 13.13. The function `kT.rate` is given at the end of this chapter.

Use the function `Q10` to plot temperature effects using the q_{10} parameter. This function computes Q10 effects for the values of `Tmax, Topt`, and `q10` parameters. Let us apply this function to feeding. Recall that T_{max} = the maximum temperature tolerance (above it feeding ceases), T_{opt} = the optimal temperature for feeding, and q_{10} = the change in feeding rate due to a 10°C change in temperature. Assign values of Tmax = 40, Topt = 25, and q10 = 1.5 are assigned. Vary the temperature from 0 to 50 degrees. The result is shown in Figure 13.13.

```
Temp <- seq(0,50); nT <- length(Temp)
Topt= 25; Tmax <- 40; q10 <- 1.5
fT <- Q10(Temp, param=c(Topt,Tmax,q10))
plot(Temp, fT, type="l", col=1,ylab="Temp effect", xlab="Temp(deg C)")
```

Now we use this basic function to explore the effect of varying each one of the three parameters. Use `Q10.mruns`. This program varies all three parameters one at a time and produces graphics for each parameter.

```
Temp <- seq(0,50); nT <- length(Temp)
# param Topt, Tmax,q10
param.nom <- c(25,40,1.5)
Topt.sens <- c(20,25,30)
Tmax.sens <- c(40,45,50)
q10.sens <- c(1,1.5,2)
np <- length(q10)
test <- Q10.mruns(Temp,param.nom,Topt.sens,Tmax.sens,q10.sens)
```

The results are shown in Figure 13.14.

Exercise 13.20

Use the Q10.mruns function to see how the Q_{10} effect varies when q_{10} = 1.1, 1.2, and 1.3. Discuss.

13.10 SEEM FUNCTIONS EXPLAINED

13.10.1 Function nut.cycle

This file has a function `f` with the ODE resulting from the multiplication of the matrix `k` of coefficients by the `X` vector and the addition of a proportional u input/output (e.g., donor) component and the independent input/output forcing terms `U`.

```
nut.cycle <- function(t,p,x){
 k <- matrix(ncol=4,nrow=4) # transfers
 k[,] <- 0
 k[1,3] <- p[1]; k[1,4] <- p[2]; k[2,1] <- p[3]
 k[3,2] <- p[4]; k[4,3] <- p[5]
 for(i in 1:4) k[i,i] <- -sum(k[,i])
 U <- c(sum(p[8:9]),0,0,0) # abs input
 uc <- c(0,0,0,p[7]) # prop input
 dX <- c(k%*%x + uc*x +U)
 return(dX)
}
nut.cycle.z <- function(t,p,x){
 # periodic impulse train
 tz <- seq(t[1],t[length(t)],p[11])
 return(tz)
}
nut.cycle.g <- function(t,p,x,tz){
 u <- rep(0,4)
 for(j in 1:length(tz))
 if(abs(t - tz[j])<0.00001){
 u[2] <- p[6]*x[2]
 u[1] <- p[10]
}
return(u)
}
```

13.10.2 Functions to Calculate Photosynthesis Response to Sunlight

The four functions PP.Smith, PP.Thornley, PP.Steele, and PP.Eilers calculate the response to sunlight by the Smith, Thornley, Steele, and Eilers–Peeters models. They are straightforward and self-explanatory.

```
PP.Smith <- function(L, Pmax, alpha){
 PP <- Pmax*alpha*L/sqrt(Pmax^2 +(alpha*L)^2)
 return(PP)
}
PP.Thornley <- function(L, Pmax, alpha, ksi){
 a <- ksi; b <- alpha*L + Pmax; c <- alpha*L*Pmax
 PP <- (1/(2*a))* (b-sqrt(b^2-4*a*c))
 return(PP)
}
PP.Steele <- function(L, Pmax, Lopt){
 PP <- Pmax*(L/Lopt)*exp(1-L/Lopt)
 return(PP)
}
PP.Eilers <- function(L, Pmax, alpha, Lopt){
 a <- 1/(alpha*Lopt^2); b <- 1/Pmax-2/(alpha*Lopt); c <- 1/alpha
 PP <- L/(a*L^2+b*L+c)
 return(PP)
}
```

13.10.3 Functions to Calculate Temperature Response

The three functions parab, kT.rate, and Q10, calculate the response to temperature using the parabolic, kT by q10, and the gamma-like q10. They are straightforward and self-explanatory.

```
# parabolic function
parab <- function(Temp,a,b){
 y <- 4*(b-Temp)*(Temp-a)/(b-a)^2
 if(y <0) y <- 0
 return(y)
}
# rate q10
kT.rate <- function(Temp,k0,q10){
   kT <- k0*q10^(Temp/10)
   return(kT)
}
# Q10 function

Q10 <- function(Temp, param=c(Topt,Tmax,q10)){

Topt <- param[1];Tmax<-param[2];q10<- param[3]
 d <- (Tmax-Topt)*(log(q10)/10)
 a <- (Tmax - Temp)/(Tmax-Topt)
 y <- a^d*exp(d-a*d)

return(y)
}
```

```
Q10.mruns <- function(Temp, param.nom, Topt.sens, Tmax.sens, q10.sens)
{

nT <- length(Temp)
mat<- matrix(1:4,2,2,byrow=T)
layout(mat,c(3.5,3.5),c(3.5,3.5),res=TRUE)
par(mar=c(4,4,1,.5), xaxs="r", yaxs="r")

# loop to change Topt, Tmax and q10
for(j in 1:3){
# determine parameter
  if(j==1) p <- Topt.sens
  if(j==2) p <- Tmax.sens
  if(j==3) p <- q10.sens
# dimension arrays according to how many temp and param values
np <- length(p); ncalc <- nT*np
X<- matrix(nrow=nT,ncol=np)
# reset to nominal
Topt<- param.nom[1]; Tmax <- param.nom[2]; q10 <- param.nom[3]
# set varying parameter and calculate
for (i in 1:np){
  if(j==1) Topt <- p[i]
  if(j==2) Tmax <- p[i]
  if(j==3) q10 <- p[i]
  X[,i] <- Q10(Temp, param=c(Topt, Tmax, q10))[,2]
}
leg.par <- c("Topt", "Tmax","q10")
```

```
nom.par <- c(rep(paste("q10=",param.nom[3]),2),"")
matplot(Temp, X, type="l", lty=1:np, col=1, ylab="Temp effect",
xlab="Temp(deg C)")
legend(0,0.5, lty=1:np, col=1, leg=paste(leg.par[j],"=",p))
mtext(side=1,line=-2,text=nom.par[j])
 if(j==1) X.Topt <- X
 if(j==2) X.Tmax <- X
 if(j==3) X.q10 <- X
} # end of j loop
return(list(X.Topt=X.Topt, X.Tmax=X.Tmax, X.q10=X.q10))
} # end function
```

14 Aquatic Ecosystems

In this chapter, we use the elements of nutrient cycles and energy flow to develop aquatic ecosystem models. First, we discuss how to model environmental variables, and these will hook to the variables of concentration, light, temperature, and so on in the elemental models studied in Chapter 13.

14.1 SOLAR RADIATION

Incoming solar radiation to Earth is 1.361 kW m^{-2} measured in units of power (kW) per unit area. This is the value at the top of the atmosphere or the extraterrestrial radiation. This value varies slightly during the year according to the distance between the sun and the earth. For our purposes, we will consider the extraterrestrial radiation a constant.

Here, we need to remind ourselves that power is a rate of energy and that 1 W is 1 J s^{-1}. These units can be converted to megajoules per day or calories per day in the following manner: 1 W m^{-2} = 1 J s^{-1} m^{-2} = 1 J s^{-1} m^{-2} (3600 × 24) s d^{-1} = 86,400 J m^{-2} d^{-1} = 0.086 MJ m^{-2} d^{-1}. Also, 1 W m^{-2} = 1 J $(4.18)^{-1}$ cal J^{-1} s^{-1} m^{-2} (3600 × 24) s d^{-1} = 20,669 cal m^{-2} d^{-1} = 2.067 cal cm^{-2} d^{-1}.

Exercise 14.1

Convert solar radiation as energy (kWh) per unit area and convert to joules per unit area and calories per unit area in cm^{-2}.

Earth's axis is tilted 23.45° with respect to the ecliptic plane (the plane determined by the movement around the sun). Sun declination, δ, varies during the year in the range –23.45° to 23.45°, and thus, the sun's elevation angle, β, varies with latitude and declination (Figure 14.1). Sun declination, $\delta(n)$, in degrees for a day, n, of the year is approximated by

$$\delta(n) = 23.45 \times \sin\left(\frac{2\pi}{365}(n-81)\right) \tag{14.1}$$

where n is the day of the year counted from January 1. Day 81 corresponds to the vernal (spring) equinox in the Northern Hemisphere, which is March 21 in a nonleap year. Note that 31 + 28 + 21 = 80 days. It plays the role of a phase change for the sine wave (Keen and Spain, 1992). The maximum solar elevation angle, β_{max}, which occurs at solar noon, varies with latitude, α, and declination in the following manner:

$$\beta_{max}(n,\alpha) = \pm\delta(n) + (90° - \varphi) \tag{14.2}$$

where declination has a positive sign for northern latitudes ($\varphi > 0$) and a negative sign for southern latitudes ($\varphi < 0$). Figure 14.2 shows declination during the year (top panel) and maximum sun elevation for several values of latitude, two in the Northern Hemisphere, and two in the Southern Hemisphere (bottom panel).

During the day, the sun's elevation angle goes from zero at sunrise to the maximum at solar noon and then decreases back to zero at sunset according to

$$\beta(H,n) = \sin^{-1}\left(\cos\varphi\cos\delta(n)\cos H + \sin\varphi\sin\delta(n)\right) \tag{14.3}$$

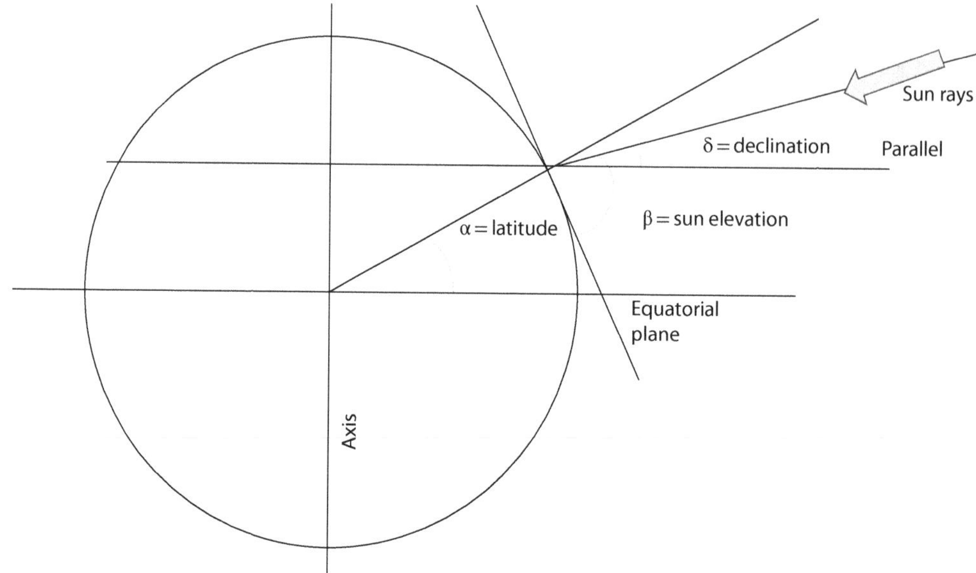

FIGURE 14.1 Angular variables: declination, latitude, and sun elevation angle.

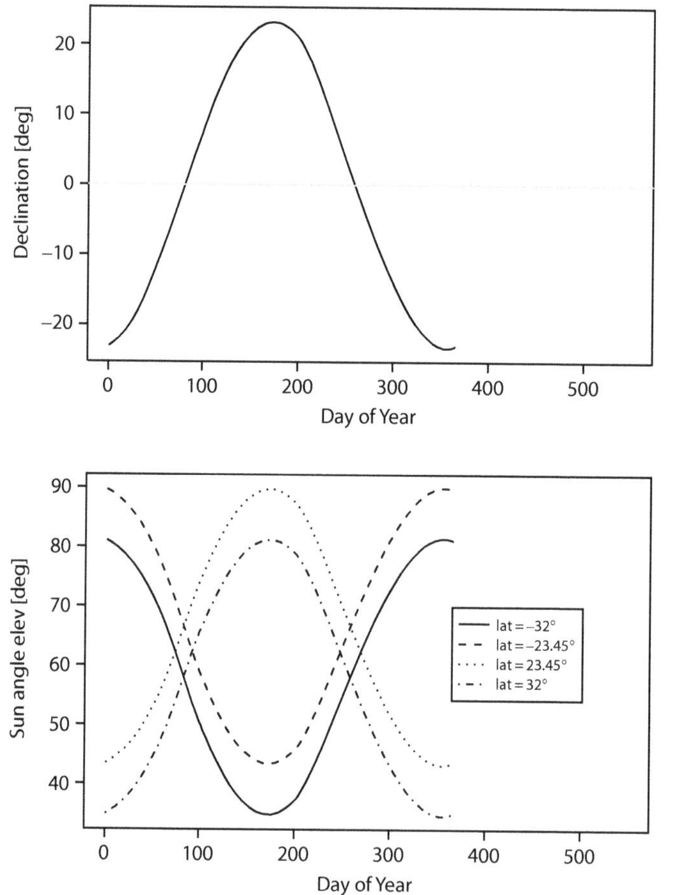

FIGURE 14.2 Declination and sun elevation angle for several latitude angles.

where H is the hour angle and equal to 15° per hour. $H = 15 \times h$, and hours h are with respect to solar noon, negative before and positive after. A simple approximate model of solar radiation during the day is to multiply the elevation angle by a factor determined from the ratio of annual average of maximum daily radiation, L_m, and the annual average of maximum sun elevation. The annual average sun elevation is for zero declination and therefore equal to $90 - \varphi$:

$$L(h,n) = \frac{L_m}{90 - \varphi} \beta(15 \times h, n) \tag{14.4}$$

Figure 14.3 shows the sun radiation during the day by latitude (top) and day of the year (bottom). We see that for higher latitudes, day length is shorter on a winter day and longer on a summer day.

When we are interested in the variation of solar radiation, $L(n)$, during the year, we can use

$$L(n) = L_m + L_a \sin\left(\frac{2\pi}{365}(n - 81)\right) \tag{14.5}$$

where L_a is the amplitude of the oscillations and can be calculated as $L_a = \pm \frac{L_m \times 23.45}{90 - \varphi}$, positive for northern latitudes and negative for southern latitudes. Examples for latitude 32.90° and two values of L_m are shown in Figure 14.4.

Exercise 14.2

Sketch a graph of $L(h)$ as a function of h from $h = 0$ to $h = 365$ days when the latitude is −32.90°. Use $L_m = 600$ W m^{-2}.

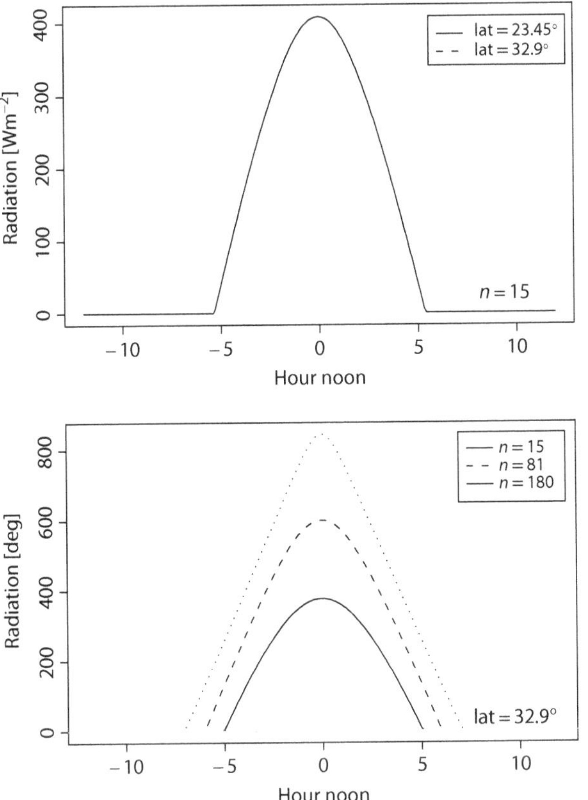

FIGURE 14.3 Solar radiation during the day by latitude and day of the year.

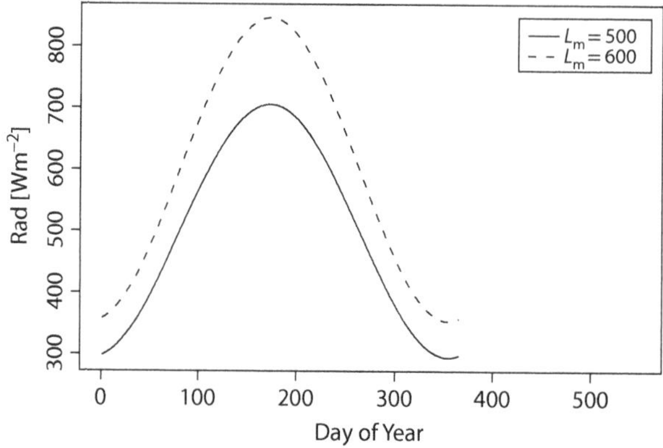

FIGURE 14.4 Sinusoidal model of maximum daily solar radiation as a function of time.

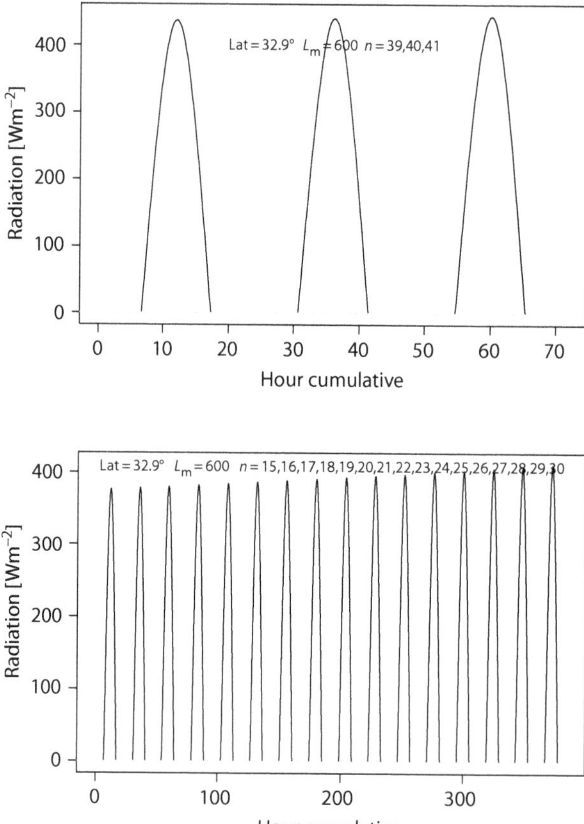

FIGURE 14.5 Solar radiation during the day for a sequence of consecutive days. Top panel: 3 days or 72 h. Bottom panel: 15 days in January illustrating modulation of the maximum value in the day with the day number.

Hourly radiation for a sequence of consecutive days is calculated by using Equation 14.4. For example, for latitude 32.90°, see Figure 14.5. The bottom panel illustrates the modulation of the maximum sunlight in a day with the day of the year. To these simple models (Equations 14.4 and 14.5), we add random variations, $L_r(h)$ and $L_r(n)$, to account for variability due to atmospheric effects and cloudiness. For example:

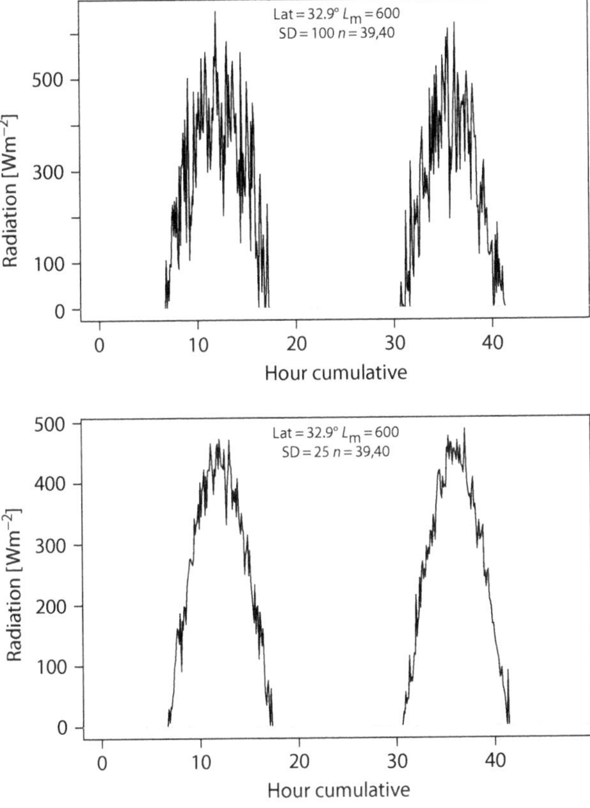

FIGURE 14.6 Solar radiation with variability during the day for two consecutive days. Top panel: standard deviation 100. Bottom panel: standard deviation 25.

$$L(h,n) = \frac{L_m}{90-\varphi}\beta(15\times t,n) + L_r(h) \tag{14.6}$$

where $L_r(h)$ is a sample at hour h from a normal distribution with mean 0 and standard deviation σ_r. The impact of the magnitude of the standard deviation is illustrated in Figure 14.6, where the sunlight in the top panel has four times the variability of the sunlight shown in the bottom panel.

These simple models would not hold for locations where sunlight intensity is greatly affected by other factors such as topography, cloudiness, and fog. Some generic models work well for many locations and are often used to drive ecosystem models (Nikolov and Zeller, 1992).

14.2 LIGHT AS A FUNCTION OF DEPTH

Light intensity decreases when it goes through water. The Beer–Lambert law states that light intensity, L, is attenuated exponentially according to

$$L(z) = L_s \exp(-kz) \tag{14.7}$$

where z is the depth, L_s is the water subsurface level, and k is the attenuation or extinction coefficient in m^{-1} (Figure 14.7). The coefficient k includes absorption by organic and inorganic

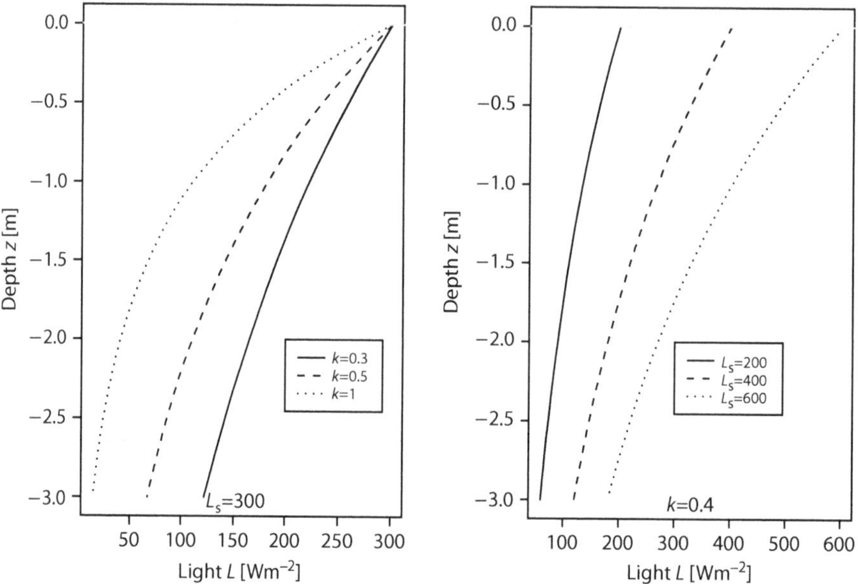

FIGURE 14.7 Exponential attenuation of sunlight in water according to the Beer–Lambert law.

compounds as well as by photosynthesis. The attenuation coefficient is estimated by regression from values of light intensity going downward (i.e., downwelling) at various values of depth. Because this is an exponential model, the process is similar to the regression method explained in Chapter 3.

A Secchi disk is a simple device to measure light attenuation in water. It consists of a 20–30 cm disk painted with a geometric pattern; it is submersed in water until it is no longer visible. The depth at which this occurs is recorded as the Secchi depth, z_s. Then, an empirical relationship between k and the inverse of z_s can be developed for specific water quality situations. This provides a quick estimation of k. For example, a relationship between the extinction coefficient and the Secchi depth is $k \sim a/z_s^b$. Values of a range from 1 to 2 and values of b from 0.75 to 1, when z_s is given in meters. For example, $a = 1.46$ and $b = 0.95$ in some marine conditions (Walker, 1982) would yield $k \sim 0.4$ for a z_s of 4 m.

Exercise 14.3

Using your knowledge of exponential models and the Beer–Lambert law, derive a linear ODE model for the attenuation of light in a water column in such a way that the solution will be the Beer–Lambert law. The derivative should be with respect to depth z.

Exercise 14.4

Assume $a = 1.56$ and $b = 0.75$. What is the extinction coefficient k for a Secchi depth of 1 m? What is the extinction coefficient for a Secchi depth of 2 m? What fraction of the subsurface light is present at 0.5 m for both cases?

Therefore, the light level reaching the algae cells depends on the vertical position (depth) of cells, so P depends on z. For example, have L as a function of depth z as shown in the top-left panel of Figure 14.8. The rate will vary with L as shown in the top-right panel of Figure 14.8 according to the

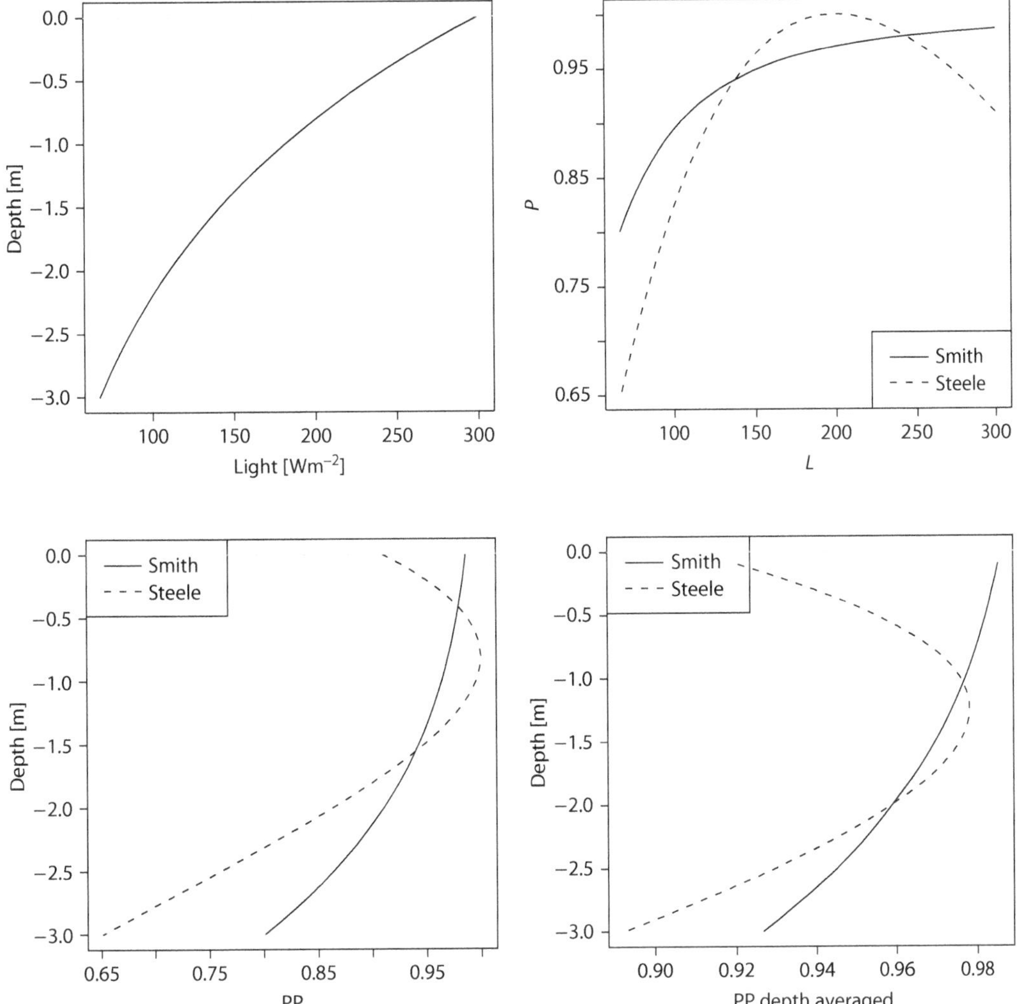

FIGURE 14.8 Sunlight as a function of depth (top left), productivity as a function of sunlight for Smith and Steele models (top right panel), productivity as a function of depth for both models (bottom left panel), and depth-average productivity as a function of depth for both models (bottom right panel).

Smith and Steele models, and combining both top panels, we obtain the variation of P with z shown in the bottom-left panel of Figure 14.8.

Note that when using the Smith model, $P(z)$ has a maximum at the water subsurface (Ginot and Herve, 1994). When using the Steele model, $P(z)$ has a maximum at some depth different than the surface; this depth depends on k (Swartzman and Kaluzny, 1987). See the lower left panel of Figure 14.8.

To get an average of primary productivity for the entire water column, it is necessary to integrate over the entire water column from the surface to the bottom of the euphotic zone. For example, denote the depth or the depth of the euphotic zone as z_e, then,

$$P(z_e) = \frac{\int_0^{z_e} P(z)\mathrm{d}z}{z_e} \tag{14.8}$$

The result depends on whether the Smith model (no photoinhibition) or the Steele model (photoinhibition) is used. Using the Smith model, we obtain the following closed expression (Ginot and Herve, 1994):

$$P(z_e) = \frac{P_{max}}{kz_e} \ln \left(\frac{L_s + \sqrt{\left(\frac{P_{max}}{\alpha}\right)^2 + L_s^2}}{L(z_e) + \sqrt{\left(\frac{P_{max}}{\alpha}\right)^2 + L(z_e)^2}} \right) \tag{14.9}$$

The results can be appreciated graphically in the lower right panel of Figure 14.8.

Using the Steele model, we obtain (Swartzman and Kaluzny, 1987)

$$P(z_e) = \frac{P_{max}\exp(1)}{kz_e} \left(\exp\left(\frac{-L(z_e)}{L_{opt}} \right) - \exp\left(\frac{-L_s}{L_{opt}} \right) \right) \tag{14.10}$$

which is displayed in the lower right panel of Figure 14.8.

Exercise 14.5

Would productivity always be higher at the top of the water column? Why? Does depth-averaged productivity always depend on the light extinction coefficient?

14.3 DISSOLVED OXYGEN AND PRIMARY PRODUCTIVITY

Of great interest in aquatic systems is the concentration of DO in water. Primary production rate is directly related to O_2 production rate because for each atom of C that is assimilated, two atoms of oxygen are released. DO rhythms during the day are affected by changes in sunlight levels. DO dynamics result from the balance of diffusion from air, production by algae, and loss by respiration (Ginot and Herve, 1994). Denote DO concentration as X, then,

$$\frac{dX}{dt} = D(X_s - X) + P - R \tag{14.11}$$

where X_s is the DO at saturation, D is the diffusion constant, R is the respiration rate, and P is the primary productivity. The inflow rate from air is proportional to the deficit $X_s - X$. The DO at saturation varies with water temperature, T_w, altitude, and salinity. At sea level for freshwater, saturation is 14.6 mg/liter at 0°C and decreases to 8.6 mg/liter at 25°C. The P is given, for example, by the depth-averaged productivity calculated in Section 14.2. For example, using the Smith model

$$\frac{dX}{dt} = D(X_s - X) + \frac{P_{max}}{kz_e} \ln \left(\frac{L_s + \sqrt{\left(\frac{P_{max}}{\alpha}\right)^2 + L_s^2}}{L(z_e) + \sqrt{\left(\frac{P_{max}}{\alpha}\right)^2 + L(z_e)^2}} \right) - R \tag{14.12}$$

The light at the surface is modeled as discussed in Section 14.1. Assuming constant cool water temperature and sunny days in winter, we can simulate Equation 14.12 for a 3-day period to obtain the results shown in Figure 14.9 corresponding to shallow turbid water. We have used the following parameter values: $P_{max} = 0.4$, $\alpha = 0.02$, $k = 1.27\ m^{-1}$, $z_e = 0.24$ m, $D = 0.01$, and $T_w = 15$°C constant ($X_s = 11.0$ mg/liter). The sunlight at the surface is modeled using Equation 14.6 with the following parameter values: $L_m = 600$, latitude $\varphi = 32.90°$, and $\sigma_r = 0$, starting at day $n = 32$. Three simulations are shown by varying respiration R by ±10% with respect to nominal $R = 0.2$. We can see

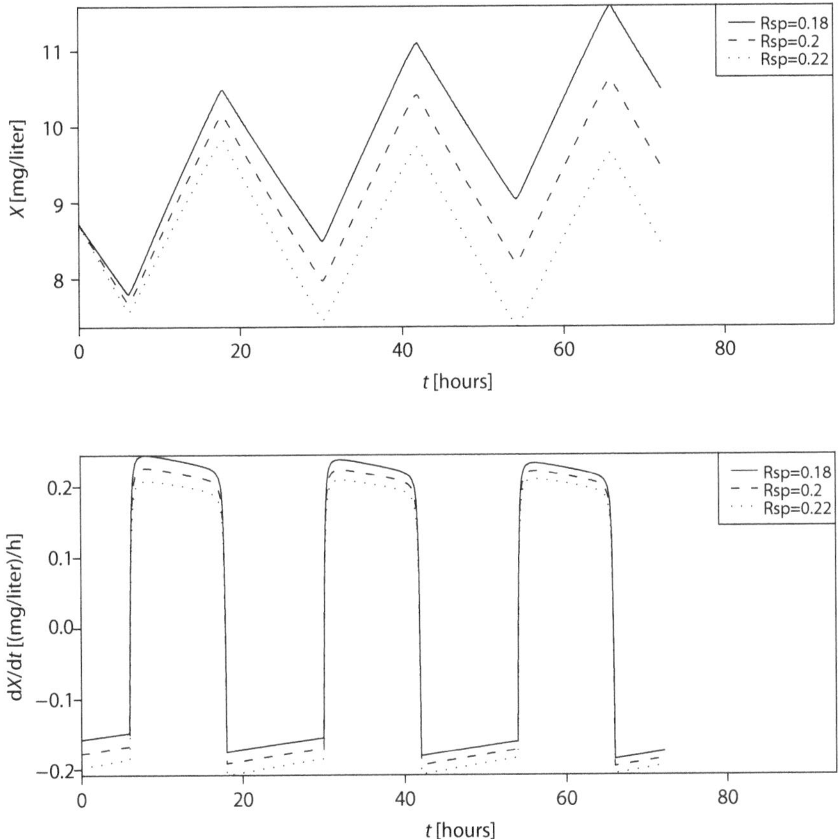

FIGURE 14.9 Dissolved oxygen over 3 days (72 h) for several values of respiration in cool water.

how the DO oscillates from a minimum to a maximum every day, and the rate of DO change has a pulsing behavior. It has low values during night hours when only diffusion and respiration count and high values during day hours when primary production is active. The overall trend is to increase or decrease according to the loss by respiration, R.

14.4 RIVER EUTROPHICATION

As an example of how to form a whole ecosystem model, we now want to couple energy flow and nutrient dynamics as a model for nutrient–phytoplankton–zooplankton in a segment or reach of a river (Swartzman and Kaluzny, 1987). This example uses very simple hydrodynamics: constant water velocity and constant cross section, that is, constant water flow rates. Agricultural and urban runoff leads to nonpoint source pollution, that is, contaminants carried by runoff waters from all over the watershed contributing to the river. Excess nutrient loads lead to eutrophication, conditioned by river flow and volume.

The model is a system of three ODEs to describe the net rate of change of densities (for algae and zooplankton) or concentrations (for nitrogen) in mg per liter. The three state variables are X_1 = nutrients (inorganic nitrogen) mg nitrogen/liter, X_2 = algae mg C/liter, and X_3 = zooplankton mg C/liter. Each equation uses a common unit: biomass as carbon (in mg) for unit consistency.

For a river segment, we balance all rates:

$$\frac{dX_1}{dt} = U_1 - U_2(X_1)X_2 + \left(I_1 - O_1(X_1)\right) \tag{14.13}$$

where U_1 is the loading by discharge, U_2 is the uptake by algae, I_1 is the inflow of nitrogen from upstream segment, and O_1 is the outflow of nitrogen to the downstream segment.

$$\frac{dX_2}{dt} = G_2(X_1, X_2) - C_2(X_2, X_3) + (I_2 - O_2(X_2)) \tag{14.14}$$

where G_2 is the growth rate of algae by primary productivity, C_2 is the consumption by zooplankton, I_2 is the inflow of algae from upstream segment, and O_2 is the outflow of algae to the downstream segment.

$$\frac{dX_3}{dt} = G_3(X_2, X_3) + (I_3 - O_3(X_3)) \tag{14.15}$$

where G_3 is the growth rate secondary productivity, I_3 is the inflow of zooplankton from the upstream segment, and O_3 is the outflow of zooplankton to the downstream segment. In Equation 14.15, we ignored the consumption by fish. We have already studied how to model some of these rates.

The simplest formulation for loading is the discharge rate divided by the volume of river reach, that is, U_1 is given as

$$U_1 = \frac{W}{V} \tag{14.16}$$

where W is the mass discharge rate = mass/time = mg/d and V is the volume of river reach in liters. This is a very simple formulation for pollutant loading that ignores the details of the mixing process. The load is mixed homogeneously and instantaneously in the river segment.

A typical form for algae uptake of nutrients is the M3 kinetics on the limiting nutrient, assumed to be inorganic nitrogen in this example. This leads to a hyperbolic response

$$U_2 = X_2 P_{max} \frac{X_1}{K_h + X_1} \tag{14.17}$$

where P_{max} is the maximum PP for saturating X_1 (assuming no limitations due to light), and K_h is the half-rate parameter. Growth rate of algae is approximated by the Smith equation, depth-averaged, extinction coefficient affected by X_2, and using the chlorophyll to carbon ratio. We use net rate, obtained by subtracting respiration from photosynthesis, where respiration is proportional to temperature.

Next, the zooplankton growth, G_3, is calculated using the M3 response on algae concentration, making it temperature dependent using Q_{10} as studied before, and scaling by assimilation efficiency. Use the net rate by subtracting respiration.

All inflows, I_i, are calculated from

$$I_i = \frac{Q}{V} X_i^u \tag{14.18}$$

where Q is the water flow rate (liter/s), V is the volume, and X_i^u is the concentration of X_i at the start of the segment. This is a **boundary condition** for the reach. The outflow of X_i is

$$O_i = \frac{Q}{V} X_i \tag{14.19}$$

Note that X_i determines the **boundary condition** for the downstream segment.

The ratio Q/V has units of T^{-1}. It is a per unit of time coefficient multiplying the gradient of X_i in the reach or per unit volume flow rate. The inverse of V/Q is the travel time or turnover time in the reach.

First, to better understand the effect of flow and load on nutrient concentration, imagine that algae and zooplankton are absent so that the model reduces to

$$\frac{dX_1}{dt} = \frac{W}{V} + \frac{Q}{V} X_1^u - \frac{Q}{V} X_1 \tag{14.20}$$

Exercise 14.6

Consider Equation 14.20 with $V = 1000$ m^3, $Q = 1000$ m^3 s^{-1}. The concentration upstream is $X_i^u =$ 1 mg/liter. For simplicity, imagine that the loading rate is zero ($W = 0$). (a) Calculate the travel time; (b) calculate nutrient concentration in the segment at equilibrium; and (c) calculate the outflow of nutrients from the segment at equilibrium.

Exercise 14.7

What is the travel time in a river reach with $V = 10^6$ m^3 and $Q = 1000$ m^3 s^{-1}? What is the outflow rate of nutrients if the concentration in the reach is 10 mg/liter?

Exercise 14.8

Explain the effect of volume dilution on water quality. Hint: Loading U_1 is lower for larger volume V.

Figure 14.10 shows an example of dynamics for $V = 27{,}500$ m^3, $W = 660$ g/s, $X_1^u = 0$ mg/liter, and two values of flow $Q = 0$ and 0.06 m^3 s^{-1}. We can see that with no flow, nitrogen increases linearly.

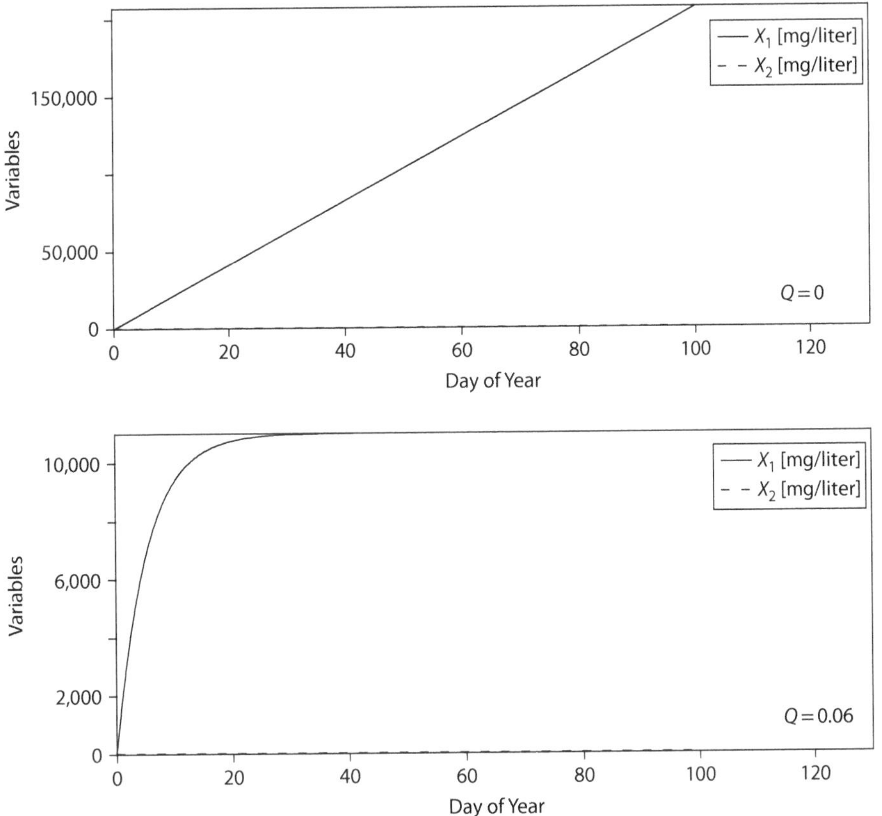

FIGURE 14.10 Nitrogen for two values of flow $Q = 0$ and 0.06 m^3 s^{-1}.

This condition is like a pond with no washout. As we add flow, the nitrogen concentration starts to level off due to washout and reaches a steady state that corresponds to $W/Q = 660/0.06 = 11{,}000$ mg/liter. The exponential transient is controlled by the ratio Q/V, which acts as a depuration rate. It has a value of $0.06/27{,}500\ \mathrm{s}^{-1} = 0.18\ \mathrm{d}^{-1}$. Its inverse is the travel time or 5.3 days.

Now, when we include algae with $P_{max} = 0.1$ (mg/liter)/d, $K_h = 1$ mg/liter, $X_2(0) = 10$ mg/liter, and $X_2^u = 10$ mg/liter, the dynamic behavior of nitrogen changes for no flow but remains similar for low flow (Figure 14.11). We can see that with no flow, nitrogen increases linearly until it starts decreasing due to algae uptake. The algae population increases exponentially. At low flow but lower loads, algae growth stabilizes (Figure 14.12). At very high flow, nutrients and algae achieve a very low steady state quickly. Figures are not shown for the sake of space.

Now, we add zooplankton and environmental drivers to work with a more complete model. An example of dynamic behavior during a year is shown in Figures 14.13 and 14.14. We can see that temperature and sunlight are modeled using sine waves parameterized by a mean, amplitude, and delay. Flow is modeled by a triangular function parameterized by onset, peak, and duration of high water flow season. This simulation is for a peak flow of 4,40,000 liter/s or 440 $\mathrm{m}^3\ \mathrm{s}^{-1}$, and the onset of peak flow is day 220.

We can see in Figure 14.13 that algae grow exponentially after the sharp increase of nutrients and then crash as the nutrients are depleted. A small peak of zooplankton follows and then there is a recovery of algae to a lower steady-state value. The effect of high flow commencing in day 220 is clear as the algae concentration decreases during the period of high flow. The values of various variables cannot be compared in magnitude because we have not scaled them according to the ratio of nitrogen to carbon or phosphorous to carbon.

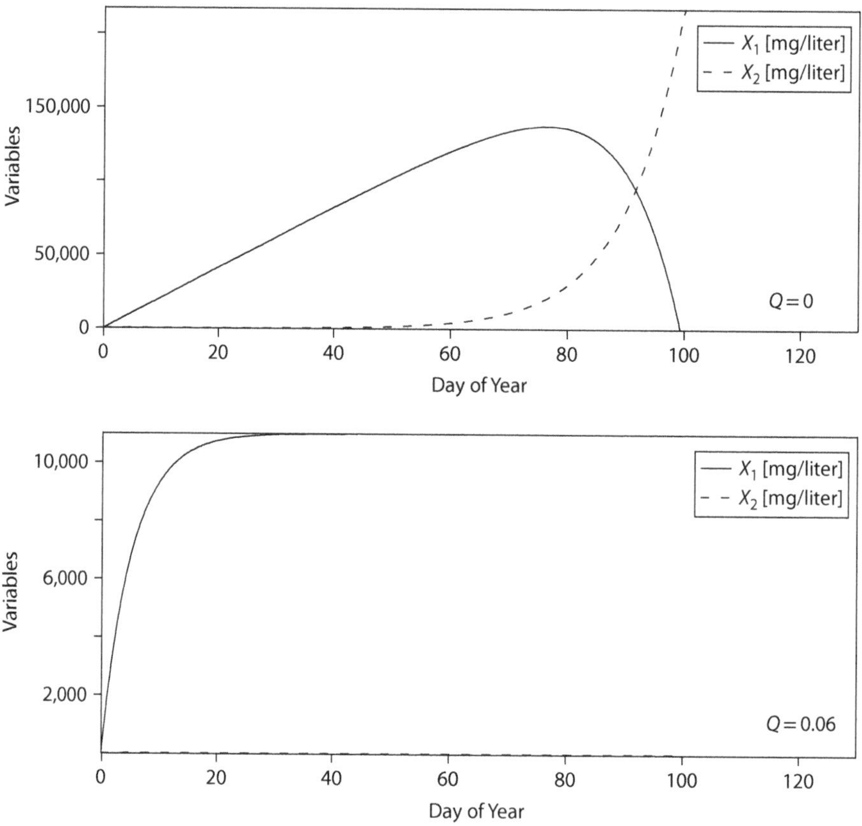

FIGURE 14.11 Nitrogen and algae for two values of flow $Q = 0$ and $0.06\ \mathrm{m}^3\ \mathrm{s}^{-1}$.

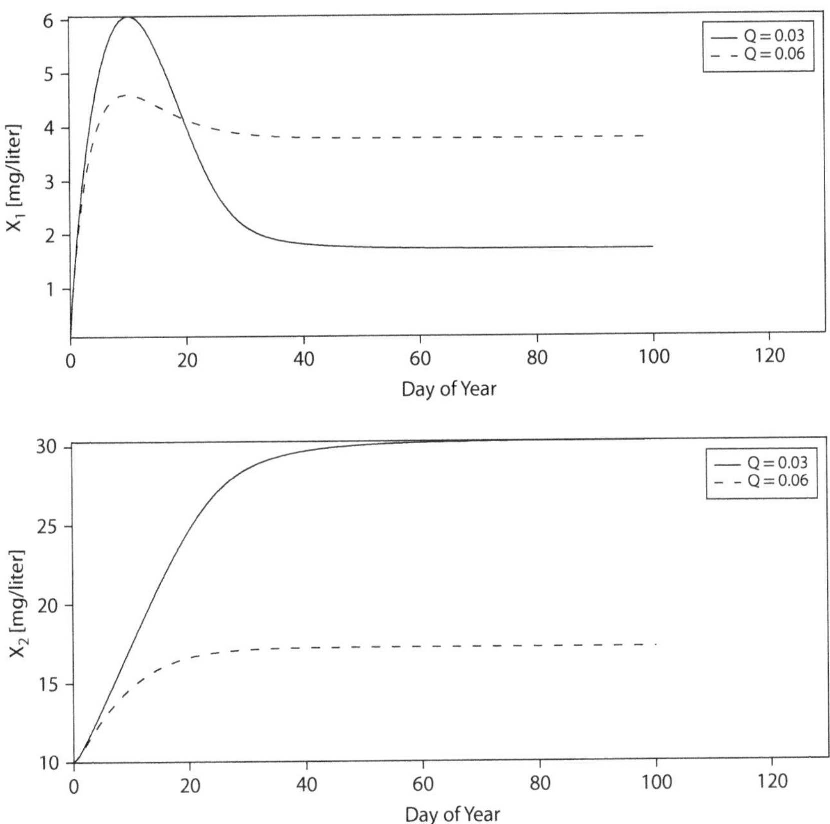

FIGURE 14.12 Nitrogen and algae for low loading and two values of flow $Q = 0.03$ and $0.06\ m^3\ s^{-1}$.

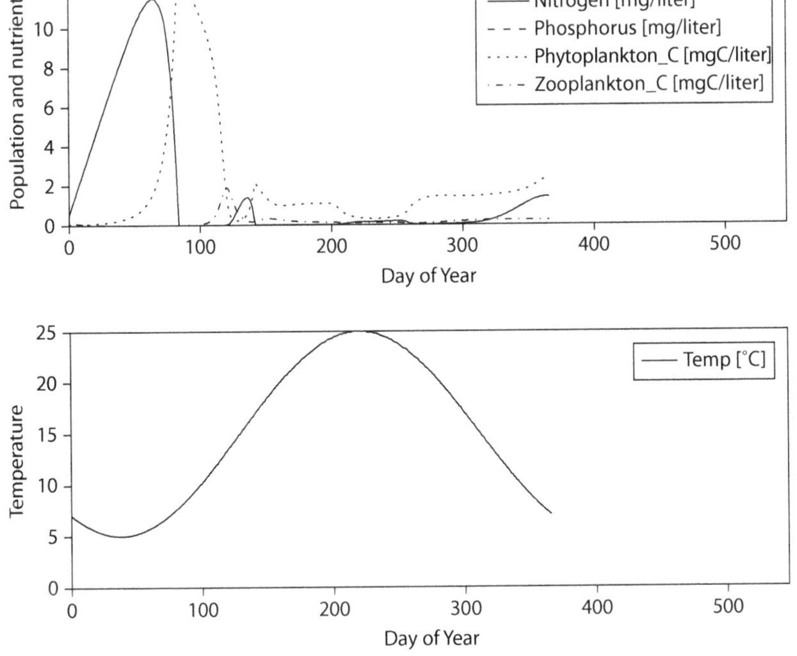

FIGURE 14.13 Nutrients, algae, and zooplankton (top) and temperature (bottom).

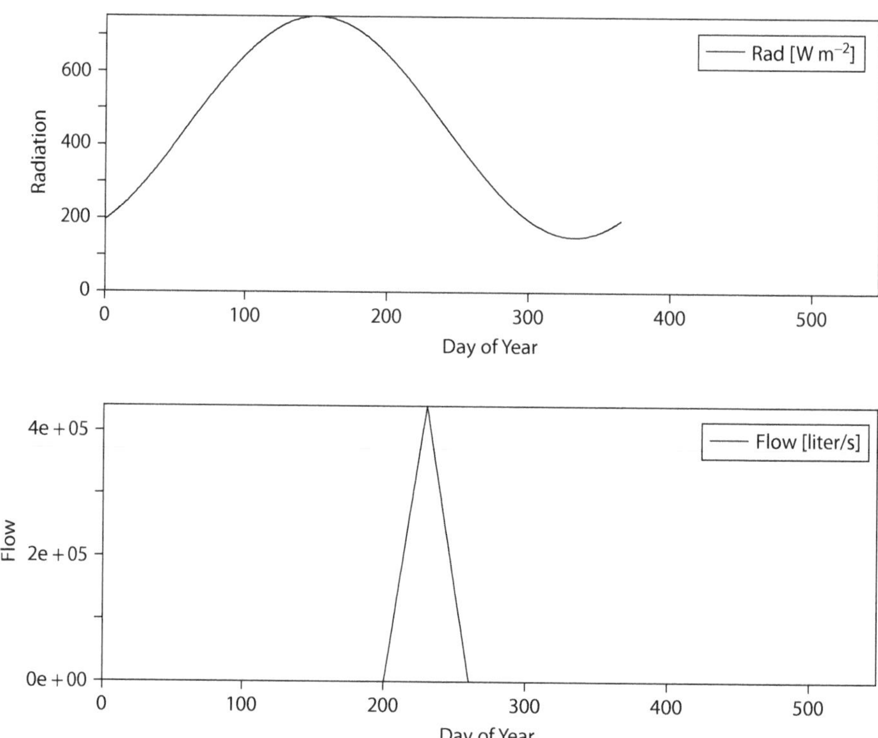

FIGURE 14.14 Radiation and flow.

14.5 COMPUTER SESSION

14.5.1 Solar Radiation

Two simple functions, `sun.dec` and `sun.elev.max`, allow for the calculation of sun declination and sun elevation for a day of the year at a given latitude. We can use them to calculate declination and elevation for all days of the year for several values of latitude, two in the Northern Hemisphere and two in the Southern Hemisphere (Figure 14.2).

```
# day of year
nday <- seq(1,365); nd <- length(nday)
dec <- sun.dec(nday)
lat = c(-32,-23.45,23.45,32); nl <- length(lat)
elev <- matrix(nrow=nd,ncol=nl)
for(i in 1:nl) elev[,i] <- sun.elev(nday,lat[i])
mat<- matrix(1:2,1,2,byrow=T)
layout(mat,c(3.5,3.5),c(3.5,3.5),res=TRUE)
par(mar=c(4,4,1,.5), xaxs="r", yaxs="r")
plot(n,dec,type="l", xlab="Day of year",ylab="Declination [deg]")
abline(h=0,col="grey")
matplot(n,elev,type="l",col=1,xlim=c(0,500), xlab="Day of year",ylab="Sun
  angle elev [deg]")
legend(300,70,lty=1:nl,leg=paste("lat=",lat),cex=0.7)
```

We can calculate the maximum daily radiation during the year at a given latitude of 32.90° for two values of average maximum radiation 500 and 600 W m^{-2} using the function sun.rad.yr and plot to obtain graphs as in Figure 14.4.

```
nday <- seq(1,365); nd <- length(nday)
Lm = c(500,600); nl <- length(Lm)
lat <- 32.90
rad <- matrix(nrow=nd,ncol=nl)
for(i in 1:nl) rad[,i] <- sun.rad.yr(nday,lat,Lm[i])
matplot(nday,rad,type="l",col=1,xlim=c(0,550), xlab="Day of
  year",ylab="Rad [W/m2]")
legend(360,600,lty=1:nl,leg=paste("Lm=",Lm),cex=0.7)
```

Now, for hourly radiation, use the function sun.rad.hr. In the following script, we calculate radiation for 1 day (15) out of 3 days of the year (15, 81, 180) for two values of northern latitude (one at the tropic 23.45° and one at medium latitude 32.90°). The annual average maximum daily is kept fixed at L_m = 600 W m^{-2}.

```
lat = c(23.45,32.90); nl <- length(lat)
nday = c(15,81,180); nd <- length(nday)
hr.noon <- seq(-12,+12,0.1); nh <- length(hr.noon)
rad <- matrix(nrow=nh,ncol=nl)
j=1
for(i in 1:nl) rad[,i] <- sun.rad.hr(nday[j],lat[i],hr.noon,Lm=600)
matplot(hr.noon,rad,type="l", col=1, xlab="Hr noon",ylab="Radiation [W/m2]")
abline(h=0,col="grey")
legend(6,200,lty=1:nl,leg=paste("lat=",lat),cex=0.7)
text(9,10,paste("nday=",nday[j]),cex=0.8)
```

The result is the top panel of Figure 14.3.

Exercise 14.9

Modify the script just given to produce a graph like the one in the top panel of Figure 14.3, but for day 81 and L_m = 500.

Exercise 14.10

Modify the script just given to produce a graph like the one in the bottom panel of Figure 14.3.

Our next step is to generate the hourly radiation for several consecutive days using the function sun.rad.hr.mult. For this purpose, we write the following lines:

```
# latitude dallas (DFW)
lat = 32.90; nday = c(39:41); Lm=600
X <- sun.rad.hr.mult(nday,lat,Lm,sdr=0,sw.plot=T)
lat = 32.90; nday = c(15:30); Lm=600
X <- sun.rad.hr.mult(nday,lat,Lm,sdr=0,sw.plot=T)
```

This produces the graphs shown in Figure 14.5, showing both sets of consecutive days. Similarly, we can include variability and compare the effects of standard deviation:

```
# hourly rad plus noise for several days
# latitude dallas (DFW)
lat = 32.90; Lm=600; nday = c(39:40)
X <- sun.rad.hr.mult(nday,lat,Lm,sdr=100,sw.plot=T)
X <- sun.rad.hr.mult(nday,lat,Lm,sdr=25,sw.plot=T)
```

This produces the graphs shown in Figure 14.6.

14.5.2 Light Attenuation in the Water Column

The following script calculates and plots sunlight as a function of depth for several values of k while keeping L_s fixed at 300 W m^{-2}. The result is shown in the left panel of Figure 14.7.

```
# Light, Beer-Lambert law
z <- seq(0,3,0.1)
nz <- length(z)
k=c(0.3,0.5,1.0); Ls=300; np <- length(k)
L <- matrix(nrow=nz,ncol=np)
for(i in 1:np)
L[,i] <- Ls*exp(-k[i]*z)
matplot(L, -z, type="l", ylab="Depth z [m]",xlab="Light L [W/
m2]",col=1)
legend(200,-2, leg=paste("k=",k), lty=1:3,cex=0.7)
mtext(side=1,line=-1, text=paste("Ls=",Ls))
abline(h=0,col="grey")
```

Exercise 14.11

Modify the script just given to produce a graph like the one in the right panel of Figure 14.7.

Our next step is to see how to estimate the extinction coefficient k of the Beer–Lambert law using light profile data. We will use a data set from three tanks in the UNT Water Research Field contained in the file named **tanks-light.txt**.

```
z(ft)  z(m)   L1c   L2b   L2c
0.0    0.00   2200  2800  2600
0.5    0.15   2000  2600  2450
1.0    0.30   1900  2100  2200
1.5    0.46   1700  2300  2000
2.0    0.61   1500  2200  1800
2.5    0.76   1700  2100  1600
3.0    0.91   1100  1900  1500
3.5    1.07   1000  1800  1400
```

The first value of depth is subsurface (right underneath the water surface). Here, the light units are langleys, but k would be independent of the light units employed. You can first obtain a quick rough estimate of k by using the depth, z_h, at which the light drops to about half of the subsurface reading (ratio = 0.5) and then calculate $k = \frac{\ln(2)}{z_h}$ (remember the exponential concepts in Chapter 3). For tank 1c, $z_h = 3$ ft = 0.91 m, then $k = 0.693/0.91 = 0.76$ m^{-1}.

We can do a better job by taking the logarithm and performing a linear regression (as explained in Chapter 3). Even better, we can do a nonlinear estimation of k. First, read the data into R. For tank 1c, we select the third column. We use the second column for depth to select depth in meters.

```
filename <- "chp14/tanks-light.txt"
z.L <- matrix(scan(filename, skip=1), ncol=5, byrow=T)
# tank 1c
z <- z.L[,2]; L <- z.L[,3]; Ls=L[1]
z.L.1c <- data.frame(z,L)
nls(L ~ Ls*exp(-k*z), data=z.L.1c, start=list(k=0.5))
```

We obtain the following:

```
Nonlinear regression model
  model:  L ~ Ls * exp(-k * z)
   data:  parent.frame
      k
0.6072285
 residual sum-of-squares:  153977.8
```

Therefore, $k = 0.61$ m^{-1}.

Exercise 14.12

What is the k value for the other two tanks? Plot the data together with the estimated Beer–Lambert estimation curve. Is it a good fit?

14.5.3 Primary Productivity as a Function of Depth

Let us use the function **PPrates**. Calculate light as a function of depth, primary productivity as a function of light, and PP in a depth-averaged mode using both the Smith (no photoinhibition) and Steele (photoinhibition) formulations.

Let us use the following parameter values to obtain Figure 14.8:

```
z <- seq(0,3,0.1)
k = 0.5; Ls=300
Pmax =1; alpha =0.02
Pmax =1; Lopt = 200
PPrates (z, param=c(k,Ls,Pmax,alpha,Lopt))
```

Exercise 14.13

Use $k = 0.3$ m^{-1} for clear water and $k = 0.8$ m^{-1} for more turbid water. Assume a depth of 2 m. What is the difference in depth-averaged PP due to the change in water turbidity? Prepare a short table with two values of k for clear and turbid water. Total PP depth averaged at the surface results for both the Smith and Steele models. For each one of these models, what is the effect of water turbidity on PP?

14.5.4 DO and Primary Productivity

The function `DO.PP.pond` corresponds to the DO dynamics model (Equation 14.11) described in Ginot and Herve (1994). It uses the function `PPT.Smith` for the depth-averaged Smith PP model, the diffusion of oxygen between air and water, and a constant respiration. Sunlight is modeled hourly using the function `sun.rad.hr`. The input file **DO-PP-inp.csv** contains the parameter values and specifies simulation for 72 hours, at latitude 32.90° for a winter day (Feb 1) or $n = 32$, with shallow turbid water ($z_d = 0.24$ m, $k = 1.27$)

```
Label,Value,Units,Description
t0,0.00,hr,time zero
tf,72.00,hr,time final
dt,0.01,hr,time step
tw,0.1,hr,time to write
Df,0.01,1/hr,Diffusion coefficient air-water
T1,15,degC,Initial temperature
dT,0,degC/h,rate of change of temp
k,1.27,1/m,Attenuation coeff in Beer-Lambert law
zd,0.24,m,depth of pond or of euphotic zone
Pmax,0.4,mg/liter/hr,maximum photosynthesis rate
alpha,0.02,mg/(literhrW/m2),initial slope of production vs light
(efficiency)
Lat,32.90,deg,Latitude
day1,32,numb,first day
Lm,600,w/m2,avg maximum daily rad
sdr,0,W/m2,variability rad
Rsp,0.2,mg/liter/h,respiration rate (water column + benthic)
X0,8.72,mg/liter,initial condition
digX,4,None,significant digits for output file
```

We will run this model varying the respiration parameter and obtain the plots shown in Figure 14.9:

```
pond <- list(f=DO.PP.pond);param <- list(plab="Rsp",pval=c(0.18,
  0.2,0.22))
t.X <- sim(pond,"chp14/DO-PP-inp.csv", param, pdfout=T)
```

Exercise 14.14

Perform sensitivity analysis on two of the parameters: What happens if you increase the diffusion coefficient (from 0.01 to 0.10)? What happens if you increase the maximum photosynthesis rate P_{max} (from 0.2 to 1.0)?

For example, we will simulate DO using the input file **DO-PP-sonde-inp.csv**, which is set up to start at hour 11.25 and end at hour 57.75. That is, starting at 11:25 a.m. on day 1 and ending at 9:45 a.m. on day 3. We do this to match a period of measurement of temperature, pH, and DO every 15 minutes as contained in the file **do-sonde.txt**. Conditions are for shallow, turbid water as before but warm water on a summer day ($T_w = 28$°C, $n = 180$). In addition, we assume a cooling trend of water at a rate of 0.1°C/h. The following lines of code will perform a simulation using the function `DO.PP` with respiration 0.25, then reads the measured data, and plots simulated together with measured for comparison (Figure 14.15, top panel).

```
# simulate
param <- list(plab="Rsp",pval=0.25)
 t.X <- sim(pond,"chp14/DO-PP-sonde-inp.csv", param, pdfout=T)$out
```

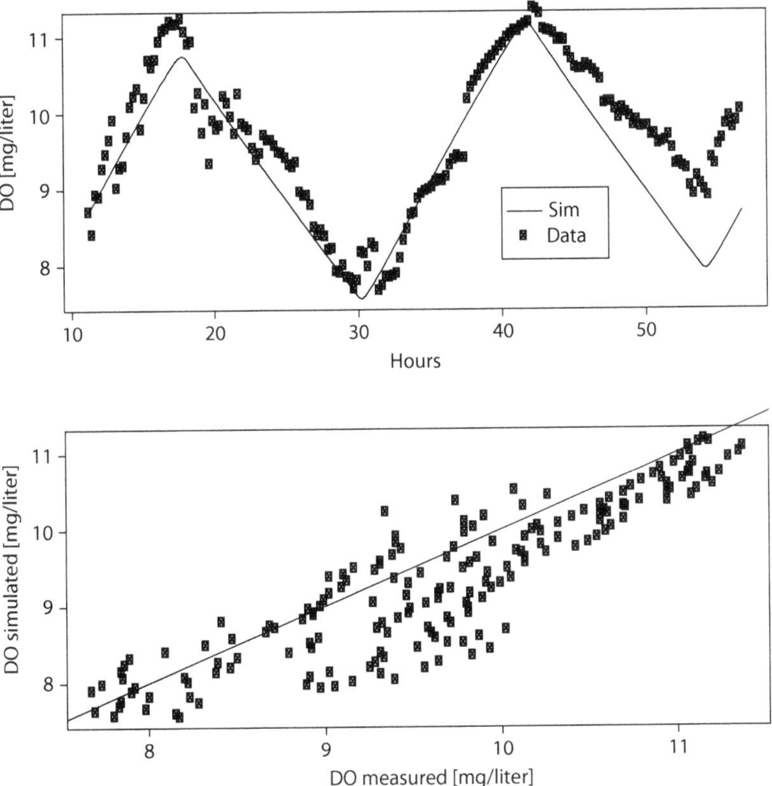

FIGURE 14.15 Simulated dissolved oxygen dynamics in warm water for 2 days in comparison to measured dissolved oxygen.

```
# read data
t.Xd <- matrix(scan("chp14/do-sonde.txt",sep=" "),ncol=4,byrow=T)
#Plot together for comparison
plot(t.X[,1],t.X[,2],type="l",xlab="Hours",ylab="DO [mg/liter]")
points(t.X[,1],t.Xd[,4],type="p")
legend(40,9,leg=c("Sim","Data"),lty=c(1,NA),pch=c(NA,1))
```

We see that DO increases during the day to a maximum of 10.5 mg/liter on the first day and 11.5 mg/liter on the second day. It decreases during the night to a minimum of about 8.2 mg/liter on the second day and 9.0 mg/liter on the third day. The model captures the main diurnal-nocturnal rhythm of DO. Additionally, we can do a scatter plot of the simulated versus measured, compare it to the line of slope 1:1, and calculate the correlation coefficient.

```
plot(t.Xd[,4],t.X[,2],xlab="DO measured [mg/liter]",ylab="DO simulated
 [mg/liter]")
abline(a=0,b=1)
cor(t.Xd[,4],t.X[,2])
```

The correlation coefficient is 0.89, and the graphic result is shown in the bottom panel of Figure 14.15. A good fit or match occurs when the points are close to the diagonal or 1:1 line. In this case, we see that the model does a reasonable job at predicting the DO.

Exercise 14.15

Vary the respiration rate and look for a better match of simulated versus measured DO. Plot the RMS of the error versus the parameter value. Use this tool to explore which respiration parameter value yields a better fit to the data.

14.5.5 River Eutrophication: Nutrient–Algae

Before we simulate the complete model, let us run a simple version with just nutrients and algae uptake. This is a function called `nut.river`. Nitrogen is controlled by load, which is calculated as discharge over volume, inflow from upstream conditions (when multiplied by flow/volume), and depuration or washout (which is the concentration multiplied by flow/volume).

First, we will use the input file **river-no-algae-inp.csv**, which contains

```
Label,Value,Units,Description
t0,0.00,d,time zero
tf,100.00,d,time final
dt,0.01,d,time step
tw,0.1,d,time to write
V,27500,m3,Volume of river reach
Q,440,m3/s,flow
W,660,g/s,N mass discharge rate
Pmax,0,(mg/liter)/d,maximum uptake rate
Kh,1,mg/liter,half-rate
X1.u,0,mg/liter,boundary condition upstream N concentration
X2.u,0,mg/liter,boundary condition upstream Algae concentration
X1.0,0.1,mg/liter,initial condition
X2.0,0,mg/liter,initial condition
digX,4,None,significant digits for output file
```

We can use it simply with values of zero flow and two low values, 0.060 and 0.60. The PDF output file is shown in Figure 14.10.

```
#  river no algae no flow and low flow,
river <- list(f=nut.river); param <- list(plab="Q",pval=c(0,0.060,0.600))
t.X <- sim(river,file="chp14/river-no-algae-inp.csv", param, pdfout=T)
```

We can verify the steady state, depuration constant, and travel times by simple calculations:

```
# verify steady state
# in file W= 660 g/s, V= 27500 m3
W=660;V=27500;Q=param$pval; Nss <- W/Q;TT <-  (V/Q)/(24*60*60)
```

```
> Nss
[1]   Inf 11000  6600
>TT
Inf 5.30 3.18
```

For no flow, the travel time is infinite and there is no steady state. But for 0.06, we get steady state concentration of 11,000 and 5.3 days of travel time.

Next, include the algae and test the same conditions: no flow and low flow. The file is **river-algae-inp.csv**, which is the same as above except for the following lines:

```
Pmax,0.1,(mg/liter)/d,maximum uptake rate
Kh,1,mg/liter,half-rate
X1.u,0,mg/liter,boundary condition upstream N concentration
X2.u,10,mg/liter,boundary condition upstream Algae concentration
X1.0,0.1,mg/liter,initial condition
X2.0,10,mg/liter,initial condition
```

We run this for the same flow conditions as above:

```
# river with algae no flow & low flow,
param <- list(plab="Q",pval=c(0,0.060,0.60))
t.X <- sim(river,file="chp14/river-algae-inp.csv", param, pdfout=T)
```

The PDF output file is shown in Figure 14.11. Finally, we lower the discharge rate by modifying the input file to be **river-algae-low-load-inp.csv**, which is the same as the previous one except for one line:

```
W,0.660,g/s,N mass discharge rate
```

Now, we run again for two values of low-flow conditions and obtain Figure 14.12:

```
# river with algae low flow, low load
param <- list(plab="Q",pval=c(0.030,0.060))
t.X <- sim(river,file="chp14/river-algae-low-load-inp.csv", param,
  pdfout=T)
```

Exercise 14.16

Repeat the previous simulation but with decreased algae growth. Test P_{max} = 0.05. Discuss.

14.5.6 River Eutrophication: Trophic Chain

Now, we work with a more complete model. Instead of adding more R code to the river function, we use a code previously written in C language and execute it from a call to a interface function `river.C` explained at the end of the chapter. We followed this approach in Chapters 10 and 11 with the `cerio.F` model. We will employ an input file is **river-sens-inp.txt**, which includes sensitivity on the parameter `base_flow` and is too long to reproduce here.

Examine the file with a text editor. The parameters and their values correspond with some changes and additions to Table 5.2 in Swartzman and Kaluzny (1987), which is inspired in values for the San Joaquin Valley, California. Most of the variables and parameters in the file are not used in this exercise (e.g., fish, events of runoff, and phosphorous dynamics).

To run the river Fortran program, first perform `dyn.load` of **seem.dll**, unless you have automated in **Rprofile.site**.

```
dyn.load(paste(.libPaths(),"/seem/libs/i386/seem.dll",sep=""))
```

Then call the `river.C` function with two arguments, folder name and file prefix corresponding to this input file:

```
fileout <- river.C("chp14","river-sens")
x <- read.plot.river.out(fileout,sens=T,pdfout=T)
```

We obtain two output files: **river-sens-out.txt** and **river-sens-dex.txt**. The latter contains the values of the sensitivity metrics for all the runs. Plots are obtained with the function **read.plot.river.out**.

The PDF output file is shown in Figures 14.13 and 14.14. Default peak flow is 4,40,000 liter/s or 440 m^3/s, and the default onset of peak flow is day 220.

Exercise 14.17

Change the onset of high flow to day 150 and a lower peak, say 50,000 liter/s^{-1} = 50 m^2s^{-1}. Plot and discuss. Use phytoplankton and zooplankton in biomass units (mgC/liter) and nitrogen in mg/liter as shown in Figure 14.13.

Exercise 14.18

The default discharge rate or load rate is 5.67 ton/day = 5670 kg/day = 5,670 10^6 mg/day. Examine the effect of changing nitrogen discharge in milligrams per day on the maximum value of phytoplankton concentration (the peak of the algae bloom). Select discharge from a low value of 57 kg/day to a high value of 5700 kg/day.

Exercise 14.19

Examine the effect of changing the water flow regime (base flow, peak flow, onset of high flow, and ending day) on the maximum value of phytoplankton concentration (the peak of the algae bloom). Base flow would affect the value of peak algae.

Exercise 14.20

Examine four scenarios (two factors and two levels). Factor 1: For two base flows (high and low), use default (60 liter/s) as low and 600 liter/s as high. Factor 2: For two nutrient discharges (low and high), use default as high and 1% of the default value as low. Produce graphs for each scenario and then a summary table of peak algae bloom for each one of four scenarios. Discuss.

14.5.7 Build Your Own

Exercise 14.21

Develop a model based on the function `DO.PP.pond`, but with two values of respiration: one for the water column and another for the bottom (benthic). Modify the input file to accommodate for an extra parameter. Partition the total respiration into these two parameters and make the water respiration parameter increase with temperature of the water while benthic respiration remains constant. Perform simulations and study how the dependence of water respiration on temperature may change the DO dynamics.

14.6 SEEM FUNCTIONS EXPLAINED

14.6.1 Solar Radiation: Daily

Declination is simply a calculation with day number as the sole argument.

```
sun.elev.max <- function(nday,lat){
 dec <- 23.45*sin((2*pi/365)*(nday-81))
 return(dec)
}
```

Then, declination is used for elevation, which includes an additional argument, latitude. Calculation of elevation depends on whether we have northern or southern latitude.

```
sun.elev.max <- function(nday,lat){
 # declination
 dec <- sun.dec(nday)
 if (lat <0) elev <- -dec + 90 - lat
 else elev <- dec + 90 - lat
 return(elev)
}
```

A third function uses the annual average of maximum radiation in the day L_m to convert maximum elevation into radiation.

```
sun.rad.yr <- function(nday,lat,Lm){
 # declination
 dec <- sun.dec(nday)
 if (lat <0) rad <- Lm - dec*Lm/(90-lat)
 else rad <- Lm + dec*Lm/(90-lat)
 return(rad)
} # end of function
```

14.6.2 Solar Radiation: Hourly

Sun elevation by hour during the day requires an argument for the hour with respect to noon, negative for hours before noon, and positive for hours after noon. Internally, this hour value is converted to hour angle before applying trigonometric expressions in radians to calculate elevation and then converting back to degrees. Finally, elevation is made zero when the calculation yields negative values.

```
sun.elev.hr <- function(nday,lat,hr.noon){
 # declination
 dec <- sun.dec(nday)
 # hr angle and trig
 nh <- length(hr.noon)
 hr.angle <- 15*hr.noon
 sin.H <- sin((pi/180)*hr.angle)
 cos.H <- cos((pi/180)*hr.angle)

 # elevation and angle
 sin.elev <- cos((pi/180)*lat)*cos((pi/180)*dec)*cos.H +
             sin((pi/180)*lat)*sin((pi/180)*dec)
 elev <- asin(sin.elev)*180/pi
 for(i in 1:nh) if(elev[i]<0) elev[i]<-0
 return(elev)
} # end of function
```

A similar function calculates radiation by hour during the day requiring an additional argument for the average of maximum daily radiation. An optional argument with default value zero is the standard deviation of noise or variability to be imposed on the hourly radiation.

```
sun.rad.hr <- function(nday,lat,hr.noon,Lm,sdr=0){
 # declination
 elev <- sun.elev.hr(nday,lat,hr.noon)
 rad <- elev
 for(i in 1:length(elev))
  if(elev[i] > 0) rad[i] <- elev[i]*Lm/(90-lat)+ rnorm(1,0,sdr)
 for(i in 1:length(rad))
  if(rad[i] < 0)  rad[i] <- 0
 return(rad)
} # end of function
```

Now, using this function, we can build another one, sun.rad.hr.mult, to calculate hourly radiation for multiple consecutive days. Its arguments are an array **nday** containing the sequence of consecutive days, latitude, average maximum radiation, optional variability, and a logical variable to plot the results.

```
sun.rad.hr.mult <- function(nday,lat,Lm,sdr=0,plot=T){

nd <- length(nday)
hr.noon <- seq(-12,+12,0.1); nh <- length(hr.noon)
nl <- max(length(lat),length(Lm))
rad <- matrix(nrow=((nh*nd)-(nd-1)),ncol=nl)
for(i in 1:nl){
   j1<- 1; j2 <- nh
   for(j in 1:nd){
    rad[j1:j2,i] <- sun.rad.hr(nday[j],lat[i],hr.noon,Lm,sdr)
    j1 <- j2; j2 <- j1+nh-1
 }
}

hr.cum <- seq(0,24*nd,0.1)
out <- data.frame(hr.cum,rad)
return(out)

if(plot==T){
 matplot(hr.cum,rad,type="l", col=1, xlab="Hr
cumulative",ylab="Radiation
   [W/m2]")
 abline(h=0,col="grey")
 days <- paste(nday[1])
 for(i in 2:nd) days <- paste(days,nday[i],sep=",")
 mtext(side=3,line=-1,paste("Lat=",lat," Lm=",Lm, " ndays=",days),
cex=0.7)
}
} # end of function
```

14.6.3 Primary Productivity versus Depth

The following two functions implement Equations 14.9 and 14.10 for the depth-averaged Smith and Steele models. In addition to the model parameters, they use sunlight at the subsurface and the attenuation coefficient as arguments.

```
# Smith and Steele -- depth averaged
PPT.Smith <- function(Ls,k,Pmax,alpha){
 y <- (Pmax/alpha)^2
 L <- Ls*exp(-k*z)
if(Ls > 0) PPTSmith <- (Pmax/(k*z)) * log( (Ls+sqrt(y+Ls^2)) /
(L+sqrt(y+L^2)) )
 else PPTSmith <- 0
return(PPTSmith)
}

PPT.Steele <- function(Ls,k,Pmax,Lopt){
 L <- Ls*exp(-k*z)
if(Ls >0) PPTSteele <- ((Pmax*exp(1))/(k*z)) * ( exp(-L/Lopt) - exp(-Ls/
   Lopt) )
 else PPTSteele <- 0
 return(PPTSteele)
}
```

We use these two functions and the ones already developed in Chapter 13 to calculate productivity versus depth.

```
PPrates.depth <- function(z, param,sw.plot=T) {
# param PP Pmax,alpha,Lopt
# param light k att coeff, Ls light subsurface

# Light, Beer's law
L <- Ls*exp(-k*z)
# Smith & Steele model vs L
PPSmith <- PP.Smith(L,Pmax,alpha)
PPSteele <- PP.Steele(L,Pmax,Lopt)
PP <- cbind(PPSmith,PPSteele)
# depth average
PPTSmith <- PPT.Smith(Ls,k,Pmax,alpha)
PPTSteele <- PPT.Steele(Ls,k,Pmax,Lopt)
PPT <- cbind(PPTSmith,PPTSteele)
out <- data.frame(z,L,PP,PPT)

if(sw.plot==T){
 mat<- matrix(1:4,2,2,byrow=T)
 layout(mat,c(3.5,3.5),c(3.5,3.5),res=TRUE)
 par(mar=c(4,4,1,.5), xaxs="r", yaxs="r")
 # Light, Beer's law
 plot(L, -z, type="l",ylab="Depth [m]",xlab="Light [W/m2]")
 abline(h=0, col="grey")
 # Smith and Steele vs. L
 matplot(L,PP, type="l", col=1, ylab="P")
 legend("bottomright",leg=c("Smith","Steele"),lty=1:2)
 abline(h=0, col="grey")
 # smith and steele vs depth
 matplot(PP,-z, type="l", ylab="Depth [m]", xlab="PP",col=1)
 legend("topleft",leg=c("Smith","Steele"),lty=1:2)
 abline(h=0, col="grey")
 # depth avg
```

```
  matplot(PPT,-z,type="l", col=1,xlab="PP Depth Averaged",ylab="Depth
    [m]")
  legend("topleft",leg=c("Smith","Steele"),lty=1:2)
  abline(h=0, col="grey")
 }
 return(out)
 }
```

14.6.4 Dissolved Oxygen and Primary Productivity: DO.PP.pond

The function DO.PP.pond is similar to the other ODE-based simulation functions we have used before. The f function to integrate is

```
DO.PP.pond <- function(t,p,x){
  # parameters
  D <- p[1]; T1 <- p[2]; dT <- p[3]
  k <- p[4]; zd <- p[5]; Pmax <- p[6]
  alpha <- p[7]; lat <- p[8]; day1 <- p[9]
  Lm <- p[10]; sdr <- p[11]; Rsp <- p[12]
  # determine day
  i <- floor(t/24); day <- day1+i
  hr <- t -24*i -12 # hr within the day
  temp <- T1 + dT*t # temp linear trend
  Xs <- 14.6 - ((14.6-8.6)/25)*temp # DO saturation
  Ls <- sun.rad.hr(day,alpha,hr,Lm,sdr)
  dX <- D*(Xs-x) + PPT.Smith(Ls,k,zd,Pmax,alpha) - Rsp
  return(dX))
}
```

14.6.5 River Eutrophication: nut.river Function

The function nut.river is similar to other ODE-based simulation functions we have used before. The f function to integrate is

```
nut.river <- function(t,p,x){
  dX <- x # define array
  V <- p[1]; Q<- p[2]; W <- p[3] # vol, flow, discharge
  Pmax <- p[4]; Kh <- p[5] # uptake M3 param
  X1.u <- p[6]; X2.u <- p[7] # boundary conditions
  # x[1] nitrogen   x[2] algae
  W.V <- (W/V)*(24*60*60) # loading convert s to d
  Q.V <- (Q/V)*(24*60*60) # depuration convert s to d
  U2 <- Pmax*(x[1]/(x[1]+Kh))*x[2] # M3 uptake
  dX[1] <- W.V + Q.V*(X1.u - x[1]) - U2
  dX[2] <- Q.V*(X2.u - x[2]) + U2
  return(dX)
 }
```

14.6.6 River Eutrophication: river.C and read.plot.river.out

The function `river.C` is almost identical to the `cerio.F` script as explained in Chapter 10. Major differences are that we call the interface `.C` instead of `.Fortran` and use filenames according to river program.

```
river.C <- function(x,fileprefix){
 y <- getwd()
 setwd(x)
 file.copy(paste(fileprefix,"-inp.txt",sep=""), "river_inp.txt")
 .C("river", package="seem")
 fileout <- paste(fileprefix,"-out.txt",sep="")
 filedex <- paste(fileprefix,"-dex.txt",sep="")
 file.rename("river_out.txt", fileout)
 file.rename("river_dex.txt", filedex)
 file.remove("river_inp.txt")
 setwd(y)
 return(paste(x,"/",fileout,sep=""))
}
```

The function `read.plot.river.out` to read and plot river output files is more complex and similar to the one employed to read and plot cerio output data files.

15 Terrestrial Ecosystems
Soils, Plants, and Water

In this chapter, we will study the water relationships between soils, plants, and atmosphere. The purpose is to model weather variables, including rain, evaporation, and transpiration, and then to model water infiltration for soil and soil moisture dynamics.

15.1 WEATHER GENERATORS

A common use of synthetic weather data in models is to replace real data when they are not readily available as it occurs when a finer temporal resolution is desired or when exploring scenarios of climate change. We encountered this situation already when we modeled solar radiation in Chapter 14. To incorporate variability, we can add random numbers to each value generated. Thus, weather becomes a stochastic variable.

Many complex models use weather generators. For example, models used in agriculture, such as the Erosion-Productivity Impact Calculator (EPIC) and Soil and Water Assessment Tool (SWAT) models. (Richardson and Nicks, 1990), generate daily values of temperature, precipitation, solar radiation, wind speed, and wind direction from the monthly values using statistical characteristics. Generators attempt to preserve the time series, correlations, and seasonality.

The simplest assumption is that of time independence. This is to generate the random component of the weather variable for the particular time step (e.g., each day) by sampling from a distribution of the variable (e.g., a normal distribution of a given mean and variance) regardless of the value of the previous time step. An example is the sinusoidal model of hourly value of solar radiation with a random component, discussed in Chapter 14.

15.1.1 Dependent Models: Markov Chain

However, in many cases, we want to make the random value depend on recent past values because the rainfall in a given day may also depend on whether it rained the previous day. Several approaches using time series and stochastic processes are available. Here, we will study one simple example using **Markov** models.

A Markov chain is a very general type of probabilistic model with a finite number of discrete states. The probability of being in a given state at time $t + 1$ depends on the state in the present time t and a conditional probability of changing to another state (Keen and Spain, pp. 319–330; Swartzmann and Kaluzny, pp. 32–39).

These conditional probabilities are called **transition probabilities**. Markov models have several applications in environmental modeling, for example, vegetation succession and weather generation. In Chapter 16, we will study the application to vegetation dynamics. Here, we will use it to generate daily rainfall values.

A Markov model consists of a transition probability matrix **P** that projects a vector of probabilities $X(t)$ through time. Each entry of the vector is a probability of being in a given state. The entries of **P** are probabilities of transition from one state to another state. The simulation is easy and proceeds exactly as we did for a projection matrix.

$$X(t+1) = \mathbf{P}X(t) \tag{15.1}$$

Consider two states to make up an $\boldsymbol{X}$ of dimension 2×1:

$$\boldsymbol{X} = \begin{bmatrix} X_1 \\ X_2 \end{bmatrix} \tag{15.2}$$

The probabilities X_1 and X_2 add up to 1 since the system can only be in one of these states. Then, $\mathbf{P}$ has dimensions of 2×2:

$$\mathbf{P} = \begin{pmatrix} p_{11} & p_{12} \\ p_{21} & p_{22} \end{pmatrix} \tag{15.3}$$

An entry p_{ij} is the probability of transitioning from j to i. Probabilities p_{11} and p_{21} have to add to 1 since the system has to transition somewhere from the source state, state 1 in this case. Likewise, p_{12} and p_{22} have to add to 1. In general, for more than two states,

$$\sum_j p_{ij} = 1 \tag{15.4}$$

After multiplying the probability vector $\boldsymbol{X}$ (state) by the matrix $\mathbf{P}$ many times, we obtain a stationary or stable probability vector. This is due to the same properties of the matrices we studied in Chapters 9 and 10. The stable distribution tells us the probability that the system will be in one of the states after a long-term run.

The stable or stationary distribution can be calculated from Equation 15.1 by making $\boldsymbol{X}(t+1) = \boldsymbol{X}(t) = \boldsymbol{X}^*$ to obtain

$$\boldsymbol{X}^* = \mathbf{P}\boldsymbol{X}^* \tag{15.5}$$

or

$$(\mathbf{P} - \mathbf{I})\boldsymbol{X}^* = 0 \tag{15.6}$$

and then we need to solve for $\boldsymbol{X}^*$. This is similar to what we have already learned from compartment models.

Take, for example, a system with two states:

$$\begin{bmatrix} X_1^* \\ X_2^* \end{bmatrix} = \begin{bmatrix} 1-p_{21} & p_{12} \\ p_{21} & 1-p_{12} \end{bmatrix} \begin{bmatrix} X_1^* \\ X_2^* \end{bmatrix} \tag{15.7}$$

Also,

$$X_2^* = 1 - X_1^*$$

therefore

$$X_1^*(1-p_{21}) + p_{12}(1-X_1^*) = X_1^* \tag{15.8}$$

and solving for X_1^*

$$X_1^* = \frac{p_{12}}{p_{21}+p_{12}} \tag{15.9}$$

Therefore, X_2^* is

$$X_1^* = \frac{p_{12}}{p_{21} + p_{12}} \quad (15.10)$$

15.1.2 Two-Step Rainfall Generation: First Step, Markov Sequence

A method used often to generate daily rainfall consists of two steps: (1) determine the occurrence of rain in a day, that is, whether a day is rainy (e.g., when the precipitation exceeds 0.2 mm); and (2) generate the precipitation amount for that day according to a distribution skewed so as to reflect the fact that daily rainfall frequency has higher values toward the lower values of rainfall.

The sequence of rainy days is generated using a Markov process (Richardson and Nicks, 1990). As an example, consider two states for a day: state 1 = dry day (it does not rain) or state 2 = wet day (it rains) (see Figure 15.1). Each day, the system can be in either a dry or a wet state.

$$\mathbf{P} = \begin{bmatrix} 0.4 & 0.2 \\ 0.6 & 0.8 \end{bmatrix} \quad (15.11)$$

Note that entry p_{21} is Pr(wet|dry) and p_{12} is Pr(dry|wet). Here, the bar denotes conditional probability. Also $p_{21} = 1 - p_{11}$ and $p_{12} = 1 - p_{22}$.

The stationary distribution is

$$X_1^* = \frac{0.2}{0.2 + 0.6} = \frac{2}{8} = \frac{1}{4} = 0.25 \quad (15.12)$$

and $X_2^* = \frac{3}{4} = 0.75$.

Therefore, as a long-term average, the probability of a day being dry is 0.25 and the probability of being wet is 0.75.

Exercise 15.1

What is the stationary probability distribution when $p_{21} = 0.6$ and $p_{12} = 0.4$?

As a long-term average, in a region, about 50% of the days are rainy in a year, and 20% of the dry days are followed by rainy days. How often are dry days followed by dry days? As 50% of the days in a year are rainy, we have that $X_2^* = 0.5$, whereas 20% of the dry days are followed by rainy days, which means that $p_{21} = 0.2$. Using Equation 15.10, we obtain

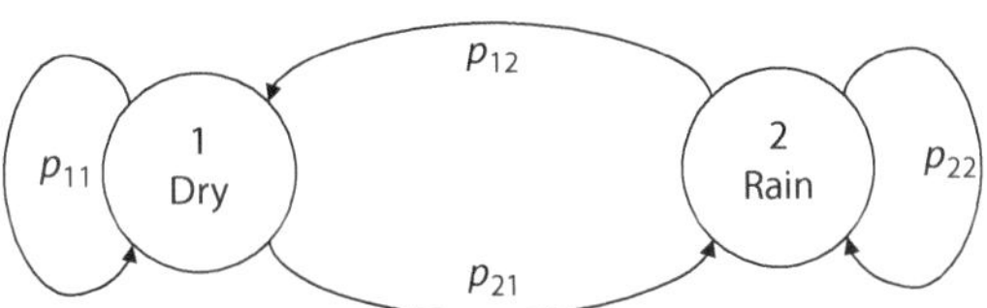

FIGURE 15.1 A Markov chain diagram for the sequence of dry or rainy days.

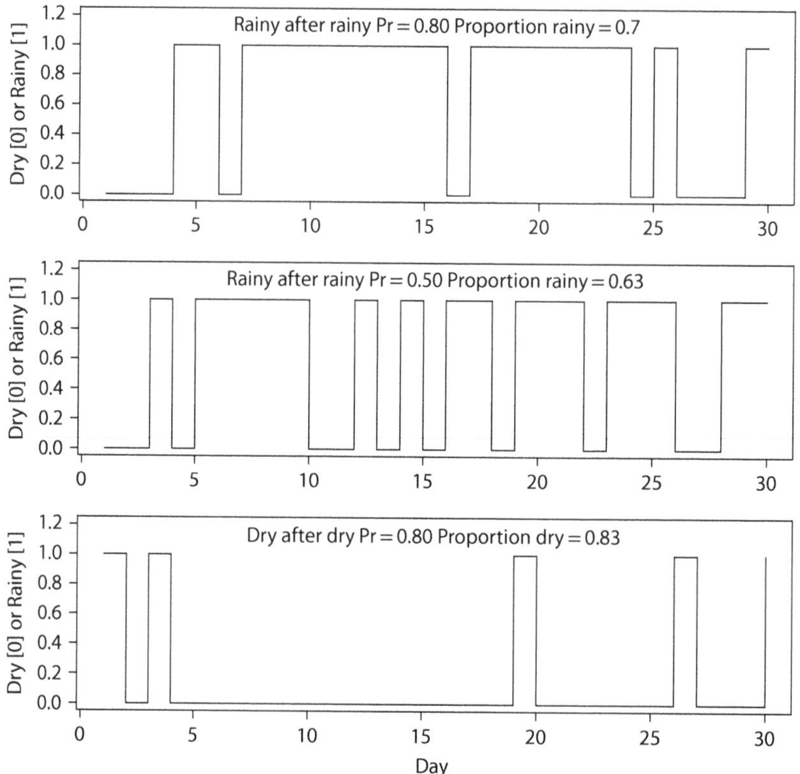

FIGURE 15.2 Examples of realizations of Markov sequences of rainy days in a month. Wet days followed by wet days with high probability (top), independent case (middle), and dry days followed by dry days with high probability (bottom).

$$p_2^* = \frac{p_{21}}{p_{21} + p_{12}} = \frac{0.2}{0.2 + p_{12}} = 0.5$$

Rearrange and then solve for p_{12} from $p_{12} + 0.2 = 2 \times 0.2$ to get $p_{12} = 0.2$, then $p_{11} = 1 - p_{12} = 0.8$. The answer is that 80% of the time dry days should be followed by dry days.

Figure 15.2 shows examples of realizations of Markov sequences for three cases: (1) rainy days are very common ($p_{22} = 0.8$), (2) rain occurrence is independent because $p_{21} = p_{12} = 0.5$, and (3) dry days are very common ($p_{11} = 0.8$). Also shown are the actual proportions obtained for rainy days and dry days.

15.1.3 Two-Step Rainfall Generation: Second Step, Amount in Rainy Days

The frequency distribution of rainfall on rainy days determines the amount of rain, once a day is selected as wet (Richardson and Nicks, 1990). Determine the distribution of rainfall for wet days using the data for the site and then use that distribution to draw samples. Daily rainfall distribution is skewed toward the low values. The distribution parameters, for example, mean and standard deviation, vary month to month according to climatic records. We will study four distributions to generate rainfall amount: exponential, Weibull, gamma, and skewed normal (Figure 15.3).

The simplest model is an exponential RV,

$$\begin{aligned} p(x) &= a\exp(-ax) \\ F(x) &= 1 - \exp(-ax) \end{aligned} \tag{15.13}$$

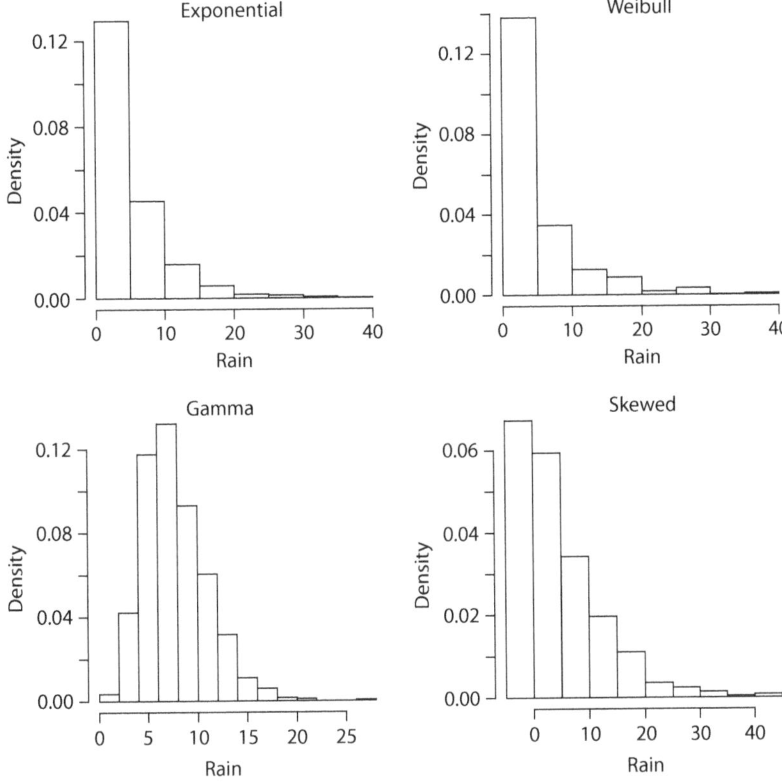

FIGURE 15.3 Examples of samples (size = 1000) generated by four different distributions: exponential with μ = scale = 5, Weibull with μ = 5, shape = 1.3, gamma with μ = 5, shape = 0.8, and skewed with μ = 5, σ = 8, γ = 2 (skew).

where x is the daily rainfall amount and has only one parameter, the rate a. Recall that both the mean and the standard deviation are equal to $1/a$. As we have learned before, this can be generated from a uniform $U(0,1)$ variate u using the inverse of the cdf:

$$x = -(1/a)\ln(u) = \mu_X \ln(u) \tag{15.14}$$

In this case, $1/a$ is the average rainfall for wet days in the month and the generated numbers will have standard deviation also equal to $1/a$.

A more complicated model is the Weibull that has the following pdf and cdf:

$$\begin{aligned} p(x) &= \left(\frac{c}{b}\right)\left(\frac{x}{b}\right)^{c-1} \exp\left(-\left(\frac{x}{b}\right)^{c}\right) \\ F(x) &= 1 - \exp\left(-\left(\frac{x}{b}\right)^{c}\right) \end{aligned} \tag{15.15}$$

with the parameters shape = c and scale = b, and mean and variance

$$\begin{aligned} \mu_X &= b\Gamma\left(1+\frac{1}{c}\right) = \left(\frac{b}{c}\right)\Gamma\left(\frac{1}{c}\right) \\ \sigma_X^2 &= b^2\left(\Gamma\left(1+\frac{2}{c}\right) - \left(\Gamma\left(1+\frac{1}{c}\right)\right)^2\right) = \left(\frac{b^2}{c}\right)\left(2\Gamma\left(\frac{2}{c}\right) - \left(\frac{1}{c}\right)\left(\Gamma\left(\frac{1}{c}\right)\right)^2\right) \end{aligned} \tag{15.16}$$

where Γ is the gamma function. Another way to write the Weibull is to use the parameter rate and power coefficients equal to the inverse of the scale and shape, $a = 1/b$, $k = 1/c$:

$$p(x) = \left(\frac{a}{k}\right)(ax)^{1/k-1}\exp\left(-(ax)^{1/k}\right)$$
$$F(x) = 1 - \exp\left(-(ax)^{1/k}\right)$$

We can see that when shape $c = 1$ or $k = 1$, this expression reduces to the exponential. A value of shape c lower than 1 (i.e., $k > 1$) will produce a pdf with higher values to the left than the exponential pdf, whereas a value of c greater than 1 ($k < 1$) will produce higher values to the right (Figure 15.4). The case $k > 1$ is of more interest for modeling rainfall amount. However, for $k > 1$, $p(0)$ does not have a finite value.

For example, when $a = 10$, $k = 1.5$,

$$\mu_X = (1/a)\Gamma(1+k) = 10\Gamma(2.5) = 10\times 1.33 = 13.3$$

and for $a = 10$, $k = 0.5$

$$\mu_X = (1/a)\Gamma(1+k) = 10\Gamma(1.5) = 10\times 0.89 = 8.9$$

For the random generation of a Weibull, we can use the inverse cdf method based on a $U(0,1)$ uniform variate u,

$$u = F(x) = 1 - \exp(-(ax)^{1/k})$$
$$1 - u = \exp(-(ax)^{1/k})$$
$$\ln(1-u) = -(ax)^{1/k}$$
$$(-\ln(1-u))^k = ax$$
$$x = (1/a)(-\ln(1-u))^k$$

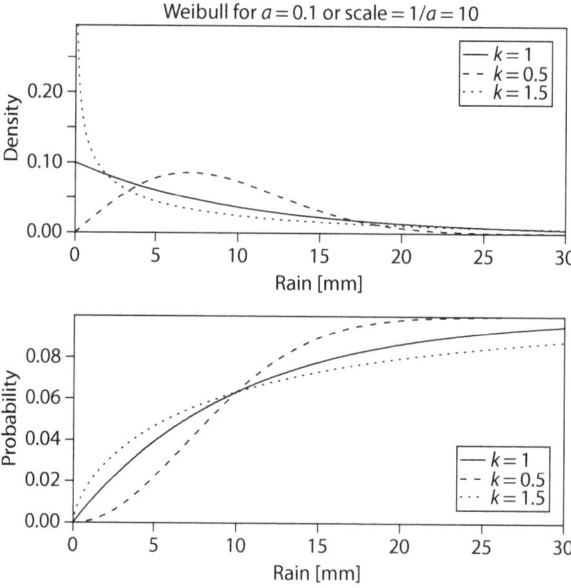

FIGURE 15.4 Weibull distribution for scale = 10 ($a = 0.1$) and three different values of shape ($1/k$). The solid line for $k = 1$ reduces to an exponential. The other two values of k are for shape greater than one and shape less than one.

or rewritten in terms of the mean daily rainfall for a month:

$$x = \left(\frac{\mu_X}{\Gamma(1+k)}\right)(-\ln(u))^k \tag{15.17}$$

This is related to the modified "exponential" distribution (Neitsch et al., 2002)

$$x = \mu_X(-\ln(u))^k \tag{15.18}$$

with exponent k that varies between 1 and 2. This coefficient controls the number of extreme events or days with high precipitation. The value $k = 1.3$ provides a distribution that fits data from many locations (Neitsch et al., 2002).

Another pdf typically employed to model rainfall amount is the gamma pdf:

$$p(x) = \frac{\left(\frac{x}{b}\right)^{c-1}\exp\left(-\frac{x}{b}\right)}{b\Gamma(c)} \tag{15.19}$$

where the parameters c and b are shape and scale, respectively. The scale is the inverse of the rate $a = 1/b$. The mean and variance are

$$\mu_X = cb \quad \sigma_X^2 = cb^2 \tag{15.20}$$

For the gamma, it is relatively easy to generate rainfall with a given mean and variance of rain since we can solve for c and b from Equation 15.20:

$$\begin{aligned} \frac{\mu_X}{b} &= c \quad \text{and} \quad c = \frac{\sigma_X^2}{b^2} \\ b &= \frac{\sigma_X^2}{\mu_X} \quad \text{and} \quad c = \frac{\mu_X^2}{\sigma_X^2} \end{aligned} \tag{15.21}$$

There is no closed-form equation for the cdf unless the shape parameter c is an integer, and in this case, $\Gamma(c) = (c-1)!$ and the pdf reduces to the Erlang pdf:

$$\begin{aligned} p(x) &= \frac{(ax)^{c-1}\exp(-ax)}{(c-1)!} \\ F(x) &= 1 - \exp(-ax)\sum_{c=0}^{c-1}\frac{(ax)^c}{c!} \end{aligned}$$

Therefore, the inverse method for random generation can only be used when the shape is an integer using c uniformly distributed variables in [0,1]; denote these variables by u_i, $i = 1, \ldots, c$:

$$x = -\left(\frac{1}{a}\right)\ln\left(\prod_{i=1}^{c} u_i\right)$$

Another pdf used to generate rainfall is based on a skewed distribution generated from a normal (Neitsch et al., 2002) used in the EPIC and SWAT models:

$$x = \mu_X + 2\frac{\sigma_X}{\gamma_X}\left[\left[\frac{\gamma_X}{6}\left(z - \frac{\gamma_X}{6}\right) + 1\right]^3 - 1\right] \tag{15.22}$$

where in addition to the mean daily rainfall for the month, we use σ_x, which is the standard deviation daily rainfall for the month, and γ_x, which is the skew coefficient of daily rainfall. In this case, z is a value from a standard normal random variable (mean = 0; standard deviation = 1). The distribution parameters, mean, standard deviation, and skewness, are varied month to month according to the climatic records.

15.1.4 Combining Dry/Wet Days with the Amount of Rain on Wet Days

The total rainfall in a month would be the sum of the rainfall amounts for the wet days, which is a random variable R, the combination of the number of wet days in the month (occurring at random according to the Markov chain model), and the rainfall amount on wet days, according to the distribution selected. The expected number of wet days is nX_2^*, where n is the total number of days in the month and X_2^* is the probability of getting a wet day given by Equation 15.10. Therefore, the expected total rainfall for the month is the expected number of wet days multiplied by the expected rainfall for a wet day:

$$\mu_R = nX_2^*\mu_X$$

For example, when using a gamma

$$\mu_R = nX_2^*cb$$

when X is a gamma, the variance of R is

$$\sigma^2_R = nX_2^*\sigma_X^2\left(1+cX_1^*\frac{p_{22}+p_{11}}{p_{21}+p_{12}}\right)$$

As a numeric example, take a month with n = 30 days and a transition matrix, given in Equation 15.11, for which we calculated $X_2^* = \dfrac{3}{4} = 0.75$. For a gamma with $c = 0.8$, $b = 6.25$, we have mean = 5, and therefore, the mean daily rainfall is $0.75\times 5 = 3.75$ mm and the mean rainfall total is

$$\mu_R = 30\times 0.75\times 5 = 112.5\,\text{mm}$$

and the variance is

$$\sigma_R^2 = 30\times 0.75\times 0.8\times 6.25^2\times\left(1+0.8\times 0.25\frac{0.4+0.8}{0.6+0.2}\right)$$
$$= 703.13\times 1.3 = 914.06$$

Figure 15.5 depicts the results of performing the two steps for a high transition probability of following a rainy day by a rainy day and a Weibull distribution with shape 1.3. These would be the conditions for a wet month. Figure 15.6, on the other hand, illustrates the results for a dry month, that is, when the probability of a dry day being followed by a dry day is high.

In general, many realizations under the same monthly conditions represent samples of the daily behavior for the month. For example, for a wet month, see Figures 15.7 and 15.8.

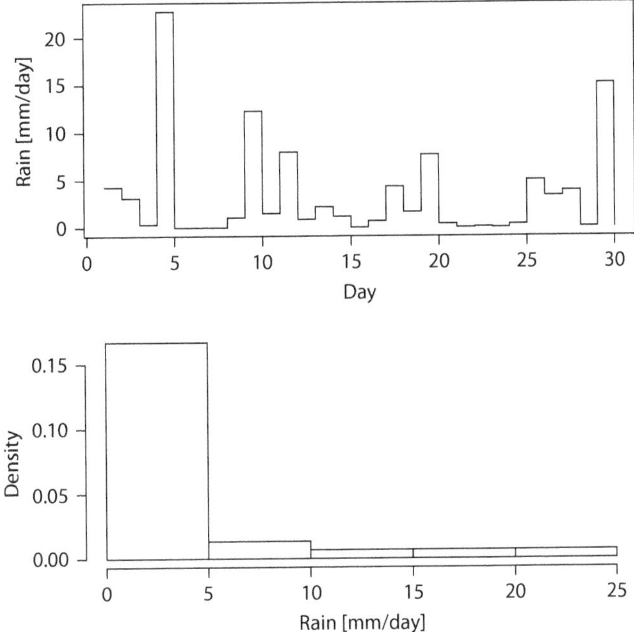

FIGURE 15.5 One realization for a wet month with a probability of rainy day following a rainy day equal to 0.8. In this simulation we have obtained a proportion of rainy days equal to 0.77. Top: sequence of daily rain values. Bottom: histogram of rain values.

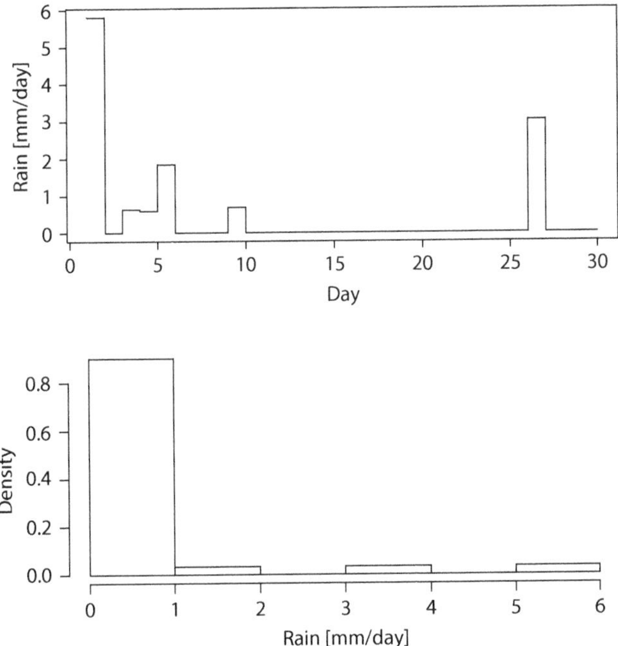

FIGURE 15.6 One realization for a dry month with a probability of dry day following a dry day equal to 0.8. In this simulation we have obtained a proportion of dry days equal to 0.8. Top: sequence of daily rain values. Bottom: histogram of rain values.

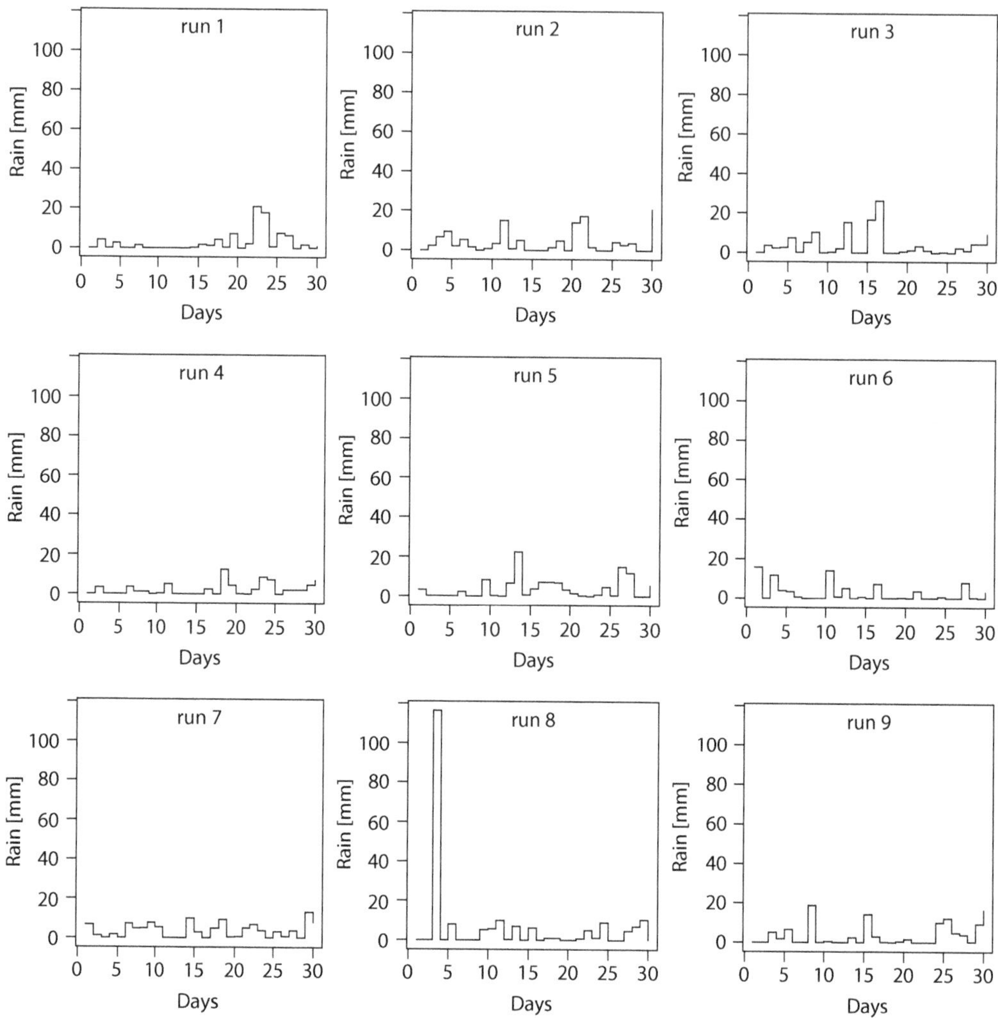

FIGURE 15.7 Nine realizations for two-step rainfall generation for a wet month. Amount drawn from Weibull density.

15.2 EVAPOTRANSPIRATION

Water balance for a terrestrial ecosystem is determined by input from precipitation and demand from evapotranspiration (ET). The latter term, evapotranspiration, is a combination of evaporation and transpiration in plants. Actually, when considering the balance of water in the soil, the input is the infiltration at the surface minus the loss by deep percolation, which will be discussed in Section 15.3. The actual ET (AET) is a fraction of the potential ET (PET) demand estimated from weather conditions, because the soil may not have enough moisture to supply the PET. There are several models to calculate PET. We will study two of these: (1) Penman and (2) Priestley–Taylor.

15.2.1 Penman

Applying the Penman method requires solar radiation, air temperature, wind speed, and relative humidity as inputs. It assumes that total evaporation is due to two terms: an energy term (driven by net solar radiation) and an aerodynamic term (driven by wind speed and relative humidity). Air temperature affects both terms.

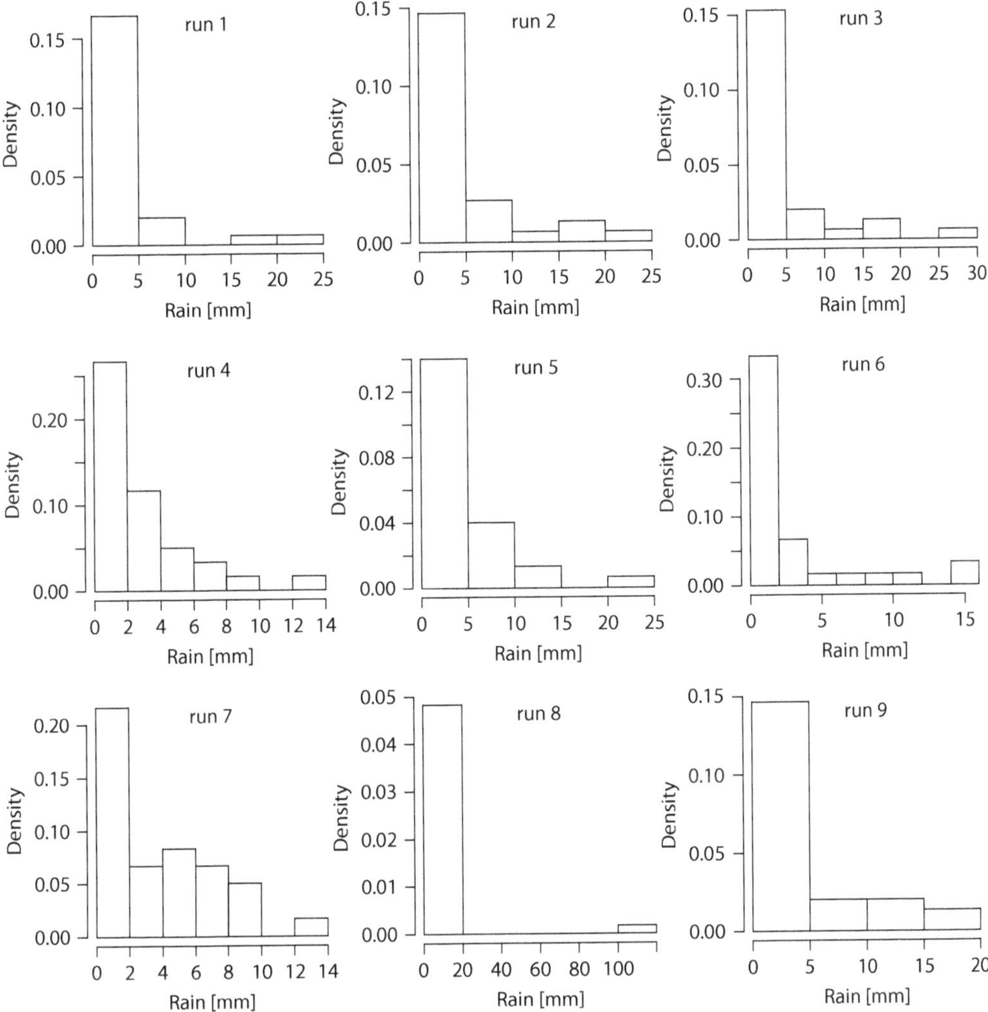

FIGURE 15.8 Histograms of nine realizations shown in Figure 15.7.

Here, we follow the calculation of the daily PET value as in the EPIC model (Williams et al., 1990). The forcing weather variables are T, air temperature in °C (mean daily); Q, solar radiation (mean daily) in MJ m^{-2}; u, wind speed at a height of 2 m in ms^{-1}; and RH, relative humidity (%). It requires values for the following parameters: α = surface albedo, about 0.23 for vegetation; γ = psycrometer constant (kPa/°C), 0.68 at 20°C and pressure 1000 mbar; and P = barometric pressure (assume 1000 mbar). Recall that Pa and mbar are units of pressure and are related by 1 mbar = 1 Hpa and 1000 mbar = 1 bar = 100 kPa.

We start by using the following submodels. Vapor pressure at saturation (in mbar) as a function of T, in kPa, is given as

$$e_s(T) = 0.1 \times \exp\left[54.88 - 5.03 \times \ln(T+273) - \frac{6791}{T+273}\right] \tag{15.23}$$

The slope of the vapor pressure curve at saturation (kPa/°C) is

$$\delta(T) = \left[\frac{6791}{T+273} - 5.03\right]\left[\frac{e_s}{(T+273)}\right]$$

The psychrometer constant is a function of pressure P (which depends on elevation H):

$$\gamma = 6.6 \times 10^{-3} P = 6.6 \times 10^{-3} \times \left[101 - 0.0115 \times H + 5.44 \times 10^{-7} H^2\right] \tag{15.24}$$

Vapor pressure is a fraction (given by relative humidity RH) of the vapor pressure at saturation in mbar:

$$e(T, P, \text{RH}) = e_s(T, P)\,\text{RH} \tag{15.25}$$

Aerodynamic resistance (s/m) is a function of wind speed at a height of $z = 2$ m:

$$r(u) = 4.72 \frac{\ln\left(\frac{2}{1.37}\right)^2}{1 + 0.54u} \tag{15.26}$$

But here, instead of using this equation, we will use a factor of wind speed at $z = 10$ m:

$$f(u) = 2.7 + 1.63u \tag{15.27}$$

Net radiation (MJ m^{-2}) is incoming radiation (discounting reflected radiation). Here, we will ignore outgoing long-wave radiation and soil heat flux:

$$R_n = Q(1 - \alpha) \tag{15.28}$$

Latent heat of vaporization in J/kg will be simplified to

$$L(T) = 2.5 - 0.0022 \times T \tag{15.29}$$

Now we combine radiation and aerodynamic terms using weighting factors for each:

$$W_R(T) = \frac{\delta(T)}{\delta(T) + \gamma} \tag{15.30}$$

$$W_A(T) = 1 - W_R(T) = \frac{\gamma}{\delta(T) + \gamma} \tag{15.31}$$

The contribution from radiation is

$$E_r = \frac{R_n}{L(T)} W_R(T) \tag{15.32}$$

and from aerodynamics is

$$E_a(T, \text{RH}, u) = f(u)(es - e) W_A(T) \tag{15.33}$$

After adding up, we get PET:

$$E = E_r + E_a \tag{15.34}$$

The top panels of Figure 15.9 are the results for the radiation and aerodynamic terms as functions of the driving variables.

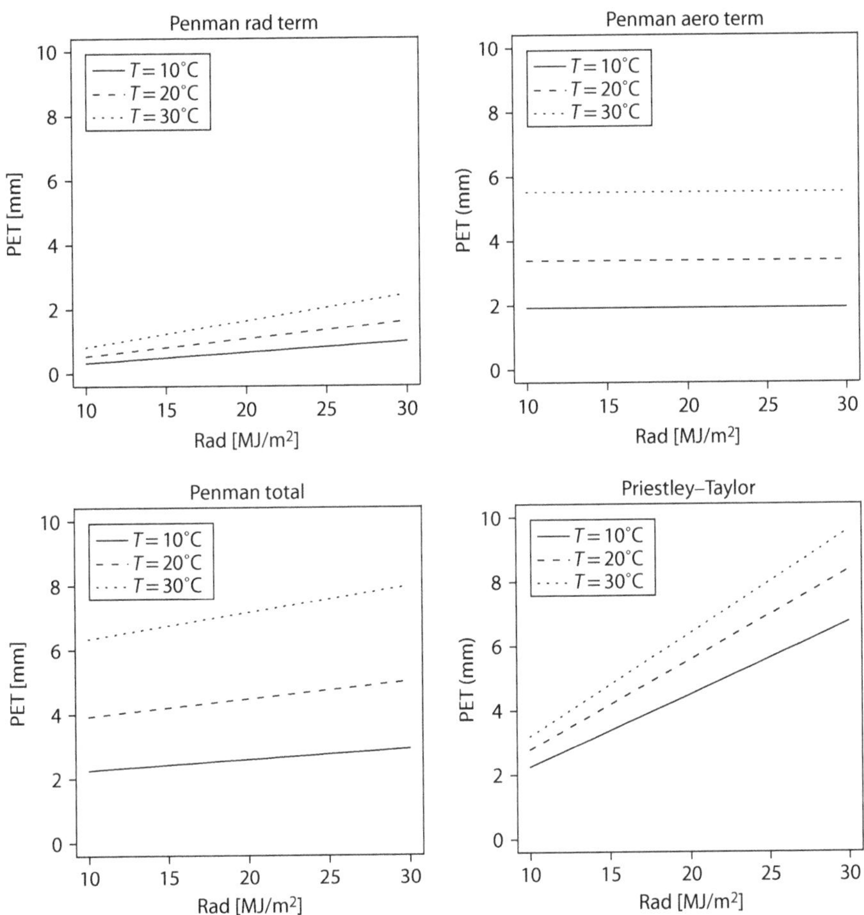

FIGURE 15.9 Top: Penman PET model as function of radiation (left) and wind speed (right), for various values of temperature at fixed RH = 70% and wind speed = 2 ms^{-1}. Bottom: Priestley–Taylor PET model as a function of radiation for several values of temperature and comparison to Penman total at RH = 70% and wind speed = 2 ms^{-1}.

15.2.2 Priestley–Taylor

This is a simplification of the Penman model, which ignores the aerodynamic term and requires only solar radiation and temperature (neither wind speed nor relative humidity).

The net radiation (mean daily) in MJ m^{-2} is calculated using the slope of the vapor pressure curve at saturation (kPa/°C):

$$\delta(T) = \exp\left[21.3 - \frac{5304}{T+273}\right]\left[\frac{5304}{(T+273)^2}\right] \tag{15.35}$$

$$W_R(T) = \frac{\delta(T)}{\delta(T)+\gamma} \tag{15.36}$$

$$h_0(Q) = \frac{2\pi}{365}Q(1-\alpha) \tag{15.37}$$

PET is estimated from

$$E = 30.6 \times h_o(Q)W_R(T) \tag{15.38}$$

The bottom panels of Figure 15.9 show how the Priestley–Taylor estimate varies with radiation for three values of temperature and how it compares with the PET estimated by Penman for the same radiation and temperature conditions, for RH = 70% and wind speed = 2 ms^{-1}.

15.3 SOIL WATER DYNAMICS

Soil moisture is an important environmental condition for terrestrial ecosystems, specifically for plant growth. Soil moisture balance or budget depends on infiltration at the surface, evaporation from the soil, transpiration in plants, and percolation to the water table. Infiltration at the surface in turn depends on precipitation, vegetation properties, soil properties, and soil water content (Hillel, 1998). Figure 15.10 is a schematic representation of the major processes.

The plant canopy intercepts a part R_c of the precipitation R (in mm/h), and therefore, the amount that reaches the soil (and that can infiltrate) is the **net or effective precipitation** R_e:

$$R_e(t) = R(t) - R_c(t) = R(t)(1 - r_c) \tag{15.39}$$

where r_c is a coefficient giving the fraction of rainfall intercepted by the canopy. The net precipitation has potentially two components: one falls through the small gaps between leaves in the canopy ("throughfall") and the other collects from leaves and flows down on the surface of the plant's stems ("stemflow"). For the sake of simplicity, we will not differentiate between these two flows for modeling purposes. Both flows are lumped in the coefficient $1 - r_c$.

Part of the net precipitation infiltrates the soil and part becomes surface **runoff**. Depending on the topography, there could be an additional inflow due to the run-on, that is to say, runoff from areas uphill to the site. Therefore, the balance of water on the surface is

$$\frac{dW_s(t)}{dt} = R_e(t) + Q_{on}(t) - Q_{off}(t) - q_{in}(t) \tag{15.40}$$

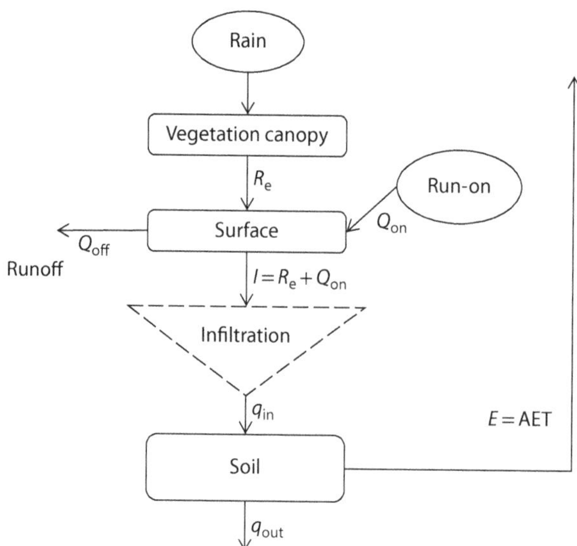

FIGURE 15.10 Ecosystem water budget processes.

where $W_s(t)$ is the water column on the surface (mm), $Q_{on}(t)$ is the runoff from uphill (mm/h) or an addition by irrigation, $Q_{off}(t)$ is the runoff (mm/h), and $q_{in}(t)$ is the infiltration (mm/h). Water at the surface can pond when the balance is positive. Starting from dry soil, runoff is zero, $Q_{off} = 0$, until the input

$$I(t) = R_e(t) + Q_{on}(t) \tag{15.41}$$

exceeds the infiltration rate q_{in}. At this point, water can pond at the surface and runoff occurs.

The infiltration rate turns out to be complicated to model since it depends on many factors such as soil texture, structure, and conditions before the rain or run-on commences. There are several models of infiltration; we will study some of these models a little later. The balance of soil water content S_w (in millimeters) is equal to the input by infiltration minus the losses E due to AET and deep percolation q_{out}:

$$\frac{dS_w(t)}{dt} = q_{in}(t) - E(t) - q_{out}(t) \tag{15.42}$$

Although the water in the soil is held against gravity, past a certain value called the water content at **field capacity**, it drains down because the suction exerted by the soil is low. This soil retention capability depends on soil texture and structure. For many soils, the suction exerted by the soil at field capacity is only about 33 kPa or 0.33 bar. When the soil is at field capacity, the drainage is very slow, and this water is available for plants and ET. The drainage flow at field capacity is about 0.05 mm/day for many soils (Nachabe, 1998).

AET depends on shading by the plant canopy, which depends on biomass (through Leaf Area Index [LAI]). AET is the fraction of PET calculated from biomass and S_w. Extractable water by plants is the water in between water content at field capacity F_c and at **wilting point** W_p, where the wilting point is ~1500 kPa or 15 bar for many soils. This suction is high enough to make it difficult to extract water by the plants.

15.4 SOIL WATER BALANCE

For functions of several variables, setting equations for the rates of change produce a partial differential equations (**PDE**). For example,

$$\frac{\partial X}{\partial t} + a\frac{\partial X}{\partial z} = 0 \tag{15.43}$$

states that the rate of change of X with respect to time t is equal to the rate of change of X with respect to position z, but scaled by a factor a. We encounter this type of equation in many environmental models because we are interested in the temporal dynamics and the variation with respect to position. For example, as we studied in Chapter 14, sunlight and DO were treated as functions of depth in the water column of a lake, and in this chapter, soil water is treated as a function of depth of the soil.

We will study a PDE for one-dimensional (1-D) (depth) soil water dynamics in saturated and nonsaturated soils. We need a PDE because the variables depend on both time and depth. Let us define the following variables and parameters:

- Two independent variables, z = depth (m) and t = time (s).
- $\theta(t, z)$ = relative soil water content as a volume fraction (adimensional [m^3 water/m^3 soil]) is a state variable dependent on both independent variables.
- q = flow density [$m^3/(s\ m^2)$] or ms^{-1}, this is water flow rate per unit of a cross-sectional area.
- $K(P)$ = hydraulic conductivity (ms^{-1}), a function of soil water potential $P(t, z)$, where $P(t, z)$ = water potential (or matric suction) (m) is an alternative state variable, dependent on both time and depth.

- $P(t, z)$ and $\theta(t, z)$ are inversely related, $\theta(P)$ is the **water retention curve**; increasing θ leads to a decrease of P.
- $h(t, z)$ = hydraulic head (m). This should be the total head, that is, the one due to matric suction and head due to gravity; $h(t,z) = -P(t,z) - z$.
- $C(P)$ = capacity (1/m) = change of water content per unit change of potential; this is the slope of $\theta(P)$, the water retention curve, or in other words the derivative of the soil water content $d\theta/dP$ with respect to potential.

Note: Suction is given in meters of water; it is often also given in bars or in kPa. Recall that 1 bar = 1020 cm = 10.20 m = 100 kPa.

Applying **continuity**, we establish that the rate of change of θ with respect to time should be accounted for changes of q with depth:

$$\frac{\partial \theta}{\partial t} + \frac{\partial q}{\partial z} = 0 \tag{15.44}$$

This is obtained by mass balance applied to a control volume. To calculate the rate of change of θ with respect to time, use the chain rule and the fact that θ is a function of P, that is to say $\theta(P)$:

$$\frac{\partial \theta}{\partial t} = \frac{\partial \theta}{\partial P}\left(\frac{\partial P}{\partial t}\right) = C(P)\frac{\partial P}{\partial t} \tag{15.45}$$

An important principle in soil water modeling is **Darcy's** law, which says that flow density is proportional to the negative difference or gradient of head:

$$q = -K(P)\frac{\partial h}{dz} \tag{15.46}$$

Note the similarity to Fick's law of diffusion for mass transport, which was discussed in Chapter 13.

Although originally developed for saturated media (as in an aquifer), Darcy's law can be used here by taking into account matric suction. In this case, the law is called the Darcy–Richards equation (Hemond and Fechner, 1994, p. 195). Recall that the head is the negative of the sum of matric suction and depth; therefore, by substituting $h = -P - z$ and taking the derivative of a sum, we obtain

$$q = -K(P)\frac{\partial(-P - z)}{dz} = K(P)\left(\frac{\partial P}{\partial z} + \frac{\partial z}{\partial z}\right) \tag{15.47}$$

which can be rewritten in the following manner:

$$q = K(P)\left(\frac{\partial P}{\partial z} + 1\right) = K(P) + K(P)\frac{\partial P}{\partial z} \tag{15.48}$$

Therefore, even though Darcy's law was originally formulated only for saturated soil, it can also be extended to nonsaturated soil by making K a function of P as we have shown here.

Exercise 15.2

Assume that the soil has achieved the same potential P_1 at all depths and that $K(P_1) = 0.001\ ms^{-1}$ for this condition. Calculate the water flow density.

Exercise 15.3

Assume a condition such that the soil becomes drier as you go deeper. Assume that you are at a depth where potential P increases 1 m for each meter of depth and that $K = 0.0001$ ms^{-1} at this potential. Calculate the water flow density.

Substituting Darcy's law in the continuity equation gives:

$$C(P)\frac{\partial P}{\partial t}+\frac{\partial}{\partial z}(K(P))+\frac{\partial}{\partial z}\left(K(P)\frac{\partial P}{\partial z}\right)=0 \tag{15.49}$$

This is a PDE with independent variables time t and depth z and one dependent variable $P(t, z)$, which can be solved numerically, after specifying $C(P)$ and $K(P)$ that are functions of the soil type and state variable itself (water potential). It requires the **initial condition** $P(0, z)$ for all depths at time 0 and a **boundary condition** $P(t, 0)$ at the surface for all times t.

Exercise 15.4

Assume two soil layers (layer 1 on top of layer 2) with potentials $P_2 = 1000$ cm and $P_1 = 100$ cm. Which layer has more water? Thickness is such that the distance between layers is 10 cm. Conductivity $K(P_1) = 0.001$ ms^{-1} for this condition. Calculate the water flow density from layer 1 to layer 2. What is the direction of flow? Hint: Use Equation 15.48 for Darcy's law in unsaturated soil and use conductivity given by the potential in layer 1.

15.4.1 Parameterization

However, we still need to specify how C and K depend on matric potential P. Therefore, we need to specify or define submodels for $C(P)$ and $K(P)$. One convenient form is the Brooks–Corey equation (Beer, 1991; van Genutchen, 1980). It specifies a model of the retention curve with P inversely related to θ_e for large values of P:

$$P(\theta_e) = P_b\theta_e^{-b} = \frac{P_b}{\theta_e^b} \tag{15.50}$$

which has two parameters: P_b = bubbling suction (cm) and b = pore-size distribution index, and θ_e = effective saturation given by

$$\theta_e = \frac{\theta - \theta_r}{\theta_s - \theta_r} \tag{15.51}$$

where θ_r = residual soil moisture and θ_s = soil moisture at saturation equal to total porosity. Effective saturation has values between 0 and 1 since $\theta \leq \theta_s$ and $\theta \geq \theta_r$. Therefore, we have introduced four parameters, P_b, b, θ_r, and θ_s. Table 15.1 presents some examples of the values of these coefficients for several soil types.

TABLE 15.1
Examples of Parameter Values

Soil Type	θ_s	θ_r	P_b (cm)	b
Sandy clay	0.43	0.109	29.17	10.4
Sandy loam	0.453	0.041	14.66	4.9
Silty loam	0.501	0.015	20.76	5.3

To determine $C(P)$, we need the inverse of Equation 15.50 expressing water content from P:

$$\theta_e(P) = \left(\frac{P}{P_b}\right)^{-1/b} = \left(\frac{P_b}{P}\right)^{1/b} \tag{15.52}$$

$C(P)$ is defined as the slope of the water retention curve. Therefore, taking the derivative we obtain

$$C(P) = \frac{d\theta}{dP} = \frac{1}{b}\left(\frac{P_b}{P}\right)^{\left(\frac{1}{b}-1\right)}\left(-\frac{P_b}{P^2}\right) \tag{15.53}$$

Now using these quantities, we express saturated conductivity in (cm/s) as follows:

$$K_s = \left[\frac{86}{(b+1)(2b+1)}\right]\left[\frac{(\theta_s - \theta_r)}{P_b}\right]^2 \tag{15.54}$$

Finally, nonsaturated or relative conductivity is a function of effective saturation:

$$K_r = \theta_e^{(3+2b)} \tag{15.55}$$

Figures 15.11 and 15.12 illustrate the submodels for the retention curve, capacity, and conductivity for sandy clay.

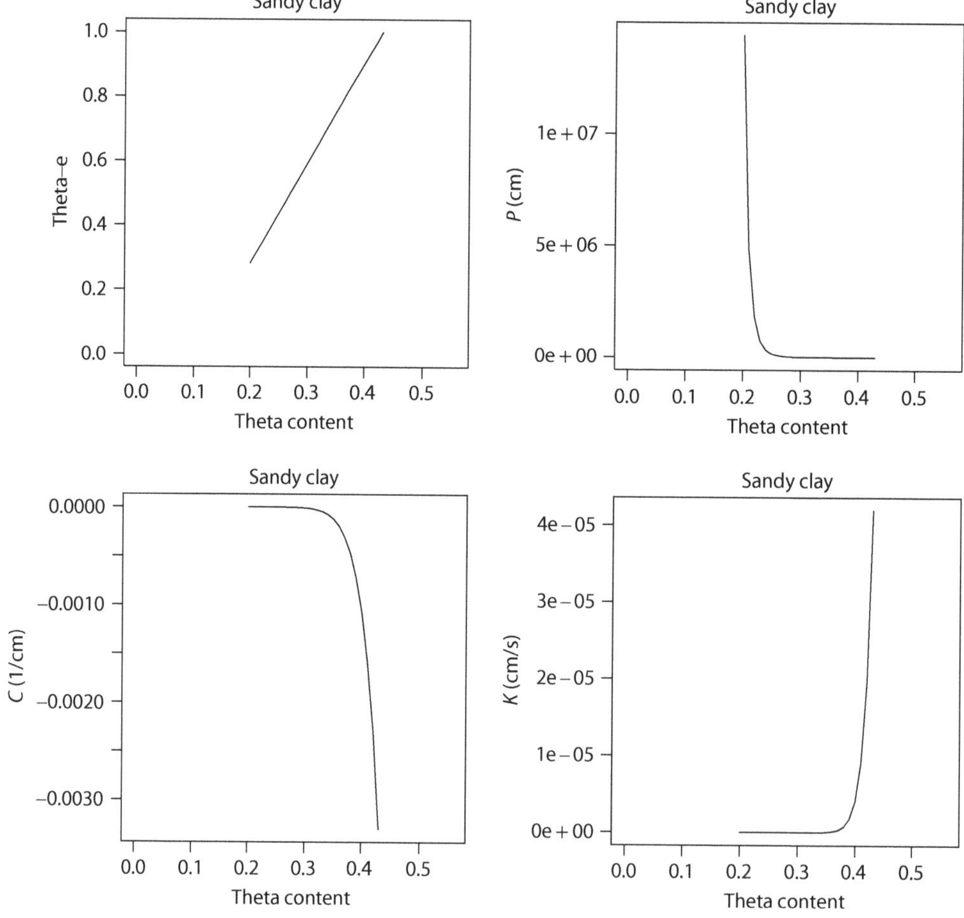

FIGURE 15.11 Soil water characteristics for sandy clay soil.

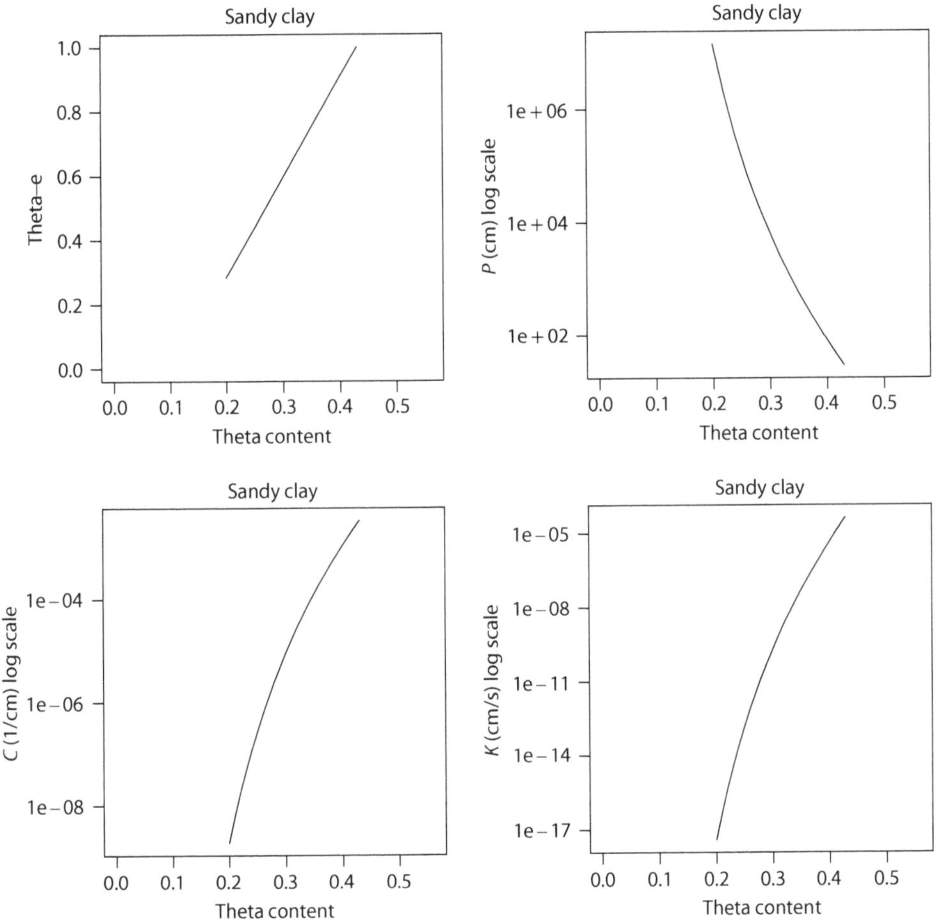

FIGURE 15.12 Soil water characteristics (semilog) for sandy clay soil.

15.4.2 Simplified Lumped ODE Models for Soil Moisture

An alternative to the PDE is a lumped approximation based on multiple discrete soil layers modeled as compartments. Each compartment or layer i has a given thickness, a homogeneous water potential P_i and functions $K_i(P_i)$ and $C_i(P_i)$. In this manner, the PDE model reduces to a system of ODEs, and it is easier to simulate. Input flow to the top layer is the infiltration of precipitation, whereas output by deep percolation occurs at the bottom layer.

For each layer i of thickness dz_i, write the discrete approximation of the **continuity equation**

$$C(P_i)\frac{dP_i}{dt} = -\frac{q_{i-1} - q_i}{dz_i} \tag{15.56}$$

where q_{i-1} is the flow density **into** layer i from layer $i - 1$ immediately above, and q_i is the flow density **out** from layer i to layer $i + 1$ immediately below.

Also, for each pair of layers $i - 1$ and i, we can write Darcy's law as

$$q_{i-1} = -K(P_i)\frac{h_{i-1} - h_i}{dz_i} \tag{15.57}$$

Again, using the head as the negative of the sum of potential and depth and by substituting, we obtain

$$q_{i-1} = -K(P_i)\frac{-P_{i-1} - z_{i-1} + P_i + z_i}{\mathrm{d}z_i} \tag{15.58}$$

The difference in depth between the two layers $z_{i-1} - z_i$ is simply the thickness $\mathrm{d}z_i$, and therefore, the above equation is rewritten as

$$q_{i-1} = K(P_i)\frac{P_{i-1} - P_i - \mathrm{d}z_i}{\mathrm{d}z_i} \tag{15.59}$$

which in turn can be rewritten as

$$q_{i-1} = -K(P_i) + K(P_i)\frac{P_{i-1} - P_i}{\mathrm{d}z_i} \tag{15.60}$$

Then, we substitute this expression for the flow density in the continuity equation to obtain an ODE for each layer:

$$\frac{\mathrm{d}P_i}{\mathrm{d}t} = \frac{-K(P_i) + K(P_i)\frac{P_{i-1} - P_i}{\mathrm{d}z_i} + K(P_{i+1}) - K(P_{i+1})\frac{P_i - P_{i+1}}{dz_{i+1}}}{C(P_i)\mathrm{d}z_i} \tag{15.61}$$

which demonstrates that the rate of change of potential at layer i depends on the potential in the layer immediately above $i - 1$ and the one immediately below $i + 1$. Therefore, the set of ODEs, $i = 1, 2, \ldots, n$ are linked to one another.

For example, consider n layers of equal thickness dz. Denote the infiltration rate by q_{in} and the percolation rate by q_{out}:

$$\frac{\mathrm{d}P_1}{\mathrm{d}t} = \frac{-(0 - q_1)}{C(P_1)\mathrm{d}z} + -q_{\text{in}} \tag{15.62}$$

$$\frac{\mathrm{d}P_2}{\mathrm{d}t} = \frac{-(q_1 - q_2)}{C(P_2)\mathrm{d}z} \tag{15.63}$$

and so on until the last layer

$$\frac{\mathrm{d}P_n}{\mathrm{d}t} = \frac{-(q_{n-1} - 0)}{C(P_n)\mathrm{d}z} + q_{\text{out}} \tag{15.64}$$

when Darcy's law is substituted for each flow density:

$$\frac{\mathrm{d}P_1}{\mathrm{d}t} = \frac{-\left(0 - K(P_2) + K(P_2)\frac{P_1 - P_2}{\mathrm{d}z}\right)}{C(P_1)\mathrm{d}z} \tag{15.65}$$

$$\frac{dP_2}{dt} = \frac{-\left(-K(P_2)+K(P_2)\frac{P_1-P_2}{dz}+K(P_3)-K(P_3)\frac{P_2-P_3}{dz}\right)}{C(P_1)dz} \tag{15.66}$$

$$\frac{dP_n}{dt} = \frac{-\left(-K(P_n)+K(P_n)\frac{P_{n-1}-P_n}{dz}\right)}{C(P_{n-1})dz} + q_{out} \tag{15.67}$$

15.5 INFILTRATION AND RUNOFF

Infiltration is water entry into the soil from the surface. It determines how much water will be available for movement and storage in the soil and therefore soil water availability for plants. It also determines the amount of surface runoff that ultimately affects stream discharge and soil erosion. Soil moisture (especially at the soil top layer or surface) affects the infiltration rate. There are many models for infiltration into the soil. Two examples are the exponential (Horton) and Green–Ampt models.

Infiltration rate is studied experimentally by measuring **cumulative infiltration** $Q_{in}(t)$, that is, how much water (in mm) has infiltrated into the soil up to a certain time t. We will assume that initially the soil is dry and the instantaneous **infiltration rate** $q_{in}(t)$ is high, and as time goes on and the soil water content increases, the infiltration rate decreases until it achieves a **steady state** or **basic** infiltration rate q_{in}^* when the soil is saturated. Note that at all times, the infiltration rate is the derivative of the cumulative infiltration $q_{in}(t) = \frac{dQ_{in}(t)}{dt}$.

15.5.1 Exponential: Horton

This model is an exponential law of **infiltration rate capacity** $f(t)$ derived from the cumulative infiltration curve. We distinguish capacity $f(t)$ from the actual infiltration rate $q_{in}(t)$, because the actual rate is also a function of water available for entry into the soil $I(t)$. Denote the initial infiltration capacity by f_0 when the soil is dry. The capacity follows an exponential decay from this value to the steady-state or final value f_c:

$$f(t) = f_c + (f_0 - f_c)\exp(-kt) \tag{15.68}$$

Parameters f_0, f_c, and k are estimated from experimental data of infiltration rate versus t. However, it is possible to derive this exponential behavior under some specific conditions of a mechanistic model from first principles (Diskin and Nazimov, 1995).

For parameters, we will use soil porosity, conductivity, and field capacity (see Figure 15.13). Infiltration capacity decreases linearly with soil water content $S_w(t)$:

$$f(t) = f_d - (f_d - f_c)\frac{S_w(t)}{\eta Z} \tag{15.69}$$

where $f(t)$ is infiltration capacity at time t, f_d is infiltration capacity when soil is dry (mm/h), f_c is capacity when soil is saturated (mm/h), η is porosity (adimensional) and equal to θ_s, and Z is depth (mm) determining saturation volume to limit infiltration (see Figure 15.14). Note that saturation will occur when $S_w = \eta Z$ and that Z is not the total soil depth but an infiltration model parameter. This is actually a simplification where we have assumed that $\theta_r = 0$. We will assume that f_c is equal to the saturated hydraulic conductivity K_s.

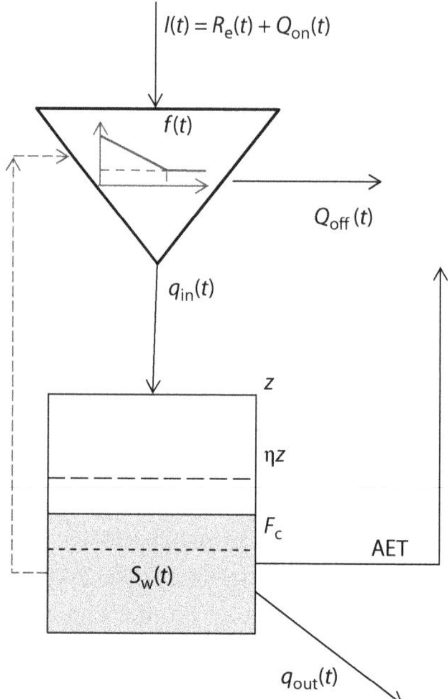

FIGURE 15.13 Infiltration model including capacity as a function of soil water content and actual rate as a function of capacity and available water at the surface.

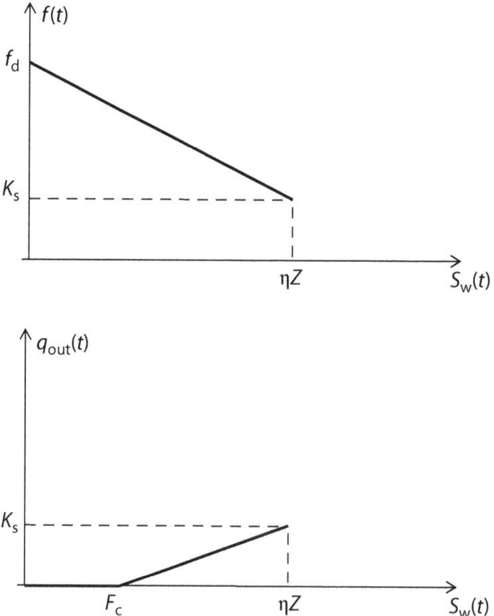

FIGURE 15.14 Infiltration capacity as a function of soil water content (top) and percolation as a function of soil water content (bottom).

Alternatively, we can formulate it as an increase in capacity with deficit with respect to saturation $\eta Z - S_w(t)$:

$$f(t) = K_s + (f_d - K_s)\frac{\eta Z - S_w(t)}{\eta Z} \tag{15.70}$$

which of course is the same as Equation 15.69. Defining effective θ_e water content (mm/mm) as $\theta_e(t) = \frac{S_w(t)}{\eta Z}$, Equation 15.70 becomes

$$f(t) = K_s + (f_d - K_s)(1 - \theta_e(t)) \tag{15.71}$$

The actual infiltration rate $q_{in}(t)$ is now a function of the available water inflow:

$$q_{in}(t) = \begin{cases} f(t) & \text{when} \quad I(t) \geq f(t) \\ I(t) & \text{when} \quad I(t) < f(t) \end{cases} \tag{15.72}$$

The difference between available input $I(t)$ and infiltration rate is the runoff $Q_{off}(t)$.

Exercise 15.5

Assume soil of depth $Z = 100$ mm, porosity $\eta = 0.43$, $K_s = 50$ mm/h, and $f_d = 200$ mm/h. Sketch a graph of infiltration capacity versus S_w. What is the soil water content at saturation? When S_w is 20 mm, what is the soil water content as a fraction θ of saturation? What is the deficit with respect to saturation? What is the infiltration capacity at this θ? Assuming that input exceeds infiltration capacity, what is the actual infiltration rate?

To calculate the infiltrated volume for start of ponding and time of ponding, use Equation 15.70. Ponding and runoff will occur when $I(t)$ exceeds capacity $f(t)$. This starts for an accumulated infiltration F_p when

$$I_c = K_s + (f_d - K_s)\frac{\eta Z - (F_p + S_w(0))}{\eta Z} \tag{15.73}$$

assuming $I(t)$ constant at rate I_c. Solving for F_p,

$$F_p = \eta Z\left(1 - \frac{I_c - K_s}{f_d - K_s}\right) - S_w(0) \tag{15.74}$$

and then, the time to ponding is $t_p = \frac{F_p}{I_c}$. For example, assume a soil with $K_s = 50, f_d = 200$ mm/h, porosity $\eta = 0.43$, and depth $= 100$ mm. Assume an initial condition $\theta = 0.3$. Using Equation 15.74 for $I_c = 70$ mm/h, we get $F_p = 100 \times 0.43\left(1 - \frac{70-50}{200-50}\right) - 100 \times 0.3 = 37.3 - 30 = 7.3\,\text{mm}$ and the time to ponding $t_p = 7.3/70 = 0.1\,\text{h}$ or 6 minutes.

This simple calculation of time of ponding assumes constant rainfall intensity that does not take into account complicated patterns of time variation of rainfall intensity. Alternative methods can be utilized (Assouline et al., 2007).

Exercise 15.6

Assume soil of depth $Z = 100$ mm, porosity $\eta = 0.43$, $K_s = 50$ mm/h, and $f_d = 200$ mm/h. For an initial condition of 20 mm, what is the infiltrated volume at ponding and time to ponding?

Percolation depends on conductivity and available water to drain, which is a function of field capacity:

$$q_{\text{out}}(t) = \begin{cases} K_s \dfrac{S_w(t) - F_c}{\eta Z - F_c} & \text{when} \quad S_w(t) \geq F_c \\ 0 & \text{when} \quad S_w(t) < F_c \end{cases} \tag{15.75}$$

where F_c is a field capacity. Note that $q_{\text{out}}(t)$ increases linearly with $S_w(t)$ from 0 when the soil is at field capacity or drier to K_s when the soil is saturated (Figure 15.14, bottom panel). Field capacity can be expressed as a coefficient c multiplied by the saturation ηZ. For example, for a saturation of $\eta Z = 50$ mm and a field capacity of $F_c = 30$ mm, the coefficient is $c = \dfrac{30}{50} = 0.6$.

Exercise 15.7

For the conditions of Exercise 15.6 and $c = 0.6$, what is the value of field capacity F_c? What is the percolation flow? What is the percolation flow when $S_w = 38$ mm?

Typically, infiltration would occur during and shortly after rainfall events. Ignoring AET loss during this time, soil water balance is now an ODE written as the balance of input given by infiltration and the output given by percolation:

$$\frac{dS_w(t)}{dt} = q_{\text{in}}(t) - q_{\text{out}}(t) \tag{15.76}$$

Using Equations 15.69, 15.72, and 15.75, when the available input $I(t)$ exceeds infiltration capacity and the soil is unsaturated

$$\frac{dS_w(t)}{dt} = f(t) - q_{\text{out}}(t) = f_d - (f_d - K_s)\frac{S_w(t)}{\eta Z} - K_s \frac{S_w(t) - F_c}{\eta Z - F_c} \tag{15.77}$$

On rearranging, we can see that it is an ODE forced by a constant term

$$\frac{dS_w(t)}{dt} = u - kS_w(t) \tag{15.78}$$

where

$$u = f_d + K_s \frac{F_c}{\eta Z - F_c} \tag{15.79}$$

$$k = \frac{f_d - K_s}{\eta Z} + \frac{K_s}{\eta Z - F_c} \tag{15.80}$$

The solution of Equation (15.78) is

$$S_w(t) = \frac{u}{k} + \left(S_w(0) - \frac{u}{k}\right)\exp(-kt) \tag{15.81}$$

With initial condition $S_w(0)$ and steady state equal to saturation, we obtain

$$S_w(\infty) = \frac{u}{k} = \eta Z \tag{15.82}$$

Therefore,

$$S_w(t) = \eta Z - (\eta Z - S_w(0))\exp(-kt) \tag{15.83}$$

The dynamics of the infiltration rate during this time is obtained by substituting Equation 15.83 into Equation 15.69:

$$\begin{aligned} q_{in}(t) &= f_d - (f_d - K_s)\frac{\eta Z - (\eta Z - S_w(0))\exp(-kt)}{\eta Z} \\ &= K_s - (f_d - K_s)\left(\frac{-(\eta Z - S_w(0))\exp(-kt)}{\eta Z}\right) \end{aligned} \tag{15.84}$$

which is a simple exponential

$$q_{in}(t) = K_s + a\exp(-kt) \tag{15.85}$$

where

$$a = (f_d - K_s)\left(\frac{\eta Z - S_w(0)}{\eta Z}\right) \tag{15.86}$$

Starting from dry soil $S_w(0) << \eta Z$, Equations 15.83 and 15.85 reduce to

$$\begin{aligned} S_w(t) &= \eta Z(1 - \exp(-kt)) \\ q_{in}(t) &= K_s + (f_d - K_s)\exp(-kt) \end{aligned} \tag{15.87}$$

This last equation is similar to Equation 15.68 describing the exponential behavior of the Horton model. This behavior is illustrated in Figure 15.15. In other words, the model described here shows Horton behavior for constant input above infiltration capacity starting from dry soil. When input does not exceed capacity at all times, actual infiltration rate is limited by either input rate or capacity. We can see in Figure 15.16 that infiltration rate is limited to the input I_c = 100 mm/h while capacity exceeds input and then is limited by capacity as it dips below I_c.

For a quick estimation of time to reach saturation starting from dry soil, we can use the time to reach a value of 5% from saturation:

$$\begin{aligned} \exp(-kT_5) &= 0.05 \\ T_5 &= -\frac{\ln(0.05)}{k} \approx \frac{3}{k} \end{aligned} \tag{15.88}$$

For example, when $k = 6$, $T_5 = 1/2$ h, but for $k = 9$, $T_5 = 1/3$ h or 20 min.

Exercise 15.8

Assume soil of depth $Z = 100$ mm, porosity $\eta = 0.43$, $K_s = 50$ mm/h, $f_d = 200$ mm/h, and field capacity coefficient $c = 0.6$. Calculate k and time to 5% of saturation starting from dry soil.

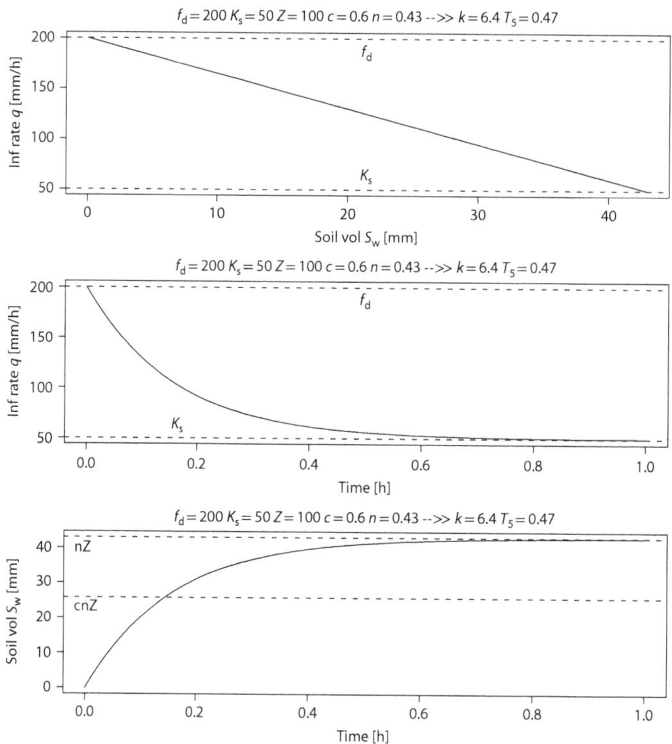

FIGURE 15.15 Exponential infiltration soil water dynamics. Top: infiltration capacity as a linear function of soil water content. Middle: exponential decay of infiltration rate versus time. Bottom: exponential increase of soil water content versus time.

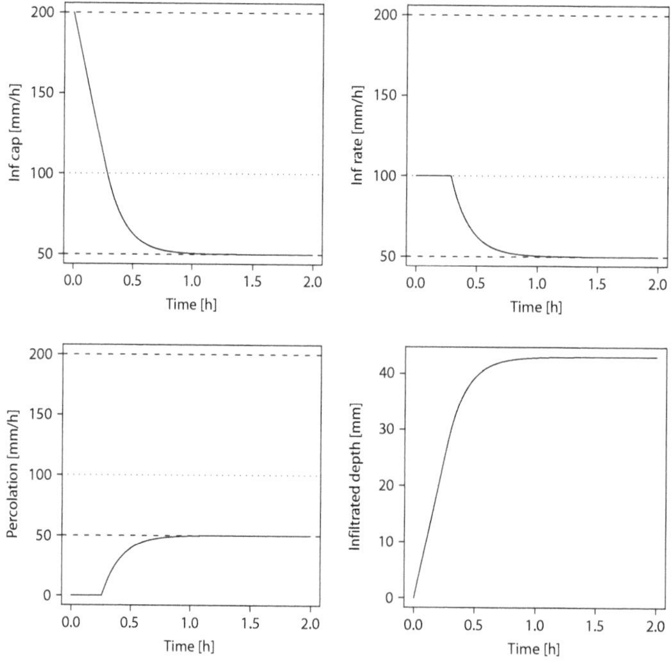

FIGURE 15.16 Exponential dynamics for constant $I_c = 100$ mm/h. Top: infiltration capacity and infiltration rate. Bottom: percolation and soil water content.

Figure 15.17 illustrates the dynamics of soil moisture once we include both infiltration and percolation processes as in Equation 15.77 and as a function or input intensity I_c. Here, S_w is the state variable X. In these simulations, we assume that I is held constant at I_c for 1.5 h and then ceases taking the value 0. We can see how a low $I_c = 10$ charges the soil with water linearly to about 25 mm, remaining constant thereafter. For higher values of I_c, the soil charges with water past field capacity, and therefore, percolation starts to operate making the charging lines curve. For these higher I_c values, the soil water drains down to the value at field capacity after cessation of the input at 1.5 h. This behavior is appreciated by considering the rates illustrated in the bottom panel of Figure 15.17. It should be noted that the percolation rate is linear with respect to the excess water above field capacity, thereby producing exponential decay.

As a simplified model of multiple layers of soil, we could assume that the flow from one layer to the next is due to percolation as a linear process. Denoting X_i as the soil content in layer i

$$\frac{dX_i(t)}{dt} = g_{i-1}(t) - g_i(t) \tag{15.89}$$

for all layers except the top for which $\frac{dX_1(t)}{dt} = q_{in}(t) - g_1(t)$. The simplest case is one where all layers have very similar characteristics leading to equal values of K_s, F_c, and ηZ.

$$\frac{dX_i(t)}{dt} = K_s \frac{X_{i-1}(t) - F_c}{\eta Z - F_c} - K_s \frac{X_i(t) - F_c}{\eta Z - F_c} \tag{15.90}$$

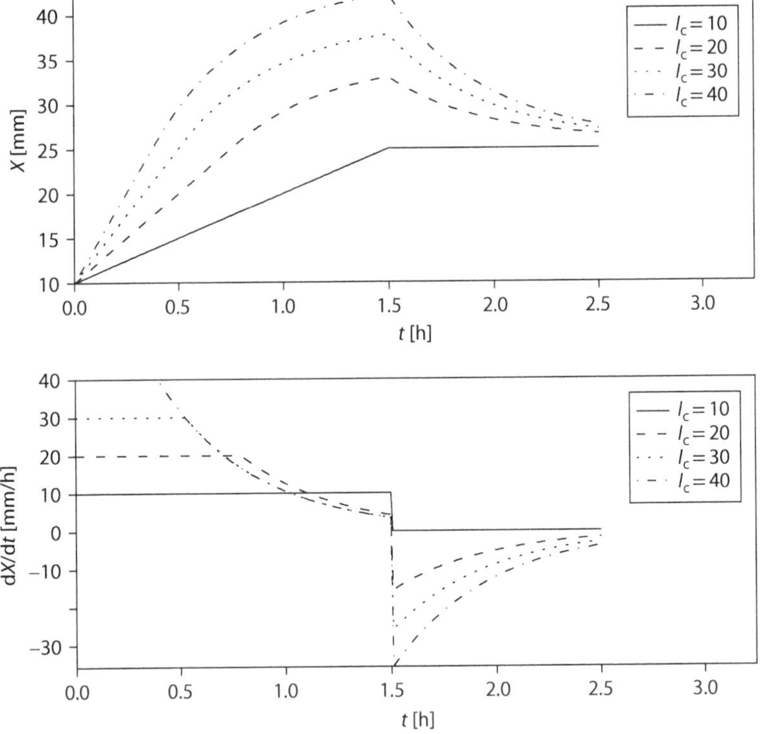

FIGURE 15.17 Exponential dynamics for several values of constant I_c during 1.5 h. Top: soil water content. Bottom: rate as balance of infiltration and percolation.

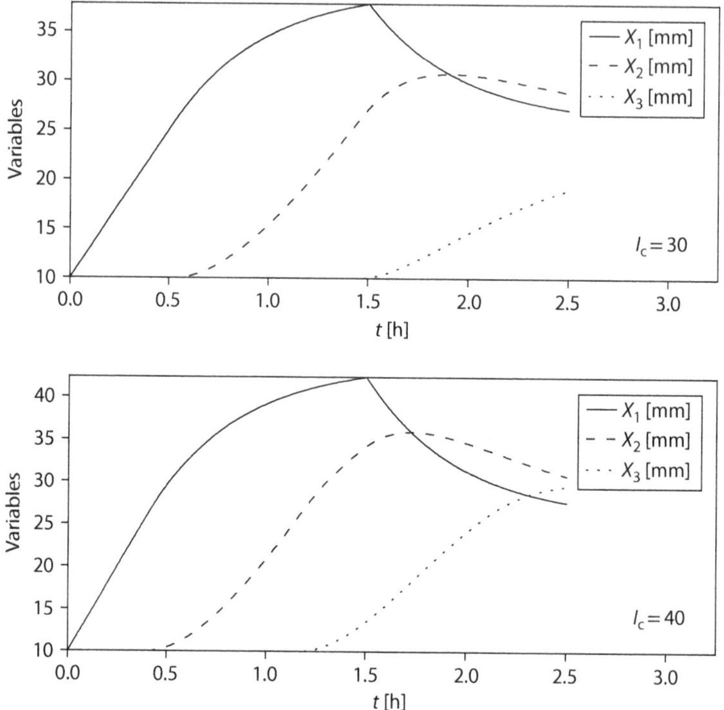

FIGURE 15.18 Soil water dynamics for three identical layers.

Figure 15.18 illustrates simulation results for three layers. Please note the time delay of about 0.5 hours for the water content of each layer to rise in response to the infiltration at the top layer.

15.5.2 Green–Ampt: Wetting Front

This is a mechanistic model based on an approximation to the movement downwards in the unsaturated soil. The main concept is that there is a **wetting front** dividing the saturated (top) and the unsaturated areas. It also assumes that there is shallow water ponding at the surface (Figure 15.19). The unsaturated area is at the initial soil moisture content. The front moves downward as time progresses, and the downward velocity is the infiltration rate. We can write Darcy's law in the form

$$q = -K_s \frac{\Delta h}{z_f} = -K_s \frac{h_0 - h_f}{z_f} \tag{15.91}$$

where q is water flow rate; K_s is saturated conductivity of the wetted area; Δh is the difference of head between the front and the surface, that is, between h_0 (head at the surface) and h_f (the head at the wetting front); and z_f the depth of the front. The head at the surface is the depth of ponding, and the head at the front is composed of depth z_f and potential P_f, and therefore,

$$q = -K_s \frac{h_0 + z_f + P_f}{z_f} = -K_s \left(1 + \frac{h_0 + P_f}{z_f}\right) \tag{15.92}$$

The negative sign makes the direction of q negative (downwards) assuming z and P are negative. We ignore the sign for simplicity and make this flow rate equal to the infiltration capacity f. Assume that h_0 is very small compared to the potential at the front:

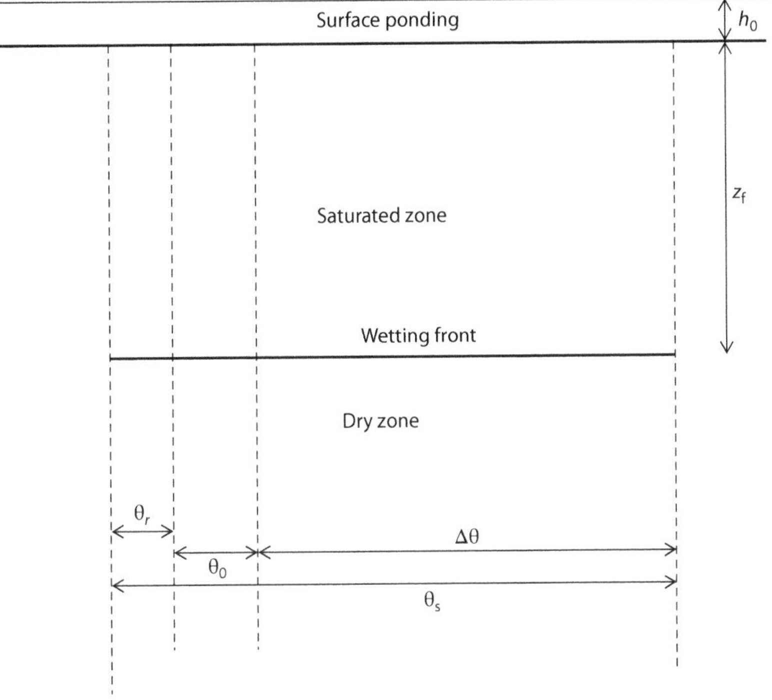

FIGURE 15.19 Schematic of Green–Ampt.

$$f(t) = K_s\left(1 + \frac{P_f}{z_f}\right) \tag{15.93}$$

The cumulative infiltration $F(t)$ should be the depth of water infiltrated in the soil, which is the product of the initial soil moisture deficit $\Delta\theta$ and the depth of the front:

$$F(t) = z_f(\eta - \theta_0) = z_f\Delta\theta \tag{15.94}$$

where ηz_f is the water content in the wetted area (above the front) and θ_0 is the initial water content (as a fraction) that prevails below the front. Solving for z_f in Equation 15.94 and substituting in Equation 15.93, we obtain

$$f(t) = \frac{dF(t)}{dt} = K_s\left(1 + \frac{P_f\Delta\theta}{F(t)}\right) \tag{15.95}$$

We can rewrite this as an ODE in which F is the dependent variable in the form

$$\frac{dF(t)}{dt} = K_s\left(1 + \frac{P_f\Delta\theta}{F(t)}\right) \tag{15.96}$$

This ODE has a solution:

$$F(t) - P_f\Delta\theta\ln\left(1 + \frac{F(t)}{P_f\Delta\theta}\right) - K_s t = 0 \tag{15.97}$$

There are various numerical methods to solve this equation (Kale and Sahoo, 2011). One practical method described in that review (also in Mailapalli et al., 2009) is a numerical integration of Equation 15.96, $F(t+\Delta t) = F(t) + \Delta F(t)$, using the following formula based on a nonstandard integration method (Ramos, 2007) and making Δt very small (e.g., 10^{-5} hours):

$$F(t+\Delta t) = F(t) + \frac{2\Delta t f}{2 - \Delta t\left(\frac{\mathrm{d}f}{\mathrm{d}F}\right)} \tag{15.98}$$

Then we calculate $\frac{\mathrm{d}f}{\mathrm{d}F}$, and substitute it along with f:

$$\Delta F(t) = \frac{2\Delta t K_s\left(1 + \frac{P_f \Delta\theta}{F(t)}\right)}{2 - \Delta t\left(-K_s \frac{P_f \Delta\theta}{F(t)^2}\right)} \tag{15.99}$$

We cannot use $F(0) = 0$ because it is in the denominator of quotients, and for practical purposes, we can make it a very small number, for example, $F(0) = 10^{-4}$. The potential at the front can be calculated from Brooks–Corey parameters P_b and b in the following manner:

$$P_f = \frac{2b+3}{b+3} P_b \tag{15.100}$$

For example, for sandy loam, $P_b = 14.66$ cm, $b = 4.9$, and by using this equation, we obtain $P_f = 23.8$ cm or 238 mm. For this soil, $K_s = 38$ mm/h, $\eta = 0.453$. We will show examples for $\theta_0 = 0.1$.

Numerical integration using Equation 15.99 varying K_s yields results as shown in Figure 15.20. The infiltrated volume F increases very rapidly at the beginning when infiltration capacity has very high values and then slows down as infiltration capacity decreases to the final K_s value. Figure 15.21 illustrates that input rate (200 mm/hour in this case) would limit actual infiltration rate and infiltrated depth because input is generally lower than the high capacity predicted at low F. From the figure, in this example, time of ponding is slightly above 0.05 hours, or about 3 minutes. We will discuss actual rates next.

For this purpose, we relate rate capacity to the input rate $I(t)$ as we did in Equation 15.72, which we repeat here for easy reference. The actual infiltration rate $q_{in}(t)$ is now a function of available water inflow:

$$q_{in}(t) = \begin{cases} f(t) & \text{when} \quad I(t) \geq f(t) \\ I(t) & \text{when} \quad I(t) < f(t) \end{cases} \tag{15.101}$$

Exceedance ponding and runoff will occur when $I(t)$ exceeds capacity $f(t)$. Using Equation 15.95, this occurs for an accumulated infiltration F_p when

$$I_c = K_s\left(1 + \frac{P_f \Delta\theta}{F_p}\right) \tag{15.102}$$

assuming $I(t)$ is constant at rate I_c and greater than K_s. Solving for $F_p(t)$,

$$F_p = \frac{P_f \Delta\theta}{\frac{I_c}{K_s} - 1} \tag{15.103}$$

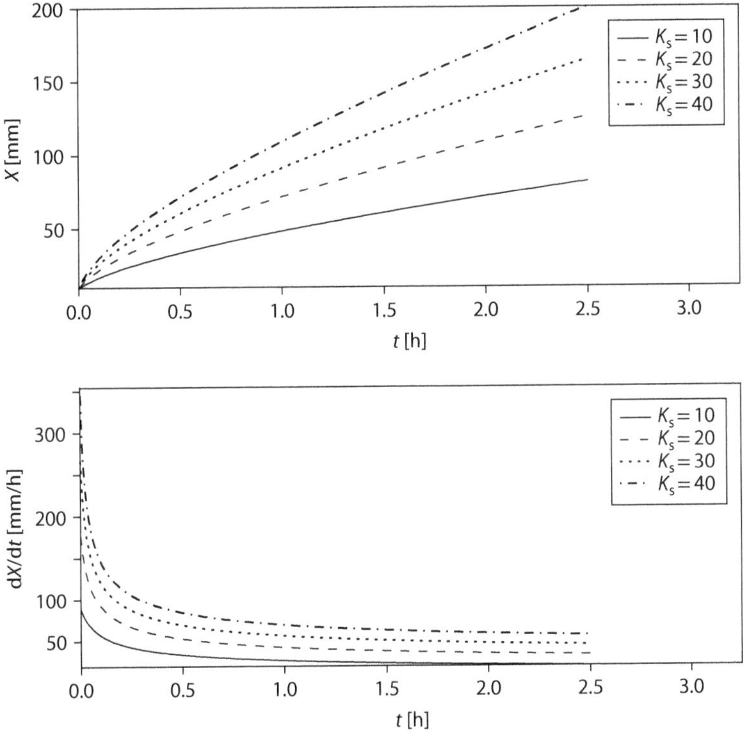

FIGURE 15.20 Green–Ampt dynamics of infiltrated depth (top) and infiltration rate capacity (bottom) for several values of parameter K_s. Here, variable X denotes F.

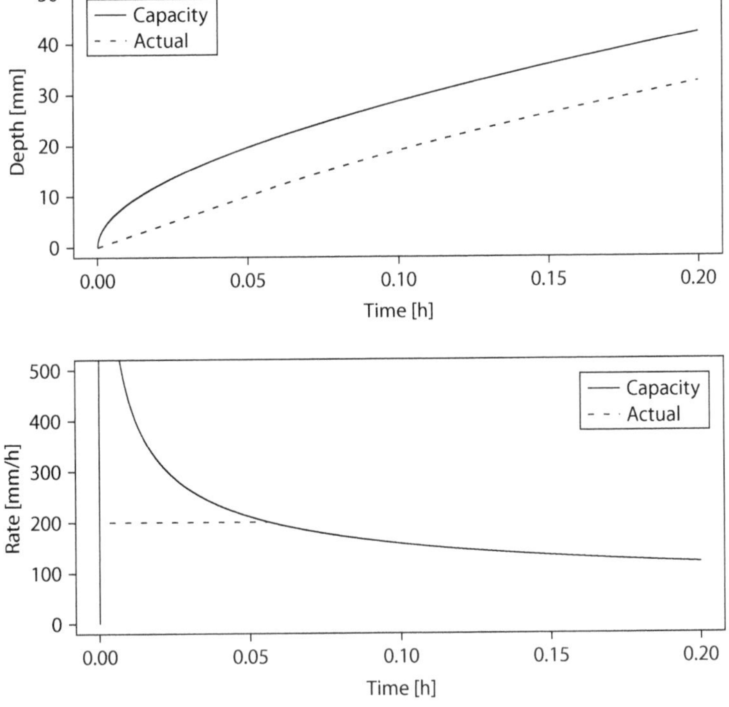

FIGURE 15.21 Green–Ampt dynamics of infiltrated depth (top) and infiltration rate (bottom) taking into account limitation by rain.

This volume will be reached when $I_c t_p = F_p$, and therefore, time to ponding is

$$t_p = \frac{F_p}{I_c} = \frac{P_f \Delta\theta}{\left(\frac{I_c}{K_s} - 1\right) I_c} \quad (15.104)$$

For example, for sandy loam soil of porosity $\eta = \theta_s = 0.453$, $P_f = 238$ mm, and $K_s = 38$ mm/h, suppose that the initial condition $\theta_0 = 0.3$; we have $P_f \Delta\theta = 238 \times (0.453 - 0.3) = 36.4\,\text{mm}$. For a rainfall intensity of 70 mm/hour, the depth of infiltration by the time to ponding is $F_p = \frac{36.4}{\frac{70}{38} - 1} = \frac{36.4}{0.84} = 43.3\,\text{mm}$ and the time to ponding is $t_p = \frac{F_p}{I_c} = \frac{43.3}{70} = 0.619\,\text{h}$ or 37 min.

Exercise 15.9

Assume sandy clay soil and an initial condition $\theta_0 = 0.3$. Use Table 15.1 to calculate P_f, K_s, and $\Delta\theta$. For an input I_c of 70 mm/hour, calculate the time to ponding and depth of infiltration by this time.

15.6 COMPUTER SESSION

15.6.1 Weather Generation: Rainfall Models

We will use the function `markov.rain`. As explained at the end of the chapter, this implements a Markov chain of dry–rainy days based on a transition probability matrix **P**. Also, for those days resulting as rainy, it draws an amount of rain at random, using function `rain.day`. This function allows to select from a set of distributions: exponential, Weibull, gamma, and skewed. Arguments are the transition matrix, the number of days, and the statistics of the daily rainfall amount: mean, standard deviation, skewness, shape, and the model pdf. The function returns the sequence of amount of rainfall one for each day as well as statistics of the sample sequence and expected values for the sequence. In addition, the function produces a graphics window with plots for the sequence and histogram of rain amounts.

As an example, let us use a month with high probability (say 0.80) that rainy days are followed by rainy days and use mean 5 and Weibull with shape 1.3.

```
amount.param=list(mu=5,std=8,skew=2, shape=1.3, model.pdf ="w")
ndays=30
# rainy followed by rainy
P <- matrix(c(0.4,0.2,0.6,0.8), ncol=2, byrow=T)
rainy1 <- markov.rain(P, ndays, amount.param)
mtext(side=3,line=-1,paste("Rainy after rainy Pr=0.80","Prop
rainy=",round(rainy1$wet.days/ndays,2)),cex=0.8)
```

The result is as shown in Figure 15.5. The figure shows the number of wet spells in the month and that most rain values are low. Let us look at the numeric results:

```
> rainy1
$x
 [1]  0.00  0.83  3.88  3.40  0.00  0.00  11.46  26.60   1.99  0.83  10.53  4.22
[13] 13.11  2.69  0.05 17.98  1.32  7.04   0.00   1.22  18.62  6.04   8.09  0.00
[25]  1.44  0.00  6.63  5.64  0.88  0.15
$wet.days
[1] 24
```

```
$expected.wet.days
[1] 22.5

$dry.days
[1] 6

$expected.dry.days
[1] 7.5

$rain.tot
[1] 154.65

$expec.rain.tot
[1] 112.5

$rain.avg
[1] 5.15

$expected.rain.avg
[1] 3.75

$rain.wet.avg
[1] 6.44

$expected.wet.avg
[1] 5
```

We see that most days were rainy, and the overall rain average of 5.15 millimeters (counting dry days as well) was below the average of 6.44 millimeters for only the wet days. We can see that the sample statistics are not too far from the expected results.

Exercise 15.10

Study the rainfall values for a contrasting condition where dry days tend to be followed by dry days. The transition matrix is c(0.8,0.6,0.2,0.4). Provide plots and numeric results. Discuss the results. Compare with Figure 15.6.

Note that every time we run this function, we obtain a different realization. We can do many realizations by writing a loop around the function as given in the following script. As an example, we will use nine realizations and obtain the results given in Figure 15.7 and their histograms in Figure 15.8.

```
amount.param=list(mu=5,std=8,skew=2, shape=1.3, model.pdf ="w")
ndays=30
# rainy followed by rainy
P <- matrix(c(0.4,0.2,0.6,0.8), ncol=2, byrow=T)
nruns <- 9

# define array
z <- matrix(ncol=nruns, nrow=ndays)

# loop realizations
for(j in 1:nruns) {
  rainy <- markov.rain(P, ndays, amount.param,plot.out=F)
```

```
  z[,j] <- rainy$x
} # end of realization loop

# plot
mat<- matrix(1:9,3,3,byrow=T)
layout(mat,rep(7/3,3),rep(7/3,3),res=TRUE)
par(mar=c(4,4,1,.5), xaxs="r", yaxs="r")
for(i in 1:nruns) {
 plot(z[,i], type="s", ylab="Rain(mm)", xlab="Days", ylim=c(0,max(z)))
 mtext(side=3,line=-1,paste("run",i),cex=0.8)
}
for(i in 1:nruns){
 hist(z[,i],prob=T,main="",xlab="Rain (mm)")
 mtext(side=3,line=-1,paste("run",i),cex=0.8)
}
```

This script simulates nine realizations producing plots of rain sequences for each realization (Figure 15.7) and histograms of rain amounts for each realization (Figure 15.8).

Exercise 15.11

Add lines of code to the previous script to calculate the sample mean of the number of wet days and the monthly averages. Use 10 and 100 sample sizes (realizations). Compare and discuss. Hint: The sample mean of the number of wet days would be the average of the number of wet days for all runs. Likewise, the sample mean of the monthly means would be the average of the mean for all runs (realizations). Also, note that this can be done in two different ways: for only the wet days of the month or for all the days of the month.

15.6.2 Evapotranspiration

Use the `pet` function to calculate PET by the Penman and Priestley–Taylor methods.

```
pet.test <- pet(0, Rad =seq(10,30), Temp=c(10,20,30), RH=70, wind=2,
  plot=T)
```

By varying temperature, RH, and wind speed, we get a sense of temperature, radiation, RH, and wind controls on PET (see Figure 15.9).

Exercise 15.12

Vary wind speed and RH to dry-windy conditions (e.g., $u = 2$ ms^{-1} and RH = 50%) and humid-calm conditions (e.g., $u = 0.2$ ms^{-1} and RH = 90%). Discuss the results. Compare the conditions. Compare the Priestley–Taylor approximation to Penman for each of these conditions.

15.6.3 Soil Water Properties

Now we use the function `soil-water`, which calculates soil water retention and water conductivity using the Brooks–Corey equation. The function is listed and explained at the end of the chapter. Arguments include a set of values of θ and parameters P_b, b, θ_r, and θ_s. In addition, we supply a `title` and a logical argument `plot` to generate graphics when = `T`. Calculations are plotted in both regular and semilogarithmic axes (the y-axis). Parameter values are given in Table 15.1. For example, for sandy clay soil,

```
# sandy clay
param=c(Pb=29.17,b=10.4, theta.r=0.109,theta.s=0.43)
theta <- seq(0.2,0.43,0.01)
sw <- soil.water(theta, param, title="Sandy Clay", plot=T)
```

This function produces two graphics windows; in the first, the plots are in regular scale for the y-axis (Figure 15.11), whereas in the second, we use log scale for the y-axis (Figure 15.12).

Exercise 15.13

Change the value of the parameters to a sandy loam soil; examples for parameter values are given in Table 15.1. Produce graphs of P, K, and C. Discuss the differences with respect to sandy clay.

15.6.4 Soil Water Dynamics: Exponential Infiltration and Percolation

We will work with a function `soilwat.1`, calling it from `sim`. The `soilwat.1` function is explained at the end of the chapter and implements Equation 15.77 as a function of input intensity and duration. The input file **chp15/sw-inp.csv** has the parameter values and initial conditions.

```
Label,Value,Units,Description
t0,0.00,hr,time zero
tf,2.5,hr,time final
dt,0.001,hr,time step
tw,0.01,hr,time to write
fd,238,mm,conductivity dry
Ks,38,mm/hr,conductivity
Z,100,mm,depth
Fc,0.6,adim,field capacity coeff
nu,0.43,porosity
Ic,100,mm,input intensity (rain+runon)
Rt,1.5,h,input(rain+runon) duration
X0,10,mm,initial condition
digX,4,None,significant digits for output file
```

In this case, we start at 10% saturation (10 mm out of 100 mm for saturation) and apply input for 1.5 hours. We run the simulations for various values of input intensity from below infiltration capacity to above infiltration capacity.

```
sw <- list(f=soilwat.1);param <- list(plab="Ic",pval=c(10,20,30,40))
t.X <- sim(sw,"chp15/sw-inp.csv", param, pdfout=T)
```

The graphical results in the PDF file include Figure 15.17, which we already discussed.

Exercise 15.14

Change the value of the initial condition to wetter conditions, say X_0 = 20 and 30 millimeters. Repeat the simulations for increasing I_c as in the example. Produce sets of graphs for each value of X_0 and discuss.

15.6.5 Infiltration Using Green–Ampt

In this section, we simulate the Green–Ampt model utilizing two sets of functions that implement the Mailapalli et al. (2009) method given in Equation 15.99 and explained at the end of the chapter. As an example, we work with a sandy loam soil, give parameter values, and use the previous function `soil.water` to calculate K_s and suction at the front P_f. Please note that we use a small step size `dtr` = 10^{-5} h. We obtain $K_s = 38$ mm/h, $P_f = 238$ mm.

```
# sandy loam
param=c(Pb=14.66,b=4.9, theta.r=0.041,theta.s=0.453)
init = 0.1 # volumetric water content m3/m3
sw <- soil.water(theta=init,param)

Ks<- sw$Ks*3600*10 # cm/s to mm/h
Pf <- sw$Pf*10 # cm to mm
porosity<- param[4]
dt=0.00001
```

First, we use a function `simr` that is just like the function `sim` except for the fact that we substitute the call to `RK4` for a call to `ramos`. For this, we use an input file **ga-ramos-inp.csv** which contains the parameter values as follows:

```
Label,Value,Units,Description
t0,0.00,hr,time zero
tf,2.5,hr,time final
dt,0.00001,hr,time step
tw,0.01,hr,time to write
Pf,238,mm,potential front
Ks,38,mm/hr,conductivity
nu,0.453,adim,porosity
init,0.1,adim,initial theta
dtr,0.00001,hr,time step
X0,30,mm,initial condition (must match 0.1 init)
digX,4,None,significant digits for output file
```

Then, we use the function `green.ampt.ramos` to define the model and call `simr` varying K_s:

```
ga <- list(f=green.ampt.ramos);param <- list(plab="Ks",pva
   l=c(10,20,30,40))
t.X <- simr(ga,"chp15/ga-inp.csv", param, pdfout=T)
```

This produces Figure 15.20, which was already discussed.

The second set of functions is function `sim.ga` and model function `green.ampt.maila`. Using the same example of parameter values, we bundle them together with porosity, initial moisture, and rain in array p, which is passed along with the time sequence `t` to the simulation function `sim.ga`.

```
rain <- 200
p <- c(Pf,Ks,porosity,init,dt,rain)
t <- seq(0,0.2,dt)
sga <- sim.ga(t,p)
```

We obtain graphics like the one shown in Figure 15.21, which was already discussed.

Exercise 15.15

Repeat the Green–Ampt simulation for sandy clay soil. Keep all other conditions the same. Produce graphs and discuss the differences with respect to sandy loam soil.

15.6.6 Soil Water Dynamics: Multiple Layers

We will work with the function `soilwat.n` calling it from `sim`. The function `soilwat.n`, explained at the end of the chapter, implements Equation 15.77 at the top layer and Equation 15.90 at layers below the top. It applies input with given intensity and duration. The input file **chp15/sw-3layer-inp.csv** has the parameter values and initial conditions. The only difference with the previous file **sw-inp.csv** is including three values for the initial condition, one for each layer.

```
X1.0,10,mm,initial condition
X2.0,10,mm,initial condition
X3.0,10,mm,initial condition
digX,4,None,significant digits for output file
```

In this case, we start at 10% saturation (10 mm out of 100 mm for saturation) for all layers and apply input for 1.5 h. We run the simulation for various values of input intensity from below infiltration capacity to above infiltration capacity.

```
sw3 <- list(f=soilwat.n);param <- list(plab="Ic",pval=c(10,20,30,40))
t.X <- sim(sw,"chp15/sw-3layer-inp.csv", param, pdfout=T)
```

The graphical results in the PDF file include Figure 15.18, which we have already discussed.

Exercise 15.16

Change the value of the initial condition to wetter conditions in the bottom two layers, for example, $X_2(0) = X_3(0) = 20$ mm. Repeat the simulations for increasing I_c as in the example. Produce sets of graphs and discuss.

15.6.7 Build Your Own

Exercise 15.17

Develop a model and simulation of infiltration based on the notion that infiltration capacity would be a nonlinear function of the deficit. When the deficit is large, the capacity is large, but as the deficit decreases, the capacity decreases exponentially with the deficit. Modify the following lines of function `soilwat.1`:

```
# slope or linear component mm/hr per mm
 linear <- (fd-Ks)/cap.soil
# infiltration cap in mm/hr
 infilt <- Ks + deficit*linear
```

Then use the modified function to perform simulations using `sim`. Compare with the linear reservoir model. You would have to modify the input file if you introduce new parameters.

15.7 SEEM FUNCTIONS EXPLAINED

15.7.1 FUNCTION RAIN.DAY

This function uses one argument, which is composed of parameters, to generate the amount of rain in a day and it is given as a list. As explained in the first few lines of the function, the list contains the mean, standard deviation, skew coefficient, shape, and the distribution to be sampled. We call the function `rain.day` once we determine that the current day is rainy.

```
rain.day <- function(param){
# mu, std and skew of daily rain used for skew
# shape for gamma, exp, and Weibull
# model.pdf "s" skewed, "w" weibull and "e" exponential, "g" gamma
  mu <- param[[1]]; std <-param[[2]]; skew <- param[[3]]
  shape <- param[[4]]; model.pdf <- param[[5]]
# calc rain
if(model.pdf=="e") {
  u <- runif(1,0,1) # generate uniform
  scale <- mu
  y <- scale*(-log(u))
}
if(model.pdf=="w") {
  u <- runif(1,0,1) # generate uniform
  scale <- mu/gamma(shape+1)
  y <- scale*(-log(u))^shape
}
if(model.pdf=="g") {
  scale <- mu/shape
  y <- rgamma(1,scale,shape)
}
if (model.pdf == "s") {
  z <- rnorm(1,0,1) # generate standard normal
  y <- mu+ 2*(std/skew)*( ( ((skew/6)*(z-skew/6)+1))^3 -1)
}
return(y)
}
```

The function returns the amount of rain generated for the day.

15.7.2 FUNCTION MARKOV.RAIN

This function implements two-step rainfall generation. Its arguments include the transition probability of the Markov chain and the number of days, which are used for the first step. In addition, one argument is the list of amount parameters, which is passed internally to the `rain.day` function described earlier.

```
markov.rain <- function(P, ndays, amount.param){

# arguments: markov matrix P & number of days
# and amount stats parameters
```

```
mu <- amount.param[[1]]
# define array with all 0
x <- rep(0,ndays); wet <- 0

# start first day with rain at random
y <- runif(1,0,1)
if(y > 0.5) {x[1] <- rain.day(amount.param); wet<- wet+1}

# loop for remaining days
for(i in 2:ndays) {

# apply markov
  y <- runif(1,0,1)
  if(x[i-1]==0) {
   if(y > P[1,1]) x[i] <- rain.day(amount.param)
  }
  else {
   if(y > P[1,2]) x[i] <- rain.day(amount.param)
  }
 if(x[i] >0) wet <- wet+1

} # end of days loop
expec.wet.days <- ndays*P[2,1]/(P[1,2]+P[2,1])
expec.dry.days <- ndays - ndays*P[2,1]/(P[1,2]+P[2,1])
dry <- ndays-wet
rain.tot<-round(sum(x),2);expec.rain.tot <- expec.wet.days*mu
rain.avg=round(mean(x),2); expec.avg <- expec.rain.tot/ndays
rain.wet.avg=round(sum(x)/wet,2); expec.wet.avg <- mu

mat<- matrix(1:2,2,1,byrow=T)
layout(mat,c(7,7),c(3.5,3.5),res=TRUE)
par(mar=c(4,4,1,.5), xaxs="r", yaxs="r")
plot(x,type="s",xlab="Day",ylab="Rain (mm/day)")
Rain <- x
hist(Rain,prob=T,main="",xlab="Rain (mm/day)")

return(list(x=round(x,2),wet.days=wet,expected.wet.days=expec.wet.days,
            dry.days=dry,expected.dry.days=expec.dry.days,
            rain.tot=rain.tot, expec.rain.tot=expec.rain.tot,
            rain.avg=rain.avg,expected.rain.avg = expec.avg,
            rain.wet.avg=rain.wet.avg,expected.wet.avg=expec.wet.avg))
}
```

15.7.3 Function pet

The function **pet** calculates the Priestley–Taylor model as a function of radiation values under various temperature values. It also calculates the Penman model (radiation term, aerodynamic term, and total) as a function of radiation for various temperature values and for fixed RH and wind speed conditions.

```
pet <- function(eleva, Rad, Temp, RH, wind, plot.out=F){

# psychrometer constant a function of elevation via pressure Kpa
psycro <- 6.8*10^(-3)*(101-0.0115*eleva+5.44*10^(-7)*eleva^2)
# convert to deg Kelvin
Temp.kelvin <- Temp + 273
nT <- length(Temp)

# radiation in MJ/m2
nRad <- length(Rad)

PETR <- matrix(nrow=nRad,ncol=nT)
PETR.taylor <- PETR; PETR.penman <- PETR
PETA.penman <- PETR.penman; PET.penman <- PETR.penman

#priestly-taylor
slope.ptaylor <- exp(21.3 - 5304/(Temp.kelvin))*(5304/(Temp.kelvin^2))
# slope vaporization in Kpa/degC
WR.taylor <- slope.ptaylor/(slope.ptaylor+psycro)
for(i in 1:nT)
PETR.taylor[,i] <- 30.6*(6.28/365)*Rad*(1-0.23)*WR.taylor[i]

# penman
satura <- 0.1* exp(54.88 - 5.03*log(Temp.kelvin)-6791/Temp.kelvin) #KPa
slope.penman <- (satura/Temp.kelvin)*(6791/Temp.kelvin -5.03)
WR.penman <- slope.penman/(slope.penman+psycro)
latent <- 2.50 - 0.0022*Temp # MJ/Kg
for(i in 1:nT) PETR.penman[,i] <- (Rad*(1-0.23)/latent[i])*WR.
penman[i]

vap.press <- satura*RH/100
wind.fact <- 2.7 + 1.63*wind
WA.penman <- 1 - WR.penman
for(i in 1:nT) PETA.penman[,i] <- WA.penman[i]*wind.fact*(satura[i]
  - vap.press[i])

PET.penman <- PETR.penman + PETA.penman

if (plot.out==T) {
  mat<- matrix(1:4,2,2,byrow=T)
  layout(mat,rep(7/2,2),rep(7/2,2),res=TRUE)
  par(mar=c(4,4,1,.5), xaxs="r", yaxs="r")

y <- PETR.penman
matplot(Rad, y, type="l", lty=1:3, ylab="PET (mm)", xlab="Rad (MJ/
  m2)",ylim=c(0,10),col=1)
title("Penman Rad Term",cex.main=0.8);legend("topleft", lty=1:3,
  leg=paste("T=",Temp,"C"))

y <- PETA.penman
matplot(Rad, y, type="l", lty=1:3, ylab="PET (mm)", xlab="Rad (MJ/
  m2)",ylim=c(0,10),col=1)
```

```
title("Penman Aero Term",cex.main=0.8);legend("topleft", lty=1:3,
  leg=paste("T=",Temp,"C"))

y <- PET.penman
matplot(Rad, y, type="l", lty=1:3, ylab="PET (mm)", xlab="Rad (MJ/
  m2)",ylim=c(0,10),col=1)
title("Penman Total",cex.main=0.8); legend("topleft", lty=1:3,
  leg=paste("T=",Temp,"C"))

y <- PETR.taylor
matplot(Rad, y, type="l", lty=1:3, ylab="PET (mm)", xlab="Rad (MJ/
  m2)",ylim=c(0,10),col=1)
title("Priestley-Taylor",cex.main=0.8);legend("topleft", lty=1:3,
  leg=paste("T=",Temp,"C"))

}

out <- data.frame(Rad,PETR.taylor,PET.penman,PETR.penman,PETA.penman)
return(out)
}
```

15.7.4 Function soil.water

This function is a straightforward application of the Brooks–Corey equations. Effective saturation is calculated as a function of θ. Then, matric suction P, capacity C, and conductivity K are calculated as a function of effective saturation. It includes the option to plot the results, but the plot script is not shown here for the sake of space. The interested reader can study it by listing the function at the console.

```
soil.water <- function(theta, param, tit="", plot=F) {

# Based on Brooks Corey
# arguments:
# theta: seq of soil water content as wetness (vol fraction)
# param:
# Pb bubbling suction in cm
# b pore size distribution index (adim)
# theta.r residual water
# theta.s total porosity
# option T or F to plot

Pb <- param[1]; b <- param[2]
theta.r <- param[3]; theta.s <- param[4]

# effective saturation
theta.e <- (theta-theta.r)/(theta.s-theta.r)

# retention curve
# P is matric suction in cm
P <- Pb/(theta.e)^b
```

```
# capacity
C <- (1/b)*(Pb/P)^(1/b-1)*(-Pb/P^2)

# Hydraulic conductivity in cm/sec
# saturated conductivity
Ks <- 86/((b+1)*(2*b+1))*((theta.s-theta.r)/Pb)^2
# Green-Ampt wetting front suction
Pf <- ((2*b+3)/(b+3))*Pb

# non saturated
K <- Ks*theta.e^(2*b+3)

# logarithmic
LP <- log(P)
LC <- log(-C)
LK <- log(K)

theta.e <- round(theta.e,2)
P <- round(P)
C <- signif(C,2)
Pf <- round(Pf)
Ks <- signif(Ks,2)
K <- signif(K,2)
LP <- round(LP)
LC <- round(LC)
LK <- round(LK)

out <- list(Ks=Ks, Pf=Pf, var.theta=data.frame(theta,theta.e,P,C,K,LP,
  LC,LK))
return(out)
}
```

15.7.5 Function soilwat.1

This function implements the combination of infiltration and percolation as in Equation 15.77 and as a function or input intensity and duration. The function is called by RK4 for simulation using the sim function.

```
soilwat.1 <- function(t,p,x){

   # infilt rate capacity linear with deficit
   # infiltration parameters in mm/hr
   fd <- p[1]; Ks <- p[2]
   # soil capacity in mm, field capacity, porosity
   Z <- p[3]; Fc.coeff <- p[4]; porosity <- p[5]
   rain.int <- p[6]; rain.dur <- p[7]
   # soil sat capacity and field capacity
   cap.soil <- Z * porosity
   Fc <- cap.soil * Fc.coeff
   # using deficit to determine infiltration cap
```

```
    deficit <- cap.soil - x
    # slope or linear component mm/hr per mm
    linear <- (fd-Ks)/cap.soil
    # infiltration cap in mm/hr
    infilt <- Ks + deficit*linear
    if (infilt > fd) infilt <- fd
    if (x >= cap.soil) f<- Ks
    else f<- infilt
    # compare to rain
    if(t <= rain.dur) rain <- rain.int else rain <- 0
    if(rain< f ) q <- rain else q <- f
    # percolation
    if(x >= Fc) g <- Ks * (x-Fc)/(cap.soil-Fc)
    else g <- 0
    dx <- q-g
    return(dx)
}
```

15.7.6 Function ramos

This function executes the numerical integration of an ODE. The function is called by the simulation function `sim` instead of `RK4`. For an autonomous model *f*, the equation is

$$x(t+\Delta t) = x(t) + \frac{2\Delta t \times f(x(t))}{2 - \Delta t\left(\dfrac{\mathrm{d}f}{\mathrm{d}x}\right)}$$

```
ramos <- function(x0, t, f, p, dt){

 # Ramos nonstandard explicit integration algorithm (EIA)
 # arguments: x0 initial condition
 #       t times for output
 #       f model
 #       df derivative of model
 #       p parameters
 #       dt time step
 # arrays to store results
    nt <- length(t); nx <- length(x0)
    ns <- floor((t[2]-t[1])/dt)
    X <- matrix(nrow=nt,ncol=nx)
 # first value of X is initial condition
    Xt <- x0; tt <- t[1]
    # loops
 for(i in 1:nt){
    X[i,] <- Xt
    for(j in 1:ns){
    tt <- tt + dt
    num <- 2*dt*f(tt,p,Xt)[1]
    den <- 2-dt*f(tt,p,Xt)[2]
    Xt <- Xt + num/den
```

```
    # X assumed positive
    for(k in 1:nx) if(Xt[k]<0) Xt[k] <- 0
   }
  } # end of run
  return(X)
} # end of function
```

15.7.7 Function green.ampt.ramos

This is utilized as a model function to define the Green–Ampt model and to respond to a call from `sim` using `ramos` that requires *f* and d*f*/d*x*.

```
green.ampt.ramos <- function(t,p,x){
 Pf<- p[1]; Ks<- p[2]
 porosity <- p[3]; init <- p[4]; Ic <- p[5]
 if(x <= 0) x <- 10^-5
 deficit <- porosity - init
 dx<- Ks*(1+ Pf*deficit/x)
 df <- -Ks*Pf*deficit/x^2
 return(c(dx,df))
}
```

15.7.8 Function simr

We use function `ramos` and substitute the call to `RK4` in `sim` for a call to ramos. The sim function is renamed as `simr`.

```
# integration (edit RK4 for ramos)
# out <- RK4(X0, t, model$f, p, dt)
out <- ramos(X0, t, model$f, p, dt)
```

15.7.9 Function green.ampt.maila

This function is an alternative to calculate the rate ΔF for the numerical integration of F in the Green–Ampt model using the Mailapalli et al. (2009) method given in Equation 15.99 and based on the Ramos algorithm. The function is called by the simulation function `sim.ga`, which will be described next.

```
green.ampt.maila <- function(t,p,x){
 Pf<- p[1]; Ks<- p[2]
 porosity <- p[3]; init <- p[4]; dtr <- p[5]
 if(x <= 0) x <- 10^-5
 deficit <- porosity - init
 num <- 2*dtr*Ks*(1+ Pf*deficit/x)
 den <- 2 - dtr*(-Ks*Pf*deficit/x^2)
 dx <- num/den
 return(dx)
}
```

15.7.10 Function sim.ga

This function implements a numerical simulation using `green.ampt.maila`, described above. It uses a simple loop for time. Arguments are time sequence and a parameter set `p`. This function is an alternative to utilize `sim` with a call to the `ramos` function instead of `RK4`.

```
sim.ga <- function(t,p){
x <- array(); f <- x; q <- x; Q <- x
x[1] <- 10^-5; f[1] <- 0; q[1] <- 0; Q[1]<-0
for(i in 2:length(t)){
# integration
dF <- green.ampt.maila(t[i],p,x[i-1])
x[i] <- x[i-1] + dF
# infiltration capacity
f[i] <- dF/dtr
rain <- p[6]
# actual rate and infiltrated depth
if(f[i] < rain) q[i] <- f[i] else q[i] <- rain
Q[i] <- Q[i-1]+q[i]*dtr
}
mat <- matrix(1:2, c(2,1),byrow=T)
layout(mat,rep(7,2),rep(7/2,2),res=TRUE)
par(mar=c(4,4,1,.5), xaxs="r", yaxs="r")

matplot(t,cbind(f,x),type="l",col=1,ylab="Capacity and
  depth",xlab="Time [h]",ylim=c(0,300))
legend("topright",lty=1:2,col=1,leg=c("Rate mm/h", "Depth [mm]"))

matplot(t,cbind(q,Q),type="l",col=1,ylab="Actual rate and
  depth",xlab="Time [h]",ylim=c(0,300))
legend("topright",lty=1:2,col=1,leg=c("Rate mm/h", "Depth [mm]"))

out <- data.frame(t,x,f,q,Q)
return(round(out,2))
} # end of sim.ga
```

15.7.11 Function soilwat.n

This function is an extension of `soilwat.1` adding the ability to calculate percolation for multiple layers. Similarly to `soilwat.1`, this function is called from `RK4` for simulation using `sim`.

```
soilwat.n <- function(t,p,x){

  # infilt rate capacity linear with deficit
  # infiltration parameters in mm/hr
  fd <- p[1]; Ks <- p[2]
  # soil capacity in mm, field capacity, porosity
  Z  <- p[3]; Fc.coeff <- p[4]; porosity <- p[5]
  rain.int <- p[6]; rain.dur <- p[7]
  n <- length(x)
```

```
 # soil sat capacity and field capacity
 cap.soil <- Z * porosity
 Fc <- cap.soil * Fc.coeff
 # using deficit to determine infiltration cap
 deficit <- cap.soil - x[1]
 # slope or linear component mm/hr per mm
 linear <- (fd-Ks)/cap.soil
 # infiltration cap in mm/hr
 infilt <- Ks + deficit*linear

 if (infilt > fd) infilt <- fd
 if (x[1] >= cap.soil) f<- Ks
 else f<- infilt

 if(t <= rain.dur) rain <- rain.int else rain <- 0
 if(rain< f ) q <- rain else q <- f
 g <- array(); dx <- array()

 # percolation from each layer (except last)
 for(i in 1:(n-1)){
 if(x[i] >= Fc) {
 if(x[i+1]>=cap.soil) g[i] <- Ks
 else g[i] <- Ks * (x[i]-Fc)/(cap.soil-Fc)
 } else     g[i] <- 0
 if(i==1) dx[i] <- q-g[i] else dx[i] <- g[i-1]-g[i]
 } # end of for

 # last layer
 if(x[n] >= Fc) g[n]<- Ks * (x[n]-Fc)/(cap.soil-Fc)
 else  g[n] <- 0
 dx[n] <- g[n-1]-g[n]

 return(dx)
}
```

16 Terrestrial Ecosystems
Vegetation Dynamics

In this chapter, we continue the study of terrestrial ecosystems by integrating plant growth to the community dynamics models discussed in Chapter 13 and to the weather and soil components we analyzed in Chapter 15. The key aspect of this integration is to make plant growth depend on individual and species interactions in their utilization of resources, such as sunlight, nutrients, and soil water availability. We limit ourselves to plant communities and do not cover other aspects of terrestrial systems.

There are many approaches to model plant communities and their integration to ecosystem processes. We will study two of these approaches: individual-based and lumped. In the first approach, although the potential responses of all individuals of the same species may be similar, the actual response of each individual may vary according to conditions and randomness. After the simulation, we calculate the statistical summaries of the species composition to describe the emergent community response. In the second approach, we aggregate beforehand all the individuals of a species into compartments or state variables. Therefore, their growth responses are the same.

16.1 INDIVIDUAL-BASED APPROACH

Individual-based models describe changes of species composition and community dynamics as an emergent property of the dynamics of the individuals (DeAngelis and Gross, 1992; Grimm and Railsback, 2005; Reuter et al., 2011). They represent an important class of models in ecological systems.

Examples in forest modeling include forest simulators such as JABOWA, FORET, and ZELIG (Botkin et al., 1972; Urban and Shugart, 1992). These models are also known by the popular term "gap models," and they calculate tree regeneration, growth, and mortality on a tree-by-tree basis. Individual trees are competing for sunlight, water, and nutrients, and they are affected by temperature. Each species has parameter values characterizing its response to restrictions of these resources or values of the environmental conditions.

16.2 GROWTH EQUATION

The volume (V) of the trunk of each tree depends on its diameter (D) and is approximately equal to the volume of a cylinder

$$V(D) = B(D) \times H(D) = \frac{\pi D^2}{4} H(D) \tag{16.1}$$

where $B(D)$ is the cross sectional area and H is the **height** (also a function of D). Diameter is measured at a standard height, H_0, and is called the **diameter at breast height** (DBH). The standard H_0 is commonly taken as 4.5 feet, or 1.37 meters. **Basal area** is defined as the cross-sectional area at breast height. This volume calculation ignores tapering and irregularities. The tree growth rate is the rate of change of volume dV/dt = volume growth rate [cm^3/year], which is expressed by a logistic or sigmoid type of equation. The rate of change is directly proportional to "leaf area" $l(D)$ (m^2) but decreases in proportion to the volume of the live wood because of maintenance costs. For simplicity, the volume of live wood can be approximated by the trunk's **surface area**, S, and therefore,

$$\frac{dV}{dt} = G_{max} \times g(\mathbf{E}(t)) \times l(D)\left[1 - \frac{S}{S_{max}}\right] \tag{16.2}$$

where G_{max} is the growth rate coefficient (cm^3 wood per m^2 of leaf area/year), S_{max} is the maximum surface area, and $g(\mathbf{E}(t)) \leq 1$ is a positive scalar function of a set of environmental or forcing variables such as air temperature, sunlight, soil moisture, and soil nutrients, which we denote by vector $\mathbf{E}(t)$. The components of this vector vary with time and can be random; we postpone their discussion until later in this chapter.

Again assuming a cylinder, the surface area is given as the circumference multiplied by the height, $S = \pi DH$. Therefore, we can rewrite Equation 16.2 as

$$\frac{dV}{dt} = G_{max} \times g(\mathbf{E}(t)) \times l(D)\left[1 - \frac{D \times H(D)}{D_{max} H_{max}}\right] \tag{16.3}$$

where H_{max} and D_{max} are the maximum height and diameter for trees of this particular species. Tree growth rate depends on the diameter: it increases when the tree is small, reaches a maximum at some intermediate point, and then declines to zero when it reaches the maximum diameter (Figure 16.1). When the tree reaches its maximum size, $DH = D_{max}H_{max}$, the growth in volume is zero; all productivity goes into maintenance. Figure 16.1 was obtained assuming H increases with D and asymptotically reaches H_{max} and that l increases with basal area. In addition, this figure shows how the growth rate varies for several values of G_{max} when conditions are optimum, that is, $g(\mathbf{E}) = 1$.

Exercise 16.1

Use Figure 16.1 to determine the diameter at which the maximum growth rate is attained for G_{max} = 1000. Calculate the basal area in m^2 for this diameter. Compare with the basal area for maximum diameter (100 cm).

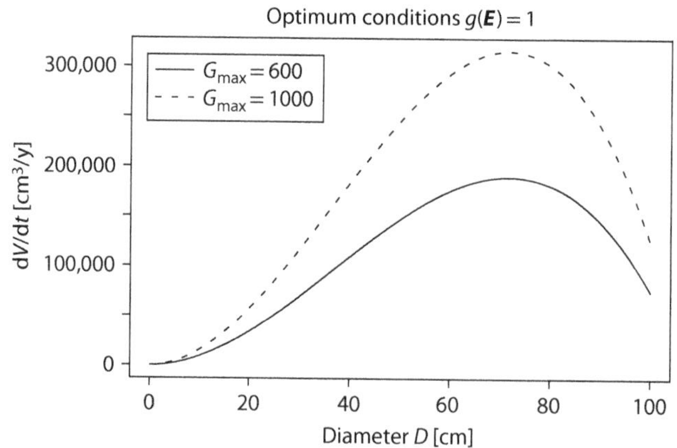

FIGURE 16.1 Volume growth rate curve for a tree as a function of diameter for two values of G_{max}, assuming optimum environmental conditions $g(\mathbf{E}) = 1$.

16.3 ALLOMETRIC RELATIONSHIPS

Gap models assume that H and l depend on D via **allometric** relationships, which vary from model to model. For example, for H versus D, models have used a parabolic form

$$H = H_0 + b_1 D + b_2 D^2 \tag{16.4}$$

or an exponential form

$$H = H_0 + (H_{max} - H_0)(1 - \exp(b_3 D))^{b_4} \tag{16.5}$$

where H_0 is the breast height defined earlier, b_1 and b_2 are the allometric coefficients of the parabolic form, and b_3 and b_4 are the coefficients for the exponential form. These coefficients are species-specific and typically assume that H is given in meters and D in centimeters. Note that $H = H_0$ when $D = 0$ for both forms, meaning that zero DBH is attained for saplings of height H_0.

H_{max} and D_{max} are estimated from the maximum values attained by the species. It is assumed that $H = H_{max}$ when $D = D_{max}$. Using a data set for pairs H, D for each species, the allometric coefficients can be calculated by regression. You can see one example of the estimation of parameters of Equation 16.5 in Figure 16.2 for trees of the green ash species growing in the bottomland forests of north Texas. We will see a program that performs this type of estimation later in this chapter in Section 16.15.

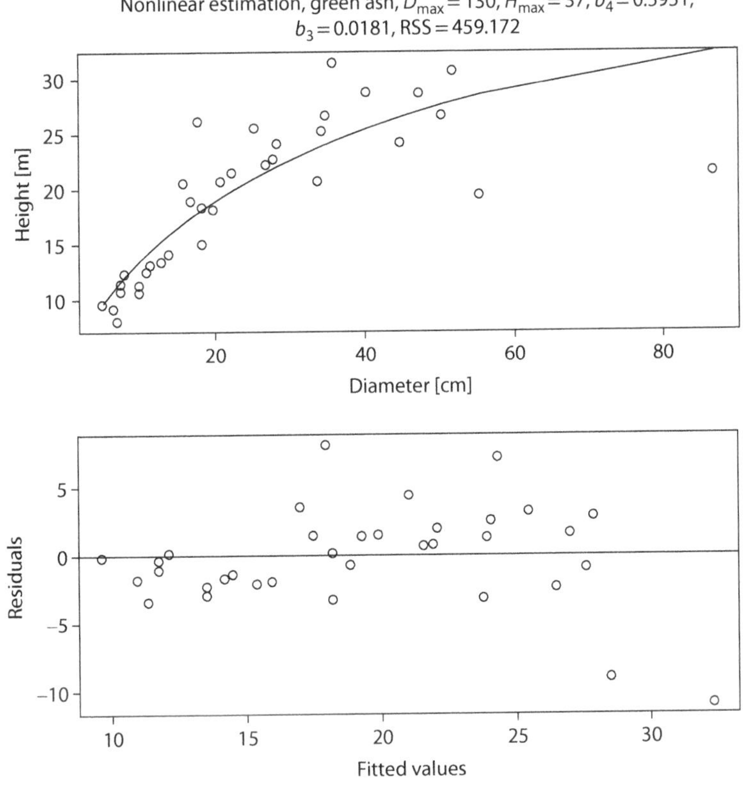

FIGURE 16.2 Example of an H and D allometric relationship.

Exercise 16.2

Use the parabolic allometric relation given by Equation 16.4 and demonstrate that $b_1 = \frac{2(H_{max} H_0)}{D_{max}}$ and $b_2 = \frac{H_{max} H_0}{D^2_{max}}$. Hint: Solve for the coefficients from a system of equations formed by the parabola evaluated at D_{max} and its derivative evaluated at D_{max}. First, $H(D)$ evaluated at D_{max} should be H_{max}. Second, the derivative at D_{max} should be zero for a maximum.

Similarly, we can calculate l from diameter D using an allometric relationship. For example, assuming that l is proportional to the basal area,

$$l(D) = c'\frac{\pi D^2}{4} = cD^2 \tag{16.6}$$

where coefficient $c = c'\frac{\pi}{4}$, l is given in m², and D in cm. Other expressions generalize this equation for a power coefficient different than 2, $l(D) = cD^d$. Yet others include an adjustment factor that takes into account the height of the crown, H_c, as a parameter, h_c:

$$l(D) = cD^d\left(1 - \frac{H_c}{H}\right) = cD^d h_c \tag{16.7}$$

Note that the crown factor, h_c, reduces to the value 1 when the foliage goes all the way down to the ground such that $H_c = 0$. Other allometric relations take into account sapwood and more complicated geometrical considerations.

16.4 DIAMETER ORDINARY DIFFERENTIAL EQUATION

Now we will convert the growth equation (Equation 16.3) into an ordinary differential equation (ODE), where D is the state variable. To do this, note that the derivative of volume is given by $\mathrm{d}V/\mathrm{d}t = \mathrm{d}(D^2H \times 100)/\mathrm{d}t$, ignoring the $\pi/4$ factor and converting H from meters to centimeters (so that we obtain V in cm³). Expand the derivative of volume $\mathrm{d}(D^2H \times 100)/\mathrm{d}t$ using the chain rule of derivatives. This rule states that the derivative of a product is the sum of two products:

$$\frac{\mathrm{d}(xy)}{\mathrm{d}t} = x\frac{\mathrm{d}y}{\mathrm{d}t} + \frac{\mathrm{d}x}{\mathrm{d}t}y \tag{16.8}$$

where $x = D^2$ and $y = H(D) \times 100$. Therefore,

$$\begin{aligned}\frac{\mathrm{d}(D^2H \times 100)}{\mathrm{d}t} &= 100 \times \left(2D(\mathrm{d}D/\mathrm{d}t)H + D^2(\mathrm{d}H/\mathrm{d}D)(\mathrm{d}D/\mathrm{d}t)\right) \\ &= D\left(\mathrm{d}D/\mathrm{d}t\right) \times 100 \times \left(2H + D(\mathrm{d}H/\mathrm{d}D)\right)\end{aligned} \tag{16.9}$$

For brevity in algebraic manipulation, denote the last factor as

$$\phi(D) = 100 \times 2H + D(\mathrm{d}H/\mathrm{d}D) \tag{16.10}$$

Solving for $\mathrm{d}D/\mathrm{d}t$, we obtain

$$\mathrm{d}D/\mathrm{d}t = (\mathrm{d}V/\mathrm{d}t)\left(D\phi(D)\right)^{-1} \tag{16.11}$$

Substitute the sigmoid Equation 16.3 for dV/dt, then

$$\frac{dD}{dt} = \frac{G_{max} g(\mathbf{E}(t)) \times l(D)\left(1 - \dfrac{DH(D)}{D_{max} H_{max}}\right)}{D\phi(D)} \tag{16.12}$$

The next steps are to plug in the expressions for l and H in terms of D so that we get the rate of diameter change dD/dt purely as a function of D. As illustrated, use the exponential model for the allometric relation H versus D (Equation 16.5) and take its derivative:

$$\frac{dH}{dD} = -(H_{max} - H_0) b_3 b_4 \exp(b_3 D)\left(1 - \exp(b_3 D)\right)^{b_4 - 1} \tag{16.13}$$

Then substitute in Equation 16.10 to obtain

$$\begin{aligned}\phi(D) &= 100 \times 2\left(H_0 + (H_{max} - H_0)(1 - \exp(b_3 D))^{b_4}\right) \\ &\quad - 100 \times (H_{max} - H_0) D b_3 b_4 \exp(b_3 D)\left(1 - \exp(b_3 D)\right)^{b_4 - 1}\end{aligned} \tag{16.14}$$

Finally, substitute this $\phi(D)$ expression in the sigmoid Equation 16.3 and again use the allometric relation H versus D (Equation 16.5) and the relation l versus D (Equation 16.6); after cancelling D in the numerator and denominator, we have

$$\frac{dD}{dt} = \frac{G_{max} g(\mathbf{E}(t)) c D \left[1 - \dfrac{D\left(H_0 + (H_{max} - H_0)(1 - \exp(b_3 D))^{b4}\right)}{D_{max} H_{max}}\right]}{200\left(H_0 + (H_{max} - H_0)(1 - \exp(b_3 D))^{b_4}\right) - 100(H_{max} - H_0) D b_3 b_4 \exp(b_3 D)\left(1 - \exp(b_3 D)\right)^{b_4 - 1}} \tag{16.15}$$

Therefore, we now have dD/dt purely in terms of the variable D and the parameters.

Exercise 16.3

Use the parabolic allometric relation (Equation 16.4) and demonstrate that $\phi(D) = 100 \times \left(2H_0 + 3b_1 D + 4b_2 D^2\right)$.

Exercise 16.4

Use Equation 16.6 and the result of the previous exercise to demonstrate that

$$\frac{dD}{dt} = G_{max} g(\mathbf{E}(t)) c D \frac{\left(1 - \dfrac{D(H_0 + b_1 D + b_2 D^2)}{D_{max} H_{max}}\right)}{100 \times \left(2H_0 + 3b_1 D + 4b_2 D^2\right)}$$

Note that Equation 16.12 is a nonlinear ODE:

$$\frac{dD}{dt} = G_{max} g(\mathbf{E}(t)) \times f(D) \tag{16.16}$$

with parameters G_{max}, whatever parameters $g(\mathbf{E}(t))$ would require, and seven other coefficients to parameterize $f(D)$ given as

$$f(D) = \frac{l(D)\left(1 - \dfrac{DH(D)}{D_{max}H_{max}}\right)}{D\phi(D)} \tag{16.17}$$

These coefficients are c, d, h_c, H_{max}, D_{max} and then either b_1, b_2 for the parabolic or b_3, b_4 for the exponential. The function $f(D)$ shows a maximum at some intermediate point D and $G_{max}g(\mathbf{E}(t))$ scales dD/dt accordingly; see the top panel of Figure 16.3. Gap models work on an annual basis and the diameter is calculated every year as a difference equation obtained using the Euler method with $\Delta t = 1$ year to solve Equation 16.16:

$$D(t+1) = D(t) + \Delta t G_{max} g(\boldsymbol{E}(t)) f(D) = D(t) + G_{max} g(\boldsymbol{E}(t)) f(D) \tag{16.18}$$

See the bottom panel of Figure 16.3. You can note from this graph that the diameter dynamics curve shows a sigmoid form with the inflexion point determined by the diameter value for which the increment shows a maximum. These graphs were produced with the parameter values in the first row of Table 16.1, for two values of G_{max}, assuming that the optimum environmental conditions $g(\boldsymbol{E}) = 1$. Once we determine the diameter for the year by this integration, the height and leaf area are calculated using allometric relations given by Equations 16.5 and 16.7 as shown in Figure 16.4.

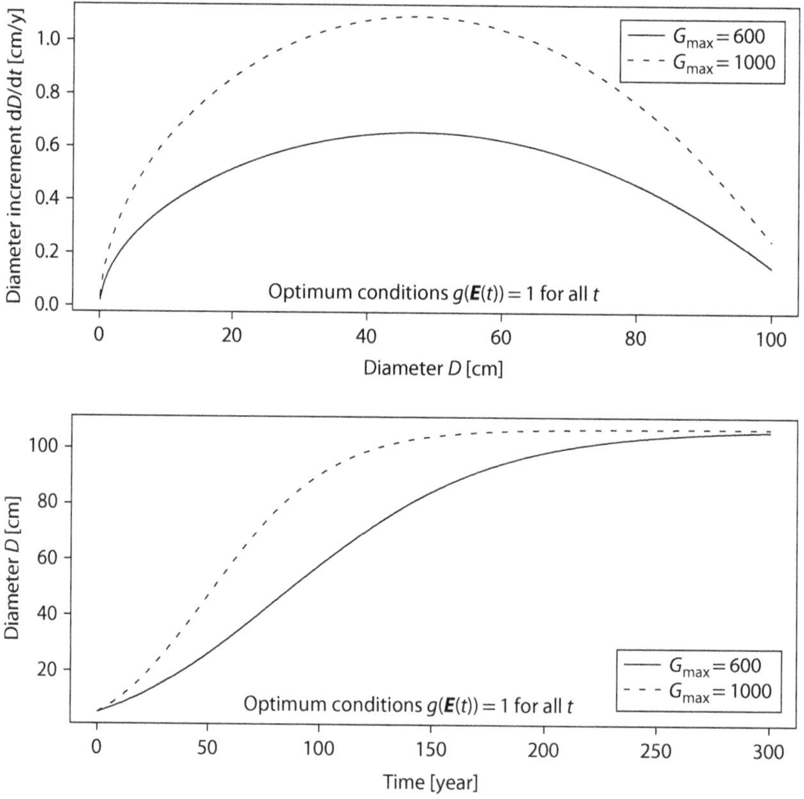

FIGURE 16.3 Top: Diameter increment as a function of diameter using both allometric relationships. Bottom: Diameter dynamics integrating the ODE by Euler.

TABLE 16.1
Parameter Examples

Species	D_{max} (cm)	H_{max} (m)	b_3 (cm^{-1})	b_4	C (m^2/cm^{-2})	h_c	d
Green ash	100	20	−0.02	0.6	0.16	0.8	2

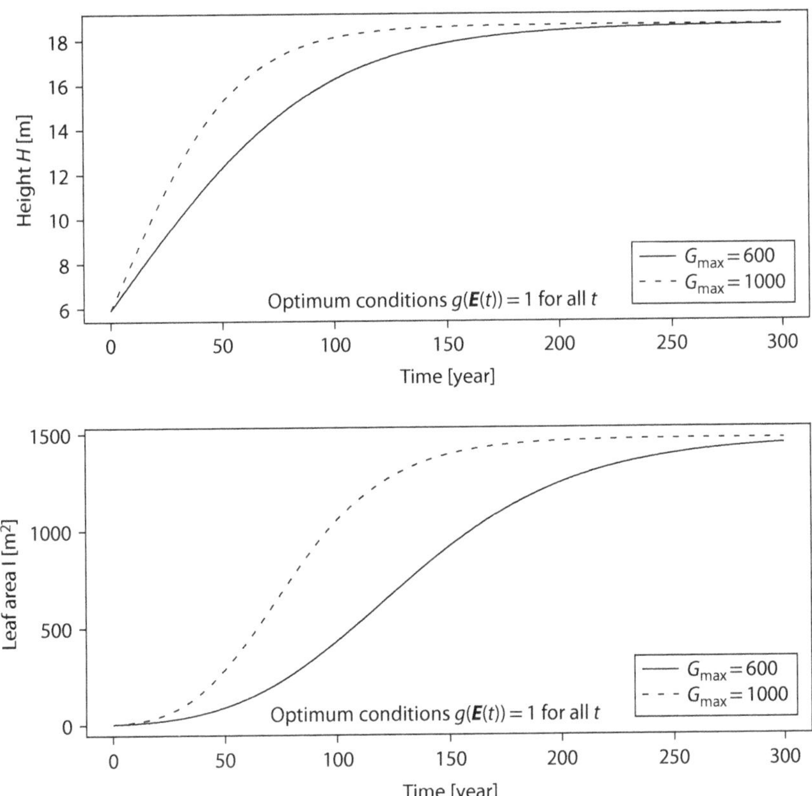

FIGURE 16.4 Top: Height dynamics obtained from the diameter dynamics by applying the H versus D allometric relation. Bottom: Leaf-area dynamics obtained from the diameter dynamics by applying the l versus D allometric relation.

Exercise 16.5

From Figure 16.3, what is the approximate maximum increment rate for $G_{max} = 1000$ and at what diameter does it occur? Determine how long it takes for the diameter to reach the asymptote for $G_{max} = 600$. From Figure 16.4, discuss why height versus time does not exhibit a sigmoid shape, whereas leaf area versus time does.

16.5 ENVIRONMENTAL CONDITIONS

Function $g(\boldsymbol{E}(t))$ affects the hypothetical optimum rate, which would be attained when conditions are optimum. Gap models commonly use a function of factors $F(E_i(t))$ = the limiting factor due to the environmental variable $E_i(t)$, for example, $g(\boldsymbol{E}(t)) = \prod_i F(E_i(t))$, or the minimum of the factors $g(\boldsymbol{E}(t)) = \min_i \left(F(E_i(t))\right)$, or a combination of both.

These factors or multipliers are normalized to have values between 0 and 1 to reduce the growth rate due to environmental conditions. Multipliers are modelled as independent factors, including sunlight (exponential), soil moisture (parabolic), soil fertility or nutrients (parabolic), and air temperature (parabolic). For example, in

$$g(\boldsymbol{E}(t)) = F_L(L(t))F_T(T(t))F_W(W(t))F_N(N(t)) \tag{16.19}$$

$E_1 = L(t)$ = solar radiation (sunlight), $E_2 = T(t)$ = temperature, $E_3 = W(t)$ = soil water, and $E_4 = N(t)$ = soil nutrients. The functions $F(E_i)$ have parameters, which depend on the species type: shade-tolerance, drought-tolerance, and cold-tolerance. As another formulation, some models use

$$g(\boldsymbol{E}(t)) = F_L(L(t))F_T(T(t))\min\left(F_W(W(t)), F_N(N(t))\right) \tag{16.20}$$

where we take the minimum or most limiting of the soil factors (moisture and fertility) and then multiply by the other factors.

Note that $g(\boldsymbol{E}(t))$ is a feedback factor from the entire tree community to the growth equation because tree growth affects environmental factors. For instance, as the trees grow, the total basal area increases and the crowding effects reduce the light, nutrients, and water.

As a specific simplistic example, consider a set $\boldsymbol{E}(t)$ of variables normalized so that each variable takes value in [0,1] by dividing into the optimum for each, E_L, E_T, E_W, and E_N as shown in Figure 16.5. Here we have used a simple parabolic expression for each factor as a function used by Phipps (1979)

$$F(E_i) = 1 - \alpha_i(1 - E_i)^2 \tag{16.21}$$

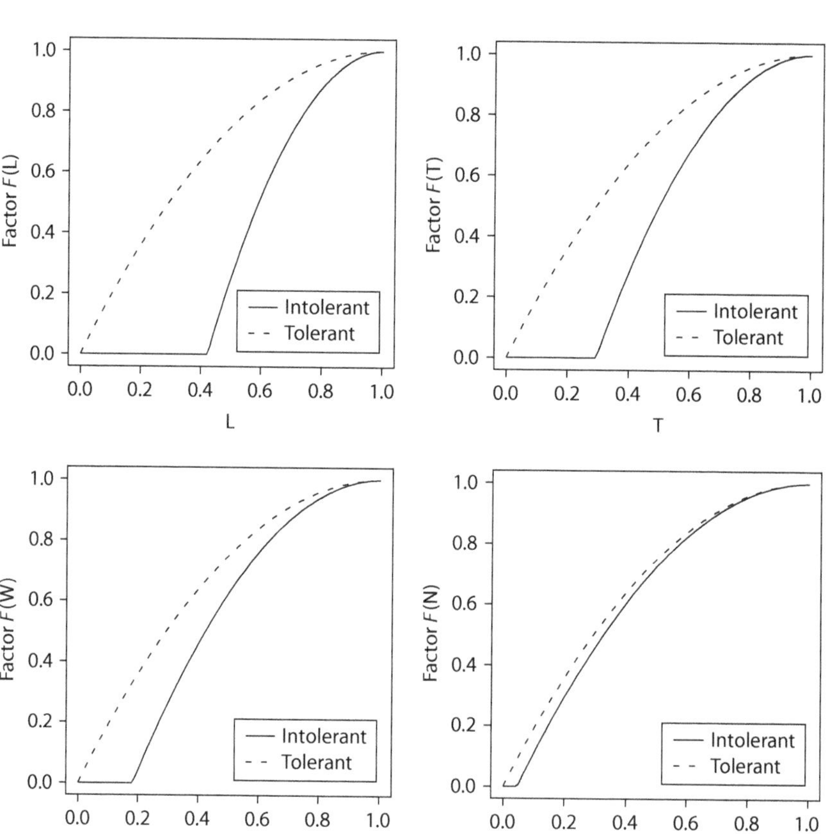

FIGURE 16.5 Hypothetical tree responses using a simple parabolic function.

There is a threshold value for E_i determined by $1-\sqrt{1/\alpha_i}$ such that the values of E_i below the threshold yield $F(E_i)$ negative and are assigned a value of zero. Figure 16.5 was produced with α_i values of 3, 2, 1.5, and 1.1 for the species labeled "intolerant" and 1, 1, 1, and 1 for the species labeled "tolerant." The intolerant species has lower tolerances for all the environmental factors. We can see that higher α corresponds to higher intolerance to a lack of the particular environmental resource.

Now, if we have $E_i(t) = 0.6$, 0.4, 0.4, and 0.8 for a particular year t, from the graphs of Figure 16.5 or applying Equation 16.21, we get for the intolerant species

$$\begin{aligned} F(E_1) &= 1-\alpha_1(1-E_1(t))^2 = 1-3(1-0.6)^2 = 0.52 \\ F(E_2) &= 1-\alpha_2(1-E_2(t))^2 = 1-2(1-0.4)^2 = 0.28 \\ F(E_3) &= 1-\alpha_3(1-E_3(t))^2 = 1-1.5(1-0.4)^2 = 0.46 \\ F(E_4) &= 1-\alpha_4(1-E_4(t))^2 = 1-1.1(1-0.8)^2 = 0.96 \end{aligned} \tag{16.22}$$

and then applying Equation 16.20,

$$g(\boldsymbol{E}(t)) = 0.52 \times 0.28 \times \min(0.46, 0.96) = 0.15 \times 0.46 = 0.07 \tag{16.23}$$

which yields only 7% of the potential growth. We see that the most limiting factor for that year was temperature stress with a value of 0.28.

Exercise 16.6

Demonstrate that the threshold for E_i is determined by $1-\sqrt{1/\alpha_i}$. Hint: Use Equation 16.21 and make it equal to zero. Then solve for E_i and use the negative root to select a value less than one.

Exercise 16.7

Estimate the $F(E_i(t))$ for each environmental variable for the tolerant species and apply Equation 16.20 to estimate $g(\boldsymbol{E}(t))$. As before use $Ei(t) = 0.6$, 0.4, 0.4, and 0.8 for a particular year t. Compare with the intolerant species calculated earlier.

To appreciate the impact of annual fluctuations of environmental conditions on growth, consider, for example, that light and nutrients are at optimal conditions $E_1 = E_4 = 1$ but that temperature and soil moisture E_2 and E_3 vary randomly from year to year. Using the simplest assumption of a uniform distribution in [0,1], we simulate the growth as shown in Figure 16.6. Here we are using two species, one intolerant and one tolerant, to all four factors. We keep allometric coefficients D_{max} and G_{max} the same for both species and compare the effect of differential tolerance on growth during a 50-year simulation. We notice how the diameter reached at the end of the simulation is about 50% greater for the tolerant species.

16.6 ESTIMATION OF G_{MAX}

We can combine Equations 16.19 or 16.20 and 16.16 to get

$$\frac{dD}{dt} = G_{max}\, g(\boldsymbol{E}(t)) f(D) \tag{16.24}$$

which we will use to estimate G_{max} as the only unknown in the expression. Assume that the data available to estimate G_{max} is a collection of the values of D for each year, which can come from tree ring measurements or annual measurements of DBH on trees of reference stands. We assume that

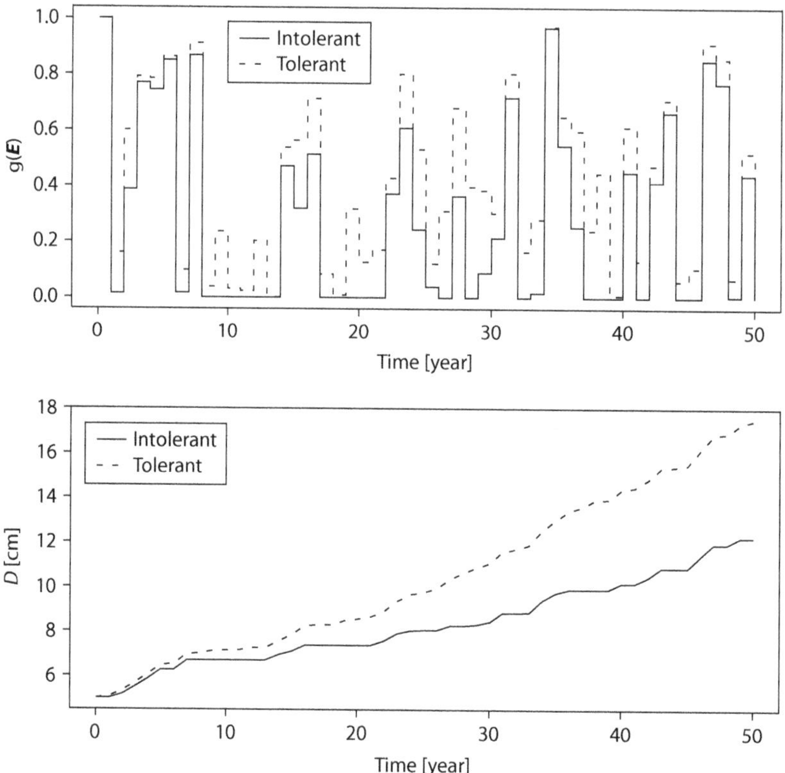

FIGURE 16.6 Impact of differential tree responses on growth using simple parabolic functions for all response functions and random environmental factors.

we can calculate dD/dt for each year from the D values. So the data set is a set of paired values D, dD. This set can be converted to pairs of $dD, f(D)$ using Equation 16.12. We will assume that there is an environmental condition in the data set that lead to maximum growth, that is, $g(\boldsymbol{E}(t)) \sim 1$. In other words, there is a pair of values, $f(D_{opt})$ and dD_{opt}, that corresponds to $g(\boldsymbol{E}(t)) \sim 1$:

$$dD_{opt} = G_{max} f(D_{opt}) \tag{16.25}$$

Note that we can solve for G_{max},

$$G_{max} = dD_{opt} / f(D_{opt}) \tag{16.26}$$

The estimation process then consists of four steps: (1) calculate the maximum of dD in the series that is, dD_{opt}; (2) determine the corresponding D_{opt} at which this occurs, D_{opt}; (3) calculate $f(D_{opt})$ using Equation 16.15; and (4) calculate G_{max} using Equation 16.26. A program to perform the estimation is given in the computer session part of this chapter. An example is shown in Figure 16.7. This approach hinges on the premise that we have good estimates of D_{max} and the allometric coefficients. Uncertainty in these parameters affects the uncertainty in the estimation of G_{max}.

16.7 SUNLIGHT

In this section and in Sections 16.8 through 16.10, we will take a closer look at environmental factors and their corresponding multipliers. Solar radiation for the site is calculated from the models such as the ones we have studied in Chapter 14; then, sunlight vertical distribution through the canopy

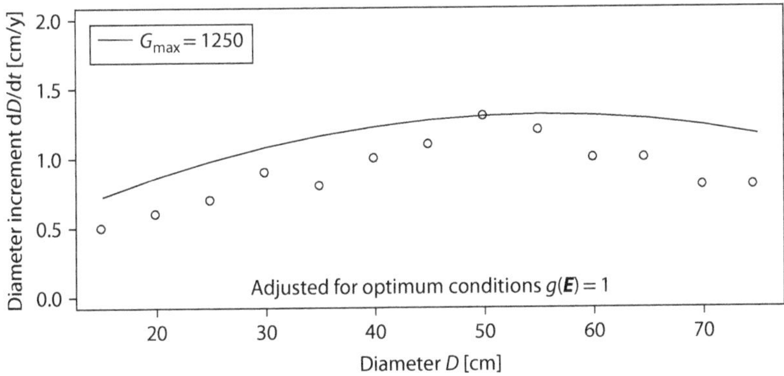

FIGURE 16.7 Example of the estimation of G_{max}.

is calculated using the Beer–Lambert law. Assume that the layer of leaves absorbs solar radiation uniformly and that a layer at height h has a leaf area of $l(h)$. The rate of light extinction is given by

$$\frac{dL(h)}{dh} = k \times l(h) \times L(h) \tag{16.27}$$

where $L(h)$ = solar radiation at height h [W m^{-2}], h = height with respect to ground level [m], and k is a constant to adjust the units; when leaf area is given in m^2, then k should be [m^{-2}] per m, that is, [m^{-3}]. When using leaf area index (LAI, m^2 m^{-2}) for $l(h)$, k would be given in m^{-1} and is approximately 0.4 m^{-1} for many forest canopies.

Integrating the light extinction from the top of the canopy at height h_t to height h is given as

$$L(h) = L_0 \exp\left(k\int_h^{h_t} l(s)\,ds\right) \tag{16.28}$$

where L_0 = solar radiation at the top [W m^{-2}] and $s > h$ is used for all heights above h. The integral is denoted by

$$l_h = \int_h^{h_t} l(s)\,ds \tag{16.29}$$

to simplify the notation in the following manner:

$$L = L_0 \exp(kl_h) \tag{16.30}$$

Most gap models calculate light extinction by adding leaf area for all trees higher than h, thereby approximating the integral by a sum.

Exercise 16.8

Consider a hypothetical special case when leaf area is constant for all h and equal to l_0, that is, a vertically homogeneous canopy. Demonstrate that l_h becomes $l_h = l_0(h_t - h)$ and that L follows an exponential $L = L_0^* \exp(-Kh)$, where L^*_0 and K are constants. Give the expressions for L^*_0 and K.

The Beer–Lambert law is applied to both direct and diffuse components of solar radiation. In spatially explicit models, the forest stand is divided into plots and there is interaction among the plots because the neighboring plots shade each other.

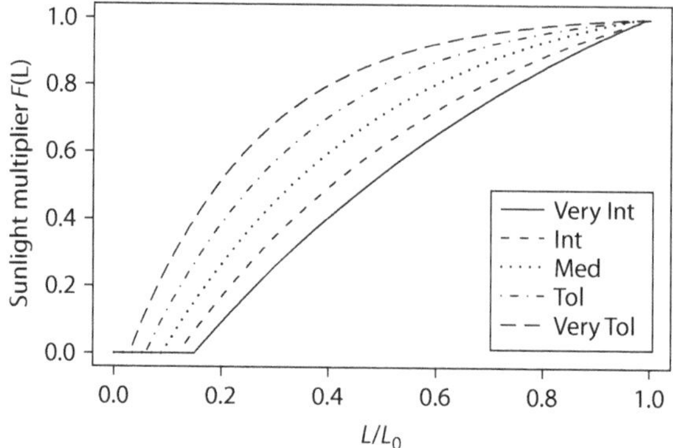

FIGURE 16.8 Five categories (very intolerant, intolerant, medium, tolerant, and very tolerant) of tree response to sunlight using Equation 16.31.

To calculate the multiplier for sunlight, use a normalized value of sunlight $\varphi = L/L_0$ in between 0 and 1 and utilize an exponential in order to make the multiplier equal to 1 at maximum light $\varphi = 1$:

$$F_L(\varphi) = c_1 \times (1 - \exp(-c_2 \times (\varphi - c_3))) \tag{16.31}$$

where c_1, c_2, and c_3 are coefficients. We see an example in Figure 16.8 that shows five categories of tolerance.

16.8 SOIL MOISTURE

Soil moisture is calculated as a balance of rainfall, run-on, runoff, ET, and percolation using the methods given in Chapter 15. Most gap models formulate the response of trees to soil moisture as a multiplier that decreases as soil moisture decreases; in other words, a declining response to dry conditions. However, excess soil moisture can also induce stress, and for the sake of generality, it is convenient to define a response function that considers the excess for both wet and dry conditions.

Define the E_{WD} proportion of wet days in a year (days when soil water is above the field capacity), and the E_{DD} proportion of dry days in a year (days when soil water is below the wilting point). Annual indices of soil moisture are in terms of the proportion of days that are not too moist ($1 - E_{\text{WD}}$) and not too dry ($1 - E_{\text{DD}}$), for example, a 50-year simulation for clay soils in the bottomland forests of north Texas; see the top panel of Figure 16.9. Here we see that excess moisture is favorable for most years (~0.8–0.9), whereas moisture deficit is not as favorable (~0.4–0.5).

As another example of response function, we use a Gaussian function for each one of these factors:

$$F_{\text{WD}}(E_{\text{WD}}) = 1 - \exp\left(-\frac{1}{2}\left(\frac{1 - E_{\text{WD}}}{\beta_{\text{WD}}}\right)^2\right) \tag{16.32}$$

where β_{WD} is a tolerance parameter for this index. An identical expression is used for E_{DD} with a parameter β_{DD}. See the bottom panel of Figure 16.9, where as an example we show a species with higher intolerance ($\beta_{\text{DD}} = 0.3$) to $1 - E_{\text{DD}}$ than to $1 - E_{\text{WD}}$ (which has $\beta_{\text{WD}} = 0.2$).

Then we can take the smaller of the two as the most limiting factor:

$$F_W(E_{\text{DD}}, E_{\text{WD}}) = \min(F_{\text{WD}}(E_{\text{WD}}), F_{\text{DD}}(E_{\text{DD}})) \tag{16.33}$$

In the example above, the minimum would be for dry days, which would be the most limiting factor.

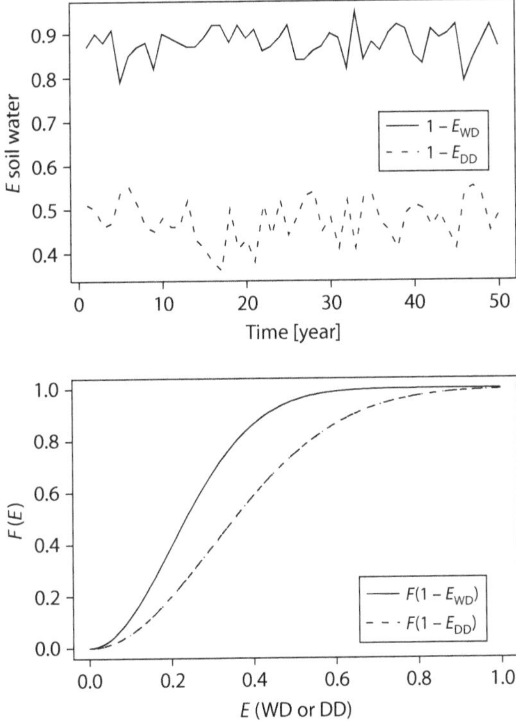

FIGURE 16.9 Top: Annual indices of soil moisture in terms of the proportion of days that are not too moist ($1 - E_{WD}$) and not too dry ($1 - E_{DD}$). Bottom: Tree response to these indices; the example is for a species more tolerant to excess moisture than to lack of moisture.

Exercise 16.9

Consider a species for which the tolerances in the example above are reversed, that is, more intolerant to excess moisture ($\beta_{WD} = 0.4$) than moisture deficit ($\beta_{DD} = 0.1$). If $1 - E_{WD}$ were to average 0.6 and $1 - E_{DD}$ to 0.5, which would be the most limiting factor? What is the average value of the growth multiplier F_w?

16.9 TEMPERATURE

Air temperature is often converted to an annual environmental factor by calculating degree-days, which is a summation of the daily average temperature above a threshold (5.56°C) over all days of the year. An example is illustrated in Figure 16.10 (top panel). A temperature parabolic response as in Chapter 13 using two parameters yields a response as illustrated in the same figure (bottom panel). For species 1 (Sp1), the minimum and maximum values are 2000 and 6000, respectively, whereas for Sp2, the parameters are 3000 and 7000, respectively. Note that for these parameter values, the parabolic response would yield similar values (~0.9) for both species, given the degree-day series on the top panel, which averages to about 4600 degree-days. For species Sp1, the site is warmer than optimum, whereas for species Sp2, the site is cooler than optimum.

Exercise 16.10

Suppose species Sp2 was to have minimum and maximum tolerances of 4000 and 7000 degree-days, respectively. Would Sp2 have higher response that Sp1 for the series shown in the top panel of Figure 16.10? Hint: Estimate responses at 4600 degree-days and compare.

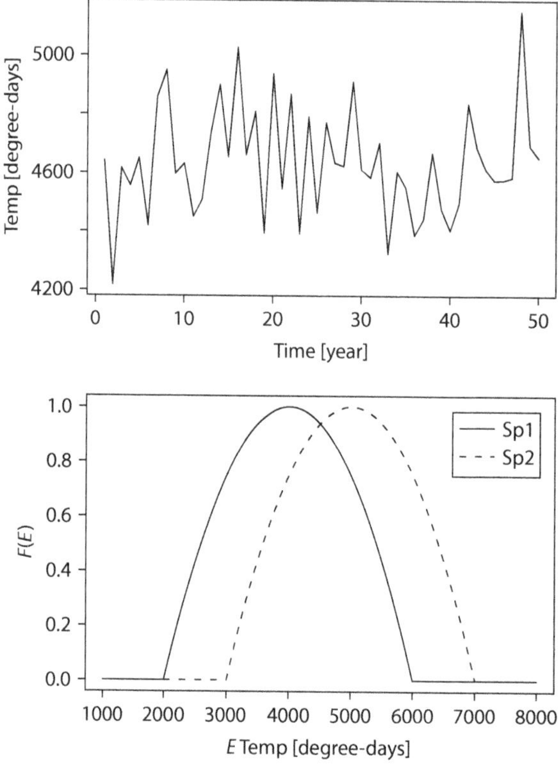

FIGURE 16.10 Top: Annual series of degree-days. Bottom: Parabolic tree response to thermal index as degree-days.

16.10 NUTRIENTS

One response function used for nutrients is the same as for sunlight Equation 16.31. We will not discuss nutrient limitation very extensively, but we should note that this response links to leaf litter and nutrient cycling as discussed in Chapter 13. Species that fix nitrogen directly from air are very tolerant to the lack of soil nutrients and their parameter values should reflect that.

16.11 TREE POPULATION DEMOGRAPHY

Most individual-based forest models (JABOWA, FORET, and ZELIG) use random establishment and mortality. For example, mortality is Poisson with two rates, one fast due to the lack of growth, and one slow due to aging. A stressed tree that has not grown in several years has a probability $\mu 1$ of dying, whereas a tree that reaches its maximum age has a probability $\mu 2$ of dying.

16.12 MODEL RESPONSES

Once we put all the preceding components together, a simulation would result on a number of trees in the stand tagged by species, diameter, and age. Each tree would have an associated height and leaf area. This information would change every simulation year as seedlings are recruited and grow and die. From this information, and given the area of the stand in ha, we can calculate the basal area [m^2/ha] and density [number of stems/ha] for each species by year. As an example, in Figure 16.11, we can examine the dynamics of the basal area for species in the cross-timbers area of north Texas. In this simulation, we have assumed a run-on water input of 20% of the rainfall to account for the inflow of

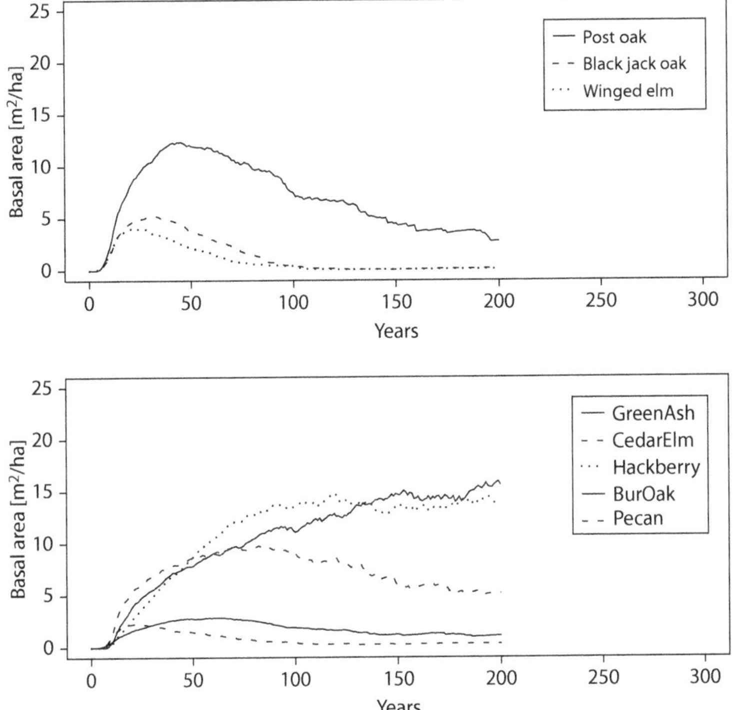

FIGURE 16.11 Basal area dynamics obtained from a gap model for species in cross-timbers area of North Texas. Top: Typical upland species. Bottom: Typical bottomland species.

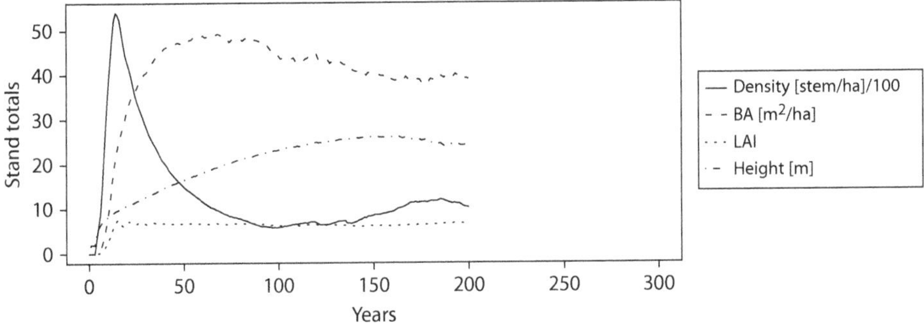

FIGURE 16.12 Stand aggregation from a gap model output.

water from uphill areas. We can see how post oak, black jack oak, and winged elm reach a maximum within 50 years into the simulation and decline to a steady state by 200 years. This behavior is similar to that of green ash, cedar elm, and pecan, but hackberry and bur oak grow at a slower pace and continue to grow by 200 years. Of course, we can aggregate the basal area and density by the level of the stand (Figure 16.12). In addition, trees can be placed in diameter categories in order to obtain tree diameter distribution every year. The basal area by species shows successional patterns when simulation starts from bare ground; fast growing and shade-intolerant species would first show an increase in the basal area and decline as the slower growing and shade-tolerant species come into the stand.

Exercise 16.11

Describe and interpret the results shown in Figure 16.12.

16.13 DETERMINISTIC AND LUMPED MODEL

Gap models can be simplified using a deterministic and lumped approach. This implies excluding random establishment and mortality or random variation of environmental factors, and basing demography on populations rather than individuals. The same equations apply and the parameter values apply to a number of trees of the same diameter, age, height, and species that collectively represent a tree population for the species. Therefore, this modeling approach is similar to a community dynamics model but with environmental constraints on growth. This approach was proposed as a simplification of JABOWA and applied to the Hubbard Brook watershed (Swartzman and Kaluzny, 1987). Figures 16.13 and 16.14 show an example of the simulation results for basal area and density, respectively. The simulated stand is 100 m^2 and we have converted the results to have units of per ha. Species 1 is an early successional species (pioneer), which is established and dominates quickly (about 80% of the basal area after 5 years), but disappears after 40 years. After that, species 3, a late successional species, starts to dominate and achieve about 90% of the basal area by 100 years. Please note that species 4, with the slowest growth, starts to pick up at about 100 years.

Exercise 16.12

Describe and interpret the results shown in Figure 16.14.

16.14 MARKOVIAN MODELS

Another approach to vegetation dynamics and succession is to use Markov models (we covered this concept in Chapter 15) and their extension, semi-Markov models, which we will explain in this chapter. In this approach, each forest stand type is a state variable and we describe the dynamic changes from type to cover type by probabilities of transition. A special case of semi-Markov models, one with exponential holding times, is similar to a compartment model.

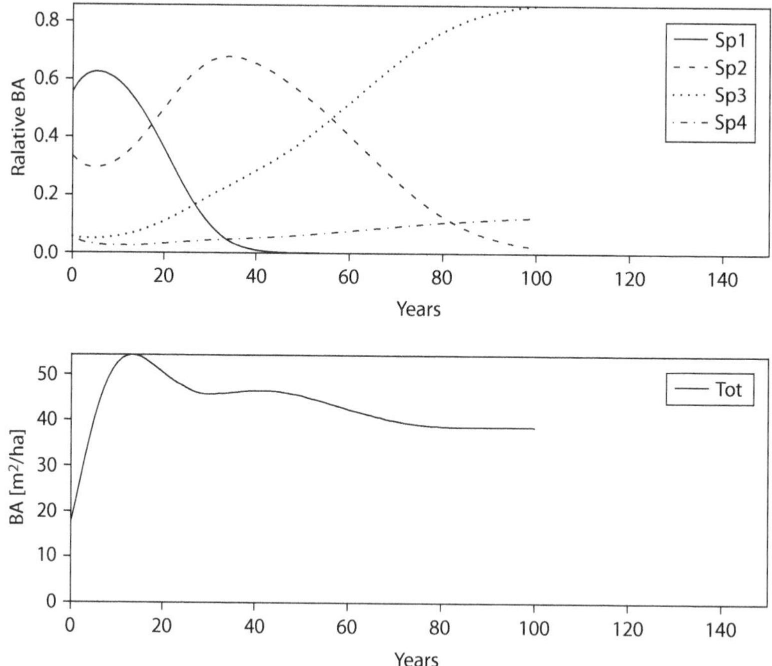

FIGURE 16.13 Forest succession: relative basal area by species (top) and total basal area (bottom).

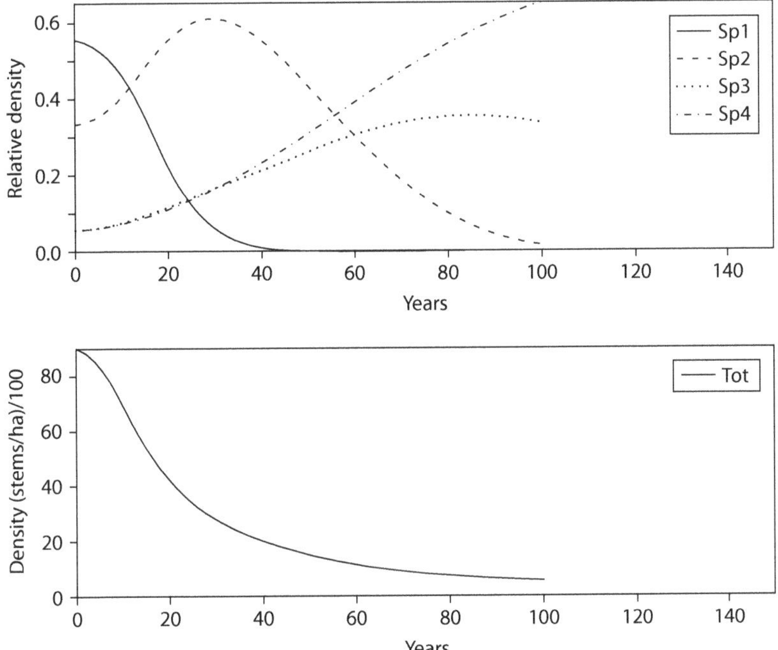

FIGURE 16.14 Forest succession: relative density by species (top) and total basal area (bottom).

There are many ways of defining the states of the Markov model. As an example, we will combine two properties, gap-creating (associated with mortality) and gap-requiring (associated with regeneration). Using these properties, we define four ecological roles to classify the tree species in the model as follows (Acevedo et al., 1996):

Role 1: Gap-creating and gap-requiring, shade-intolerant trees that grow to a large size
Role 2: Gap-creating and non-gap-requiring, shade-tolerant trees that grow to a large size
Role 3: Non-gap-creating and gap-requiring, shade-intolerant trees that grow to relatively small size
Role 4: Non-gap-creating and non-gap-requiring, shade-tolerant trees that grow to relatively small sizes

16.14.1 Markov

In a probability transition matrix, the entries are probabilities of transition among several states at each time step. State variables represent occupancy probabilities or fractions of space occupied by each forest type at time *t*. The steady state depends on the transition probabilities.

Let us consider, for example, a Markov chain with transitions of a forest plot among the four roles defined in the previous section (Acevedo et al., 1996). We will estimate the long-term patterns of regeneration and succession in a canopy gap. A gap-size forest plot makes transitions among several states defined by the dominance of one of the four roles (Figure 16.15).

At any particular time *t*, a tree species belonging to either one of the four roles dominates the plot. Denote by $p_i(t)$ the probability that the plot is occupied by role *i* at time *t*. The total canopy space covered by a collection of gap-size plots will be distributed among the roles according to the proportions X_is, which should be approximately equal to the p_is. Logically, we will require that

$$\sum_{i=1}^{4} X_i = 1 \tag{16.34}$$

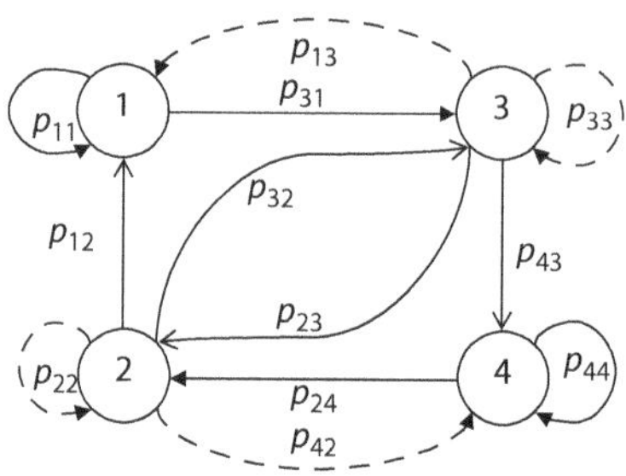

FIGURE 16.15 Markov model of forest succession based on gap creation upon mortality and shade tolerance.

that is, the total canopy cover consists of the sum of proportions X_i of each role i in the canopy. The probability p_{ij} is associated with the transition from role j to role i. Some transitions are unlikely, and therefore, their associated transition probabilities have very low values. For example, since role 3 requires gaps for regeneration and roles 3 and 4 do not produce them, we assume that p_{33} and p_{34} have low values. Likewise, p_{14} and p_{13} are negligible because roles 4 and 3 do not produce the gaps required by role 1.

Some other probabilities are also small and depicted as dashed lines because shade-tolerant species are likely to be outcompeted in gaps. Thus, p_{21}, p_{22}, p_{41}, and p_{42} have low values. Since roles 3 and 4 are considered to be of low stature relative to role 1 and 2, a two-layer canopy vertical structure is implicitly assumed in the model, even though the states of the model do not explicitly consider the relative proportions in a mixed canopy. A more detailed model can be written to consider the states as composed of role i in the upper canopy and role j in the lower canopy (Acevedo et al., 1995a).

Several scenarios can be analyzed using this model, for example, extremes of shade-tolerance and gap-creation characteristics. These conditions will tend to make the previous set of low-value transition probabilities almost negligible. Therefore, the transition matrix **P** is

$$\mathbf{P} = \begin{bmatrix} p_{11} & p_{12} & 0 & 0 \\ 0 & 0 & p_{23} & p_{24} \\ 1-p_{11} & 1-p_{12} & 0 & 0 \\ 0 & 0 & 1-p_{23} & 1-p_{24} \end{bmatrix} \quad (16.35)$$

which has only four independent parameters, for example, p_{11}, p_{12}, p_{23}, and p_{24}. These are described as follows: p_{11}, probability of role 1 reoccupying the gaps produced by themself; p_{12}, probability of role 1 reoccupying the gaps produced by mortality of role 2; p_{23}, probability of role 2 species reoccupying the space previously occupied by role 3; and p_{24}, probability of role 2 species reoccupying the space previously occupied by role 4.

A simulation that uses matrix $\mathbf{P} = \begin{bmatrix} 0.5 & 0.4 & 0 & 0 \\ 0 & 0 & 0.9 & 0.5 \\ 0.5 & 0.6 & 0 & 0 \\ 0 & 0 & 0.1 & 0.5 \end{bmatrix}$ yields the results shown in Figure 16.16. We can note that the fraction of space occupied by each role X_i fluctuates, but settles down to a steady state within 10 years of simulation.

The steady-state values, that is,

$$X_i(t+1) = X_i(t) = X_i^* \quad i = 1,\ldots,4 \quad (16.36)$$

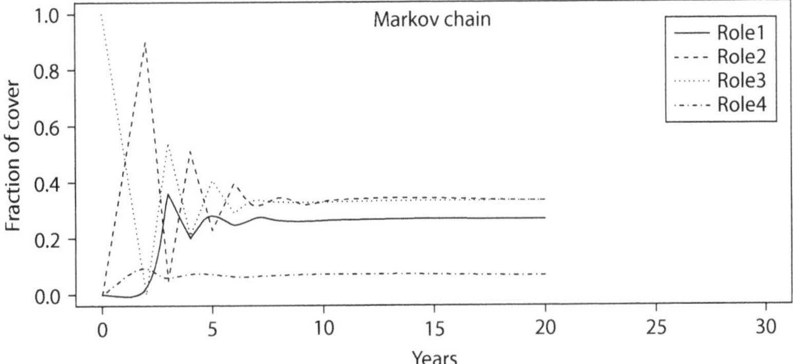

FIGURE 16.16 Markov chain dynamics with transitions among forest cover types.

are obtained as

$$\begin{aligned} X_1^* &= p_{12}p_{24}D^{-1} \\ X_2^* &= X_3^* = (1-p_{11})p_{24}D^{-1} \\ X_4^* &= (1-p_{23})(1-p_{11})D^{-1} \end{aligned} \tag{16.37}$$

where D is given as

$$D = (1-p_{11})(2p_{24}+1-p_{23}) + p_{12}p_{24}$$

We can estimate some general patterns of forest composition. For example, if p_{11} is high so that $p_{31} = 1 - p_{11} \approx 0$, then the equilibrium tends to approximately $X^* = [1,0,0,0]^T$, that is, a forest dominated by role 1. (Recall that subscript T denotes transpose; we use it here to convert to a column vector). On the contrary, if p_{11} is low so that $p_{31} = 1 - p_{11} \approx 1$, then the equilibrium contains all four roles in proportions $X^* = [p_{12}p_{24}, p_{24}, p_{24}, 1 - p_{23}]^T$. The value of p_{11}, discriminating colonization of gaps created by role 1, controls the dominance of role 1. As another example, a high value of p_{23}, indicating few transitions to role 4, would produce an equilibrium $X^* = [p_{11}p_{24}, (1 - p_{11})(2p_{24}), (1 - p_{11})(2-p_{24}), 0]^T$ with a low proportion of role 4.

Succession modeling using the steady state derived from a Markov chain to infer the long-term distribution does not account for the time spent in a given state before making a transition. In the next section, we address this limitation by using semi-Markov models.

Exercise 16.13

Calculate the steady state for the matrix $\mathbf{P} = \begin{bmatrix} 0.5 & 0.4 & 0 & 0 \\ 0 & 0 & 0.9 & 0.5 \\ 0.5 & 0.6 & 0 & 0 \\ 0 & 0 & 0.1 & 0.5 \end{bmatrix}$. Demonstrate that it is equal to $X^* = [0.27, 0.33, 0.33, 0.07]^T$. Confirm by examining the steady state in Figure 16.16.

16.14.2 Semi-Markov

The calculations in the previous section do not consider the different longevities and the growth rates of the species. We can extend the previous approach by using semi-Markov models. In these models, transitions depend not only on the source state but also on the time spent in the source state. This is the **holding time** of the transition, and it is a random variable with a given probability density function.

The basic Markov process underlying the semi-Markov process is called the "embedded chain" and corresponds to the basic model of Figure 16.15 with the associated transition probability matrix **P**. The holding time densities h_{ij}s, that is, the probability densities for the time spent in making the transition from role j to i, for every pair of roles i, j will be important in determining the transients and the steady state.

To analyze this model, a convenient form to use for the holding time density is the Erlang density (Acevedo, 1980a; Acevedo et al., 1996; Hennessey, 1980; Lewis, 1977):

$$h_{ij}(\tau) = \frac{d_{ij}^{k_{ij}}\tau^{k_{ij}-1}\exp(-d_{ij}\tau)}{(k_{ij}-1)!} \tag{16.38}$$

which has two parameters, d_{ij} and k_{ij}. The first one, d_{ij}, is a first-order rate corresponding to a Poisson process and the second one, k_{ij}, is an integer representing the order of the function. In this expression, τ is the time to make the transition. For $k_{ij} > 1$, the probability of making the transition in zero time is zero. A first-order Erlang density $k_{ij} = 1$ is an exponential. Semi-Markov with exponential holding time is the same as a continuous time Markov process and the same as a compartment model.

The mean and variance of this pdf are equal to

$$\mu_{ij} = k_{ij}/d_{ij} \quad \sigma^2{}_{ij} = k_{ij}/(d_{ij})^2 \tag{16.39}$$

Statistics of interest, for example, occupancy probabilities, are calculated from the available semi-Markov methodology (Howard, 1971) in the following manner. The waiting time density, that is, the probability density of the time spent in role j, before making a transition to any one of the other roles is given by

$$W_j(\tau) = \sum_{i=1}^{4} p_{ij}h_{ij}(\tau) \quad j = 1,2,\ldots,4 \tag{16.40}$$

as the sum of all the holding time densities corresponding to the transitions out of state j. The mean waiting time in role j is then the mean of $W_j(\tau)$ and denoted as M_j,

$$M_j = \sum_{i=1}^{4} p_{ij}\mu_{ij} \quad j = 1,2,\ldots,4 \tag{16.41}$$

or the sum of all products of the probability p_{ij} and the mean k_{ij}/d_{ij} of the $h_{ij}(t)$ density. In turn, the mean time between transitions M is a weighted sum of the mean waiting times in each role:

$$M = \sum_{j=1}^{4} X_j^* M_j \tag{16.42}$$

where the weights used in the average are the stationary proportions of the embedded chain as given in the previous section. The steady-state occupancy probabilities, and therefore, the stationary proportions X^{**}_is for all the roles can be calculated as

$$X_j^{**} = X_j^* M_j M^{-1} \quad j = 1,2,\ldots,4 \tag{16.43}$$

where X^{**}_i represents the fraction of space occupied by the role i after a sufficiently long time has elapsed.

Under the assumption of zero probabilities for the transition shown in the dashed lines in Figure 16.15, that is, the extreme-role case already discussed in the previous section, the application of the previous equations yields

$$\begin{bmatrix} X_1^{**} \\ X_2^{**} \\ X_3^{**} \\ X_4^{**} \end{bmatrix} = \begin{bmatrix} p_{12}p_{24}(a_{11}+a_{31}) \\ (1-p_{11})p_{24}(a_{12}+a_{32}) \\ (1-p_{11})p_{24}(a_{23}+a_{43}) \\ (1-p_{11})(1-p_{23})(a_{44}+a_{24}) \end{bmatrix} M^{-1}D^{-1} \tag{16.44}$$

where $a_{ij} = p_{ij}\mu_{ij}$ for short and M is given by

$$M = \begin{bmatrix} p_{12}p_{24}(a_{11}+a_{31}) + (1-p_{11})p_{24}(a_{12}+a_{32}) \\ +(1-p_{11})p_{24}(a_{23}+a_{43}) + (1-p_{11})(1-p_{23})(a_{44}+a_{24}) \end{bmatrix} D^{-1} \tag{16.45}$$

Note that for each one of the eight transitions, three parameters require values: the rate and order for each one of the eight time lags and one transition probability. As a hypothetical example, assume the following values for the transition probabilities and the lag orders:

$$\mathbf{P} = \begin{bmatrix} 0.5 & 0.4 & 0 & 0 \\ 0 & 0 & 0.9 & 0.5 \\ 0.5 & 0.6 & 0 & 0 \\ 0 & 0 & 0.1 & 0.5 \end{bmatrix} \qquad \mathbf{k} = \begin{bmatrix} 2 & 2 & 0 & 0 \\ 0 & 0 & 2 & 2 \\ 2 & 2 & 0 & 0 \\ 0 & 0 & 2 & 2 \end{bmatrix} \tag{16.46}$$

Instead of specifying the rates, it is easier to start with the means of the lags, estimated from the longevities of the tree species represented by each role,

$$\boldsymbol{\mu} = \begin{bmatrix} 80 & 130 & 0 & 0 \\ 0 & 0 & 50 & 100 \\ 80 & 130 & 0 & 0 \\ 0 & 0 & 50 & 100 \end{bmatrix} \tag{16.47}$$

and then compute the rates d_{ij} as k_{ij}/μ_{ij},

$$\mathbf{d} = \begin{bmatrix} 0.025 & 0.0154 & 0 & 0 \\ 0 & 0 & 0.04 & 0.02 \\ 0.025 & 0.0154 & 0 & 0 \\ 0 & 0 & 0.04 & 0.02 \end{bmatrix} \tag{16.48}$$

Exercise 16.14

Evaluate the stationary states according to Equation 16.44 and demonstrate that we get $\boldsymbol{X}^{**} = [0.24, 0.49, 0.19, 0.08]^T$. Compare to $\boldsymbol{X}^*$ for the embedded process calculated in the previous section. This exercise should demonstrate the importance of the holding time in determining the stationary distribution.

A convenient method for the simulation is to numerically integrate a set of first-order ODE that emulate the Erlang function as we approached distributed delays in Chapter 10 (Acevedo, 1980a; Acevedo et al., 1996). The transition from state j to state i is composed of a sequence of transitions among intermediate states, at the rate d_{ij}. The transition from the state j to the first intermediate state is affected by the probability p_{ij}. Therefore, for the transition from j to i, we can write

$$\frac{dy_n^{ij}}{dt} = d_{ij}[y_{n-1}^{ij}(t) - (t)] \quad n = 1,2,\ldots,k_{ij} \tag{16.49}$$

$$\frac{dy_0^{ij}}{dt} = -d_{ij}y_0^{ij}(t) + p_{ij}\sum_{n=1}^{4} d_{jn}y_{k_{jn}}^{jn}(t) \tag{16.50}$$

where the variables

$$y_0^{ij}, y_1^{ij}, \ldots, y_{k_{ij}}^{ij} \tag{16.51}$$

denote the intermediate states in the transition from j to i. All the proportions in the intermediate states of this transition add up to the total proportion of the state j undergoing that transition. Therefore, the sum over all destination states i will give the total proportion in state j,

$$X_j = \sum_{i=1}^{4}\sum_{n=0}^{k_{ij}} y_n^{ij} \quad j = 1,\ldots,4 \tag{16.52}$$

This equivalence has been referred to as the linear chain trick (McDonald, 1978), catenary system (van Hulst, 1979), or pseudo-compartments (Matis et al., 1992). This is precisely how we approached distributed delays in Chapter 10. These equations are easy to solve numerically. For example, for the values of the parameters given earlier, the results shown in Figure 16.17 are

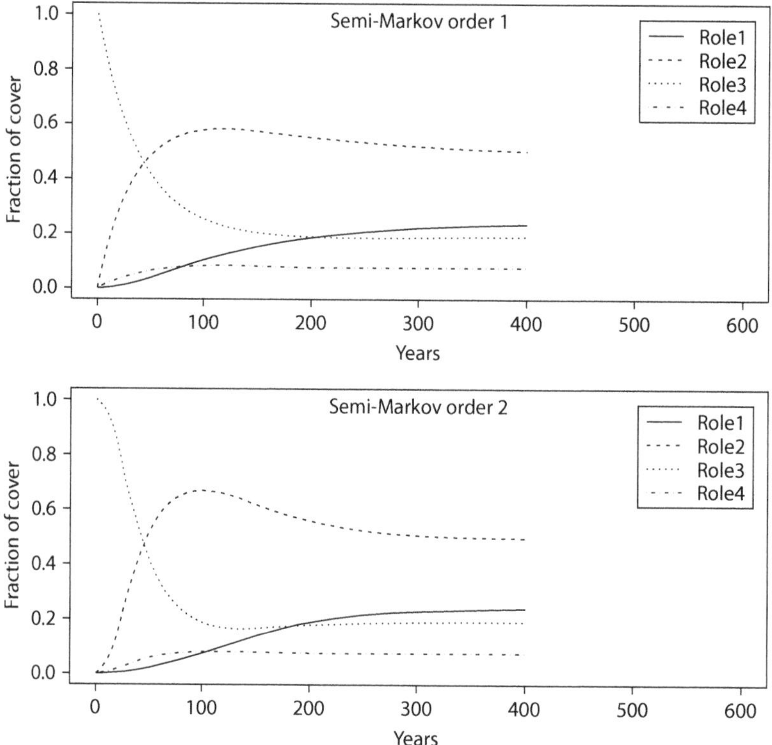

FIGURE 16.17 Semi-Markov model with transitions among forest cover types illustrating the effect of holding time in each transition. Top: Erlang order 1. Bottom: Erlang order 2.

obtained by assuming the initial condition $[0,0,1,0]^T$, that is, all plots are dominated by role 3. We can appreciate the impact of the order of the Erlang pdf on the dynamics.

In Markov models, there is the potential for expanding the state description to include environmental and other biotic factors important in ecosystem dynamics. Many examples of environmental models include space and time. Bi-dimensional, spatially explicit models operate on a grid or lattice made of square cells laid out on a map. Adding a third dimension like, for example, depth or height, would lead to a cubic grid. The individual-based approach applies to landscape by making topographic position (elevation, slope and its aspect) and soils affect the environmental constraints (temperature, precipitation, and radiation). Each gap–model plot corresponds to a homogeneous parcel of landscape. The semi-Markov approach applies to the landscape level by making the parameters (probabilities and holding times) depend on the elevation, slope and its aspect, soils, and so on. In this case, the semi-Markov model can be parameterized from the gap model (Acevedo et al., 2001; Acevedo et al., 1995a). The structure could affect the dynamics, for example, the proximity of landscape parcels with abundance of a cover type could affect dispersal and establishment patterns.

Similarly, the prairie or grassland succession model based on compartments we studied in Chapter 12 extends to ecosystem dynamics by making the rates vary for different soil types and topographic positions. Thus, the semi-Markov parameters or transfer rates would be dependent on terrain. An example is introducing a fixed delay to account for the changes in soil litter during succession (Bledsoe and Van Dyne, 1971).

16.15 COMPUTER SESSION

16.15.1 H versus D Allometric Relation

This is an example of how to estimate the parameters of an allometric relation. It invokes the function `D.H.est` explained at the end of this chapter.

```
# Change this according to directory and input file.
filedata <- "chp16/greenash.txt"
# Change filename and name for title according to species as needed.
species.name <- "Green ash"
# Input file has 2 columns, H and DBH, with header
# listed in ascending order of DBH. Maximum H and DBH on the second row
D.Hmax <- scan(filedata, skip=1, nlines=1)
D.H <- read.table(filedata, skip=2); names(D.H) <- c("D","H")
pdf("chp16/greenash-DHallom-out.pdf")
d.h <- D.H.est(D.H,D.Hmax,species.name,pdfout=T)
dev.off()
```

The results in **d.h** show the values of b3 and b4, height estimates, and residual sum of square errors.

```
> d.h
$b3
  b3
0.0181

$b4
  b4
0.5951
```

```
$H.est
[1]  9.68 11.01 11.41 11.81 11.81 12.19 13.60 13.60 14.24 14.55 15.44
16.00
[13] 17.05 17.54 18.01 18.24 18.24 18.91 19.34 19.95 21.09 21.62 21.96
22.13
[25] 23.81 23.95 24.09 24.37 25.51 26.54 27.06 27.65 27.93 28.54 32.41

$RSS.est
[1] 459.172
```

The resulting figure is the one shown in Figure 16.2, which we already discussed.

Exercise 16.15

Use the file **chp16/cedarelm.txt** to determine the coefficients b3 and b4. Produce the graph as shown in Figure 16.2.

16.15.2 Growth Equation

This is an example of calculating the growth rate using the function `dD.dt.analysis` as explained at the end of this chapter. This function uses `dD.rate`, also given at the end of this chapter. Here we first define a set of D values spanning the interval from 0 to $D_{max} = 100$, a set of two values of G, and allometric parameters valued as in the first row of Table 16.1 discussed already. In addition, we use an initial condition of $D_0 = 5$ cm and a time sequence from 0 to 300 years.

```
pdf("chp16/dD-figure.pdf")
Dmax <-100; D <- seq(0,Dmax,0.1)
G <- c(600,1000)
param.H <- c(20, -0.02,0.6); param.l <- c(0.16,0.8,2)
D0 <- 5; t <- seq(0,300)
rates.var <- dD.dt.analysis(D,G,Dmax,param.H,param.l,D0,t,pdfout=T)
dev.off()
```

The results are as shown in Figures 16.3 and Figure 16.4, which we have already discussed.

Exercise 16.16

Consider cedar elm and use the values of the coefficients b_3 and b_4 from the previous exercise. For simplicity, use the leaf area allometric coefficients as given in the first row of Table 16.1. Produce the graph as in Figures 16.3 and 16.4. Discuss.

16.15.3 Environmental Factors

This is an example of calculating and plotting the environmental response functions using `env.factors.analysis`, which is explained at the end of this chapter. Here we will use as an example the simple parabolic response (Equation 16.21) implemented as the function `parab1`, which is also given at the end of this chapter. Here we first define a list `FE` of response functions (all `parab1`), an array of `Eopt` (all 1), and a list of arrays of the parameters. The arrays are of dimension 2 because we have two species; the second member of the array is always made equal to 1 in order to have a tolerant response to all the factors by species 2.

```
pdf("chp16/gE-figure.pdf")
  nsp <- 2
  splab <- c("Intolerant","Tolerant")
  FE <- list(parab1,parab1,parab1,parab1)
  Eopt <- c(1,1,1,1)
  param<- list(c(3,1), c(2,1), c(1.5,1), c(1.1,1))
  env.factors.analysis(nsp,splab,FE,Eopt,param)
  dev.off()
```

The result is shown in Figure 16.5, which we have already discussed.

Exercise 16.17

Vary the parameters from the example such that species 1 and 2 are labeled "Sp1" and "Sp2," such that species 1 is tolerant to the lack of E1 and E2, but intolerant to the stress in E3 and E4 with α = 3; and species 2 is intolerant to stress in E1 and E2 with α = 2, but tolerant to stress in E3 and E4.

16.15.4 Impact of g(E) on Growth

In this example, we use the function `ge.growth` and call it using two species, one intolerant and one tolerant to all four factors. We keep the allometric coefficients, D_{max} and G_{max}, the same for both the species and compare the effect of differential tolerance on growth.

```
pdf("chp16/gE-D-figure.pdf")
  nsp <- 2; splab <- c("Intolerant","Tolerant")
  Dmax <-rep(100,nsp); Gmax <- rep(1000,nsp)
  param.H <- list(c(20, -0.02,0.6),c(20, -0.02,0.6))
  param.l <- list(c(0.16,0.8,2),c(0.16,0.8,2))
  D0 <- rep(5,nsp); t <- seq(0,50); nt <- length(t)
  FE <- list(parab1,parab1,parab1,parab1)
  Eopt <- c(1,1,1,1)
  param.E<- list(c(3,2,1.5,1.1), rep(1,4))
  Ef <- list(runif,runif,runif,runif)
  param.Ef <- list(c(Eopt[1],Eopt[1]),c(0,Eopt[2]),c(Eopt[3],Eopt[3]),
    c(0,Eopt[4]))
  t.GD <- ge.growth(nsp,splab,FE,Eopt,param.E,Ef,param.Ef,Dmax,Gmax,param.H,
    param.l,D0)
  dev.off()
```

The result is shown in Figure 16.6, which we already discussed.

Exercise 16.18

As in the previous exercise, vary the parameters from the example such that species 1 and 2 are labeled "Sp1" and "Sp2," such that species 1 is tolerant to low E1 and E2, but intolerant to low E3 and E4 with α = 3; and species 2 is intolerant to low E1 and E2 with α = 2, but tolerant to low E3 and E4. Now keep everything else the same as in the example and perform several simulations of 50, 100, and 200 years. Compare the impact of tolerance on the diameter growth. Discuss.

16.15.5 Estimation of G_{MAX}

In this example, we use a series of dD measurements in the diameter categories of 5 cm from 15 to 75 cm, values of D_{max}, and values of allometric coefficients for height and leaf area. A call to the function `dD.dt.calibra` yields Figure 16.7, which was already discussed.

```
dD <- c(0.5,0.6,0.7,0.9,0.8,1.0,1.1,1.3,1.2,1.0,1.0,0.8,0.8)
D <- seq(15,75,5)
Dmax <-100
param.H <- c(40, -0.015,0.6); param.l <- c(0.16,0.6,2.2)
dD.dt.calibra(D,dD,Dmax,param.H,param.l)
```

16.15.6 Sunlight Response

Here we use the function `expon.alf` given at the end of this chapter for exploring the behavior of the tree response to sunlight given by Equation 16.31. Five categories of tolerance are established. The results are shown in Figure 16.8, which was already discussed.

```
clab <- c("Very Int","Int","Med","Tol","Very Tol")
c1<- c(1.58,1.26,1.13,1.05,1.02)
c2<- c(1.19,1.79,2.44,3.29,4.16)
c3<- c(0.15,0.12,0.09,0.06,0.03)
L <- seq(0,1,0.01); nl <- length(L)
ALF <- matrix(ncol=5,nrow=nl)
for(i in 1:5)ALF[,i] <- expon.alf(L, c(c1[i],c2[i],c3[i]))
matplot(L,ALF,type="l",col=1,xlab="L/L0",ylab="Sunlight Multiplier F(L)")
legend("bottomright",leg=clab, lty=1:5,col=1)
```

16.15.7 Soil Moisture Response

We can practice the use of the function `gauss.smf` explained at the end of this chapter. Use the following lines of code, where we define the tolerance parameters to be 0.2 and 0.3 for β_{WD} and β_{DD}, respectively:

```
Elab <- c("F(1-E_WD)","F(1-E_DD)")
x <- seq(0,1,0.01)
E <- matrix(nrow=length(x),ncol=3)
y <- matrix(nrow=length(x),ncol=4)
for(i in 1:2) E[,i] <- x
p <- c(0.2,0.30)
for(i in 1:length(x))
y[i,] <- gauss.smf(E[i,],p)
matplot(x,y,type="l",col=1,xlab="E (WD or DD)",ylab="F(E)")
legend("topleft",col=1,lty=1:2,leg=Elab)
```

16.15.8 Individual-Based Model

As an example, we will work with a modification of the ZELIG model and upland and bottomland forests in the Greenbelt Corridor of the Elm Fork of the Trinity River (GBC) in north Texas. The upland stands are dominated by Post oak and Elm (Acevedo et al., 1997) and the bottomland by Hackberry (Holcomb, 2001). Biological information is provided by the input file **gbc-species-inp.txt**, whereas the site-specific information on soil and weather variables is provided in the **gbc-site-inp.txt** input file. An additional input file is **gbc-control-inp.txt** used to define the simulation controls.

The control file defines a 200-year run; it specifies to write to output files every 50 years, except the tracer output, which is for every year. The species input file has the parameter values for the species. The first three species are more abundant on the upland sites, whereas the last five are more abundant on the bottomland sites. The site input file has three types of soils. We will use type 1 (Ovan clay) as indicated by the matrix at the end of the file. More explanation about the input files is given by the documents file in the folder.

Because zelig is precompiled from a Fortran program, we use a function **zelig.F** and a function **read.plot.zelig.out** in R to read and plot the output file. Recall that we should load the `seem.dll` unless we have automated in `Rprofile.site` as explained before in Chapters 10, 11, and 14.

```
dyn.load(paste(.libPaths(),"/seem/libs/i386/seem.dll",sep=""))
```

The function **zelig.F** call includes two arguments: the folder, the and a prefix for input and output filenames. As explained in Chapters 10, 11 and 14, to be able to use a Fortran or C program, first load the `seem.dll` by using `dyn.load` unless you have automated the loading in the Rprofile.site.

```
spp <- c("Post oak", "Black jack oak", "Winged elm", "GreenAsh",
     "CedarElm", "Hackberry", "BurOak", "Pecan")
grp1 <- c(1:3); grp2 <- c(4:8) #upland 3 spp and bottomland 5 spp
fileout <- zelig.F("chp16","gbc")
x <- read.plot.zelig.out(fileout,spp, grp1, grp2, pdfout=T)
```

You should see two output files: **gbc-print-out.txt** and **gbc-tracer-out.txt**. More explanation on the output files is given by the documents file in the folder. Use a text editor to look at the output files. In particular, look at DBH distribution and stand aggregates for each simulation interval.

The file **gbc-tracer-out.txt** contains the following variables for each year: stand aggregates (density, biomass, standard deviation of biomass, basal area, mean LAI, mean canopy height) and basal area per species (eight in this case). The function `read.plot.zelig.out` reads the tracer file and makes plots of the total aggregated stand variables (Figure 16.12) and basal area by species (Figure 16.11), which were already discussed.

Exercise 16.19

Examine the diameter distribution in the stand at the end of the simulation run.

16.15.9 Deterministic and Lumped JABOWA

In this example, the state variables are basal area and number per species. The species are **cherry, birch, maple**, and **spruce**. We simulate a scenario of 100 years successional dynamics after a clear-cut. The initial conditions are 50, 30, 5, and 5 for number of trees and 5 cm DBH, and age of 1 year for all species. The parameters and their values are adapted from the work by Swartzman and Kaluzny (1986).

A default data set is in the file **forsucc-inp.txt**, which is too long to reproduce here entirely, but can be examined with a text editor. We extract the initial conditions:

```
dbh0[1]   5.00
dbh0[2]   5.00
dbh0[3]   5.00
dbh0[4]   5.00
x_unit    [cm]
```

```
num0[1]   50
num0[2]   30
num0[3]   5
num0[4]   5
num_unit [trees]
age0[1]   1.00
age0[2]   1.00
age0[3]   1.00
age0[4]   1.00
```

We use the seem function forsucc.C which call a precompiled program forsucc written in C. As we have done in other occasions when we have a precompiled simulation program, we the function forsucc.C includes calls to manage files for this particular program, and in this case a call to the function .C(). We do not need to dyn.load the **seem.dll** because we did it in the 16.15.8 section (unless you have closed the R program since then and have not automated the dyn.load in Rprofile.site). Once the run is made, we call another function `read.plot.forsucc.out` to read the output file and plot.

```
fileout <- forsucc.C("chp16","forsucc") # forsucc is fileprefix
x <- read.plot.forsucc.out(fileout, sens=F, pdfout=T)
```

The results are shown in Figures 16.13 and 16.14, which we have already discussed. The batch and R functions are at the end of this chapter.

Exercise 16.20

Generate a figure of basal area as a function of time for a late successional scenario. For this, edit the input file to modify the initial conditions as (10, 5, 3, and 3) for numbers, (20, 15, 0.5, and 0.5) for diameters, and (15, 15, 3, and 3) for age. Run and plot the figure (it should be similar to the one in Figure 6.6, page 162, in the work of Swartzman and Kaluzny, 1986); explain the successional sequence obtained for this scenario.

16.15.10 Semi-Markov

The seem function semi.F is an interface for the program semi written in Fortran and which simulates semi-Markovian dynamics assuming Erlang densities for the holding times. We use two different input files for forest succession based on roles, one, **rolesm-inp.txt,** is for the first-order Erlang and the other, **roless-inp.txt**, is for the second-order Erlang. For example, the contents of the latter are

```
  4
0.01        10       40.00
0.000      0.000     1.000      0.000
0.500      0.400     0.000      0.000
0.000      0.000     0.900      0.500
0.500      0.600     0.000      0.000
0.000      0.000     0.100      0.500
2          2         0          0
0          0         2          2
2          2         0          0
0          0         2          2
8.00       13.00     0.00       0.00
0.00       0.00      5.00       10.00
```

```
8.00      13.00     0.00      0.00
0.00      0.00      5.00      10.00
0.00      0.00      0.00      0.00
0.00      0.00      0.00      0.00
0.00      0.00      0.00      0.00
0.00      0.00      0.00      0.00
```

The first line defines the number of and these are the four roles. The second line has the dt, time to print, and final time in decades. The next line is the initial condition, and then four matrices of 4 × 4: **P, k, μ**, and a matrix of fixed delays, which we do not use here in this example. The values for the parameters are the ones given earlier in this chapter.

Because the simulator semi is a Fortran program, we use the functions semi.F to call semi and **read.plot.semiout** to read and plot the output file. We do not need to dyn.load the **seem.dll** because we did it in the 16.15.8 section (unless you have closed the R program since then and have not automated the dyn.load in Rprofile.site). The function semi.F includes two arguments: the folder and a prefix corresponding to input and output filenames.

```
spp <- c("Role1", "Role2", "Role3", "Role4")
fileprefix <- "rolesm"; label <- "Semi-Markov Order 1"
fileout <- semi.F("chp16",fileprefix)
x <- read.plot.semi.out(fileout, spp, label, pdfout=T)
```

After executing we produce the file **roless-out.txt**. This output file has the following: time (in decades), fraction in each one of the four states, and the total (1.00). After executing `read.plot.dynlayer.out`, we obtain the results as shown in Figure 16.16 and in the bottom panel of Figure 16.17.

The other input file **rolesm-inp.txt** defines a semi-Markov with exponential holding times, which makes it the same as a compartment model:

```
spp <- c("Role1", "Role2", "Role3", "Role4")
fileprefix <- "roless";label <- "Semi-Markov Order 2"
fileout <- semi.F("chp16",fileprefix)
x <- read.plot.semi.out(fileout, spp, label, pdfout=F)
```

The results should match Figure 16.16 and the top panel of Figure 16.17.

Exercise 16.21

Explore the effect of a third-order Erlang pdf. To modify the parameter values, edit the ***-inp.txt** file (be careful to preserve the position of the data because it is read with a fixed format). Instructions on the input file are contained in the readme file.

16.15.11 Build Your Own

Exercise 16.22

Develop a semi-Markov model for five states. Four of these are the four roles already studied and the fifth state is a canopy gap or opening. Only those roles that start in gaps should have nonzero transition probabilities from gap state. Only those roles that create gaps should have nonzero transition probabilities to the gap state. Draw a transition graph, write a matrix, assign parameter values, calculate the steady-state values, and develop a simulation using **semi**.

16.16 SEEM FUNCTIONS EXPLAINED

16.16.1 Function DH.allom

It calculates the height, given the diameter and parameters H_{max}, b_3, and b_4 according to Equation 16.5.

```
# height vs diameter alometric relation
DH.allom <- function(D,param){
H0 <- 1.37
Hmax <- param[1]; b3 <- param[2]; b4 <- param[3]
H <- H0 + (Hmax-H0)*(1 - exp(b3*D))^b4
return(H)
}
```

16.16.2 Function D.H.est

This function performs a set of linear regressions on a log-transformed H versus D allometric relation in order to determine the best coefficient b_3 value for each one of a set of values of b_4 (Delgado et al., 2005). The log-transformed version of Equation 16.5 is found in the following manner. First separate the terms in D and H:

$$\frac{H - H_0}{H_{max} - H_0} = (1 - \exp(b_3 D))^{b_4}$$

Raise both sides to power $1/b_4$ and rearrange:

$$\left(1 - \frac{H - H_0}{H_{max} - H_0}\right)^{1/b_4} = \exp(b_3 D)$$

Then take the natural logarithm of both sides:

$$y = \ln\left(\left(1 - \frac{H - H_0}{H_{max} - H_0}\right)^{1/b_4}\right) = b_3 D$$

Now we can perform linear regression of y versus D.

Then the b_3, b_4 pair with the smallest residual sum-of-squares error is selected to perform nonlinear regression, where the starting values are the pair selected above. Arguments are a data frame with D and H values and an array with D_{max} and H_{max} values. This function uses the previous function DH.allom.

```
D.H.est <- function(D.H,D.Hmax,species.name,pdfout=F,linregout=F) {

# Assign variables from arguments and constant breast height.
D <- D.H$D; H <- D.H$H; Dmax<- D.Hmax[1]; Hmax <- D.Hmax[2]
H0 <- 1.37
# define set of b4 and allocate
b4 <- seq(0.2,1.8,0.1); n <- length(b4)
b3 <- array(); RSS <- array()
# Loop through different b4 values and perform linear regression on log
# transformed data for each.
```

```
for (i in 1:n){
# Transforming variables; use negative of log to see fitted line as
# positive.
Hlog <- - log(1 - ((H-H0)/(Hmax-H0))^(1/b4[i]))
# Starting linear regression
allom.lm <- lm(Hlog ~ D -1)
b3[i] <- allom.lm$coeff
# Predicted height. Note: negative sign used in exp since -log was used above.
H.est <-DH.allom(D,param=c(Hmax,-b3[i],b4[i]))
RSS[i] <- sum((H.est-H)^2)

if(linregout==T){
# Graph scatter plot and fitted line
  if(pdfout==F) win.graph()
  mat<- matrix(1:2,2,1,byrow=T)
  layout(mat,c(7,7),c(3.5,3.5),res=TRUE)
  par(mar=c(4,4,1,.5), xaxs="r", yaxs="r")
  plot(D,H, xlab="Diameter (DBH, cm)", ylab="Height (m)"); lines(D,H.est)
  title(paste("Linear Estimation,",species.name,",","Dmax=",Dmax,",",

  "Hmax=",Hmax,",","b3=",round(b3[i],4),",","b4=",round(b4[i],4),",","
  RSS =",round(RSS[i],4)),cex.main=0.5)
  plot(fitted(allom.lm), resid(allom.lm),xlab="Fitted Values",
  ylab="Residuals")
  abline(h=0)
  }
}
# Find the linear regression that resulted in the minimum residual
# sum-of-squares error.
# Select this b3, b4 pair as starting values of non linear regression.
for(i in 1:n){
 if(identical(RSS[i],min(RSS))){startb3 <- b3[i]; startb4 <- b4[i]}
}
# Nonlinear Regression.
nl.est <- nls(H ~DH.allom(D,param=c(Hmax,-b3,b4)), data= D.H, start=
list(b3=startb3, b4=startb4))
# Graph scatter plot with fitted line and the residuals vs fitted
# values.
b3.est <- coef(nl.est)[1];b4.est <- coef(nl.est)[2]
H.est <- DH.allom(D,param=c(Hmax,-b3.est,b4.est))
RSS.est <- round(sum((H.est-H)^2),4)
b3.est <- round(b3.est,4);b4.est <- round(b4.est,4)
if(pdfout==F) win.graph()
 mat<- matrix(1:2,2,1,byrow=T)
 layout(mat,c(7,7),c(3.5,3.5),res=TRUE)
 par(mar=c(4,4,1,.5), xaxs="r", yaxs="r")
plot(D, H, xlab="Diameter [cm]", ylab="Height [m]")
lines(D, H.est)
title(paste("Non Linear Estimation,", species.
name,",","Dmax=",Dmax,",",
 "Hmax=", Hmax,",","b4=", b4.est,",","b3=", b3.est,",","RSS=",RSS.est),
 cex.main=0.5)
```

```
plot(H.est, resid(nl.est), xlab="Fitted Values", ylab="Residuals")
abline(h=0)
H.est <- round(H.est,2)
return(list(b3=b3.est,b4=b4.est,H.est=H.est,RSS.est=RSS.est))
}
```

16.16.3 FUNCTION Dl.ALLOM

This function calculates the leaf area by applying Equation 16.7, which requires three parameters.

```
Dl.allom <- function(D,param){
 c <- param[1]; hc <- param[2]; d <- param[3]
 l <- c*D^d*fhc
 return(l)
}
```

16.16.4 FUNCTION dD.RATE

This function calculates the increment in the diameter using Equations 16.12 and 16.14 together with allometric functions **DH.allom** and **Dl.allom** already defined. It requires a value or a set of diameter plus specification of G and seven parameters. Besides D_{max}, three of these are for the H versus D relation and three others for the l versus D relation.

```
dD.rate <- function(D,G,Dmax,param.H,param.l){
 H <- DH.allom(D,param.H)
 l <- Dl.allom(D,param.l)
 Hmax <- param.H[1]; b3 <- param.H[2]; b4 <- param.H[3]
 nume <- l*(1 - (D*H)/(Dmax*Hmax))
 phi <- 100*(2*H-D*(Hmax-H0)*b3*b4*exp(b3*D)*(1-exp(b3*D))^(b4-1))
 deno <- D*phi
 dD <- G*nume/deno
return(dD)
}
```

16.16.5 FUNCTION dD.DT.ANALYSIS

This function allows for analyzing the d*D*/d*t* increment for a given set of values of G. It uses the function **dD.rate** given above. Besides the values of G, it requires a set of values of D, seven parameters as described above in the function **dD.rate**, initial D_0 condition for the numerical integration, and the time sequence t.

```
dD.dt.analysis <- function(D,G,Dmax,param.H,param.l,D0,t,pdfout=F){

mat<- matrix(1:2,2,1,byrow=T)
layout(mat,c(7,7),c(3.5,3.5),res=TRUE)
par(mar=c(4,4,1,.5), xaxs="r", yaxs="r")

dD.dt <- matrix(nrow=length(D),ncol=length(G))
for(i in 1:length(G))
dD.dt[,i] <- dD.rate(D,G[i],Dmax,param.H,param.l)
```

```
matplot(D,dD.dt,type="l",col=1,xlab="Diameter D [cm]",ylab="Diameter
increment dD/dt [cm/y]")
legend("topleft",leg=paste("G=",G),lty=1:length(G),col=1)

Dt <- matrix(nrow=length(t),ncol=length(G))
for (j in 1:length(G)){
Dt[1,j] <- D0
for(i in 2:length(t))
Dt[i,j] <- Dt[i-1,j] + dD.rate(Dt[i-1,j],G[j],Dmax,param.H,param.l)
}
matplot(t,Dt,type="l",col=1,xlab="Time [yr]",ylab="Diameter D [cm]")
legend("bottomright",leg=paste("G=",G),lty=1:length(G),col=1)

if(pdfout==F) win.graph()
 mat<- matrix(1:2,2,1,byrow=T)
 layout(mat,c(7,7),c(3.5,3.5),res=TRUE)
 par(mar=c(4,4,1,.5), xaxs="r", yaxs="r")
 Ht <- DH.allom(Dt,param.H)
 matplot(t,Ht,type="l",col=1,xlab="Time [yr]",ylab="Height H [m]")
 legend("bottomright",leg=paste("G=",G),lty=1:length(G),col=1)
 lt <- Dl.allom(Dt,param.l)
 matplot(t,lt,type="l",col=1,xlab="Time [yr]",ylab="Leaf area l [m2]")
 legend("bottomright",leg=paste("G=",G),lty=1:length(G),col=1)
 return(list(D.dD=data.frame(D,dD.dt),t.Xt =data.frame(t,Dt,Ht,lt)))
}
```

16.16.6 Function parab1

This is a simple function to evaluate a simple parabolic response to an environmental variable according to Equation 16.21; it takes the value 1 when $E = E_{opt}$ and uses the argument param for α. Negative values are assigned zero. Although the function is not supposed to receive values of E leading to F greater than 1, we assign one to F values above 1 for caution.

```
parab1 <- function(E,Eopt,param){
ne <- length(E)
F <- 1.0-param*(Eopt-E)^2
 for(i in 1:ne) if(F[i]<0) F[i] <- 0
 for(i in 1:ne) if(F[i]>Eopt) F[i] <- 1
return(F)
}
```

16.16.7 Function env.factors.analysis

This function is designed to plot the environmental factors for several species. Their number and labels are arguments `nsp` and `splab`. Then four response functions (e.g., `parab1`) are passed as a list `FE`, `Eopt` as an array, and param for the functions as a list of arrays.

```
env.factors.analysis <- function(nsp,splab,FE,Eopt,param){
 # parabolic
 Elab <- c("L","T","W","N")
 mat<- matrix(1:4,2,2,byrow=T)
```

```
layout(mat,c(3.5,3.5),c(3.5,3.5),res=TRUE)
par(mar=c(4,4,1,.5), xaxs="r", yaxs="r")
for(k in 1:4){
 E <- seq(0,Eopt[k],0.01*Eopt[k]); ne <- length(E)
 F <- matrix(nrow=ne,ncol=nsp)
 for(j in 1:nsp) F[,j] <- FE[[k]](E,Eopt[k],param[[k]][j])
 matplot(E,F,type="l",col=1,xlab=Elab[k],ylab=paste("Factor
 F(",Elab[k],")"))
 legend("bottomright",leg=splab, lty=1:nsp,col=1)
 }
 }
```

16.16.8 Function ge.growth

This function performs a simulation of random environmental fluctuation and uses tree response functions to calculate $g(\boldsymbol{E})$ according to Equation 16.20. Then $G_{max}g(\boldsymbol{E})$ is used in conjunction with the allometric parameters and D_{max} to simulate tree diameter growth using `dD.rate` to implement the Euler method. Arguments include the number and label for species (`nsp, splab`), functions `FE` for tree response, `Eopt` optima for environmental variables, parameters `param.E` for tree response by species, functions `Ef` and parameter `param.Ef` to generate random samples of environmental variables, arrays for D_{max} and G_{max}, allometric parameters `param.H` and `param.l`, and initial diameter `D0` for simulation. The output includes graphs of $g(\boldsymbol{E})$ and D versus time for each species.

```
ge.growth <- function(nsp,splab,FE,Eopt,param.E,Ef,param.
Ef,Dmax,Gmax,param.H,param.l,D0){
E <- matrix(nrow=nt,ncol=4)
G <- matrix(nrow=nt,ncol=nsp); D <- G
F <- structure(1:(nt*nsp*4), dim=c(nt,nsp,4))
D[1,1:nsp] <- D0
E[1,] <- 1; F[1,,] <- 1; G[1,] <- 1
for(i in 2:nt){
   for(k in 1:4) {
      E[i,k] <- Ef[[k]](1,param.Ef[[k]])
      for(j in 1:nsp) F[i,j,k] <- FE[[k]](E[i,k],Eopt[k],param.E[[j]][k])
 } #k
 for(j in 1:nsp) {
  G[i,j] <- F[i,j,1]*F[i,j,2]*min(F[i,j,3],F[i,j,4])
  D[i,j] <- D[i-1,j] + dD.rate(D[i-1,j],Gmax[j]*G[i,j],Dmax[j],
  param.H[[j]],param.l[[j]])
 } #j
} #t

mat<- matrix(1:2,2,1,byrow=T)
layout(mat,c(7,7),c(3.5,3.5),res=TRUE)
par(mar=c(4,4,1,.5), xaxs="r", yaxs="r")
matplot(t,G,type="s",xlab="Time [yr]",ylab="gE",col=1)
legend("topleft", lty=1:nsp, leg=splab)
matplot(t,D,type="l",xlab="Time [yr]",ylab="D [cm]",col=1)
legend("topleft", lty=1:nsp, leg=splab)
return(list(round(E,2),round(F,2),round(data.frame(t,E,G,D),2)))
}
```

16.16.9 Function dD.dt.calibra

This function performs an estimation of G_{max}, given a set of D and dD values together with D_{max} and allometric coefficients. It uses Equation 16.26 and plots the corresponding growth curve for reference.

```
dD.dt.calibra <- function(D,dD,Dmax,param.H,param.l){
mat<- matrix(1:2,2,1,byrow=T)
layout(mat,c(7,7),c(3.5,3.5),res=TRUE)
par(mar=c(4,4,1,.5), xaxs="r", yaxs="r")

fmax <- max(dD,na.rm=T)
Dopt <- D[which(dD==fmax)]
fopt <- dD.rate(Dopt,1,Dmax,param.H,param.l)

Gmax <- round(fmax/fopt,0)
matplot(D,dD, type="p",col=1,pch=1,xlab="Diameter D [cm]",ylab="Diameter
increment dD/dt [cm/y]",ylim=c(0,2))
dD.dt <- array()
dD.dt <- dD.rate(D,Gmax,Dmax,param.H,param.l)
lines(D,dD.dt,col=1)
legend("topleft",leg=paste("Gmax=",Gmax),lty=1,col=1)
mtext(side=1,line=-1,"Adjusted for Optimum conditions g(E)=1")
return(Gmax)
}
```

16.16.10 Function expon.alf

This function implements the response given by Equation 16.31. Arguments are a value of a set of values of environmental variable E and an array p containing the three parameters.

```
expon.alf <- function(E,p){
ne <- length(E)
F <- p[1]*(1.0-exp(-p[2]*(E-p[3])))
 for(i in 1:ne) if(F[i]<0) F[i] <- 0
 for(i in 1:ne) if(F[i]>1) F[i] <- 1
return(F)
}
```

16.16.11 Function gauss.smf

This function applies Equations 16.32 and 16.33 to two moisture indices and calculates the most limiting factor. Arguments are the indices and the tolerance parameters. Both are given as arrays with two components.

```
gauss.smf <- function(x,p) {
# calculates response to soil water conditions
# WetDay: x2, # DryDay: x3
y <- x
for(i in 1:2){
 if(x[i]<0) x[i] <- 1
```

```
  y[i]<- 1-exp(-(x[i]/p[i])^2.00/2.00)
 }
 # combine as limiting factor
 smf<- min(y[1],y[2])
 return(c(y,smf))
 }
```

16.16.12 Functions forsucc.C, smi.F, and zelig.F

These functions are similar to the other interface functions to Fortran and C programs. we have explained and used before. First, forsucc.C is

```
forsucc.C <- function(x,fileprefix){
y <- getwd()
  setwd(x)
  file.copy(paste(fileprefix,"-inp.txt",sep=""), "forsucc_inp.txt")
  .C("forsucc", package="seem")
  fileout <- paste(fileprefix,"-out.txt",sep="")
  filedex <- paste(fileprefix,"-dex.txt",sep="")
  file.rename("forsucc_out.txt", fileout)
  file.rename("forsucc_dex.txt", filedex)
  file.remove("forsucc_inp.txt")
  setwd(y)
  return(paste(x,"/",fileout,sep=""))
}
```

Then semi.F

```
semi.F <- function(x,fileprefix){
y <- getwd()
  setwd(x)
  file.copy(paste(fileprefix,"-inp.txt",sep=""), "semi_inp.txt")
  .Fortran("semi", package="seem")
  fileout <- paste(fileprefix,"-out.txt",sep="")
  file.rename("semi_out.txt", fileout)
  file.remove("semi_inp.txt")
  setwd(y)
  return(paste(x,"/",fileout,sep=""))
}
```

is simpler, since we only change the directory as specified by argument 1 and apply the directive file given in argument 2.

16.16.13 Function read.plot.zelig.out

This function and the next two are too lengthy to reproduce here in their entirety, but they are similar to the other functions we have used to read and plot output files from precompiled programs. In this case, it produces two figures, one for stand aggregates and one for basal area by species.

and finally, zelig.F follows the same principles but it is longer because there are more files involved

```
zelig.F <- function(x,fileprefix){
  y <- getwd()
  setwd(x)
  file.copy(paste(fileprefix,"-control-inp.txt",sep=""), "z_control_inp.
txt")
  file.copy(paste(fileprefix,"-site-inp.txt",sep=""), "z_site_inp.txt")
  file.copy(paste(fileprefix,"-species-inp.txt",sep=""), "z_species_inp.
txt")
  .Fortran("zelig", package="seem")
  fileout1 <- paste(fileprefix,"-tracer-out.txt",sep="")
  fileout2 <- paste(fileprefix,"-print-out.txt",sep="")
  file.rename("z_tracer_out.txt", fileout1)
  file.rename("z_print_out.txt", fileout2)
  file.remove("z_control_inp.txt")
  file.remove("z_site_inp.txt")
  file.remove("z_species_inp.txt")
  setwd(y)
  return(paste(x,"/",fileout1,sep=""))
}
```

16.16.14 Function read.plot.forsucc.out

This function produces two figures, one for basal area and the other for density.

16.16.15 Function read.plot.semi.out

This function produces two figures, one for Markov chain dynamics and one for semi-Markov dynamics.

Bibliography

Abrams, P. A., and J. D. Roth. 1994. The effects of enrichment of three species food chains with nonlinear functional responses. *Ecology* 75:1118–1130.

Acevedo, M. F. 1998. Modeling stress in ecosystems: Potential application to space exploration. *SAE Transactions–Journal of Aerospace* 107:656–666.

Acevedo, M. F. 1980a. "Electrical network simulation of tropical forests successional dynamics." In *Progress in Ecological Engineering and Management by Mathematical Models*, ed. D. Dubois, 883–892. Liège, Belgium: Centre Belge d'etudes et de documentation (Belgian Center of Studies and Documentation).

Acevedo, M. F. 1980b. "Nonequilibrium ecology; chronic and impulsive disturbances." In *Paradigms in Changing Times*, eds. R. Ragade and J. Dillon, 72–80. Louisville, Kentucky: Society for General Systems Research.

Acevedo, M. F. 1981. "Modeling ecosystems subject to sudden and periodic disturbances." In *Applied Systems and Cybernetics*, Vol. IV, ed. G. Lasker, 1972–1931. New York: Pergamon Press.

Acevedo, M. F. 2012. *Data Analysis and Statistics for Geography, Environmental Science & Engineering*. Boca Raton, FL: Taylor & Francis/CRC Press.

Acevedo, M. F., and R. Raventós. 2003. Dinámica y Manejo de Poblaciones: modelos unidimensionales (Population dynamics and management: one dimensional models). Alicante, Spain: Publicaciones de la Universidad de Alicante.

Acevedo, M. F., and W. T. Waller. 2000. Modelling and control of a simple trophic aquatic system. *Ecological Modelling* 131:269–284.

Acevedo, M. F., S. Pamarti, M. Ablan, D. L. Urban, and A. Mikler. 2001. Modeling forest landscapes: Parameter estimation from gap models over heterogeneous terrain. *Simulation* 77:53–68.

Acevedo, M. F., D. P. Smith, and M. Ablan. 1997. "Vegetation dynamics in North-Central Texas: A prospectus for landscape scale modeling." In eds. D. Lyons, and P. Hudak, *Geographic Perspectives on the Texas Region* 115–124. Washington, DC: Association of American Geographers.

Acevedo, M. F., D. L. Urban, and M. Ablan, M. 1995a. Transition and gap models of forest dynamics. *Ecological Applications* 5:1040–1055.

Acevedo, M. F., D. L. Urban, and H. H. Shugart. 1996. Models of forest dynamics based on roles of tree species. *Ecological Modelling* 87:267–284.

Acevedo, M. F., T. W. Waller, D. Smith, D. Poage, and P. McIntyre. 1995b. Cladoceran population responses to stress with particular reference to sexual reproduction. *Non Linear World* 2:97–129.

Anosov, D. V., and V. I. Arnold (eds.). 1988. *Dynamical Systems I*. New York: Springer-Verlag.

Arnold, V. I. 1978. *Ordinary Differential Equations*. Cambridge: MIT Press.

Assouline, S., J. S. Selker, and J. Y. Parlange. 2007. A simple accurate method to predict time of ponding under variable intensity rainfall. *Water Resources Research* 43:W03426.

Beer, T. 1991. *Hydrological Tables. Applied Environmetrics*. Australia: Computational Mechanics.

Bledsoe, L. J., and G. M. Van Dyne. 1971. "A compartment model simulation of secondary succession." In *Systems Analysis and Simulation in Ecology*, Vol. 1, ed. B. C. Patten, 479–511. New York: Academic Press.

Bolker, B. M., 2008. *Ecological Models in R*. Princeton, NJ: Princeton University Press.

Botkin, D. B., J. F. Janak, and J. R. Wallis. 1972. Some ecological consequences of a model of forest growth. *Journal of Ecology* 60:849–873.

Breck, J. E., 1988. Relationships among models for acute toxic effects: Applications to fluctuating concentrations. *Environmental Toxicology* and *Chemistry* 7:775–778.

Brewer, J. W. 1979. "Toward optimal impulsive control of agroecosystems." In *Theoretical Systems Ecology*, ed. E. Halfon, 401–417. New York: Academic Press.

Cadzow, J. A. 1973. *Discrete-Time Systems. An Introduction with Interdisciplinary Applications*. Englewood Cliffs, NJ: Prentice Hall.

Calabrese, E. J., and L. A. Baldwin. 1993. *Performing Ecological Risk Assessments*. Chelsea, MI: Lewis Publishers.

Carr, J. R. 1995. *Numerical Analysis for the Geological Sciences*. New York: Prentice Hall.

Caswell, H., 2001. *Matrix Population Models: Construction, Analysis, and Interpretation*, 2nd ed. Sunderland, MA: Sinauer.

Colinvaux, P., 1993. *Ecology 2*. New York: Wiley.

Cox, P. M., C. Huntingford, and R. J. Harding. 1998. A canopy conductance and photosynthesis model for use in a GCM land surface scheme. *Journal of Hydrology* 212–213:79–94.

Crouse, D. T., L. B. Crowder, and H. Caswell, H. 1987. A stage-based population model for loggerhead sea turtles and implications for conservation. *Ecology* 68:1412–1423.

Cushing, J. M. 1977. *Integrodifferential Equations and Delay Models in Population Dynamics*. New York: Springer-Verlag.

Dambacher, J. M., H. -K. Luh, H. W. Li, and P. A. Rossignol. 2003. Qualitative stability and ambiguity in model ecosystems. *American Naturalist* 161:876–888.

Davis, J. C., 2002. *Statistics and Data Analysis in Geology*. 3rd ed. New York: Wiley.

DeAngelis, D. L., and L. J. Gross. 1992. *Individual-Based Models and Approaches in Ecology: Populations, Communities, and Ecosystems*. New York: Chapman-Hall.

Delgado, L. A., M. F. Acevedo, H. Castellanos, H. Ramirez, and J. Serrano. 2005. Allometric relations and growth patterns for tree species in the Imataca forest reserve, Venezuela. *Interciencia* 30:275–283.

Diskin, M. H., and N. Nazimov. 1995. Linear reservoir with feedback regulated inlet as a model for the infiltration process. *Journal of Hydrology* 172:313–330.

Eilers, P. H. C., and J. C. H. Peeters. 1993. Dynamic behaviour of a model for photosynthesis and photoinhibition. *Ecological Modelling* 69:113–133.

Ellner, S. P., and J. Guckenheimer. 2006. *Dynamic Models in Biology*. Princeton, NJ: Princeton University Press.

Ford, A. 1999. *Modeling the Environment*. Washington, DC: Island Press.

Gates, D. M. 1993. *Climate Change and its Biological Consequences*. Sunderland, MA: Sinauer.

Ginot, V., and J. C. Herve. 1994. Estimating the parameters of dissolved oxygen dynamics in shallow ponds. *Ecological Modelling* 73:169–187.

Gotelli, N., 2008. *A Primer of Ecology*, 4th ed. Sunderland, MA: Sinauer Associates.

Grimm, V., and S. F. Railsback. 2005. *Individual-Based Modeling and Ecology*. Princeton, NJ: Princeton University Press.

Gurney, W. S. C., E. McCauley, R. M. Nisbet, and W. W. Murdoch. 1990. The physiological ecology of Daphnia: A dynamic model of growth and reproduction. *Ecology* 71:716–732.

Gurney, W. S. C., R. M. Nisbet, and S. P. Blythe. 1986. "The systematic formulation of models of stage structured populations." In *The Dynamics of Physiologically Structured Populations*, Vol. 68, eds. J. A. J. Metz and O. Dickmann, 474–494. New York: Springer Verlag.

Hallam, T. G. 1986a. "Community dynamics in homogeneous environments." In *Mathematical Ecology*, ed. T. G. Hallam and S. A. Levin, 241–285. New York: Springer-Verlag.

Hallam, T. G., 1986b. "Population dynamics in a homogeneous environment." In *Mathetmatical Ecology*, eds. T. G. Hallam, and S. A. Levin, 61–94. New York: Springer-Verlag.

Hanson, F. B., and H. C. Tuckwell. 1981. Logistic growth with random density independent disasters. *Theoretical Population Biology* 19:1–18.

Hardisty, J., D. M. Taylor, and S. E. Metcalfe. 1993. *Computerised Environmental Modelling*. Chichester, England: Wiley.

Harte, J., 1988. *Consider a Spherical Cow: A Course in Environmental Problem Solving*. Mill Valley, CA: University Science Books.

Hemond, H. F., and E. J. Fechner. 1994. *Chemical Fate and Transport in the Environment*. New York: Academic Press.

Hennessey, J. C. 1980. An age dependent, absorbing semi-Markov model of work histories of the disabled. *Mathematical Biosciences* 51:283–304.

Hillel, D. 1998. *Environmental Soil Physics*. New York: Academic Press.

Hirsch, M. W., and S. Smale. 1974. *Differential Equations, Dynamics Systems, and Linear Algebra*. New York: Academic Press.

Holcomb, S. S. 2001. *An Examination of the Riparian Bottomland Forest in North Central Texas through Ecology, History, Field Study, and Computer Simulation, Environmental Science*. Denton, TX: University of North Texas.

Howard, R. A. 1971. *Dynamic Probabilistic Systems. Vol II. Semi-Markov and Decision Processes*. New York: Wiley.

Huggett, R. J. 1993. *Modelling the Human Impact on Nature*. Oxford: Oxford University Press.

Jones, O., R. Maillardet, and A. Robinson. 2009. *Introduction to Scientific Programming and Simulation Using R*. Boca Raton, FL: Taylor & Francis/CRC Press.

Jopp, F., H. Reuter, and B. Breckling (eds.). 2011. *Modelling Complex Ecological Dynamics: An Introduction into Ecological Modelling for Students, Teachers & Scientists*. Berlin, Heidelberg: Springer.

Jorgensen, S., 1988. *Fundamentals of Ecological Modelling*. Amsterdam: Elsevier.

Kale, R. V., and B. Sahoo. 2011. Green-Ampt infiltration models for varied field conditions: A revisit. *Water Resources Management* 25:3505–3536.

Kaufman, D. G., and C. M. Franz. 2000. *The Biosphere: Protecting Our Global Environment*, 4th ed. Dubuque, IA: Kendall/Hunt.

Keen, R. E., and J. D. Spain. 1992. *Computer Simulation in Biology: A BASIC Introduction*. New York: Wiley-Liss.

Kirchner, T. B. 1992. *TIME-ZERO The Integrated Modeling Environment. Reference Manual*. Quaternary Software.

Kooijman, S. A. L. M. 1993. *Dynamic Energy Budgets in Biological Systems*. Cambridge: Cambridge University Press.

Kot, M. 2001. *Elements of Mathematical Ecology*. Cambridge: Cambridge University Press.

Lean, D. R. S. 1973. Phosphorus dynamics in lake water. *Science* 179:678–680.

Levins, R., 1974. The qualitative analysis of partially specified systems. *Annals of the New York Academy of Sciences* 231:123–138.

Lewis, E. R., 1977. *Network Models in Population Biology*. New York: Springer Verlag.

Lewis, M. A., and H. G. Othmer. 1997. "Qualitative theory of ordinary differential equations." In *Case Studies in Mathematical Modeling—Ecology, Physiology, and Cell Biology*, eds. H. G. Othmer, F. R. Adler, M. A. Lewis, and J. C. Dallon, J. C., 357–380. Upper Saddle River, NJ: Prentice Hall.

Macedo, M. F., J. G. Ferreira, and P. Duarte. 1998. Dynamic behaviour of photosynthesis-irradiance curves determined from oxygen production during variable incubation periods. *Marine Ecology Progress Series* 165:31–43.

Mailapalli, D. R., W. W. Wallender, R. Singh, and N. S. Raghuwanshi. 2009. Application of a nonstandard explicit integration to solve Green and Ampt infiltration equation. *Journal of Hydrologic Engineering* 14:203–206.

Mancini, J. L. 1983. A method for calculating effects, on aquatic organisms, of time varying concentrations. *Water Research* 17:1355–1362.

Matis, J. H., W. E. Grant, and T. H. Miller. 1992. A semi-Markov process model for migration of marine shrimp. *Ecological Modelling* 60:167–184.

Maurer, B. A. 1999. *Untangling Ecological Complexity: The Macroscopic Perspective*. Chicago: The University of Chicago Press.

May, R. M. 1973. Qualitative stability in model ecosystems. *Ecology* 54:638–641.

McDonald, N. 1978. *Time Lags in Biological Models*. New York: Springer Verlag.

Morford, S. L., B. Z. Houlton, and R. A. Dahlgren, R. A. 2011. Increased forest ecosystem carbon and nitrogen storage from nitrogen rich bedrock. *Nature* 477:U78–U88.

Nachabe, M. H., 1998. Refining the definition of field capacity in the literature. *Journal of Irrigation and Drainage Engineering* 124:230–232.

Neitsch, S. L., J. G. Arnold, J. R. Kiniry, J. R. Williams, and K. W. King. 2002. *Soil and Water Assessment Tool. Theoretical Documentation Version 2000*. Tempe, TX: Grassland, Soil And Water Research Laboratory, Agricultural Research Service.

Nikolov, N., and K. F. Zeller. 1992. A solar radiation algorithm for ecosystem dynamic models. *Ecological Modelling* 61:149–168.

Nisbet, R. M., and W. S. C. Gurney. 1986. "The formulation of age-structured models." In *Mathematical Ecology*, eds. T. G. Hallam, and S. A. Levin, 95–115. New York: Springer-Verlag.

Nisbet, R. M., W. S. C. Gurney, W. W. Murdoch, and E. McCauley. 1989. Structured population models: a tool for linking effects at individual and population level. *Biological Journal of the Linnean Society* 37:79–99.

Nisbet, R. M., E. McCauley, W. S. C. Gurney, W. W. Murdoch, and A. M. de Roos. 1997. "Simple representations of biomass dynamics in structured populations." In *Case Studies in Mathematical Modeling–Ecology, Physiology, and Cell Biology*, eds. H. G. Othmer, F. R. Adler, M. A. Lewis, and J. C. Dallon, 61–79. Upper Saddle River, NJ: Prentice Hall.

O'Connor, D. J., J. P. Connolly, and E. J. Garland. 1989. "Mathematical models—fate, transport, and food chain." In *Ecotoxicology: Problems and Approaches*, eds. S. A. Levin, M. A. Harwell, J. R. Kelly, and K. D. Kimball, 221–242. New York: Springer-Verlag.

Pastorok, R. A. 2002. *Ecological Modeling in Risk Assessment: Chemical Effects on Populations, Ecosystems, and Landscapes*. Boca Raton, FL: Lewis Publishers.

Phipps, R. L. 1979. Simulation of wetlands forest vegetation dynamics. *Ecological Modelling* 7:257–288.

Ramos, H. 2007. A non-standard explicit integration scheme for initial-value problems. *Applied Mathematics and Computation* 189:710–718.

Reuter, H., B. Breckling, and F. Jopp (eds.). 2011. "Individual-based models." In *Modelling Complex Ecological Dynamics. An Introduction into Ecological Modelling for Students, Teachers & Scientists*, 163–178. Berlin, Heidelberg: Springer.

Richardson, C. W., and A. D. Nicks. 1990. "Weather generator description." In *EPIC-Erosion/Productivity Impact Calculator: 1 Model Documentation*, eds. A. N. Sharpley, and J. R. Williams, 93–104. Durant, OK and Temple, TX: United States Department of Agriculture, Agricultural Research Center. Technical Bulletin No: 1768.

Ricklefs, R. E., and G. L. Miller. 2000. *Ecology*, 4th ed. New York: W.H. Freeman.

Shaffer, M. J., L. Ma, and S. Hansen (eds.). 2001. *Modeling Carbon and Nitrogen Dynamics for Soil Management*. Boca Raton, FL: CRC Press.

Soetaert, K., and P. M. J. Herman. 2009. *A Practical Guide to Ecological Modelling Using R as a Simulation Platform*. New York: Springer.

Stevens, M. H. H., 2009. *A Primer of Ecology with R*. New York: Springer.

Swartzman, G. L., and S. Kaluzny. 1987. *Ecological Simulation Primer*. New York: MacMillan.

Thomann, R. V. 1989. "Deterministic and statistical models of chemical fate in aquatic systems." In *Ecotoxicology: Problems and Approaches*, eds. S. A. Levin, M. A. Harwell, J. R. Kelly, and K. D. Kimball, 245–277. New York: Springer-Verlag.

Thornley, J. H. M. 2002. Instantaneous canopy photosynthesis: Analytical expressions for sun and shade leaves based on exponential light decay down the canopy and an acclimated non-rectangular hyperbola for leaf photosynthesis. *Annals of Botany* 89:451–458.

Tuljapurkar, S. and H. Caswell (eds.). 1997. *Structured-Population Models in Marine, Terrestrial, and Freshwater Systems*. New York: Chapman-Hall.

Urban, D. L., and H. H. Shugart. 1992. "Individual-based models of forest succession." In *Plant Succession: Theory and Prediction*, eds. D. C. Glenn-Lewin, R. K. Peet, and T. T. Veblan, 249–292. New York: Chapman and Hall.

van Genutchen, M. T. 1980. A closed-form equation for predicting the hydraulic conductivity of unsaturated soils. *Soil Science Social of America Journal* 44:892–898.

van Hulst, R. 1979. On the dynamics of vegetation: succession in model communities. *Vegetation* 39:85–96.

Walker, J. C. G. 1991. *Numerical Adventures with Geochemical Cycles*. Oxford: Oxford University Press.

Walker, T. A. 1982. Use of a Secchi disc to measure attenuation of underwater light for photosynthesis. *Journal of Applied Ecology* 19:539–543.

Williams, J. R., C. A. Jones, and P. T. Dyke. 1990. "The EPIC model." In *EPIC-Erosion/Productivity Impact Calculator: 1 Model Documentation*, eds. A. N. Sharpley, and J. R. Williams, 3–92. Durant, OK and Temple, TX: United States Department of Agriculture, Agricultural Research Center, USDA Technical Bulletin No: 1768.

Zeigler, B. P., H. Praehofer, and T. G. Kim. 2000. *Theory of Modeling and Simulation. Integrating Discrete Event and Continuous Complex Dynamic Systems*, 2nd ed. San Diego: Academic Press.

Index

A

B

C

T

U

V

W

Z